Supplements to the 2nd Edition of

RODD'S CHEMISTRY OF CARBON COMPOUNDS

ELSEVIER SCIENCE PUBLISHERS B.V.
Molenwerf 1
P.O. Box 211, 1000 AE Amsterdam, The Netherlands

Distributors for the United States and Canada:

ELSEVIER SCIENCE PUBLISHING COMPANY INC.
52, Vanderbilt Avenue
New York, N.Y. 10017

Printed in The Netherlands

Supplements to the 2nd Edition of

RODD'S CHEMISTRY OF CARBON COMPOUNDS

VOLUME I

ALIPHATIC COMPOUNDS
★

VOLUME II

ALICYCLIC COMPOUNDS
★

VOLUME III

AROMATIC COMPOUNDS
★

VOLUME IV

HETEROCYCLIC COMPOUNDS
★

VOLUME V

MISCELLANEOUS
GENERAL INDEX
★

Supplements to the 2nd Edition (Editor S. Coffey) of

RODD'S CHEMISTRY OF
CARBON COMPOUNDS

A modern comprehensive treatise

Edited by
MARTIN F. ANSELL
Ph.D., D.Sc. (London) F.R.S.C. C. Chem.
Department of Chemistry, Queen Mary College,
University of London (Great Britain)

Supplement to

VOLUME IV   HETEROCYCLIC COMPOUNDS

Part A:
Three-, Four- and Five-Membered Heterocyclic
Compounds with a Single Hetero-Atom in the Ring

ELSEVIER
Amsterdam – Oxford – New York – Tokyo   1984

VI

CONTRIBUTORS   TO   THIS   VOLUME

ROBERT LIVINGSTONE, B.Sc., Ph.D., F.R.S.C.

Department of Pure and Applied Chemistry, The Polytechnic,

Queensgate, Huddersfield, HD1 3DH

RAYMOND E. FAIRBAIRN, B.Sc., Ph.D., F.R.S.C.

Formerly of Research Department,

Dyestuffs Division,

I.C.I. (INDEX)

PREFACE TO SUPPLEMENT IVA

The publication of this volume continues the supplementation of the second edition of Rodd's Chemistry of Carbon Compounds thus keeping this major work of reference up to date. This supplement covers the Chapters in volume 4A and thus starts the supplementation of volume four of the second edition. Although the Chapters in this book stand on their own, it is intended that each one should be read in conjunction with the parent chapter in the second edition.

At a time when there are many specialist reviews, monographs and reports available, there is still in my view an important place for a book such as "Rodd", which gives a broader coverage of organic chemistry. One aspect of the value of this work is that it allows the expert in one field to quickly find out what is happening in other fields of chemistry. On the other hand a chemist looking for the way into a field of study will find in "Rodd" an outline of the important aspects of that area in chemistry together with leading references to other works to provide more detailed information.

As editor I have been fortunate in that Dr Livingstone agreed to supplement the Chapters he wrote for the second edition, and as I expected he has produced a very readable critical assessment of the areas of chemistry covered. As an organic chemist I have enjoyed reading the chapters and profited from learning of advances in the fields of chemistry that are outside my own special interests.

This volume has been produced by direct reproduction of the manuscript. I am most grateful to Dr Livingstone for all the care and effort which he and his secretary put into the production of the manuscript, including the diagrams. I am confident that readers will find this presentation acceptable. I also wish to thank the staff at Elsevier for all the help they have given me and for seeing the transformation of authors manuscripts to published work.

Martin F. Ansell

CONTENTS

VOLUME IVA

Heterocyclic Compounds; Three-, Four- and Five-Membered Heterocyclic Compounds with a Single Hetero-Atom in the Ring

Chapter 1. Compounds with Three- and Four-Membered Heterocyclic Rings
by R. LIVINGSTONE

Chapter 2. Compounds Containing a Five-Membered Ring with One Hetero Atom of Group VI; Oxygen
by R. LIVINGSTONE

*Chapter 3. Compounds with Five-Membered Ring having One Hetero
Atom from Group VI; Sulphur and its Analogues*
by R. LIVINGSTONE

*Chapter 4. Compounds Containing a Five-Membered Ring with One
Hetero Atom from Group V; Nitrogen*
by R. LIVINGSTONE

Chapter 5. Compounds Containing Five-Membered Rings with One Hetero Atom from Group V: Nitrogen: Fused Ring Compounds
by R. LIVINGSTONE

Chapter 6. Other Five-Membered Ring Compounds with One Hetero Atom in the Ring from Groups III, IV and V
by R. LIVINGSTONE

OFFICIAL PUBLICATIONS

B.P.	British (United Kingdom) Patent
F.P.	French Patent
G.P.	German Patent
Sw.P.	Swiss Patent
U.S.P.	United States Patent
U.S.S.R.P.	Russian Patent
B.I.O.S.	British Intelligence Objectives Sub-Committee Reports
F.I.A.T.	Field Information Agency, Technical Reports of U.S. Group Control Council for Germany
B.S.	British Standards Specification
A.S.T.M.	American Society for Testing and Materials
A.P.I.	American Petroleum Institute Projects
C.I.	Colour Index Number of Dyestuffs and Pigments

SCIENTIFIC JOURNALS AND PERIODICALS

With few obvious and self-explanatory modifications the abbreviations used in references to journals and periodicals comprising the extensive literature on organic chemistry, are those used in the World List of Scientific Periodicals.

LIST OF COMMON ABBREVIATIONS AND
SYMBOLS USED

A	acid
Å	Ångström units
Ac	acetyl
a	axial; antarafacial
as, $asymm$.	asymmetrical
at	atmosphere
B	base
Bu	butyl
b.p.	boiling point
C, mC and μC	curie, millicurie and microcurie
c, C	concentration
C.D.	circular dichroism
conc.	concentrated
crit.	critical
D	Debye unit, 1×10^{-18} e.s.u.
D	dissociation energy
D	dextro-rotatory; dextro configuration
DL	optically inactive (externally compensated)
d	density
dec. or decomp.	with decomposition
deriv.	derivative
E	energy; extinction; electromeric effect; Entgegen (opposite) configuration
E1, E2	uni- and bi-molecular elimination mechanisms
E1cB	unimolecular elimination in conjugate base
e.s.r.	electron spin resonance
Et	ethyl
e	nuclear charge; equatorial
f	oscillator strength
f.p.	freezing point
G	free energy
g.l.c.	gas liquid chromatography
g	spectroscopic splitting factor, 2.0023
H	applied magnetic field; heat content
h	Planck's constant
Hz	hertz
I	spin quantum number; intensity; inductive effect
i.r.	infrared
J	coupling constant in n.m.r. spectra; joule
K	dissociation constant
kJ	kilojoule

LIST OF COMMON ABBREVIATIONS

k	Boltzmann constant; velocity constant
kcal	kilocalories
L	laevorotatory; laevo configuration
M	molecular weight; molar; mesomeric effect
Me	methyl
m	mass; mole; molecule; *meta-*
ml	millilitre
m.p.	melting point
Ms	mesyl (methanesulphonyl)
[M]	molecular rotation
N	Avogadro number; normal
nm	nanometre (10^{-9} metre)
n.m.r.	nuclear magnetic resonance
n	normal; refractive index; principal quantum number
o	*ortho-*
o.r.d.	optical rotatory dispersion
P	polarisation, probability; orbital state
Pr	propyl
Ph	phenyl
p	*para-*; orbital
p.m.r.	proton magnetic resonance
R	clockwise configuration
S	counterclockwise config.; entropy; net spin of incompleted electronic shells; orbital state
S_N1, S_N2	uni- and bi-molecular nucleophilic substitution mechanisms
S_Ni	internal nucleophilic substitution mechanisms
s	symmetrical; orbital; suprafacial
sec	secondary
soln.	solution
symm.	symmetrical
T	absolute temperature
Tosyl	*p*-toluenesulphonyl
Trityl	triphenylmethyl
t	time
temp.	temperature (in degrees centigrade)
tert.	tertiary
U	potential energy
u.v.	ultraviolet
v	velocity
Z	zusammen (together) configuration

LIST OF COMMON ABBREVIATIONS

α	optical rotation (in water unless otherwise stated)
$[\alpha]$	specific optical rotation
α_A	atomic susceptibility
α_E	electronic susceptibility
ε	dielectric constant; extinction coefficient
μ	microns (10^{-4} cm); dipole moment; magnetic moment
μ_B	Bohr magneton
μ_g	microgram (10^{-6} g)
λ	wavelength
ν	frequency; wave number
χ, χ_d, χ_μ	magnetic, diamagnetic and paramagnetic susceptibilities
$\sim$	about
(+)	dextrorotatory
(−)	laevorotatory
(±)	racemic
$\ominus$	negative charge
$\oplus$	positive charge

Chapter 1

COMPOUNDS WITH THREE- AND FOUR-MEMBERED HETEROCYCLIC RINGS

R. LIVINGSTONE

1. *Introduction*

A number of novel routes to three- and four-membered heterocycles are reported and the existence of some related species have been identified as intermediates in the formation of other heterocycles.

M.O. calculations on oxirene, thiirene, and 2-azirine show that these formally antiaromatic analogues of the cyclopropenyl anion should be stable and may therefore occur as stable intermediates in reactions (M.J.S. Dewar and C.A. Ramsden, Chem. Comm., 1973, 688). Evidence has been present for their existence in some instances.

Ring-forming reactions are important and common processes in heterocyclic chemistry, and a set of rules found useful, on an empirical basis, to predict the relative facility of ring forming reactions has been described (J.E. Baldwin, *ibid.*, 1976, 734).

A number of bicyclic ring systems containing a three- or four-membered ring with one heteroatom have been synthesised.

2. *Three-membered Rings Containing An Oxygen Atom*

(a) *Oxiranes (ethylene oxides)*

A number of properties and reactions of oxiranes appear in Vol. ID under the various headings concerned with epoxides. Also the preparation and properties of arene oxides are to be found in Vol. IIIF,G.

(i) *Synthesis*

In addition to existing methods oxiranes have been prepared by a number of new methods, including the following, some of which can be regarded as novel.

(1) Peroxyacetyl nitrate (violently explosive when condensed as a pure liquid) reacts with simple olefins quantitatively in chloroform or benzene to give the corresponding oxiranes (K.R. Darnall and J.N. Pitts, Jr., *ibid.*, 1970, 1305).

$$\text{Me C(:O)·O·ONO}_2$$
$$+ \xrightarrow[\text{or} \quad C_6H_6]{\text{CHCl}_3} \quad R_2\!\!\triangle\!\!R_2 + \text{MeONO} + \text{MeNO}_2 + (CO_2)$$
$$R_2C=CR_2$$

(2) The reaction of an olefin with iodine in the presence of water and a suitable oxidising agent such as iodic acid, or oxygen catalysed by nitrous acid gives an iodohydrin. The latter will form an oxirane readily in high yield on treatment with bases, for example calcium hydroxide, or by the action of moist sodium aluminate suspended in ethylene dichloride (J.W. Cornforth and D.T. Green, J. chem. Soc., C, 1970, 846).

$$\text{C=C} \xrightarrow[\substack{\text{NaNO}_2, \\ \text{M H}_2\text{SO}_4 \\ 70°}]{\text{I}_2,\ \text{H}_2\text{O},} \quad -\overset{\text{OH}}{\underset{|}{\text{C}}}-\overset{\text{I}}{\underset{|}{\text{C}}}- \xrightarrow{\text{base}} \quad \triangle$$

(3) Treatment of 2,2-dimethyl-3-hydroxypropyl 2-bromoisobutyrate with sodium hydride in 1,2-dimethoxyethane gives the oxirane derivative 2,2,6,6-tetramethyl-1,4,8--trioxaspiro[2.5]octane (M.S. Newman and E. Kilbourn, J. org. Chem., 1970, 35, 3186).

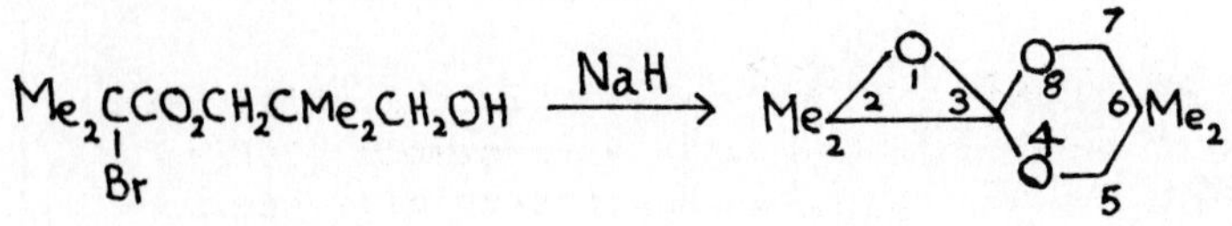

$$\underset{\overset{|}{Br}}{Me_2C}CO_2CH_2CMe_2CH_2OH \xrightarrow{NaH}$$

The epoxy ketene ketal is very stable towards basic
reagents, but surprisingly after boiling with methyllithium
in ether, or with sodium methoxide in methanol at room
temperature, it is converted into 2,2-dimethyl-3-hydroxy-
propyl 2-hydroxyisobutyrate $Me_2COH.CO.OCH_2.CMe_2-CH_2OH$.
This conversion occurs readily on treatment with either
aqueous acid or boron trifluoride etherate. Treatment
of the spiro compound in benzene or anisole with one
equivalent of aluminium chloride at room temperature
affords 3-chloro-2,2-dimethylpropyl 2-hydroxyisobutyrate.

(4) Boiling a mixture of 1,2-dibromo-3,4-bis-
-(diphenylmethylene)cyclobutene, an alcohol, tetrahydrofuran and potassium hydroxide (or sodium alkoxide) yields a
cyclobutadiene epoxide { 1-alkoxy-3-(bromodiphenylmethyl)-
-4-(diphenylmethylene)cyclobut-2-en[1,2-b]oxirane} (R =
Me, Et, Pr[n], Pr[i], Bu[n], or C_6H_{11}) (F. Toda and
K. Akagi, Chem. Comm., 1970, 764).

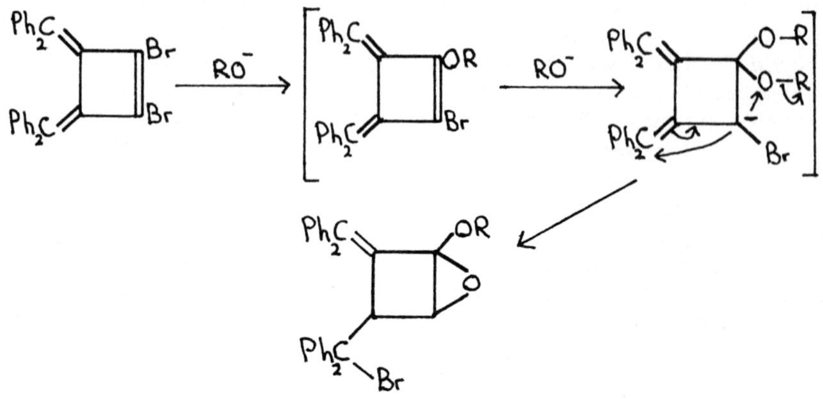

(5) The addition of 4-nitroperbenzoic acid to methyl-
enecyclopropane in methylene chloride at -10° gives oxa-
spiropentane, which is converted into cyclobutanone (95%)
together with traces of 2-methylpropenal and methyl vinyl
ketone by the addition of a catalytic amount of lithium
iodide (J.R. Salün and J.M. Conia, Chem. Comm., 1971,
1579).

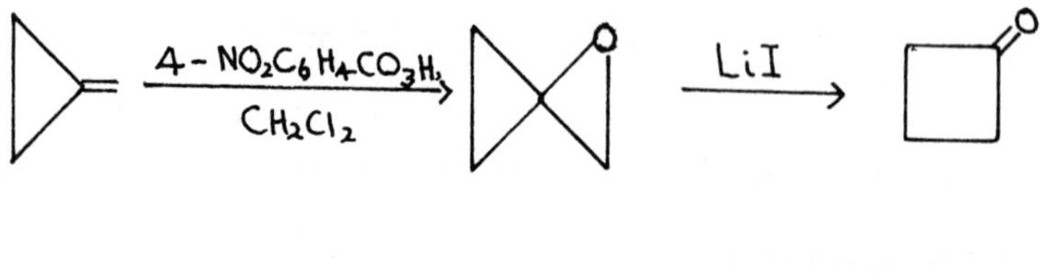

The formation of other oxaspiropentanes and related
compounds is reported along with their thermal reactivity,
and their conversion to cyclobutanones with Lewis acids
(D.H. Aue, M.J. Meshishnek, and D.F. Shellhamer,
Tetrahedron Letters, 1973, 4799).

(6) Dehydration of *erythro-* and *threo-*
$RCH(OH)CR^1(OH)CR^2=CR^3R^4$ by sodium hydride and
toluenesulphonyl chloride in tetrahydrofuran gives the

corresponding *cis-* and *trans*-2-alkenyl-3-aryl-oxiranes
(1) and (2) stereospecifically (J.C. Paladini and J.
Chuche, Bull. Soc. chim. Fr., 1974, 187). 1-Chloro-1-
phenylbut-3-yne may be converted into *trans*-2-ethynyl-3-
-phenyloxirane. Thermal ring-cleavage of the vinyloxiranes
(1) and (2) gives the dihydrofurans (3). The reaction
is accompanied by *cis-trans* isomerisation of the oxirane
ring, and the stereoselectivity of the reaction is greater
with the E than with the Z configuration. The reactions
are rationalised on the basis of a disrotatory
electrocyclisation of a vinylic carbonyl ylide with *cis-*
trans isomerisation of the double band (*ibid.*, p.197).

(1)   (2)   (3)

R^1	R^2	R^3	R^4	R^5
Ph	H	H	H	H
Ph	Me	H	H	H
Ph	H	H	Me	H

R^1	R^2	R^3	R^4	R^5
Ph	H	H	H	Me
Ph	H	H	Ph	H
2-Furyl	Me	Me	H	H

2,3-Diisopropenyloxirane on gas phase racemisation and
rearrangement produces 2-isopropenyl-4-methyl-2,3-dihydro-
furan (Paladini and R.J. Crawford, Canad., J. Chem., 1974,
52, 2098).

   (7) Oxiranes may be formed by a route involving
highly nucleophilic selenocarbanion intermediates.
β-Hydroxyalkyl phenyl selenides are obtained by the
reaction of selenoacetals with *n*-butylithium and trapping
the resulting carbanions with carbonyl compounds. The
selenides with methyl iodide in the presence of silver
tetrafluoroborate readily form β-hydroxyalkylselenonium
salts, which with potassium *tert*-butoxide in DMSO afford
the oxiranes *via* the betaine (W. Durmont and A. Krief,

6

Angew. Chem. internat. Edn., 1975, _14_, 350).

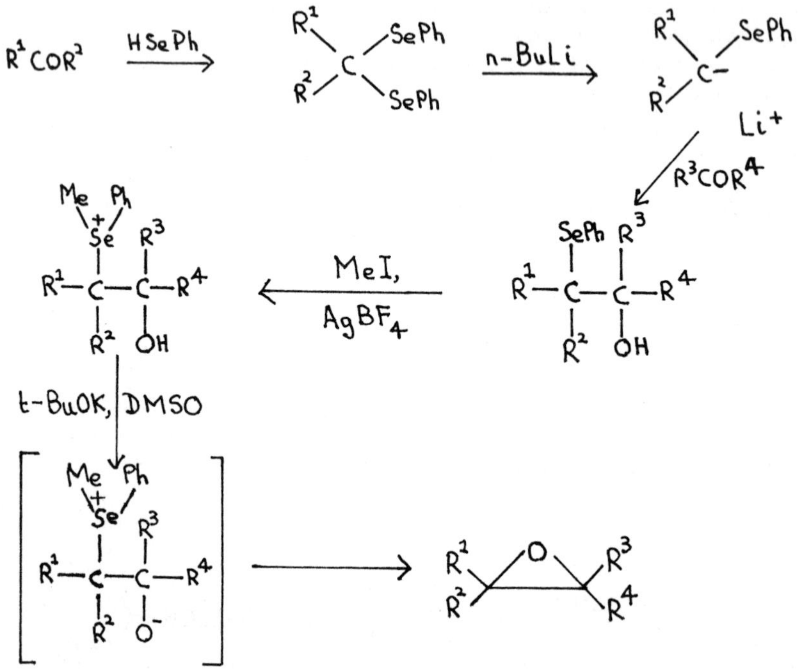

The use of methyl selenoacetals offer several advantages over phenylacetals. These include (a) that methyl selenoacetals, even of aldehydes, can be prepared in high yields, (b) the nucleophilicity of α-methyl selenocarbanions, even of highly substituted ones is so high that they can react with a large variety of carbonyl compounds, (c) the nucleophilicity of the selenium atom in the β-hydroxy methyl selenides is higher than in the phenyl analogues and allows the synthesis of the corresponding selenonium salt with methyl iodide or dimethyl sulphate (expensive silver tetrafluoroborate is not required) without concomitant rearrangement.
Several examples are reported (D. Van Ende, Dumont and Krief, _ibid._, p.700).

Treatment of an equimolar solution of cyclopropyl-diphenylsulphonium fluoroborate and a ketone or aldehyde, e.g.,cyclopentanone, cyclohexanone, benzaldehyde, with solid potassium hydroxide in dimethylsulphoxide followed by flash distillation of the crude extract in *vacuo* yields pure oxaspiropentanes (M.J. Bogdaanowicz and B.M. Trost, Tetrahedron Letters, 1972, 887).

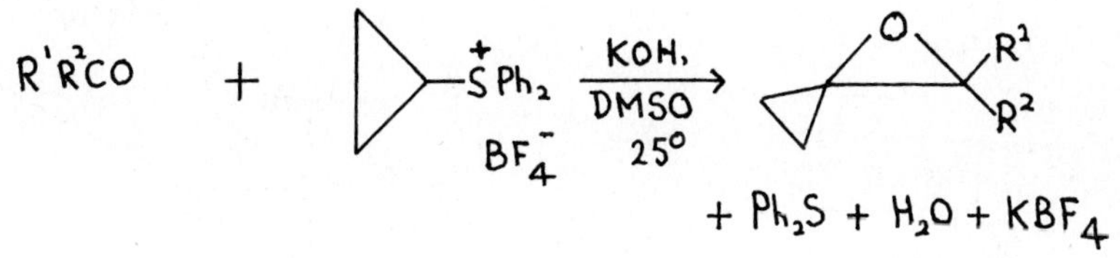

(8) 3-*tert*-Butyl-2-methyleneoxirane (1-*tert*-butyl-allene oxide) is obtained by the route shown below.

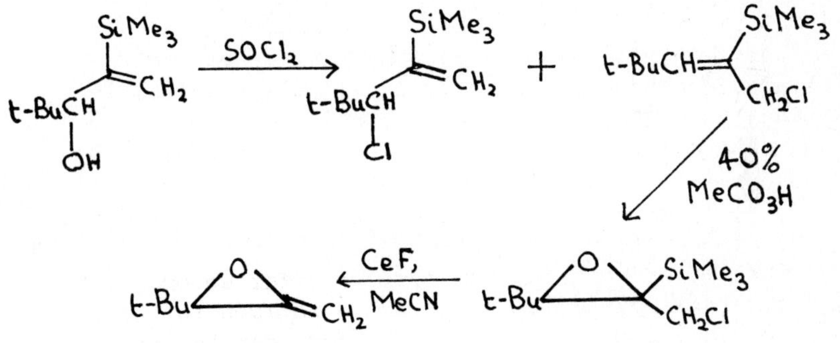

1-Alkyl-, 3-alkyl-, and 3-aryl- substituted allene oxides, when similarly generated, suffer regiospecific nucleophilic epoxide opening to give substituted ketones.   1-Aryl- and 1,1-dialkyl-substituted allene oxides, once generated, are believed to undergo facile isomerisation to cyclopropanones (T.H. Cahn and B.S. Ong, J. org. Chem., 1978, 43, 2994).

(9) The nitrosochloroethylenes (4), react with
cyclopentadiene to yield a bicyclic compound containing
an oxirane and an aziridine ring, *via* (4+2) cycloadduct
intermediates (5) which are detectable at -60°
(H.G. Viehe *et. al.*, J. Amer. chem. Soc., 1977, <u>99</u>, 2340).

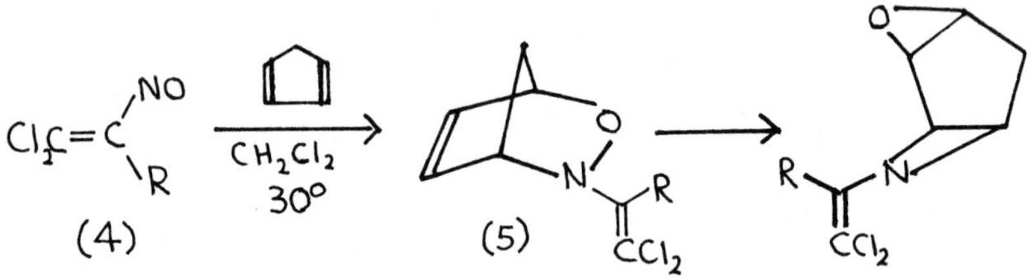

A successful synthesis of a stable oxiranimine has
been claimed (E. Ziegler, G. Kollenz, and W. Ott, Ann.,
1976, 2071).

(ii) Properties and Reactions

(1) *n*-Butyllithium in tetrahydrofuran at -78°
metallates epoxyethyltriphenylsilane(2-triphenylsilyl-
oxirane) in high yield and exclusively on the oxirane
carbon α to silicon. The resulting 1-triphenylsilyl-1,2-
-epoxyethyllithium retains its stereochemical and
structural integrity at this temperature, and it can be
quenched with various reagents to form oxirane derivatives
in high yield (J.J. Eisch and J.E. Galle, J. Amer. chem.
Soc., 1976, <u>98</u>, 4646).

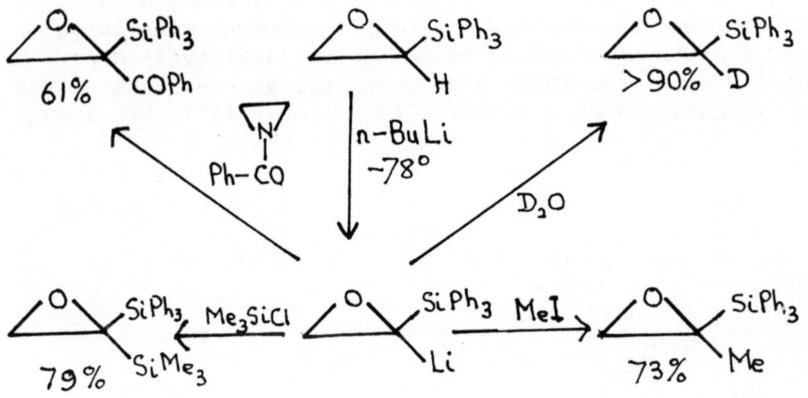

(2) An example of the *cis*-opening of an oxirane is given when dieldrin is treated with acetic anhydride--sulphuric acid. At least four products are obtained, one in a 11-14% yield is a *cis*-diacetate resulting from the *cis*-opening of the epoxide ring of dieldrin. The *endo-exo* configuration of dieldrin and the *exo* orientation of the oxirane ring suggest that the rearward approach of the bulky AcO⁻ ion would be sterically hindered, thus favouring approach from the front. The possibility of σ-bond participation would also favour the approach of AcO⁻ ion from this side. The corresponding *trans*-diacetate from the expected ring opening has not yet been isolated from the reaction products (A.S.Y. Chau and W.P. Cochrane, Chem. and Ind., 1970, 1568).

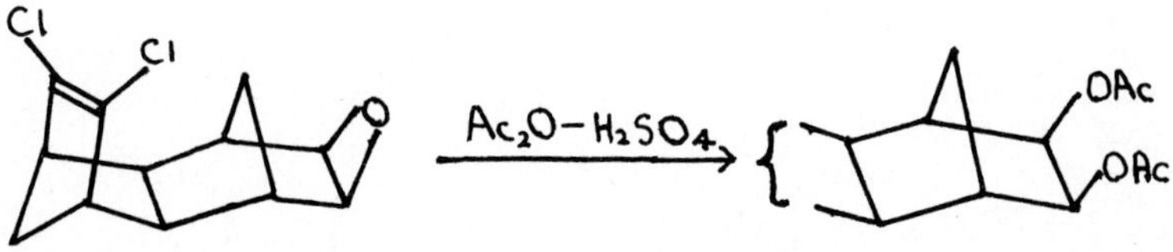

Dieldrin

(3) The lithium diethylamide induced epoxide-allylic
alcohol rearrangement of *cis*- and *trans*-4- *tert*-butylcyclo-
hexene oxide is shown, through appropriate deuterium
substitution, to occur by a *syn*-elimination process. Thus
cis-3-deuterio-*trans*-5-*tert*-butylcyclohexene oxide is
converted into *trans*-5-*tert*-butyl-cyclohexen-2-ol with
loss of deuterium, while *trans*-3 -butylcyclohexen-2-ol with
-butylcyclohexene oxide yields *cis*-5-*tert*-butylcyclohexen-
2-ol-3-d with loss of a proton (R.P. Thummel and
B. Rickborn, J. Amer. chem. Soc., 1970, 92, 2064).

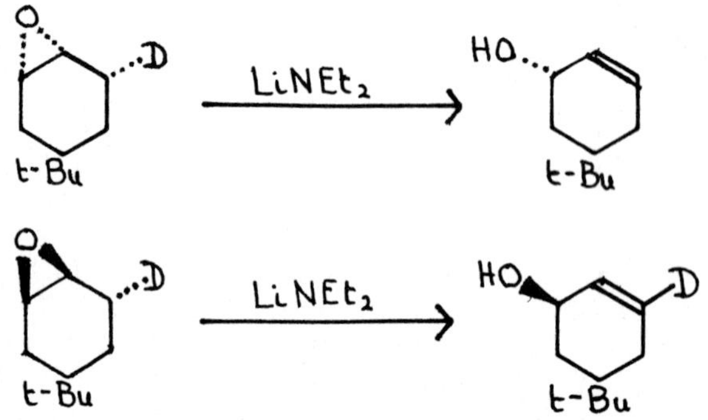

Another reagent diethylaluminium 2,2,6,6-tetramethylpiper-
idide allows the above transformation to take place under
milder conditions. Thus can be achieved a useful regiospecific isomerisation of epoxides to allyclic alcohols,
for example (E)-cyclododecene oxide can be converted into
(E)-cyclodec-2-en-1-ol (A. Yasuda *et. al.*, *ibid.*, 1974,
96, 6513).

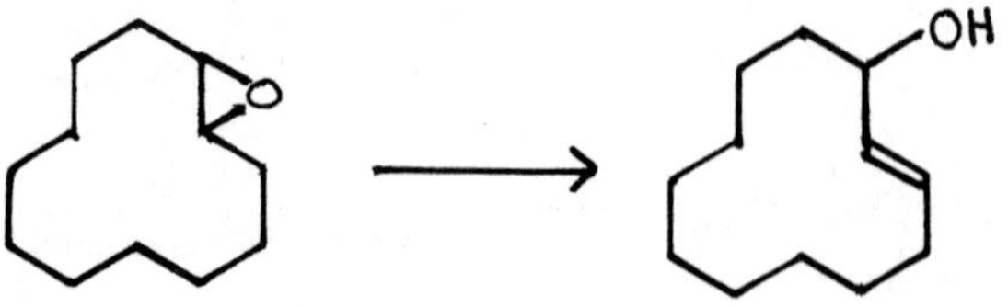

(4) The structure of cycloadducts from 2-cyano-*trans*-
-stilbene oxide with acetylenic and olefinic dipolarophiles
confirm the electrocyclic C-C ring scission of the oxirane
to a carbonyl ylide, while the cycloadditions of 3-cyano-
-*trans*- and *cis*-stillbene oxide to dimethyl fumarate
establish the conrotatory course.   The carbonyl ylides
give with acetylenes dihydrofurans, and with olefins
tetrahydrofurans (A. Dahmen *et. al.*, Chem. Comm., 1971,
1192).

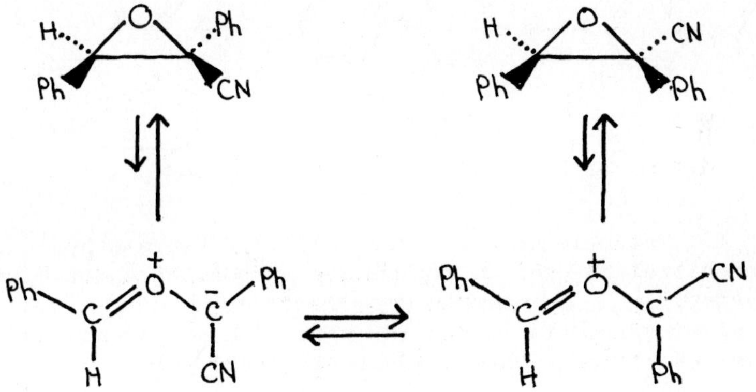

With carbonyl compounds the carbonyl ylides afford
dioxolanes (A. Robert, J.-J. Pommeret, and A. Foucaud,
Tetrahedron Letters, 1971, 231).

(5) In a unifying concept of the mechanism of the
ozonolysis of olefins a peroxy epoxide is suggested as
a key intermediate (P.R. Story *et. al.*, J. Amer. chem.
Soc., 1971, 93, 3044).

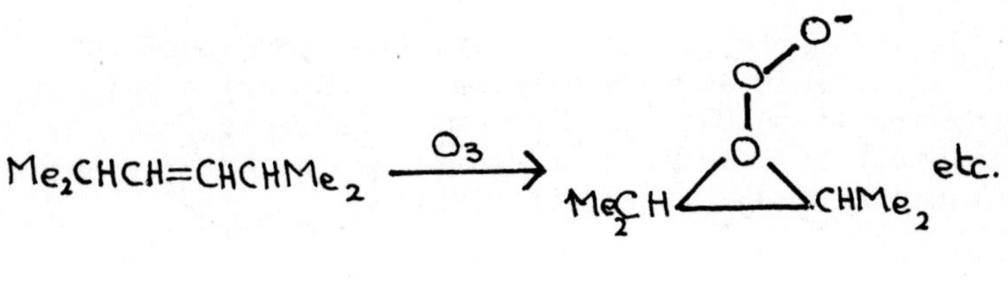

(6) 9,10-Dihydro-9,10-epoxyphenanthrene undergoes photo-rearrangement to the previously unknown 1,2:3,4-dibenzoxepin. The mechanism may be considered as an analogue of a photochemical Berson-Willcott rearrangement in which a [1,5] suprafacial shift of oxygen is involved (N.E. Brightwell and G.W. Griffin, Chem. Comm., 1973, 37).

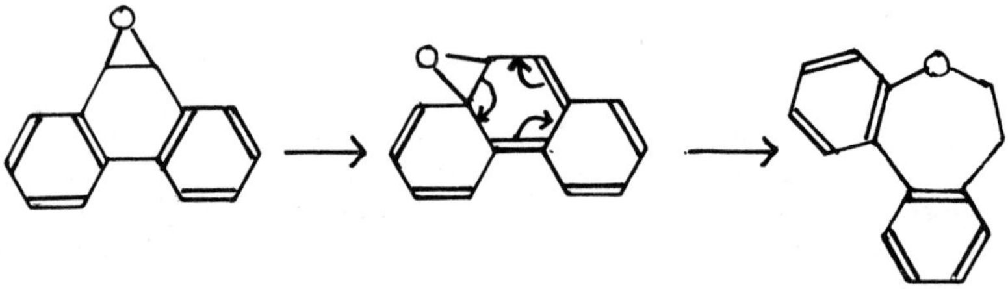

(7) The bridged anthracene, 1,4,7,8-tetramethyl-dibenzobarrelene epoxide rearranges in boiling chloroform to 1-acetyl-1,4,5-trimethyldibenzocycloheptatriene, rather than eliminating the bridge (H. Hart, J.B.-C. Jiang, and M. Sasaoka, J. org. Chem., 1977, 42, 3840).

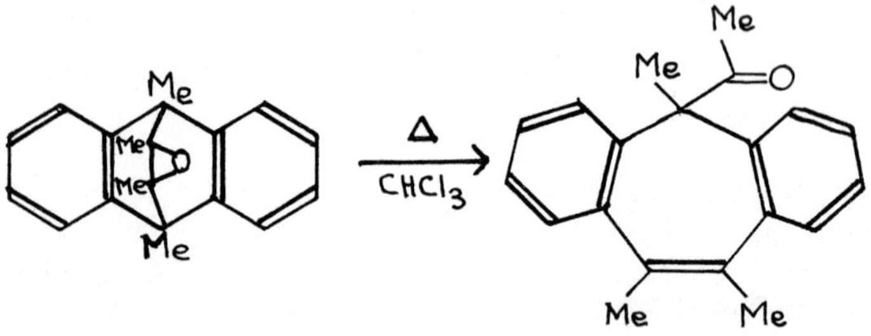

(8) The *syn*-oxabicyclo[5.1.0]octa-2,5-dienes (6) and (7) are liquid at room temperature and are a pair of heterocycles in mutual equilibrium *via* a rapid Cope rearrangement (H. Klein, W. Kursawa, and W. Grimme, Angew. Chem. internat. Edn., 1973, 12, 580).

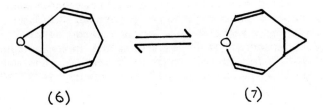

(6)   (7)

Selective catalytic hydrogenation of 9-oxabicyclo-
[6.1.0]nona-2,4,6-triene (8) protected by the 1,1,1-
-trifluoro-2,4-pentanedionatorhodium group yields 9-oxa-
bicyclo[6.1.0]nona-2,6-diene (9), which is stable at room
temperature, but rearranges at 90° to 4-oxabicyclo[5.2.0]-
nona-2,5-diene (10). The Cope rearrangement occurs only
in one direction (W. Grimme and K. Seel, *ibid.*, p.507).

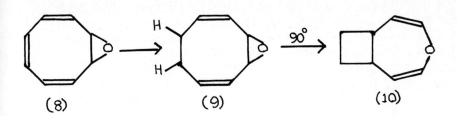

(8)   (9)   (10)

(9) *cis*-2-Vinyl-3-ethynyloxirane in benzene or carbon
tetrachloride on heating in a sealed tube rearranges
cleanly, *via* two Cope rearrangements to *cis*-2-ethynylcyclo-
propan-2-carboxaldehyde (N. Manisse and J. Chuche, J. Amer.
chem. Soc., 1977, 99, 1272).

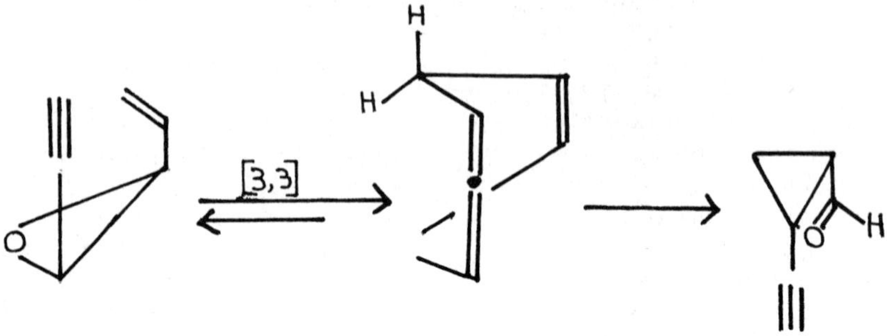

The rate of hydration 2-pyridyloxirane is greatly increased by the addition of divalent transition-metal ions, the magnitude of the catalytic effect decreasing in the order Cu(II)>Co(II)>Zn(II)>>Mn(II). The reaction with Cl⁻, Br⁻ and MeO⁻ is 100% regiospecific for β-attack (R.P Hanzlik and W.J. Michaely, Chem. Comm., 1975, 113).

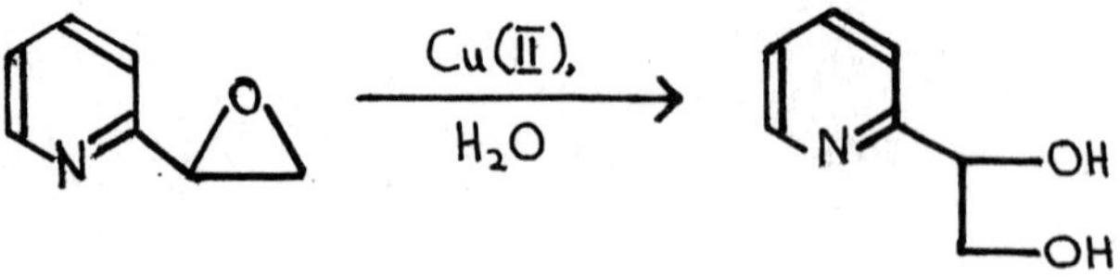

(b) *Oxiran-2-one*

The thermal decomposition of bis(α-naphthoyl)peroxide in benzene and carbon tetrachloride at 60° is a mixed spontaneous and induced radical reaction. Experiments with radical scavengers indicate the occurence of a large amount of cage recombination, the major part of which is proposed to give an α-lactone as an intermediate (J.E. Lettler and R.G. Zepp, J. Amer. Chem. Soc., 1970, 92, 3713). The α-lactone, 3,3-di-*tert*-butyloxiran-2-one (11) has been prepared by the action of ozone on

di-*tert*-butylketene in fluorotrichloromethane (Freon II)
at -78°. Evidence that the solution contains the
α-lactone is indicated by the following facts: (a) ammonia
gas gives a quantitative precipitation of di-*tert*-butyl-
glycine, (b) on warming the solution to -20° a polyester
(91%) is formed as a white precipitate, and (c) when hexafluoroacetone is added to the Freon II solution at -78°
and the mixture brought to room temperature, two
rearrangement products, namely 1-*tert*-butyl-1,2,2-trimethyl-
ethene and 3-*tert*-butyl-3,4,4-trimethyloxetan-2-one are
isolated after distillation. The polyester is presumably
formed by participation of the dipolar form (12).

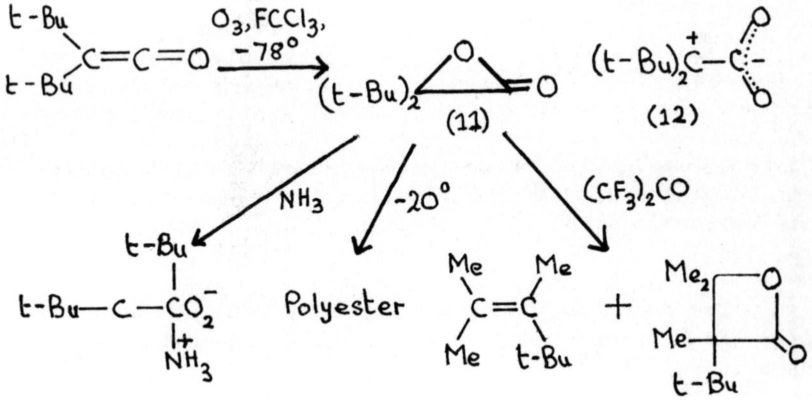

Ozone with diphenylketene does not afford a stable oxiran-
-2-one, a polyester being formed immediately at -78°
(R. Wheland and P.D. Bartlett, *ibid.*, p.6057).

   Photolysis of di-n-butylmalonylperoxide involves
photodecarboxylation to produce an α-lactone intermediate.
The dipolar form (14) has been trapped with methanol to
give the α-methoxy acid (W. Adam and R. Rückaschel, *ibid.*,
1971, 93, 557).

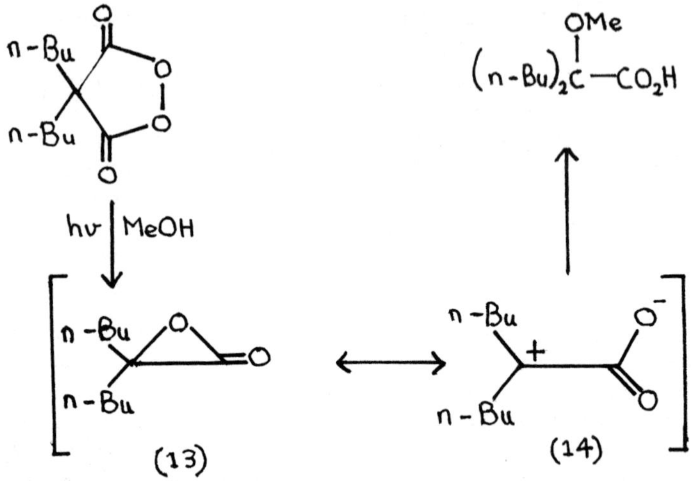

Di-n-butylmalonylperoxide irradiated as the neat liquid at 77K yields 3,3-di-n-butyloxiran-2-one (13) rather than the zwitterion (14) (O.L. Chapman *et. al., ibid.*, 1972, 94, 1365).

The photolysis of 2-vinyloxiranes and pentacarbonyl-iron has been studied (R. Aumann and R. Froehlich, Angew. Chem., 1974, 86, 309).

(c) Oxirene

Attempts have been made to obtain oxirene (1) by thermal reverse Diels-Alder fragmentation of (2). There appears to be some evidence that fragmentation of (2) to oxirene (1) and dimethyl phthalate occurs but the reaction pathway is more complex. At 300-400° the cyclohepta-trienecarboxaldehyde derivative (3) is obtained, but flash thermolysis at 600-800° produces some ketene, the expected rearrangement product of oxirene (E. Lewars and G. Morrison, Tetrahedron Letters, 1977, 501, Canad. J. Chem., 1977, 55, 966).

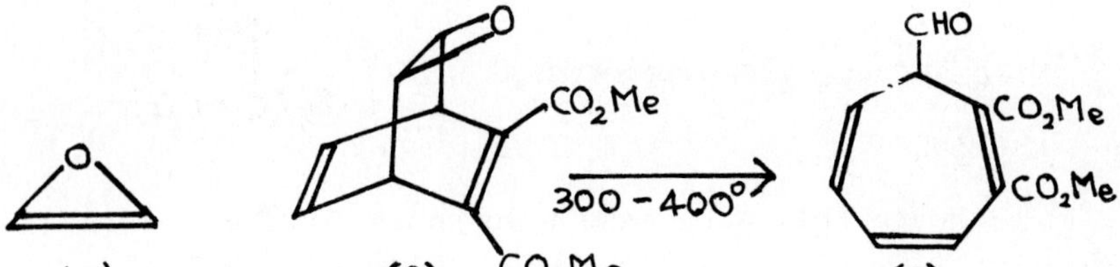

Evidence in favour of the oxocarbene $\rightleftharpoons$ oxirene equilibrium has been presented and a simple chemical test devised to indicate the participation of oxirenes in the photochemical and thermal decomposition of α-diazoketones (S.A. Matlin and P.G. Sammes, Chem. Comm., 1972, 11). Studies of the oxocarbene-oxirene equilibrium on the pathway to the Wolff rearrangement show a 13-16% oxirene participation in the formation of ketene on photolysis of 1-diazoethan-2-one (K.-P. Zeller, Angew. Chem. internat. Edn., 1977, 16, 781; Tetrahedron Letters, 1977, 707; Tetrahedron, 1977, 33, 453).

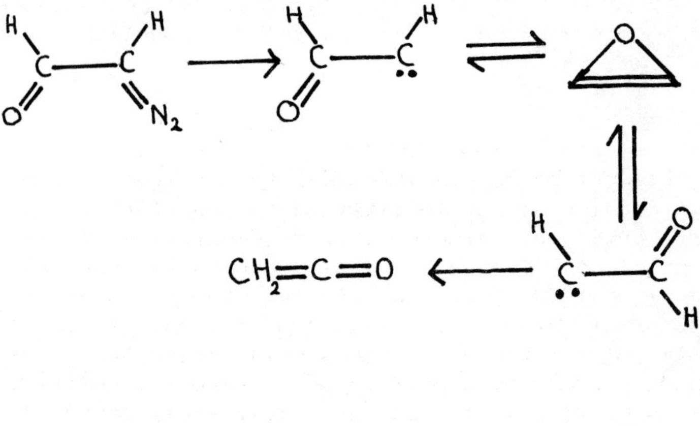

3. Three-membered Rings Containing One Sulphur Atom

(a) Thiiranes (ethylene sulphides)

(i) Synthesis

Although thiiranes have been known for many years, synthetic approaches to the rings system are somewhat limited.

(1) In general, oxiranes can be converted to thiiranes by the use of thiocyanate ion, thiourea, or thioamides and related compounds. Thiiranes may be conveniently and rapidly synthesised from the corresponding oxiranes and triphenyl- or tri-n-butyl-phosphine sulphide in benzene in the presence of trifluoroacetic acid. In analogy to the thiocyanate and thiorea reactions, it is proposed that the reaction follows the general mechanism (T.H. Chan and J.R. Finkenbine, J. Amer. chem. Soc., 1972, 94, 2880).

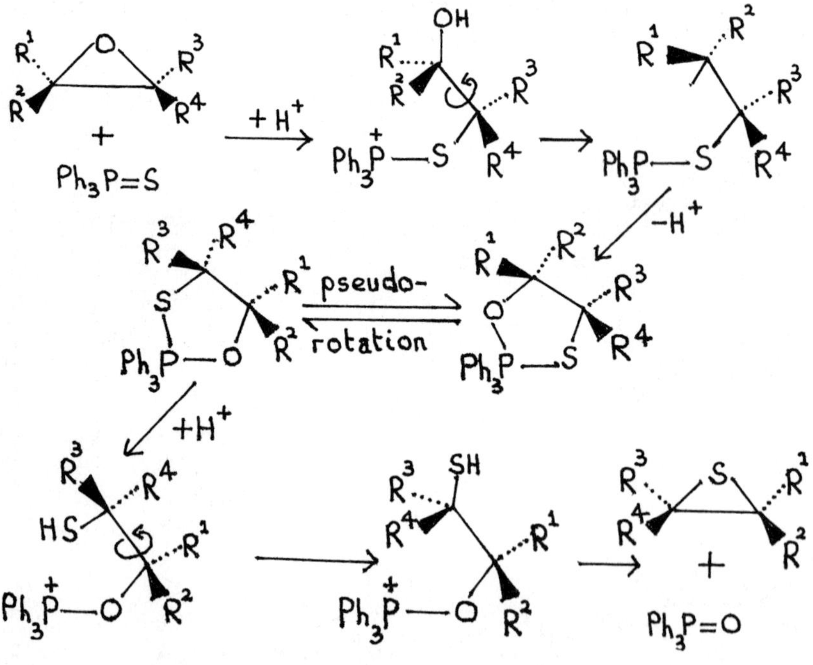

(2) Treatment of ether solutions of certain alicyclic β-iodothiocyanates with methanolic potassium hydroxide at room temperature affords thiiranes in acceptable yields. It is assumed that trans addition of iodine thiocyanate

to the carbon-carbon double bond occurs, followed by preferential hydrolysis of the thiocyanate moiety by base and rapid ring closure to the thiirane.

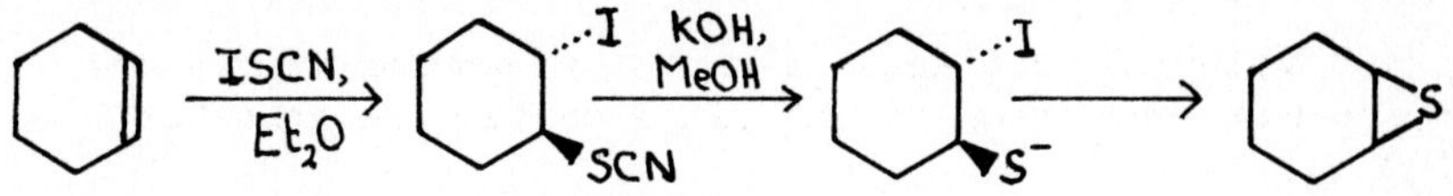

The procedure does not appear suitable when applied to acyclic olefins (J.C. Hinshaw, Tetrahedron Letters, 1972, 3567).

(3) 3-Methylbenzothiazole-2-thione in the presence of trifluoroacetic acid converts oxiranes into thiiranes stereospecifically. The reaction is rapid and almost quantitative. By analogy with the thiocyanate and thiourea reactions the following mechanism has been proposed, and supported by the evidence that *cis*-stilbene oxide is converted exclusively into *cis*-stilbene sulphide (V. Caló, *et. al.*, Chem. Comm., 1975, 621).

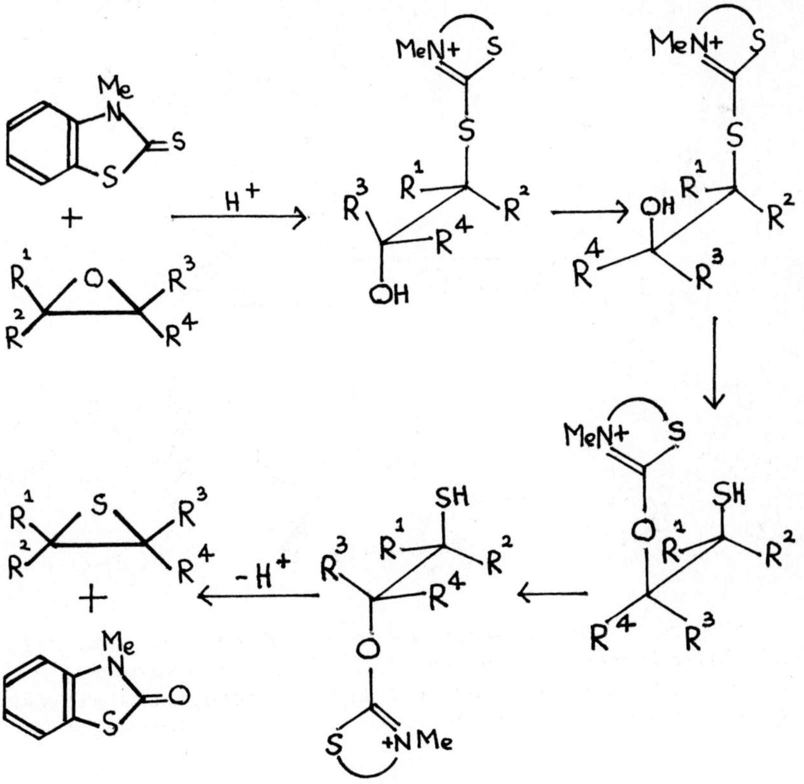

(4) Thiiranium hexachloroantimonates are obtained in high yield (85-90%) by the addition of the appropriate alkene to a solution of methyl(bismethylthio)sulphonium hexachloroantimonate in methylene chloride at 0° or in sulphur dioxide at -60°.

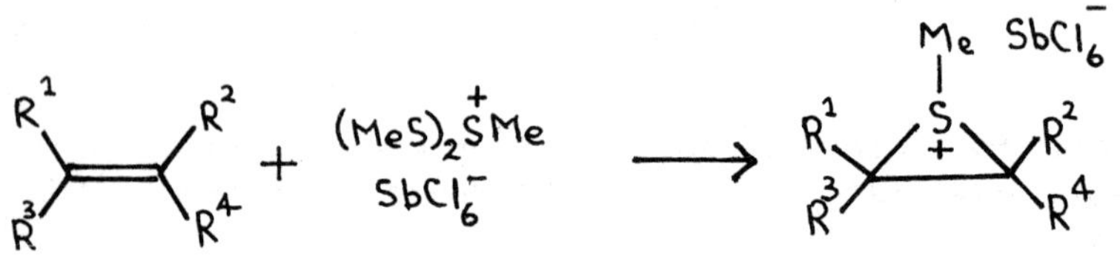

Examples are known where R^1, R^2, R^3, R^4 are either Me or H; $R^1 = R^4 = Et$, $R^2 = R^3 = H$; and $R^1R^2 = -(CH_2)_6-$, $R^3 = R^4 = H$ (G. Capozzi $et.$ $al.$, Tetrahedron Letters, 1975, 2603).

(5) Metallation of 2-(alkythio)-2-oxazolines followed by addition of an aldehyde or ketone lead to a thiirane (60-70% yield).

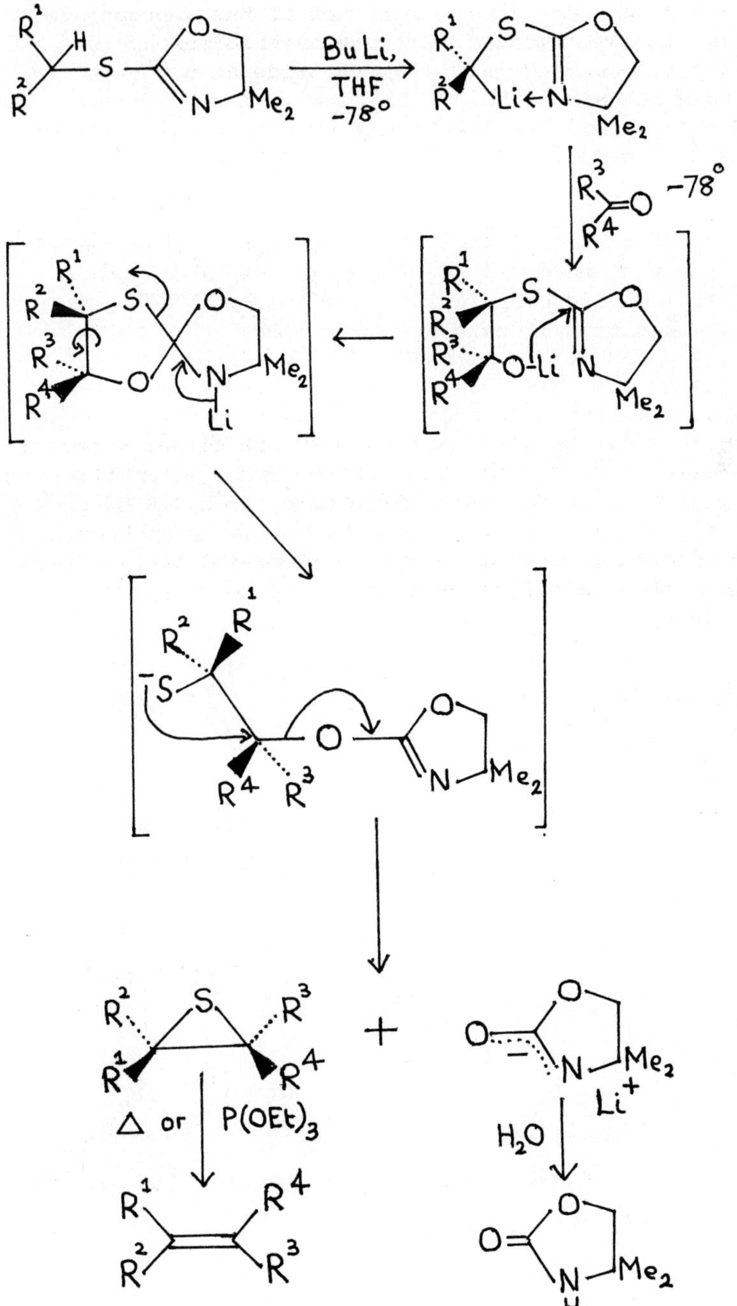
BuLi,
THF
-78°
-78°
Li
O-Li
Me2
△ or
P(OEt)3
H2O
Li+

The process is also useful for the direct synthesis of
alkenes and dienes by extrusion of the sulphur from
thiiranes. In many cases a high degree of stereo-
selectivity is observed in the alkene formation. An
asymmetric synthesis of chiral thiiranes has also been
achieved providing these substances in 19-32% enantiomeric
excess (A.I. Meyers and M.E. Ford, J. org. Chem., 1976,
<u>41</u>, 1735).

     (6) Treatment of methyl and other alkyl thiocyanates
with a variety of bases, including proton-specific bases
such as lithium diisopropylamide in various solvents,
followed by addition of carbonyl compounds failed to
produce any detectable thiiranes. An alteration of the
reagent design was therefore needed and the CN group was
replaced by a system less prone to act as a leaving group
but which retained an electrophilic multiple bond to serve
as an "alkoxide trap". The lithium alkylthio-thiazoline
(1, R = H or Ph) reacts with aldehydes and ketones to give
alcohols (2, R = H or Ph), which when treated with base,
acid, or heat give thiiranes (C.R. Johnson *et. al.*,
Tetrahedron Letters, 1975, 2865).

(1)   (2)
R = H (75%)
R = Ph (80%)

$80°$ R = H

R = Ph
NaOMe, MeOH
$25°$

(73%)   (100%) Ph

It is believed that a variety of other types of sulphur stabilised carbanionic reagents should be capable of effecting such transformations.

(7) When the dry sodium salt of toluene-p-sulphonyl-hydrazone of 2,2,4,4-tetramethylthietanone is heated slowly to 150° in vacuo, 2,2-dimethyl-3-dimethylmethylenethiirane is the only volatile product formed. The thiirane is desulphurised on heating with hexamethylphosphoramide in benzene to afford tetramethylallene and treatment with 70% perchloric acid in ether gives a crystalline 1,4--dithiane (A.G. Hortmann and A. Bhattacharjya, J. Amer. chem. Soc., 1976, 98, 7081).

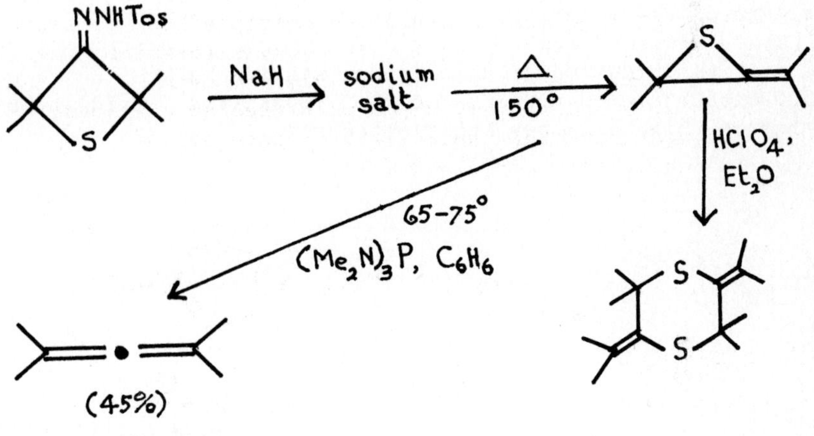

Methylenethiirane (allene episulphide) is obtained on the pyrolysis of (3) or the anthracene adduct (4), and on heating the tosylhydrazide salt (5) (E. Block *et. al.*, *ibid.*, 1978, 100, 7436).

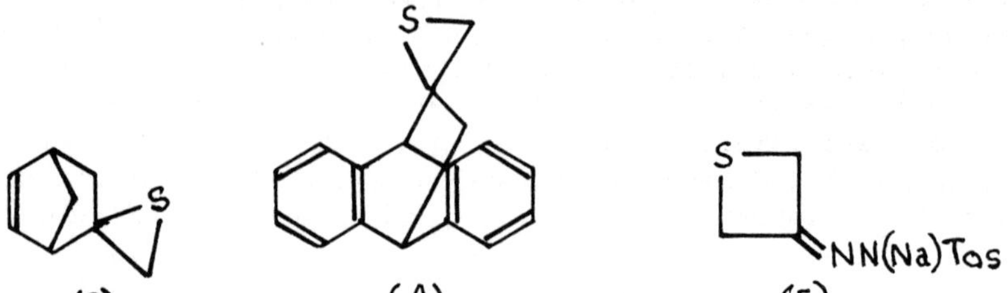

(3)   (4)   (5)

Experimental evidence indicates that allene episulphide
is thermodynamically more stable than cyclopropanethione,
in apparent contrast to the order of stability in the
allene oxide-cyclopropanone system.

(8) Oxidation of thioketenes bearing bulky
substituents with nitrones of the 1-pyrroline 1-oxide type
affords thiiran-2-ones (thiolactones).

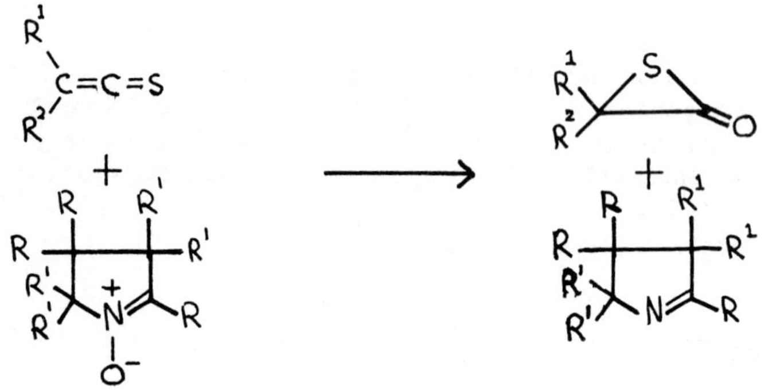

R^1 = CHMe$_2$, R^2 = t-Bu;
R^1 = R^2 = t-Bu;
R^1R^2 = -CMe$_2$-(CH$_2$)$_3$CMe$_2$-

Structural proof is provided by X-ray structure
determination of thiirane-2-one [R^1R^2 = -CMe$_2$-(CH$_2$)$_3$-CMe$_2$-].

The same compound on heating undergoes qualitative
decarbonylation to a thioketone. It is relatively stable
to hydrolysis and solvolysis, probably as a result of the
bukly alkyl group; on heating decarbonylation occurs
to form the thioketone. Addition products are obtained
with methanol and benzylamine (E. Schaumann and U. Behrens,
Angew. Chem. internat. Edn., 1977, 16, 722).

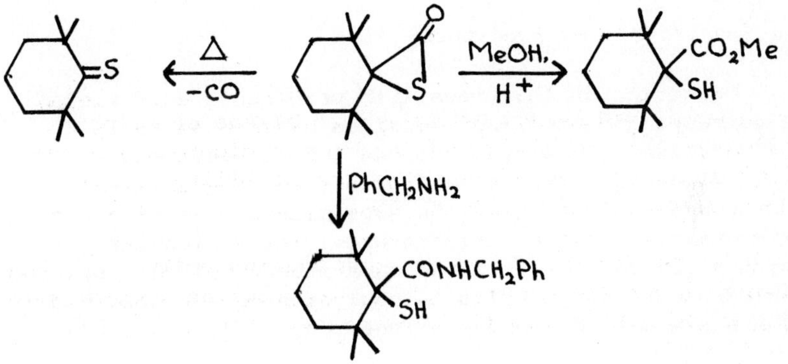

(9) Equimolar amounts of tosyl isothiocyanate and
diphenyldiazomethane react in anhydrous ether at 0° with
nitrogen evolution to give tosyl 2,2-diphenylthiiranimine
(67%). Also obtained is 4,4-diphenyl-5-tosylimino-1,2,3-
-trithiolene (1%).

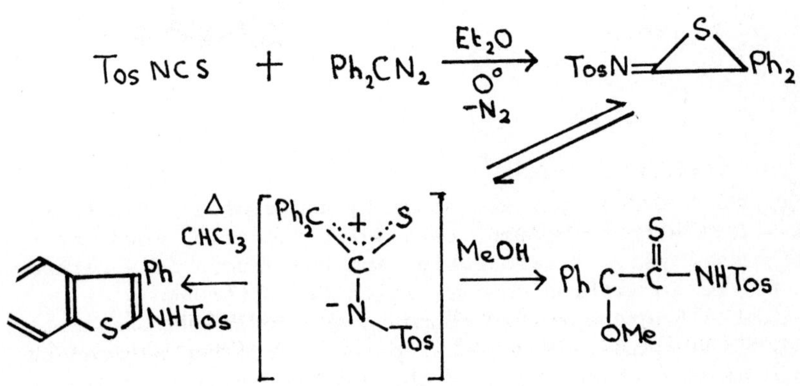

X-Ray structure analysis of the thiiranimine shows that
the four atoms S, C-2, C-3, and N are almost coplanar.
In chemical reactions it is cleaved at the unusually long
S-C-2 bond. Thus a chloroform solution of the thiiranimine
left at room temperature for a number of days yields a
benzo[b]thiophene quantitatively. Methanolysis gives tosyl
2-amino-1,1-diphenyl-1-methoxyethan-2-thione in
quantitative yield (G. L'abbé *et. al.*, Angew. Chem.
internat. Edn., 1978, <u>17</u>, 195).

(ii) Properties and Reactions

Oxidation of thiiranes with perbenzoic acid yields
the corresponding thiirane 1-oxides having the anti-
configuration with respect to the substituents and
sulphinyl oxygen. Thiirane 1-oxides bearing an alkyl
substituent and oxygen on the same side of the ring are
quite unstable at room temperature. Intramolecular
hydrogen abstraction from the substituent followed by ring
opening occurs and allylic thiosulphinates are obtained
(K. Kondo and A. Negishi, Tetrahedron, 1971, <u>27</u>, 4821).

$$R^1 \overset{S}{\triangle} R^2 + PhCO_3H \xrightarrow[\text{2. dry } NH_3]{\text{1. } CH_2Cl_2, O^\circ} R^1 \overset{\overset{O^-}{\underset{|}{S}}}{\triangle} R^2 + PhCO_2NH_4$$

R^1, R^2 = H, Me, Ph

Thiirane 1-oxides may decompose by two routes. The first
is a facile rearrangement to a sulphenic acid, when the
stereochemistry is favourable, and the second is a pathway
of higher activation energy, which leads through a
partially stereospecific route to an olefin and presumably
sulphur monoxide. The latter pathway involves a diradical
intermediate which explains the non-stereospecific olefin
formation due to bond rotation in this intermediate

(J.E. Baldwin, G. Höfle, and Se Chun Choi, J. Amer. chem. Soc., 1971, 93, 2810).

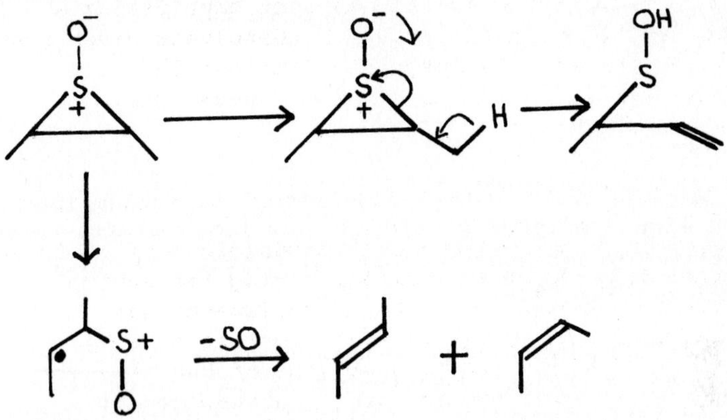

The rate of exchange of the trifluoromethyl groups between different positions (automerisation) of perfluorotetramethyl Dewar thiophene and its *exo*-S-oxide (a thiirane 1,1-oxide) have been compared and a marked difference found between the rates of intramolecular exchange. The former being slow and the latter extremely rapid, (C.H. Bushweller, J.A. Ross, and D.M. Lemal, *ibid.*, 1977, 99, 629).

(b) Thiirenes

The trimethylthiirenium ion is quantitatively formed by a reaction of methyl(bismethylthio)sulphonium hexachloroantimonate with excess but-2-yne at about -80°C in liquid sulphur dioxide.

$$MeC \equiv CMe$$
$$+ \ +$$
$$(MeS)_2 \overset{+}{S} Me \ SbCl_6^-$$

$$\xrightarrow{\ SO_2,\ -80° \ }$$

$$+ \ MeSSMe$$

1,2,3-Trimethyl- and 1-methyl-2,3-diethylthiirenium ions stable in liquid sulphur dioxide at low temperatures have been observed by nmr spectroscopy (G. Capozzi *et. al.*, Chem. Comm., 1975, 248).

1,2,3-Trimethylthiirenium hexachloroantimonate has been obtained in the solid state but it decomposes above -40° and is very sensitive to moisture. 1-Methyl-2,3-di- -*tert*-butylthiirenium hexachloroantimonate is much more stable and can be obtained as white crystals at room temperature (*idem.* Tetrahedron Letters, 1977, 911). The thiirenium ion at 173K shows a pyramidal conformation, (R. Destro, T. Pilali, and M. Simonetta, Chem. Comm., 1977, 576).

The preparation and characterisation of thiirene has been reported. Irradiation of argon matrix-isolated 1,2,3-thiadiazole with light of λ = 232-280 nm produces ethynyl mercaptan and thioketene along with a substance which it is suggested must be thiirene or species derived from thiirene (A. Krantz and J. Laureni, J. Amer. chem. Soc., 1977, 99, 4842).

$$+ \ H_2C=C=S \ + \ HC \equiv CSH$$

6-Methoxycarbonyl-1,2,3-benzothiadiazole on heating yields nitrogen and 3-methoxycarbonylbenzothiirene, which subsequently forms 2,7- and 2,8-dimethoxycarbonylthianthrene.

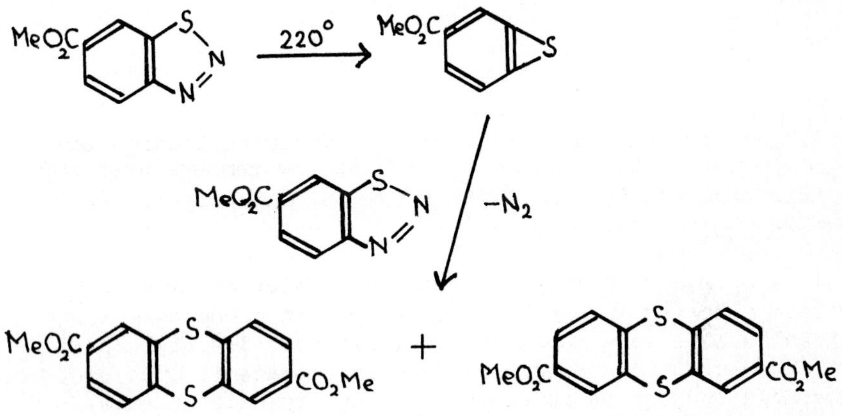

It is not clear whether benzothiirene exists as a transition state or in the ground state (T. Woolridge and T.D. Roberts, Tetrahedron Letters, 1977, 2643). Different views excluding the intervention of a benzothiirene as a reaction intermediate have been expressed (L. Benati, P.C. Montevecchi and G. Zanardi, J. org. Chem., 1977, 42, 575).

The above results have been criticised but supported, and a much clearer indication of the presence of thiirene obtained by the photolysis of 4-methyl-5-ethoxycarbonyl- and 5-methyl-4-ethoxycarbonyl-1,2,3-thiazole. Of the expected eight ir absorption bands of the parent thiirene molecule seven have been located and a tentative assignment of them made. Thiirene, trifluoromethylthiirene, and benzothiirene are highly unstable, but electron-withdrawing substituents exert a marked stablising effect on the 4π-electron ring system: ethoxycarbonylmethylthiirene is stable up to at least 73K (M. Torres et. al., ibid., 1978, 43, 2490).

Evidence indicates that benzothiirenes rather than thioketocarbenes are intermediates in the production of dibenzodithianes by the thermolysis of sodium σ-bromo-benzothiolates, when two products are obtained. This contrasts with the thermolysis of sodium σ-halogeno-phenoxides, which appear to decompose *via* a ketocarbene to give only one dibenzodioxane (J.I.G. Cadogan, J.T. Sharp and M.J. Trattles, Chem. Comm., 1974, 900).

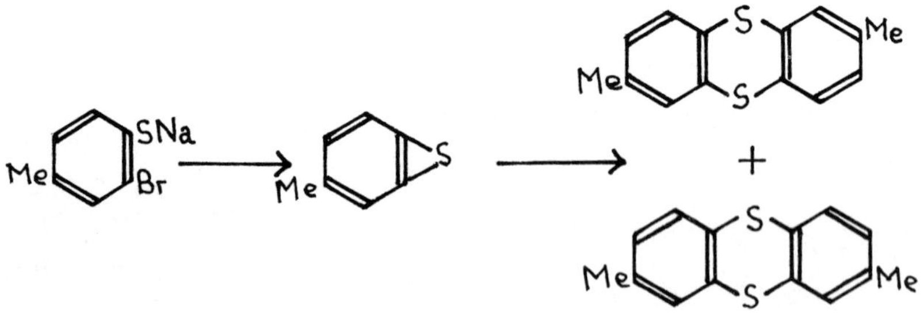

The results of a non empirical SCF-MO calculation on the linear and cyclic structures of the ion $C_2H_3S^+$ indicate that the cyclic form is 65.9 kcal/mole more stable than the linear form.

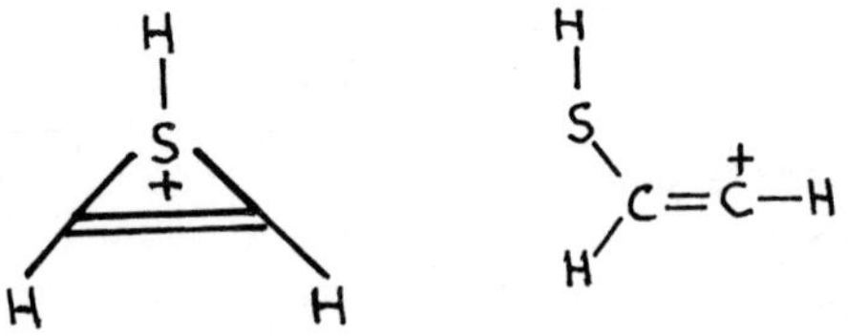

The stability of the bridged species depends greatly on the electronegativity of the peripheral atoms or group, which is in agreement with the increased stability of the thiirenium ion associated with the substitution of a hydrogen with a methyl (A.S. Denes and I.G. Csizmadia, *ibid.*, 1972, 8).

　　Treatment of 4,5-disubstituted 1,2,3-thiadiazoles
with di-iron nonacarbonyl gives thioxocarbene complexes,
(e.g. 1), but when the substituents are different,
formation of the cross-over products (2) and (3) occurs,
implicating thiirene derivatives as symmetrical
intermediates (P.G. Mente and C.W. Rees, *ibid.*, p.418;
T.L. Gilchrist, Mente and Rees, J. chem. Soc., Perkin I,
1972, 2165).

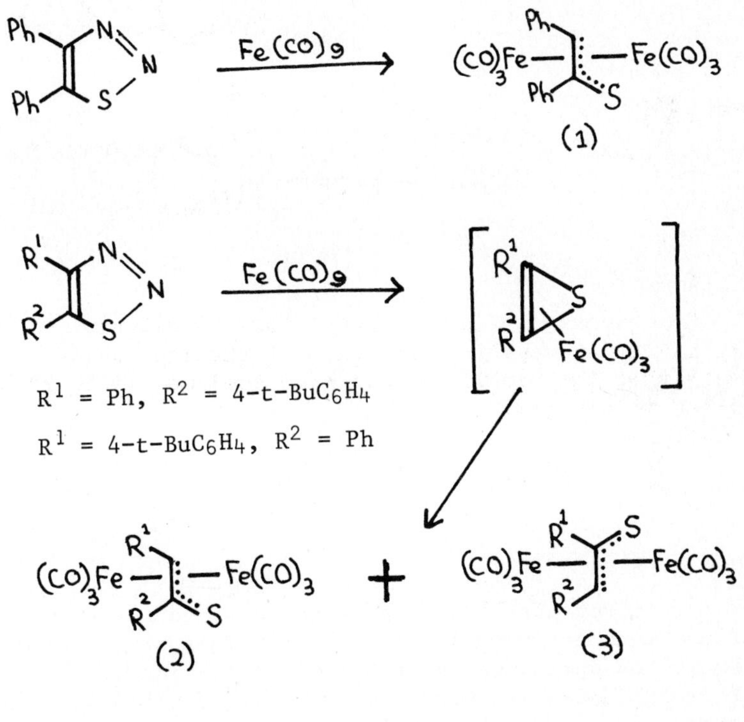

R^1 = Ph, R^2 = 4-t-BuC_6H_4

R^1 = 4-t-BuC_6H_4, R^2 = Ph

　　Treatment of α,α'dibromodibenzyl sulphone with
triethylamine in boiling methylene chloride yields 2,3-
diphenylthiirene 1,1-dioxide.　On heating at its melting
point or boiling for several hours in benzene it decomposes
to diphenylacetylene and sulphur dioxide.　Reaction with
hydrazine gives desoxybenzoin azine, with hydroxylamine

desoxybenzoin oxirine, phenylmagnesium bromide affords
diphenylacetylene and the salt of benzenesulphinic acid.
The reduction of the double bond using aluminium amalgam
in wet ether −45° gives *cis*-2,3-diphenylthiirane 1,1-
-dioxide.

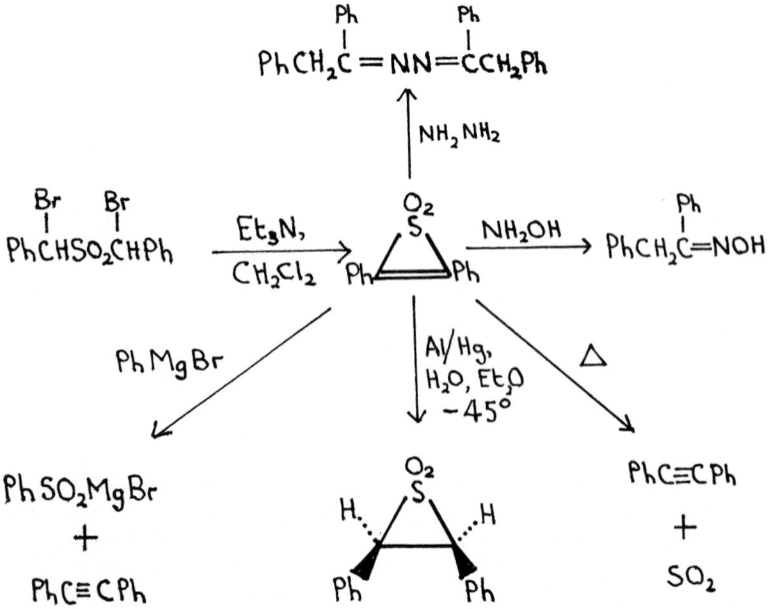

It is not possible to extend the above reaction
(Ramberg-Bäcklund reaction) to the synthesis of the alkyl-
-substituted analogues since α,α'-dialkyl- and α,α'-
-dibromosulphones are too weakly acidic to react with an
organic base such as triethylamine. They react with
aqueous sodium hydroxide, but the thiirene dioxide which
are clearly formed as intermediates do not survive the
reaction conditions. 2-Methylthiirene 1,1-dioxide is
obtained by dehydrobrominating 2-bromo-2-methylthiirane
1,1-dioxide, formed from α-bromoethanesulphonyl bromide
and triethylamine in the presence of diazomethane.

In order to obtain 2,3-dimethylthiirene 1,1-dioxide an alcohol-free solution of diazoethane must be used to avoid destruction of the intermediate thiirane and a stronger base such as 1,5-diazabicyclo[4.3.0]non-5-ene is required for the elimination (L.A. Carpino, J. Amer. chem. Soc., 1971, 93, 476).

The electronic structure of thiirene 1,1-dioxide has been investigated by uv photoelectron spectroscopy, and the results indicate that it is less aromatic than cyclopropenone or tropone (C. Müller, A. Schweig, and H. Vermeer, *ibid.*, 1975, 97, 982). The reaction of 2,3- -diphenylthiirene 1,1-dioxide with enamines provides novel acyclic and cyclic systems (M.H. Rosen and G. Bonet, J. org. Chem., 1974, 39, 3805).

4. Three-membered Rings With One Selenium Atom

Although seleniranes have never been isolated, they have been observed as transient intermediates in the reaction between selenium and olefins (A.B. Callear and W.J.R. Tyerman, Proc. chem. Soc., 1964, 296; Trans. Faraday Soc., 1965, 61, 2395; 1966, 66, 371; Tyerman *et. al.*, J. Amer. chem. Soc., 1966, 88, 4277). Treatment of alkenes and alkynes with areneselenyl hexafluorophosphates

or hexafluoroantimonates affords seleniranium and
selenirenium salts respctively (G.H. Schmid and
D.G. Garratt, Tetrahedron Letters, 1975, 3991).

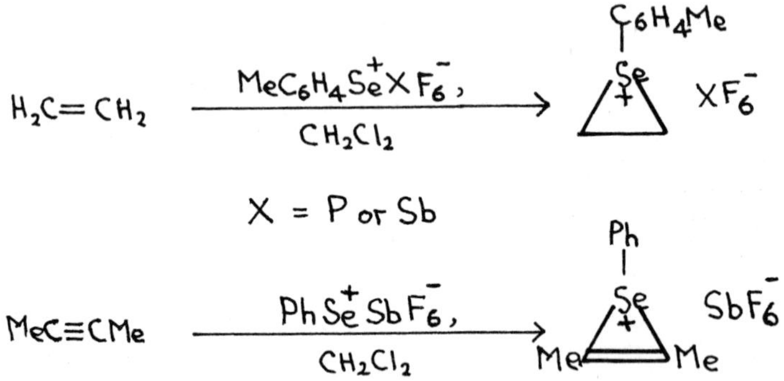

$$X = P \text{ or } Sb$$

5. Three-membered Rings Containing One Nitrogen Atom

(a) Aziridines (ethyleneimines)

Aziridines are of interest as biological alkylating
and anticancer agents.

(i) Synthesis

(1) Evidence has been presented for the existence
of *O*-nitrenes and for the facile addition of alkoxynitrenes
to alkenes, thus providing a new route to pyramidally
stable aziridines. The oxidation of methoxyamine with
lead tetracetate in the presence of excess tetramethyl-
ethene at −50° gives 1-methoxy-2,2,3,3-tetramethyl-
aziridine, also obtained by cyclisation, with sodium
methoxide in methanol, of the product resulting from the
addition of *N*-chlorosuccinimide to equivalent amounts of
2,2,3,3-tetramethylethene and methoxyamine in
dichloromethane.

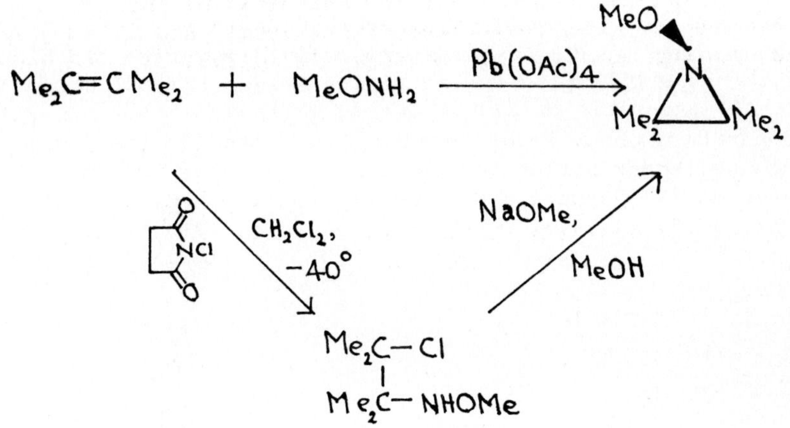

$$Me_2C=CMe_2 \quad + \quad MeONH_2 \xrightarrow{\ Pb(OAc)_4\ }$$

$$\begin{array}{c} Me_2C-Cl \\ | \\ Me_2C-NHOMe \end{array}$$

1-Alkoxyaziridines exhibit an appreciable energy barrier to nitrogen inversion, and by analogy with *N*-halogeno-, *N*-amino-aziridines, and oxaziridines exhibit pyramidal stability at room temperature. A favourable combination of inductive and electrostatic factors effectively stabilise the pyramidal configuration of those aziridines in which the ring nitrogen is attached to halogen, nitrogen, or oxygen atom (S.J. Brois, J. Amer. chem. Soc., 1970, **92**, 1079).

(2) Treatment of olefins with positively charged halogen and cyanamide results in addition of halogen and cyanoamine across the double bond. The halogenocyanoamine cyclises spontaneously or in the presence of alkali to cyanoaziridines (K. Ponsold and W. Ihn, Tetrahedron Letters, 1970, 1125).

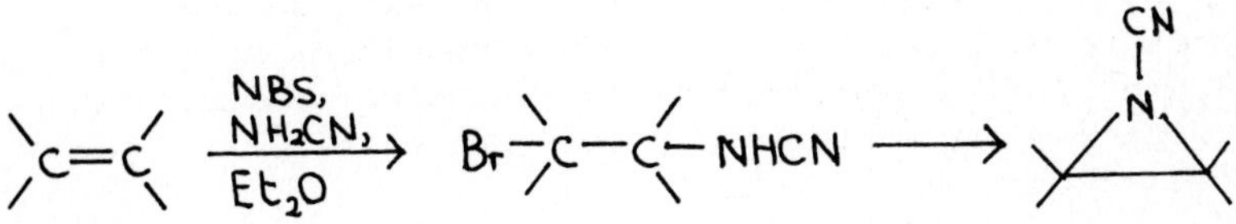

(3) Treatment of the α-bromoamidine (1) with potassium *tert*-butoxide in ether results in a 1,3 elimination, which proceeds with both high regio- and high stereo-selectivity to the aziridine imines (2) and (3). At -40° the ratio of 2:3 is 86:14, while at room temperature about equal amounts of (2) and (3) are formed. Valence isomerisation between (2) and (3) has not been detected (H. Quast and E. Schmitt, Angew. Chem. internat. Edn., 1970, 9, 381).

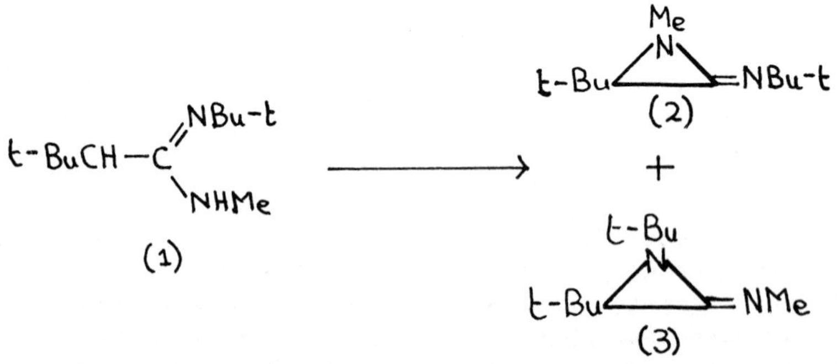

(4) 1-Phthalimidoaziridines, with a conjugating substituent - a carbonyl group or an aromatic ring, when irradiated in the presence of an olefin, undergo an exchange reaction which results in the formation of a new phthalimido aziridine incorporating the added olefin (T.L. Gilchrist, C.W. Rees, and E. Stanton, J. chem. Soc., C, 1971, 988).

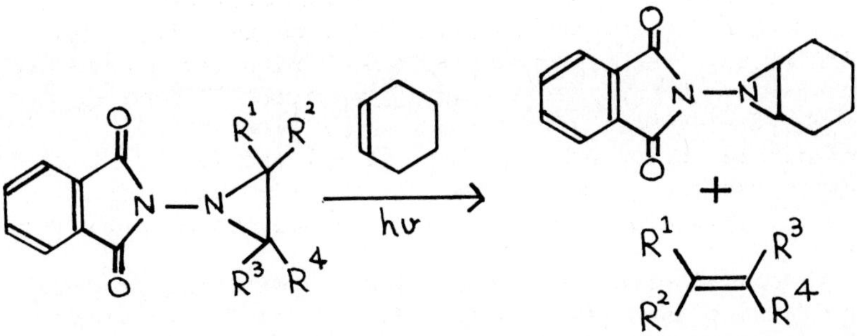

(5) The reaction of amines and formaldehyde with
an aminating agent, hydroxylamine-O-sulphonic acid gives
N-alkylglycinonitriles (A.H. Lawrence *et. al.*, Tetrahedron
Letters, 1972, 2025), and not as previously stated 2-alkyl-
-2,4-diazabicyclo[1.1.0]butanes (A.A. Dudinskaya *et. al.*,
Tetrahedron, 1971, <u>27</u>, 4053).

(6) Photolysis of a solution of ethyl azidoformate
and 1,2-dimethylcyclobutene in methylene chloride gives
N-ethoxycarbonyl-1,4-dimethyl-5-azabicyclo[2.1.0]pentane,
which is stable and remains unchanged on boiling in carbon
tetrachloride for 24h (J.N. Labows, Jr., and D. Swern,
Tetrahedron Letters, 1971, 4523).

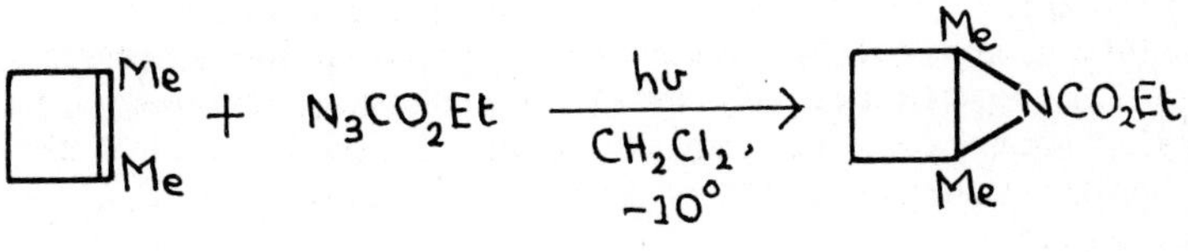

(7) An investigation has been made into the
involvement of 1-azirine in the formation of aziridines
by the reaction between Grignard reagents and oximes
(R. Chaabouni, A. Alurent, and P. Mison, *ibid.*, 1973,
1343).

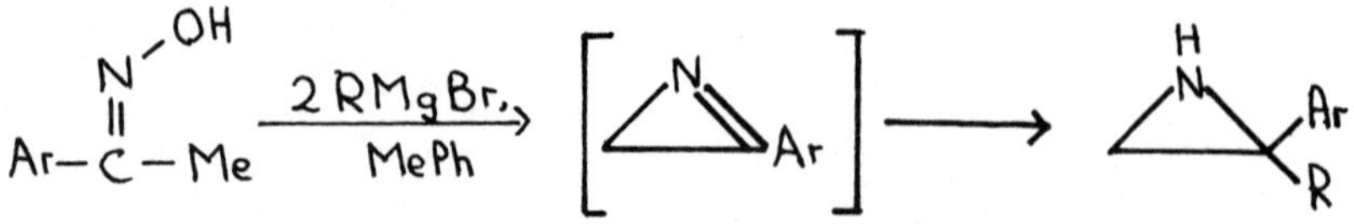

The reaction between oximes, RCHPh.CPh=NOH(R=Et or CHMe$_2$) and Grignard reagents, R^{1}MgBr (R^1 = Me or Et) to yield aziridines is stereospecific only when the difference between the steric hindrance of the R and Ph groups is significant (R. Bartrick and Laurent, Compt. rend., 1974, 279, C, 289). Treatment of the *trans*-phenyl oximes with an excess of *iso*-butylmagnesium bromide gives the aziridine (60-65%).

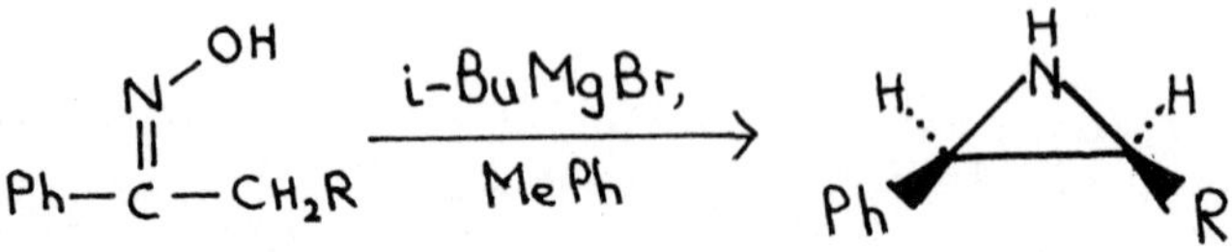

R = Me or Ph

2-Phenyl-3-isopropylaziridine is prepared regiospecifically (40%) from a mixture of the *cis*- and *trans*-isomers of benzyl isopropyl ketoxime (Y. Diab, Laurent, and Mison Tetrahedron Letters, 1974, 1605; Bull. Soc., chem. Fr., 1974, 2202).

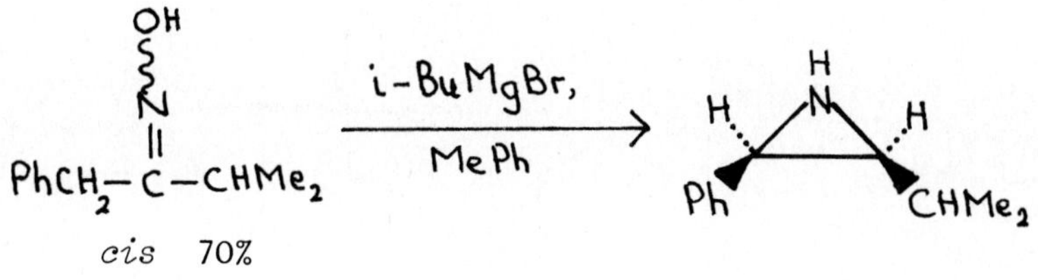

The preparation of azirdines by the reaction between a
Grignard regagent or organolithium and a ketoxime is
limited and a more reliable versatile reaction involves
a ketone dimethylhydrazone methiodide with a Grignard
reagent (G. Alvernhe *et. al.*, Tetrahedron Letters, 1975,
335).

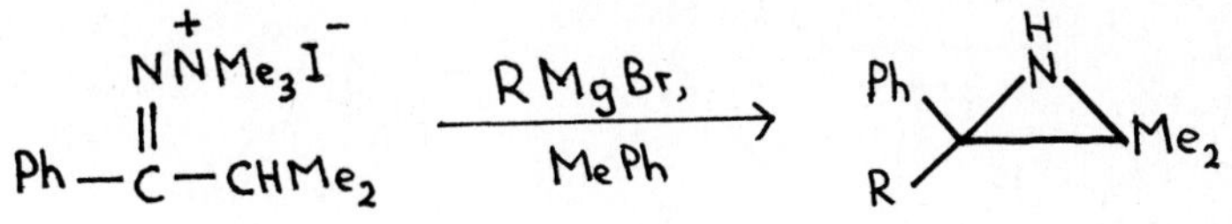

(8) Prolonged heating of 1,1-dibromo-2,2-dimethyl-
cyclopropane with *tert*-butylamine yields the bromoallyl-
amine (4), which with sodamide in liquid ammonia gives 1-
tert-butyl-2-isopropylideneaziridine (5) (Quast and W. Risler,
Angew. Chem. internat., Edn., 1973, 12, 414).

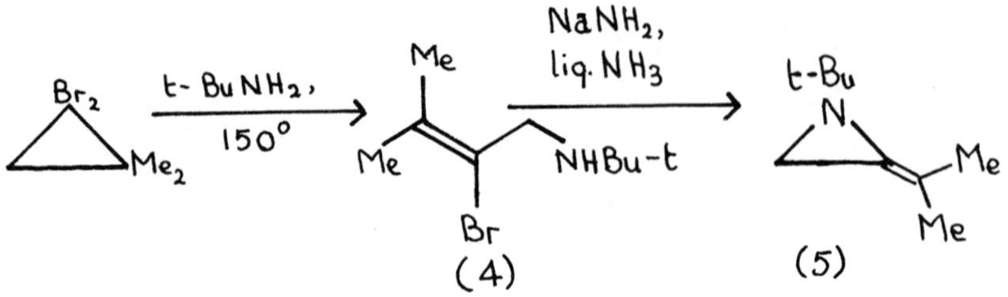

(4)   (5)

Nmr investigations have presented proof of the methylene-
aziridinecyclopropanimine valence isomerisation. The
methyleneaziridine (5) decomposes slowly at 120° and
rapidly at 190° to give quantitatively the olefin and
nitrile, which can arise only by cheletropic decomposition
of the cyclopropanimine intermediate (6).

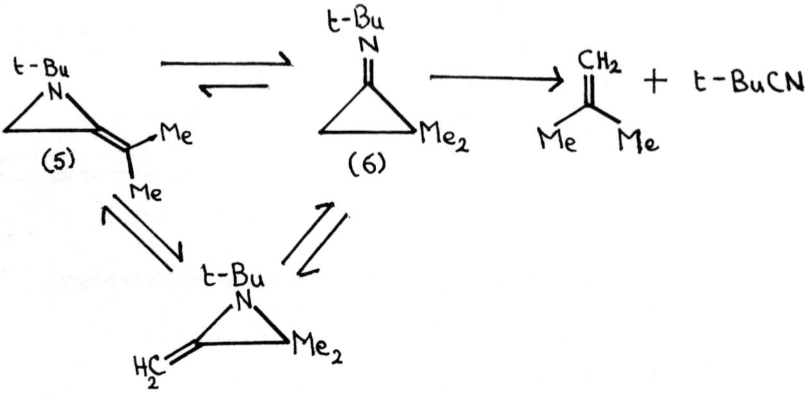

Alkyl diazoacetates react with N,N'diisopropylcarbodiimide
in the presence of transition metal salts, for example,
copper (II) trifluoromethanesulphonate (copper triflate)
or rhodium (II) acetate, to give 2-alkoxycarbonyl-1-
-isopropyl-3-isopropyliminoaziridine (A.J. Hubert et. al.,
Tetrahedron Letters, 1976, 1317).

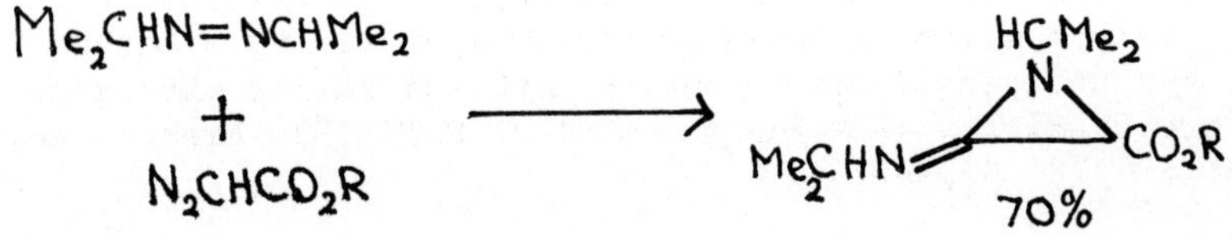

$$Me_2CHN=NCHMe_2$$
$$+$$
$$N_2CHCO_2R$$

⟶

(structure with 70%)

The triethylamine-catalysed decomposition of
N-(4-nitrobenzenesulphonoxy)urethane produces ethoxycarbonyl nitrene, which reacts with allene and 1,1-dimethyl-
allene to give by 1,2-cycloaddition 2-methylene-1-ethoxy-
carbonylaziridine and 2-isopropylidene-1-ethoxycarbonyl-
aziridine respectively. The latter compound on heating
isomerises to 5,5-dimethyl-2-ethoxy-4-methylene-2-oxazoline
(E.M. Bingham and J.C. Gilbert, J. Org. Chem., 1975, **40**,
224). Tetramethyallene fails to produce an adduct with the
nitrene.

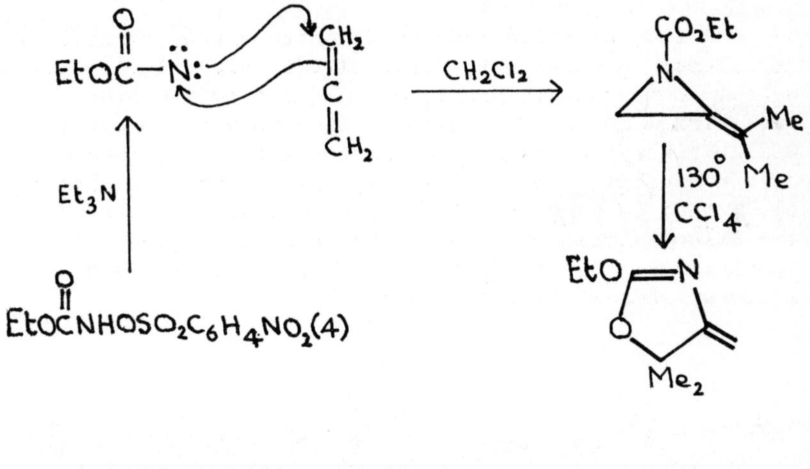

(reaction scheme with CH_2Cl_2, $130°$, CCl_4)

(9) 2-Iodoalkyl azides undergo a facile reaction
with aryl- and alkyl-dichloroboranes to produce β-iodo
secondary amines, which on treatment with base ring close
to give the corresponding *N*-aryl- and *N*-alkyl-aziridines

in good yield. The stereocheimstry of the original 2-
-iodoalkyl azide is maintained, thus providing a method
of synthesising N-aryl- and N-alkyl-aziridines with known
stereochemistry (A.B. Levy and H.C. Brown, J. Amer. chem.
Soc., 1973, <u>95</u>, 4067).

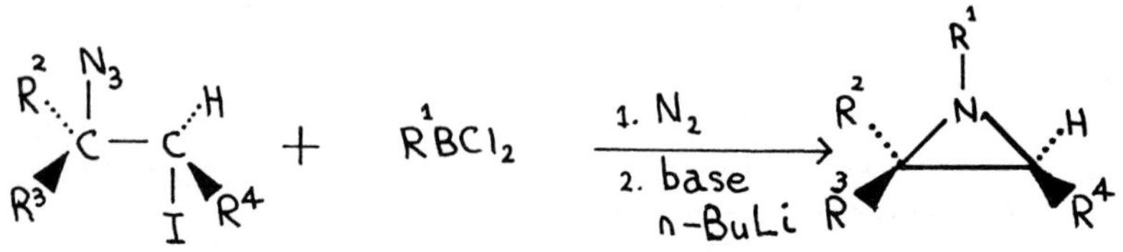

(10) Separable diastereomeric *cis*-1-menthyl 2-phenyl-
-3-aziridinecarboxylates are prepared by the Gabriel
reaction, thus permitting the introduction of asymmetry
in their subsequent conversion to amino acids (J.W. Lown,
T. Itoh, and N. Ono, Canad. J. Chem., 1973, <u>51</u>, 856).

A reagent which adds Br/N_3 units directiospecifically
to alkenes (according to Markownikoff rule) to give azido
β-bromo compounds is prepared *in situ* from N-bromosuccinimide and sodium azide in dimethoxyethane/water.
Trisubstituted olefins react faster with Br/N_3 than di-
or mono-substituted olefins. The aziridines are formed
by lithium tetrahydridoaluminate-mediated cyclisation of
the azido β-bromo compound. The reaction proceeds stereospecifically (D. van Ende and A. Krief, Angew. Chem.
internat. Edn., 1974, <u>13</u>, 279).

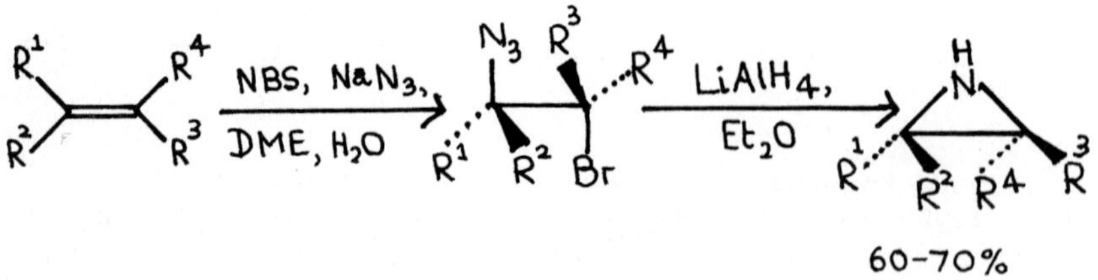

60-70%

(11) 1,2-Diphenyl-1-azaspiro[2,2]pentane is obtained by irradiation of the thermal adduct of phenyl azide and benzylidenecyclopropane. It thermally rearranges to cyclo-butanimine, and with hydrogen chloride and methanol respectively, gives the adducts (7) and (8). The former is reconverted to the spiro compound by the DMSO anion (J.K. Crandall and W.W. Conover, Chem. Comm., 1973, 33; J. org. Chem., 1974, **39**, 63).

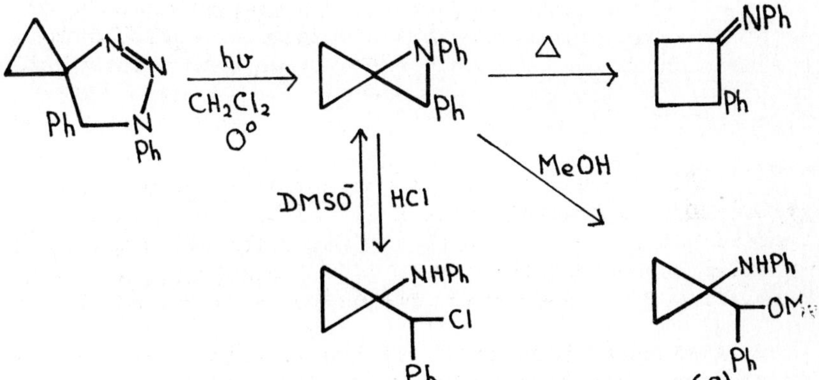

1-Phenyl and the 1-methoxycarbonyl derivatives of 1-azaspiropentane are also available. Nitrene addition to 1,3,3-trimethylcyclopropene at 0° fails to yield 2-azabicyclobutane (D.H. Aue, R.B. Lorens, and G.S. Helwig, Tetrahedron Letters, 1973, 4795), but the analogous direct addition of methoxycarbonylnitrene to cyclobutenes provides a convenient synthesis of 5-azabicyclo[2.1.0]pentanes (Aue, H. Iwahashi, and D.F. Shellhamer, *ibid.*, p.3719).

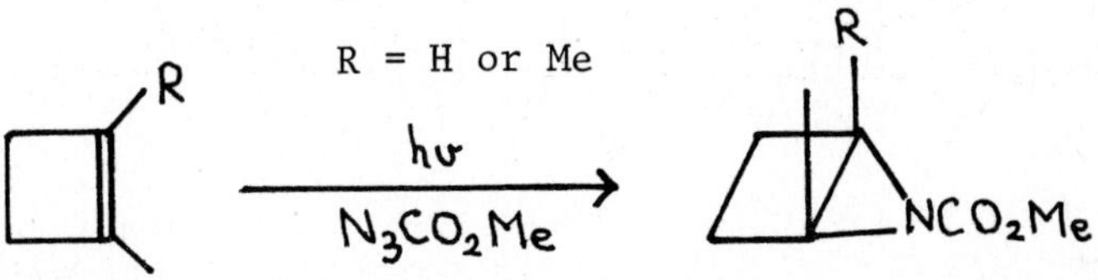

(12) 2-Formylaziridines are obtained by the reaction of the appropriate α,β-dibromoaldehyde with the necessary amine, and subjecting the resulting anil to mild acid hydrolysis (L. Wartski and A. Sierra-Escudero Compt. rend., 1974, **279**, C, 149).

$$R^1CHBr \cdot CHBr \cdot CHO \xrightarrow{R^2NH_2} \underset{CH=NR^2}{R^1 \overset{\overset{R^2}{N}}{\triangle}} \xrightarrow{H^+, H_2O} R^1 \overset{\overset{R^2}{N}}{\triangle} CHO$$

(13) In an extension of the Wenker synthesis aziridines are prepared by the simultaneous action of triphenylphosphine, carbon tetrachloride, and triethylamine on β-amino-alcohols (R. Alpel and R. Kleinstueck, Ber., 1974, <u>107</u>, 5).

$$HOCHR^1CHR^2NHR^3 \xrightarrow[Et_3N]{Ph_3P, CCl_4,} R^1 \overset{\overset{R^3}{N}}{\triangle} R^2$$

(14) α,α-Dichloranils obtained by the chlorination of the anil using *N*-chlorosuccinimide, on treatment with lithium tetrahydridoaluminate, cyclise to give 1,2-disubstituted aziridines (N. De Kimpe *et. al.*, Synth., Comm., 1975, <u>5</u>, 269).

$$RCH_2CH=NBu^t \xrightarrow{NCS} RCCl_2CH=NBu^t \xrightarrow{LiAlH_4} \overset{\overset{t-Bu}{N}}{\triangle} R$$

R = Me, Et, Pr, i-Pr, Bu, EtMeCH

(15) The reaction between the stereoisomeric nitronic esters (9) and (10) with benzoylacetylene is stereo-selective, each isomer leads to different aziridines. Thus the reaction permits the synthesis under kinetic control of 1-methoxyaziridine invertomers (R. Grée and R. Carrié, Chem. Comm., 1975, 112).

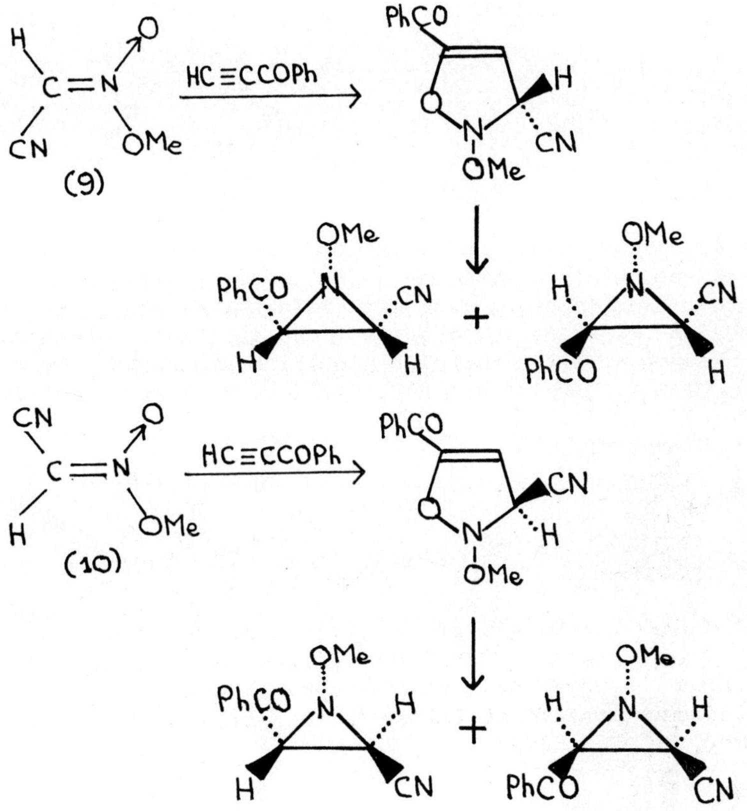

(16) *cis*-1-Amino-2,3-diphenylaziridine is obtained from *meso*-hydrobenzoin dimesylate with 95% hydrazine (H. Paulsen and D. Stoye, Angew. Chem. 1968, 80, 120).

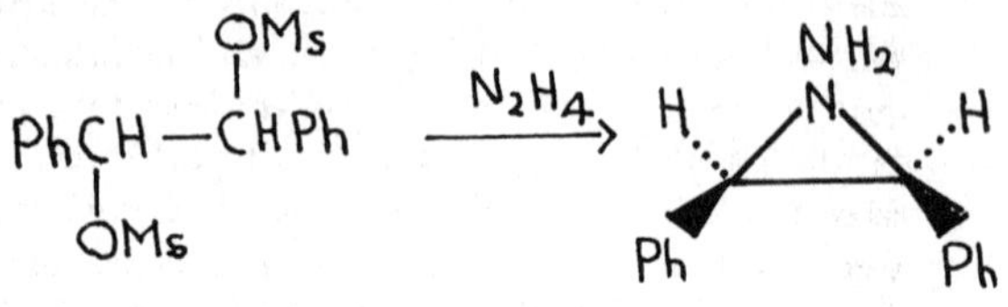

The *trans*-isomer cannot be prepared by the same technique. It can be obtained from the adduct formed by the oxidation of the *N*-aminophthalimide with lead tetraacetate in a methylene chloride solution of *trans*-stilbene. Treatment of the adduct with ethanolic hydrazine yields the *trans* — isomer (L.A. Carpino and R.K. Kirkley, J. Amer. chem. Soc., 1970, <u>92</u>, 1784).

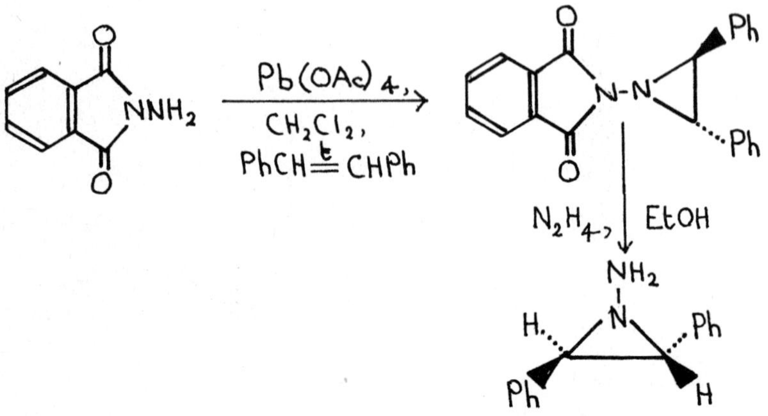

(17) Tricyclic compound containing an oxirane and an aziridine ring (see p. 8).

(ii) Properties and Reactions

Barriers to nitrogen inversion have been studied in a number of 5-membered cyclic amines by variable temperature nmr spectroscopy and the substituent effects observed in these compounds compared with those obtained in aziridine derivatives, which are much larger (J.M. Lehn and J. Wagner, Tetrahedron, 1970, 26, 4227). Comparison between aziridine derivatives and derivative of some other heterocycles containing one nitrogen atom have also been made (H. Kessler and D. Leibfrit, Tetrahedron Letters, 1970, 4289, 4293, 4297).

It has been found that phthalimidoaziridines (1) show temperature-dependent [1]H-nmr spectra. The coalescence temperatures for several 2- and 3- substituted phthalimido-aziridines have been found to lie in the range 75-125°. However, in the case of 2,3-dichloro- derivatives $(R^1 = R^3 = Cl)$ the coalescence temperatures are below 0° (D.J. Anderson and T.L. Gilchrist, J. chem. Soc., 1971, 2273).

(1)

Conformation analysis of 1,1′-bisaziridyl by vibrational spectroscopy has been recorded (P. Rademacher and W. Lüttke, Angew. Chem. internat. Edn., 1970, 9, 245).

Oxidation by ozone of 1-*tert*-butylaziridine in methylene chloride provides a solution of 1-*tert*-butyl-axiridine 1-oxide stable up to 0° (J.E. Baldwin *et. al.*, J. Amer. chem. Soc., 1971, 93, 4082).

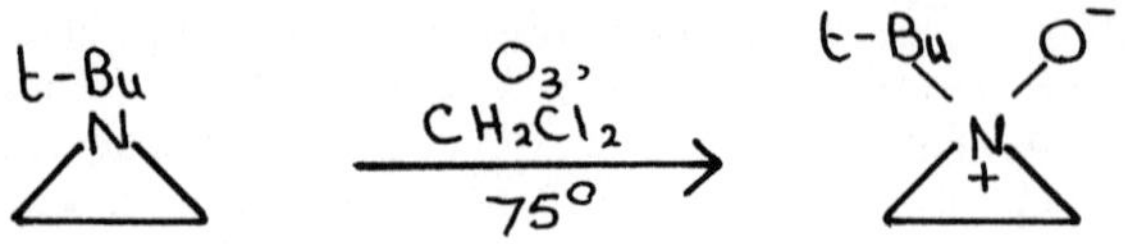

3-Aroylaziridines react with a variety of imines in boiling benzene to yield imidazolidines, *via* the intermediate ylide, formed by thermal cleavage of the C-C bond. The orientation of the [2 + 3] cycloaddition of the azomethine ylide has been demonstrated (J.W. Lown, J.P. Moser, and R. Westwood, Canad. J. Chem., 1969, <u>47</u>, 4335; 1970, <u>48</u>, 1682).

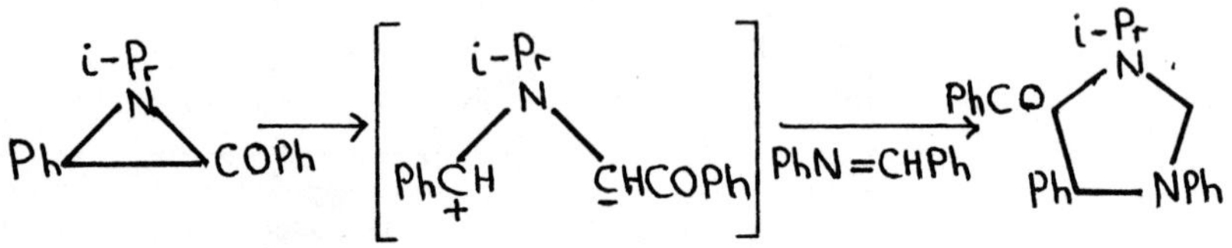

The *cis-* and *trans-* isomeric aziridines (2) and (3) on heating establish equilibria with small concentrations of the azomethine ylides by conrotatory ring opening. The *trans*-isomer combines stereospecifically even with weak dipolarophiles, while in the case of the less reactive *cis*-isomer the isomerisation to the *trans*-isomer competes with the 1,2-dipolar cycloaddition except for dipolarphiles of highest activity (R. Huisgen and H. Mäder, J. Amer.

chem. Soc., 1971, _93_, 1777).

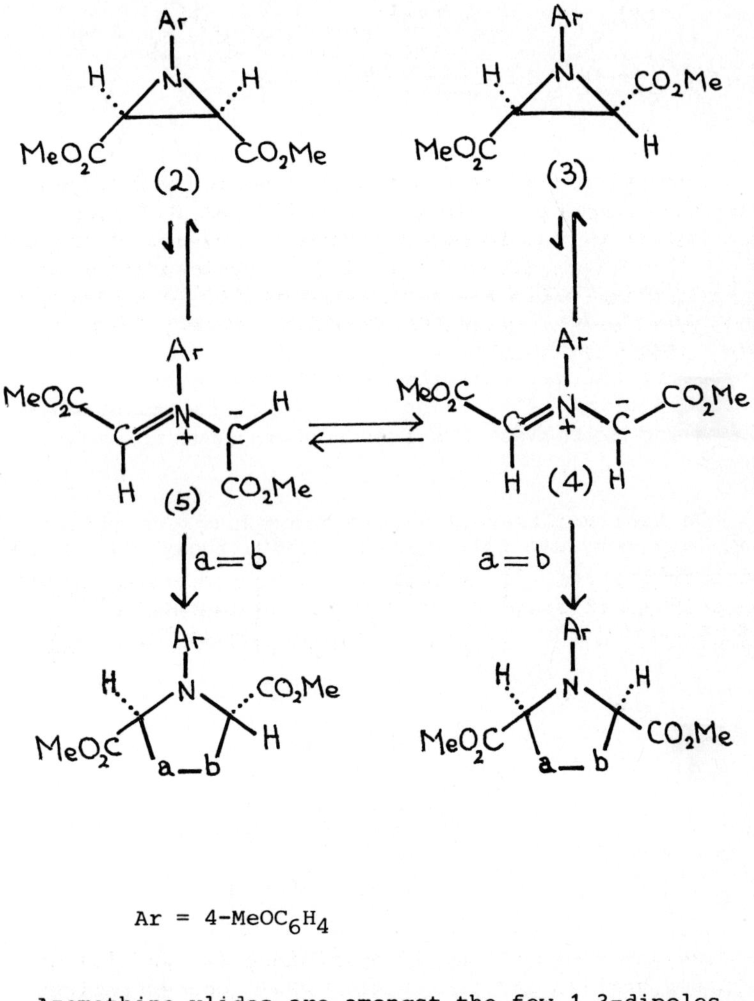

Ar = 4-MeOC$_6$H$_4$

Azomethine ylides are amongst the few 1,3-dipoles that equal ozone in their ability to attack an aromatic double band, thus the reaction of dimethyl 1-(4-methoxyphenyl)aziridine-2,3(_trans_)-dicarboxylate (3) with an excess of phenanthrene gives an adduct (6) (Huisgen and W. Scheer, Tetrahedron Letters, 1971, 481).

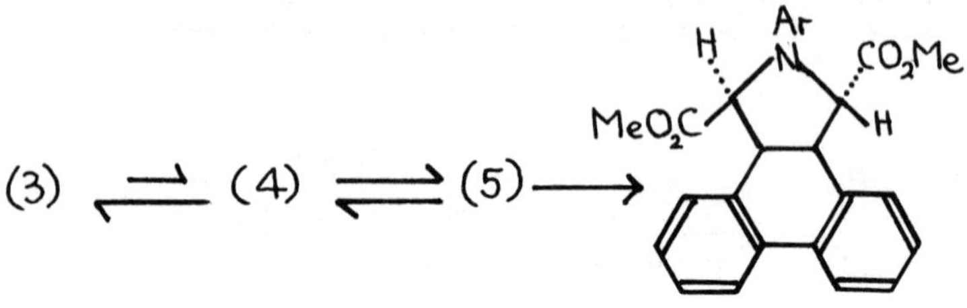

$$(3) \; \underset{\longleftarrow}{\longrightarrow} \; (4) \; \rightleftharpoons \; (5) \longrightarrow$$

(6)

The structure of the adduct (6) leaves no doubt that the azomethine ylide. (5) is the reactive species. The *exo,exo*—substituted azomethine ylide (4) exists in a conrotatory thermal equilibrium with the aziridine. Anthracene gives a 1:1 adduct and two stereoisomeric 1:2 adducts and napthalene two tris-adducts, but no mono-adduct.

On heating dimethyl 1-(4-methoxyphenyl)aziridine-2,3-(*cis*)-dicarboxylate with diethyl azodicarboxylate, a triązo-lidine derivative (7) is formed with *trans*-located ester groups in positions -3 and -5 (E. Brunn and Huisgen, *ibid.*, p.473).

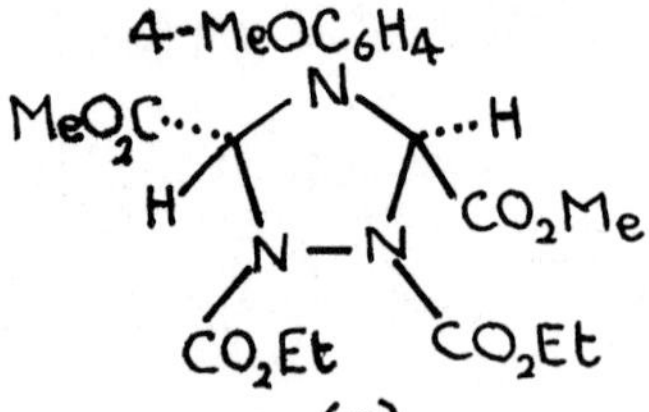

(7)

1-Alkyl-2-aroyl-3-arylaziridines, with electron-releasing substituents in the 2- or 3-positions react through the ylide with substituted 4-nitrosophenols to

give 3,5-dihydro-2H-pyrrolo[3,4-d]oxazoles (Lown and
M.H. Akhtar, J. chem. Soc., Perkin I, 1972, 1459).

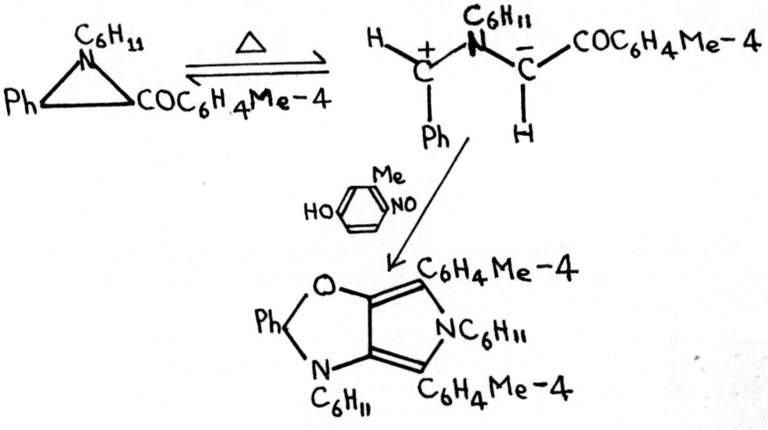

Also the azomethine ylide arising from dimethyl 1,3-
-diphenylaziridine-2,2-carboxylate adds preferentially
to the C=O bond of ketenes to give 5-methyleneoxazolidines
with two ester groups on C-4 (F. Texier, R. Carrié and
J. Jaz, Chem. Comm., 1972, 199).

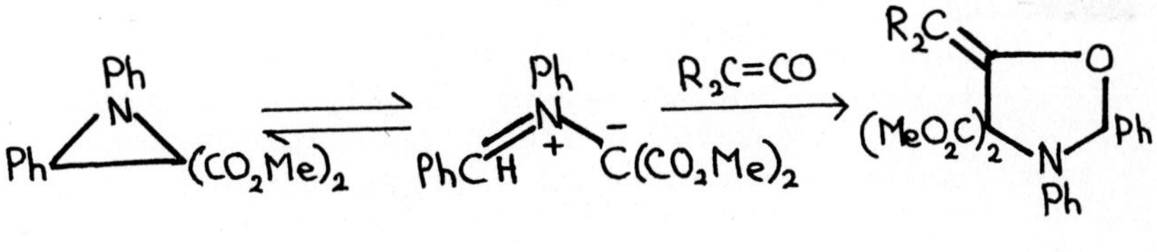

The thermal rearrangement of aziridinyl ketones (8)
and (9) to *N*-alkyl-3-arylpyrroles differs dramatically
from previously observed three-membered ring 'ene'
reactions, in that the *trans*-isomer reacts at a faster
rate than the corresponding *cis*-form. In previous examples
only the *cis*-isomer reacted. The simplest mechanism is
based on the fact that aziridines readily undergo thermal

cleavage to azomethine ylides by conrotation of the
substituent groups. With the *cis*-isomer there is steric
interaction between the nitrogen substituent and one or
other of the 2,3-substituents (A. Padwa *et. al.*, J. Amer.
chem. Soc., 1973, <u>95</u>, 7168).

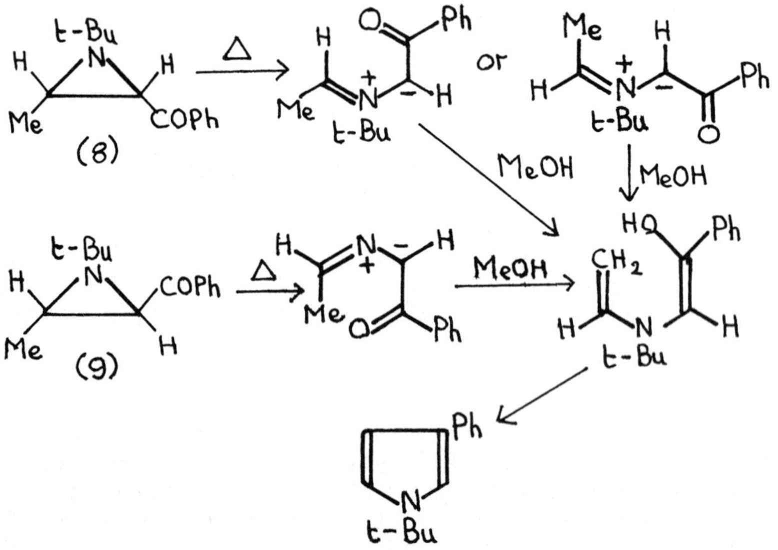

2-Arenoylaziridines undergo thermal 1,3-dipolar
addition to aromatic aldehydes and chloral *via* azomethine
ylides to give exclusively 4-arenoyloxazolidines, for
example 2-benzoyl-1-isopropyl-3-(3-nitrophenyl)aziridine
reacts with 2,4-dinitrobenzaldehyde to yield the
oxazolidines (10) (G. Dallas, Lown, and Moser, J. chem.
Soc., C, 1970, 2383).

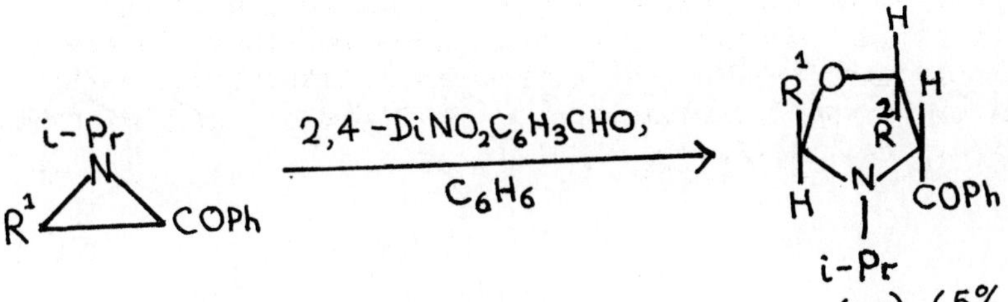

$$R^1 = 3\text{-}NO_2C_6H_4 \qquad\qquad R^2 = 2,4\text{-}DiNO_2C_6H_3$$

For the deamination of 2-(aminomethyl)aziridines there are two mechanistic pathways, a carbene and a carbonium ion route. The former is associated with the formation of alkynes and 3-pyrroline products and the latter with β-(nitrosamine)ketones and esters of acetic acid (G. Szimies, Ber., 1973, 106, 3695).

The aziridine {1-cyclohexyl-6-(cyclohexylimino)-
-1,1a,6,6a-tetrahydro-1a-phenylindeno[1,2-b]azirine} (11) undergoes a formally disallowed valence tautomerism to an isoquinolinium imine, which may be trapped as an azomethine ylide in a series of 1,3-dipolar cycloadditions. Although the thermal valence tautomerism (11) to (12) is formally a forbidden process, substantial driving force for this process is provided by the relief of ring strain in (11) and the gain in resonance energy in (12) (Lown and K. Matsumoto, Chem. Comm., 1970, 692).

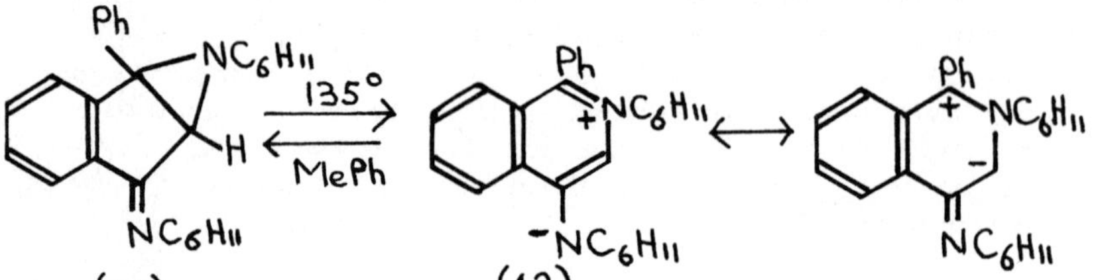

Reactions of aziridines involving ring expansion have been reviewed (F.N. Galdysheva, A.P. Sineokov, and V.S. Etlis, Russ. chem. Rev., 1970, 39, 118).

Oxidation of *trans*-1-amino-2,3-diphenylaziridine using activated manganese dioxide in methylene chloride at 0° affords only *trans*-stilbene, whereas oxidation of the *cis*-isomer gives a mixture of 85% *trans*- and 15% *cis*-stilbene (L.A. Carpino and R.K. Kirkley, J. Amer. chem. Soc., 1970, 92, 1784).

3-Phenyl-1-azabicyclo[1,1,0]butane is extremely sensitive to decomposition under acid conditions to give 3-hydroxy-3-phenylazetidine and other azabicyclobutanes undergo facile addition of a variety of reagents to yield 3-substituted azetidines *via* cleavage of the 1,3-bond (J.L. Kurz *et. al.*, *ibid.*, 1970, 92, 5008).

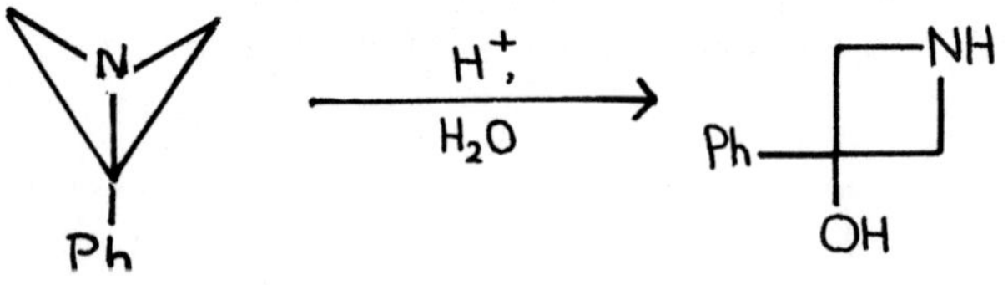

Thermolysis of a number of 1-alkyl-2-aroyl-3-methyl-
aziridines affords 1-alkyl-3-arylpyrroles in high yield.
Kinetic studies show that *trans*-1-*tert*-butyl-2-benzoyl-3-
-methylaziridine rearranges 40 times more rapidly than
the corresponding *cis*-isomer at 70°. A mechanism
involving conrotatory opening of the aziridine ring to
an azomethine ylide accounts for the observed rearrangement
patterns (Padwa, D. Dean, and T. Oine, J. Amer. chem. Soc.,
1975, <u>97</u>, 2822).

The products obtained by the thermal isomerisation of
1-substituted 2-vinylaziridines (13) depends on the nature
of the substituents in positions 1 and 3. In most cases
with 3-phenyl derivatives 2-pyrrolines are obtained and
with either a methyl group or no substituent at C-3,
depending on the nature of the substituent at C-1 they
give an imine or a mixture of 2- and 3-pyrrolines
(D. Borel, Y. Gelas-Mialhe, and R. Vessiere, Canad. J.
Chem., 1976, <u>54</u>, 1590).

(13)

2,2-Di(methoxycarbonyl)-1,3-diphenylaziridine reacts
with the bromomalononitrile anion on heating in the presence
of lithium perchlorate to give an azetidine (see p. 114)
(M. Vaultier and Carrié, Tetrahedron Letters, 1978, 1195).

The synthesis of the stable 1-nitroaziridines,
3β-acetoxy-5β,6β-1-nitroaziridinylcholestane and 10-methyl-
-1,9-(1-nitroaziridino)decalin has been described
(M.J. Haire and G.A. Boswell, Jr., J. org. Chem., 1977,
42, 4251). The oxidation of some 1-aminoaziridines has
been investigated (L. Hosch, N. Eggo, and A.S. Dreiding,
Helv., 1978, 61, 795).

The methyleneaziridine-cyclopropanimine rearrangement
proceeds stereospecifically. Kinetic data on the
rearrangement of the optically-active 2-methylene-1-methyl-
-3-trimethylsilylaziridine has been reported (H. Quast
and C.A. Weise Vélez, Angew. Chem. internat. Edn., 1978,
17, 213).

The thermal rearrangement of aziridinimines (14)
and (15) has been studied kinetically, but their facile
decomposition competes with the rearrangement (Quast and
P. Schäfer, Tetrahedron Letters, 1977, 1057).

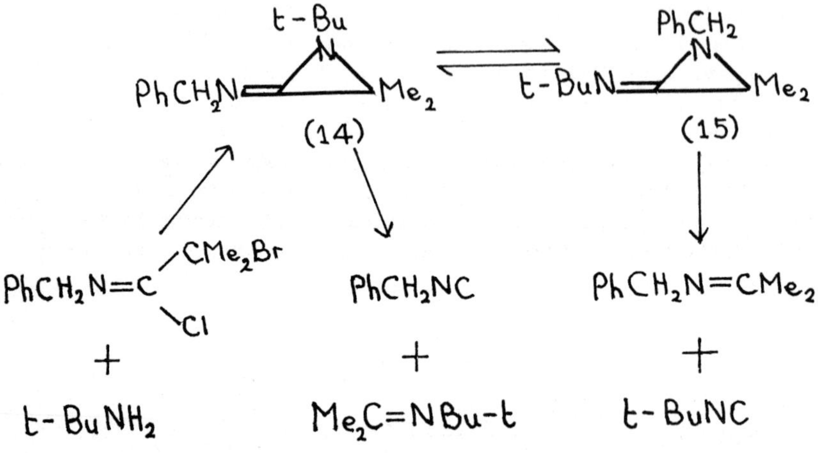

(iii) Halogenoaziridines

The addition of dichlorocarbene to diethyl
azodicarboxylate gives an open chain compound, which reacts
further with dichlorocarbene to yield an aziridine

(D. Seyferth and H.-m. Shih, J. Amer. chem. Soc., 1972, 94, 2508).

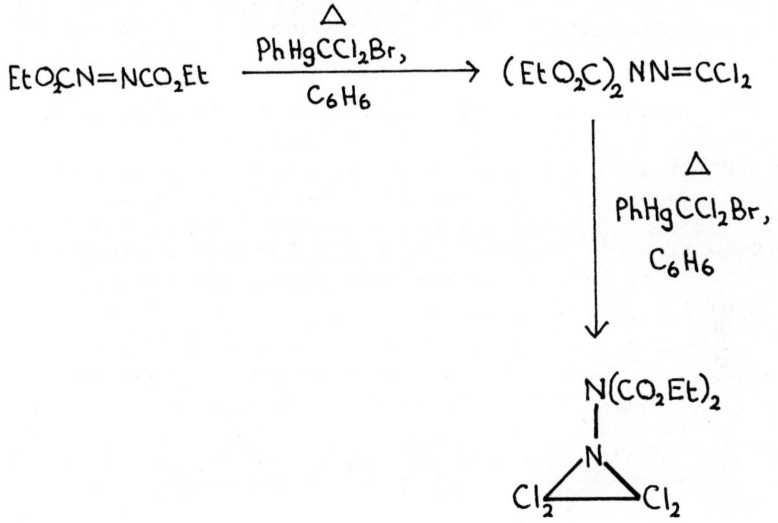

$$EtO_2CN=NCO_2Et \xrightarrow[C_6H_6]{\overset{\triangle}{PhHgCCl_2Br,}} (EtO_2C)_2NN=CCl_2$$

$$\downarrow \overset{\triangle}{\underset{C_6H_6}{PhHgCCl_2Br,}}$$

Phenyl(bromodichloromethyl)mercury transfers CCl_2 to alkyl- and aryl-carbonimidoyl dichlorides to give 1-alkyl- and 1-aryl-2,2,3,3-tetrachloroaziridines in fair yield (Seyferth, W. Tronich, and Shih, J. org. Chem., 1974, 39, 158).

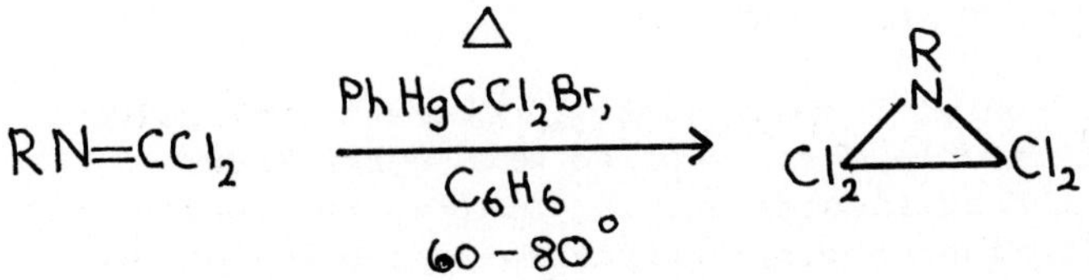

$$RN=CCl_2 \xrightarrow[\substack{C_6H_6 \\ 60-80°}]{\overset{\triangle}{PhHgCCl_2Br,}}$$

The tetrachloroaziridines are much more stable thermally than monochloro- or *gem*-dichloro- aziridines, but rearrange when heated at 180°, except in the case of 1-isopropyltetrachloroaziridine.

$$RN=C \begin{smallmatrix} Cl \\ CCl_3 \end{smallmatrix}$$

from

$$Cl_2 \underset{N-R}{\triangle} Cl_2 \xrightarrow{180°}$$

R = Ph, $c-C_6H_{11}$

Reduction of 1-acyl-2-chloro-3,3-dimethyl-2-phenyl-
-aziridines with lithium tetrahydriodoaluminate affords
3,3-dimethyl-2-phenylaziridine.

$$Me_2 \underset{N-COR}{\triangle} \overset{Ph}{\underset{Cl}{}} \xrightarrow[Et_2O]{LiAlH_4,} Me_2 \underset{N-H}{\triangle} Ph$$

The related 1-benzoylaziridine reacts with phenylmagnesium
bromide yielding 3,3-dimethyl-2,2-diphenylaziridines.
The acylchloroaziridines may be converted into
functionalised (azido, acetoxyl)1-acyl-aziridines and
oxazolines (A. Hassner, S.S. Burkes, and J. Cheng-fan,
J. Amer. chem. Soc., 1975, **97**, 4692).

(b) Aziridinones (α-lactams)

1-*tert*-butyl-3-phenyl-3-trifluoromethylaziridinone
is unusually stable when compared with other known
trisubstituted aziridones and is obtained by treating the
N-*tert*-butylamide of 3,3,3-trifluoro-2-phenylpropanoic
acid with *tert*-butyl hypochlorite and potassium *tert*-
butoxide (E.R. Talaty and C.M. Utermoehlen, Tetrahedron
Letters, 1970, 3321).

$$\underset{CF_3}{\overset{|}{PhCHCONHBu-t}} \longrightarrow \underset{CF_3}{Ph} \underset{N-t-Bu}{\triangle} O$$

The preparation of three optically pure steroidal
α-lactams has been reported (S. Sarel, B.A. Weissman, and
Y. Stein, *ibid.*, 1971, 373). Optically active aziridin-2-
-ones (1) are synthesised from the corresponding
benzyloxycarbonyl L-amino acid by dehydration in THF at
-20 to -30° using carbonyl dichloride, thionyl chloride,
or phosphoryl chloride (M. Miyoshi, Bull, chem. Soc. Japan,
1973, 43, 212).

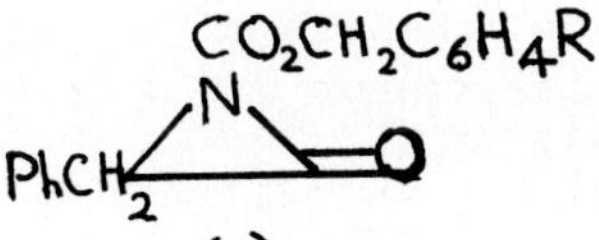

(1)

R = 4-H, Br, Cl

Ring fission of the aziridinones takes place
exclusively at the carbonyl-nitrogen bond to give
L-acylamino acid derivatives.   The reaction has been used
successfully in peptide synthesis (*idem*, *ibid.*, p.1489).

Aziridinones with phenyl isocyanate yield 1:1 adducts
by the trapping of 1,3-dipolar species derived from the
aziridinones (M. Kakimoto, S. Kajigaeshi, and S. Kanemas,
Chem. Letters, 1976, 47).

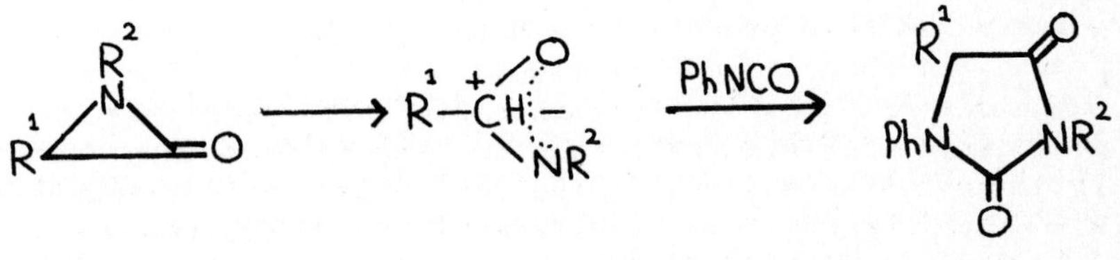

(c) Azirines

The chemistry of 1-azirine is well documented

(L.A. Paquette, "Principles of Modern Heterocyclic Chemistry", W.A. Benjamin, Inc., New York, 1968, p.11), but a 2-azirine has not yet been prepared and this has been rationalised by noting its antiaromatic character. However, it has figured in a number of investigations, for example, its involvement in the formation of aziridines by the reaction between a Grignard reagent and an oxime, (p.38), the interception of transitory 1-azirines by cyclopentadienones during the thermal decomposition of certain vinyl azides resulting in the formation of 3H--azepines (D.J. Anderson and A. Hassner, J. org. Chem., 1973, 38, 2565). The mchanism of the azepine formation has been discussed (A. Hanner and D.J. Anderson, J. Amer. chem. Soc., 1972, 94, 8255).

(i) 1-Azirines (2H-Azirines): synthesis

Reaction of alkylsubstituted α-chloroenamines with sodium azide affords the previously unknown 2-amino-1--azirines (M. Rens and L. Ghosez, Tetrahedron Letters, 1970, 3765).

4,5-Disubstituted 1-phthalimide-1,2,3-triazoles on pyrolysis in the vapour phase give 1-azirines as the primary isolable products. 4-Methyl-5-phenyl- and 5-methyl-4-phenyl-1-phthalimido-1,2,3-triazole give identical mixtures of 1-azirines and their pyrolysis products, indicating that they are formed through a common intermediate, considered to be the antiaromatic 2-methyl--3-phenyl-1-phthalimido-2-azirine (T.L. Gilchrist, G.E. Gymer, and C.W. Rees, J. chem. Soc., Perkin I, 1973, 555).

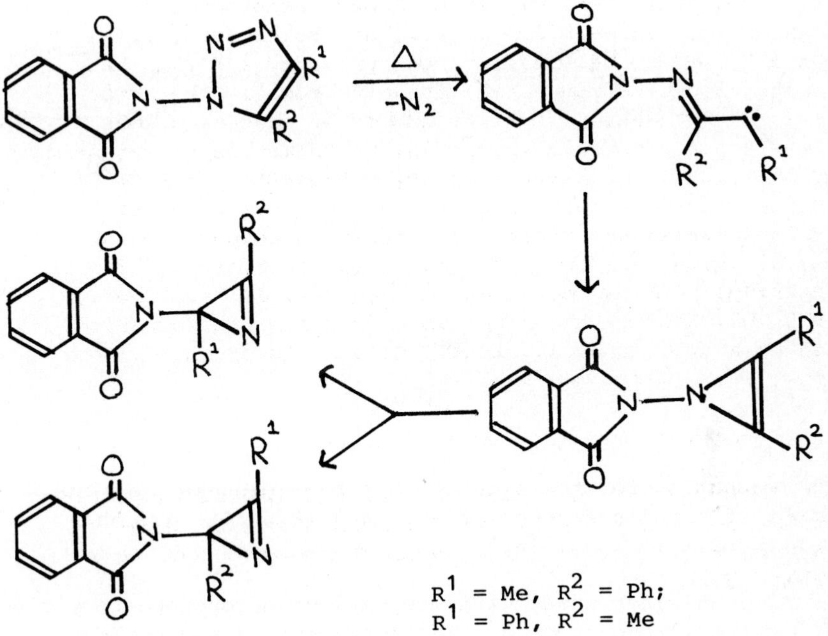

$$R^1 = Me, \quad R^2 = Ph;$$
$$R^1 = Ph, \quad R^2 = Me$$

(ii) Properties and reactions of 1-azirines

2-Phenylazirine with dimethylsulphonium methylylide gives 2-phenyl-1-azabicyclobutane, the first heterocyclic bicyclobutane to be isolated and characterised. The azabicyclobutane system readily undergoes proton-initiated hydration to yield 3-phenylazetidin-3-ol and treatment in anhydrous methanol with a trace of perchloric acid affords 3-methoxy-3-phenyl-azetidine (A.G. Hortmann and D.A. Robinson, J. Amer. chem. Soc., 1972, 94, 2758) (see p. 54).

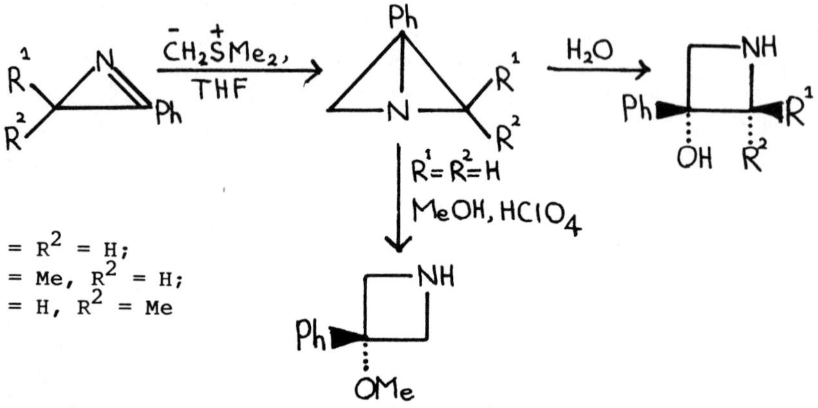

$R^1 = R^2 = H$;
$R^1 = Me$, $R^2 = H$;
$R^1 = H$, $R^2 = Me$

The mechanism for the cleavage has been discussed in the light of kinetic isotope effect data (B.K. Gillard and J.L. Kurz, *ibid.*, p.7199).

1-Azirines react with diphenylcyclopropenone to yield 2,3-diphenyl-4-pyridones. The formation of the pyridones can be seen to involve nucleophilic attack of the weakly basic azirine nitrogen on the electrophilic cyclopropenone ring followed by an intramolecular Cope cyclisation. (Hassner and A. Kascheres, J. org. Chem., 1972, 37, 2328).

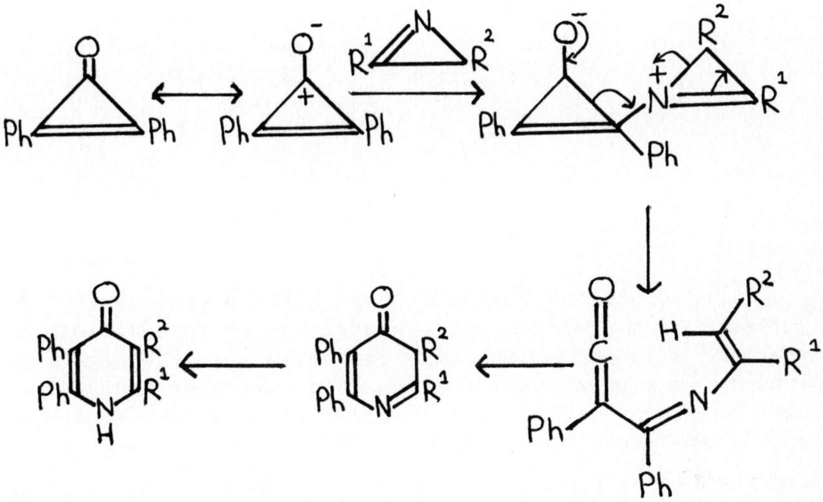

1-Azirines react with triphenylcyclopropenyl bromide to yield pyridine derivative in good to modest yield (R.E. Moerck and M.A. Battiste, Tetrahedron Letters, 1973, 4421).

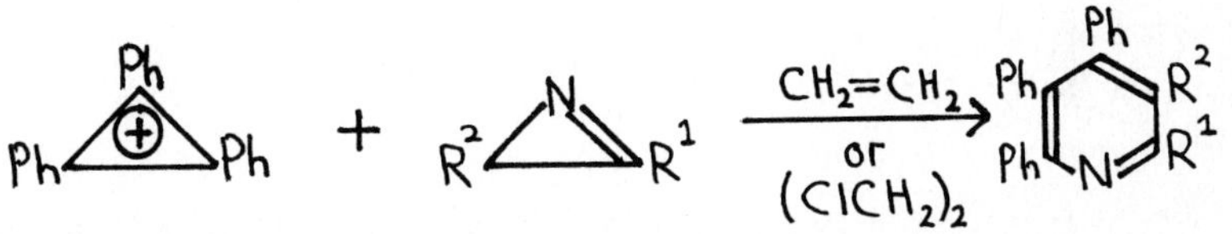

$R^1 = R^2 = H$;
$R^1 = Ph$, $R^2 = H$;
$R^1 = Ph$, $R^2 = CO_2Et$

Cyclopentadienones and 1-azirines react in boiling toluene with loss of carbon monoxide to give 3H-azepines (Hassner and D.T. Anderson, J. org. Chem., 1974, 39, 3070).

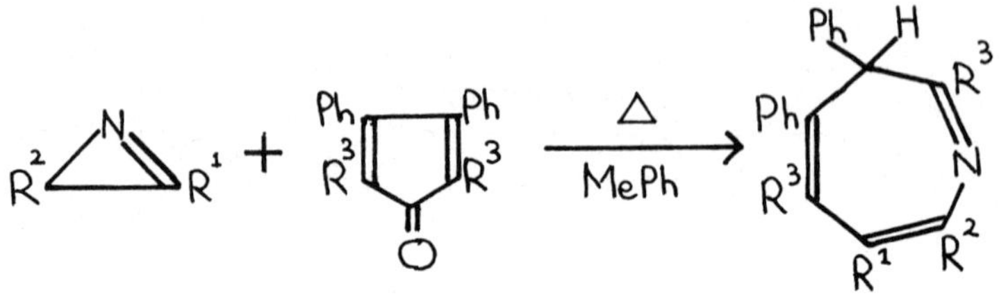

The photoconversion of substituted arylazirines into methoxyimines proceeds *via* a nitrile ylide intermediate and reaction with deuteriomethanol indicates that the electron density is highest on the disubstituted rather than the trisubstituted carbon atom of the nitrile ylide (A. Padwa and J. Smolanoff, Chem. Comm., 1973, 342).

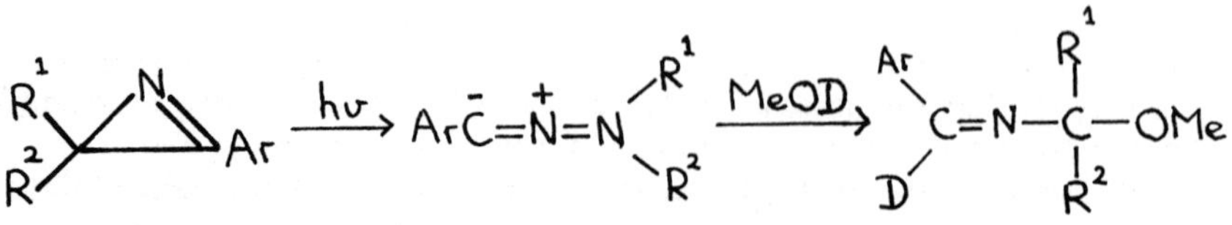

R^1, R^2 = H, Me or Ph;
Ar = Ph or $C_{10}H_7$

2-Naphthyl-substituted 1-azirines on uv irradiation undergo ring opening to yield nitrile ylide intermediates, which can be trapped with electron-deficient olefins to produce 1-pyrolines (Padwa and S.I. Wetmore, J. org. Chem., 1974, 39, 1396).

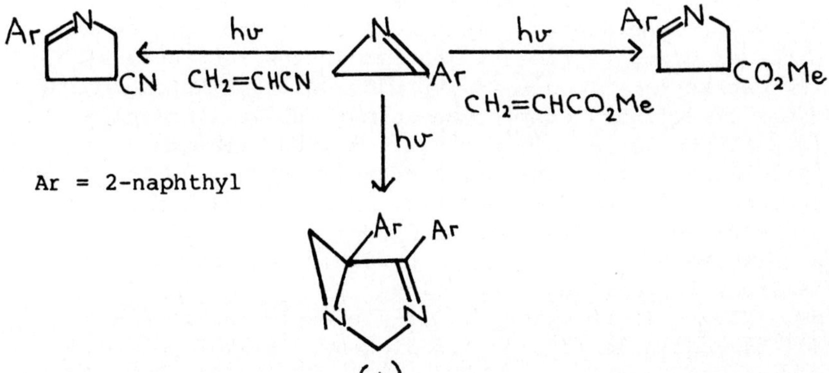

Ar = 2-naphthyl

(1)

Irradiation of the 1-azirine in the presence of electron-
-rich acyclic or cyclic olefins produces no photoadduct,
but instead affords a dimer, 4,5-di(2-naphthyl)1,3-diaza-
bicyclo[3.1.0]hex-3-ene (1).

Acyl chlorides add to 3,3-dimethyl-2-phenylazirine
in benzene at room temperature to give 1-acyl-2-chloro-3,3-
-dimethyl-2-phenylaziridines (Hassner, Burkes, and Cheng-
fan, J. Amer. chem. Soc., 1975, 97, 4692).

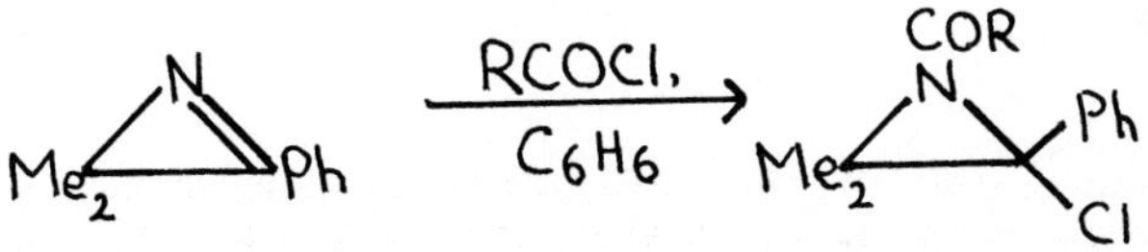

Substituted 1-azirines undergo photochemical ring
opening to form nitrile ylides. These 1,3-dipoles can
be trapped with a variety of dipolarophiles to produce
five-membered heterocycles, for example, ketenes give
oxazolines (H. Heimgartner et. al., Chimia, 1972, 26, 424),
aldehydes also give oxazolines, carbon dioxide affords

oxazolin-5-ones (H. Giezendannes *et. al.*, Helv., 1972,
55, 745), isocyanates, isothiocyanates, and isocyanides
give oxazoles, thiazoles, and imidazoles respectively
(B. Jackson *et. al.*, *ibid.*, p.916), diethyl ketomalonate
gives oxazolines, ethyl cyanoformate gives imidazoles
(Jackson *et. al.*, *ibid.*, p.919), styrene affords 1-
pyrrolines, norbornene gives adducts (Padwa, D. Dean, and
Smolanoff, Tetrahedron Letters, 1972, 4087), ketones give
3-oxazolines (P. Claus *et. al.*, Helv., 1974, 57, 2173;
A. Orahovats *et. al.*, *ibid.*, p.2626), and esters give
5-alkoxy-3-oxazolines (P. Gilgen *et. al.*, *ibid.*, 1975,
58, 1739). Irradiation of 2,3-diphenyl-1-azirine and
1,4-benzoquinone affords 1,3-diphenyl-2H-isoindole-4,7-
dione (Gilgen *et. al.*, *ibid.*, 1974, 57, 2364).

In the absence of an added dipolarophile the photochemically produced ylides react with a further molecule
of the original 1-azirine to yield a bicyclic compound,
for instance, 2-phenyl-1-azirine affords 4,5-diphenyl-1,3-
-diazabicyclo[3.1.0]hex-3-ene (N. Gakis, *et. al.*, Helv.,
1972, 55, 748).

Similarly *endo-* and *exo*-2,4,5,6-tetraphenyl-1,3-diaza-
bicyclo[3.1.0]hex-3-ene are obtained from 2,3-diphenyl-1-
-azirine (Padwa *et. al.*, J. Amer. chem. Soc., 1972, 94,
1395; Chem. Comm., 1972, 409). On further irradiation,
the 1,3-diazabicyclohexene system undergoes photochemical
ring opening to an ene di-imine intermediate (Padwa and
Wetmore, Jr., Chem. Comm., 1972, 1116). Thermal and
photochemical reactions of diazabicyclo compounds
containing an aziridine ring have been studied (T. DoMinh
and A.M. Trozzolo, J. Amer. chem. Soc., 1972, 94, 4048;
Padwa and L. Gehrlein, *ibid.*, p.4933; Padwa and E. Glazer,

ibid., p.7788).

Oxazoles, isoxazoles, imidazoles, pyrazoles and pyrroles are available by intramolecular 1,3-dipolar cyclo-additions of 1-azirines with unsaturated side-chains effected either by photolysis or thermolysis. Photolysis of 3-formyl-2-phenylazirine (2) affords 2-phenyloxazole, whereas thermolysis gives 3-phenyloxazole; similarly the anil (3) gives 1,2-diphenylimidazole and 1,3-diphenyl-pyrazole, and the vinyl substituted 1-azirines (4) give 2,3-disubstituted pyrroles and 2,5-disubstituted pyrroles.

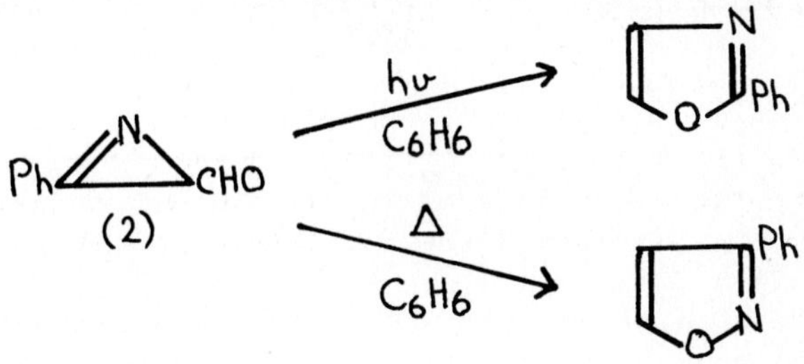

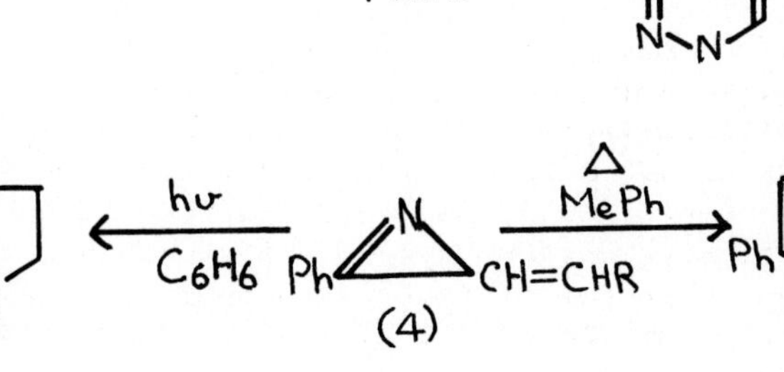

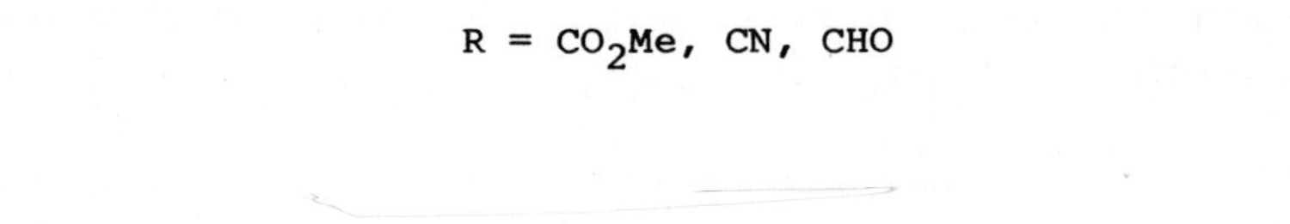

$R = CO_2Me,\ CN,\ CHO$

The photorearrangement proceeds *via* a transient nitrile ylide intermediate, which can be trapped with external dipolarophiles, whereas the thermal transformation can be rationalised in terms of an equilibration of the 1-azirine with a transient vinyl nitrene, which subsequently rearranges to the final product (Padwa, Smolanoff, and A. Tremper, Tetrahedron Letters, 1974, 29).

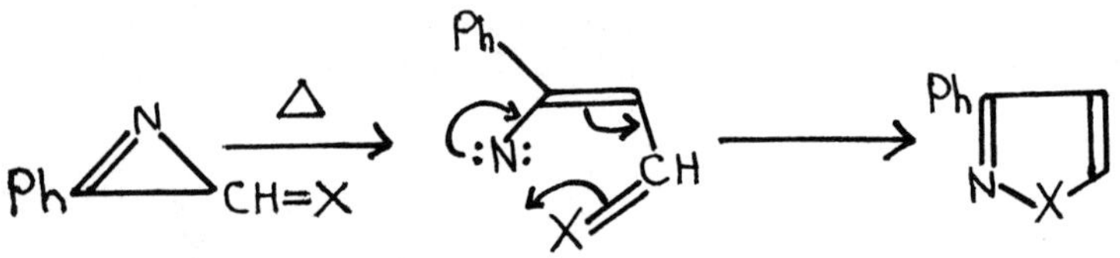

X = O, NPh, CHR

Photocyclisation of styryl-substituted 1-azirines affords substituted benzazepines as the major product (Padwa and Smolanoff, *ibid.*, p.33). The scope of the thermal and photochemical ring expansion reactions of a number of 2-vinyl-substituted 1-azirines has been examined (Padwa, Smolanoff, and Tremper, J. Amer. chem. Soc., 1975, 97, 4682).

1-Azirines react regiospecifically with thiobenzoyl isocyanate in *p*-xylene at room temperature to give exclusively [4 + 2] cycloaddition products. Controlled thermolysis of the adducts yield thiadiazepinones, which on prolonged thermolysis extrude elemental sulphur to yield a pyrimidine derivative (V. Nair and K.H. Kim, Tetrahedron Letters, 1974, 1487).

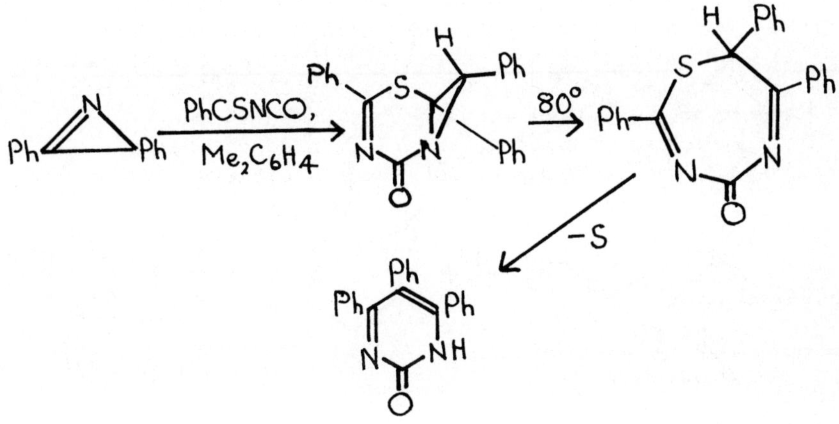

The irradiation of 3-allyl-3-methyl-2-phenyl-1-azirine (5) produces 2-azabicyclo[3.1.0]hex-2-ene (8) *via* an unusual 1,1-cycloaddition reaction. It is found that the short term irradiation of (5) produces a 1:1 mixture of (7) and (8). On further irradiation, (7) is quantitatively isomerised to (8). 3-Allyl-2-methyl-3--phenyl-1-azirine (6) on irradiation affords a quantitative yield of (8) and no significant quantity of (7). Evidence is presented in favour of a biradical intermediate with partial dipolar character (Padwa and P.H.J. Carlsen, J. Amer. chem. Soc., 1976, 98, 2006).

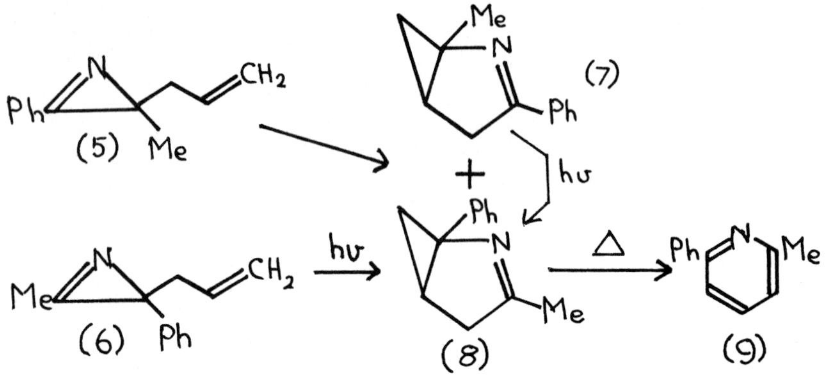

2-Azabicyclo[3.1.0]hex-2-ene (8) on heating yields 2-methyl-
-6-phenylpyridine (9) (*idem.*, *ibid.*,1975, 97, 3862). The
thermolysis of 3-allyl-3-methyl-2-phenyl-1-azirine (5)
gives 1-methyl-2-phenyl-3-azabicyclo[3.1.0]hex-2-ene (7)
and 3-methyl-2-phenylpyridine. The thermal transformation
can be explained in terms of an equilibration of the
1-azirine with a transient vinyl nitrene, which
subsequently rearranges to the final azabicyclohexene (7)
(*idem.*, J. org. Chem., 1976, 41, 180).

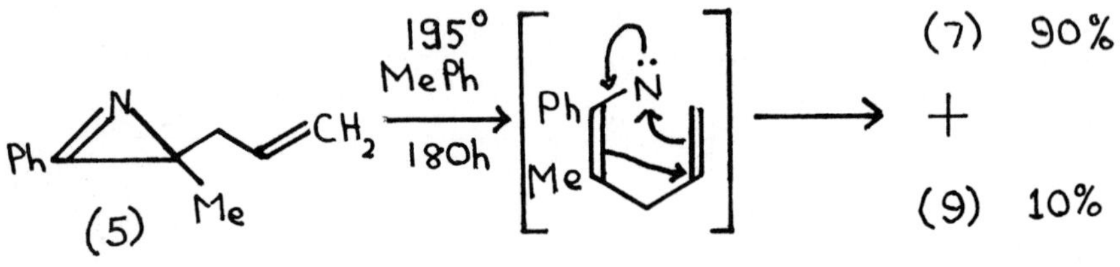

A 1,1-cycloaddition occurs when 2-(2-allylphenyl)-3,3-
-dimethyl-1-azirine (10) is irradiated to give a single
product, a benzobicyclo[3.1.0]hex-2-ene (11), which is
readily hydrolysed to acetone and the corresponding amine
(Padwa *et. al.*, J. Amer. chem. Soc., 1976, 98, 1048).

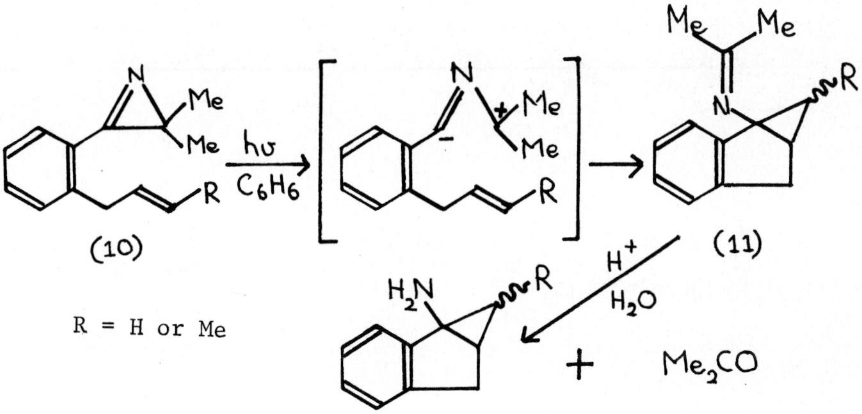

(10)

R = H or Me

(11)

Thermolysis of but-3-enyl substituted 1-azirines
gives substituted pyridines and biphenyl derivatives (Padwa
and N. Kamigata, Chem. Comm., 1975, 789).   2-Formyl-3-
-phenyl-1-azirine-N-allylimine on photolysis yields 1-allyl-
-2-phenylimidazole, whereas thermolysis affords exclusively
1-allyl-3-phenylpyrazole (Padwa, Smolanoff, and Tremper,
J. org. Chem., 1976, 41, 543).

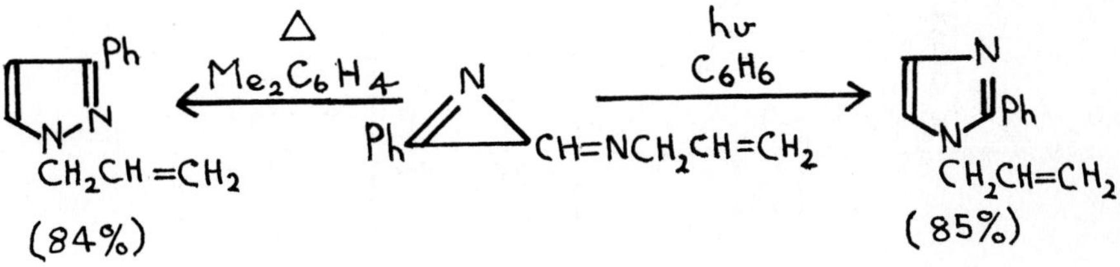

(84%)

(85%)

Reaction of 2-formyl-3-phenyl-1-azirine with 3-methoxy-
carbonyl-2-propenylidene-1-triphenylphosphorane affords
a mixture containing 2-phenyl-7-methoxycarbonyl-1H-azepine
(40%) as well as an azirinyl diene.   2-Dimethylamine-3,3-
-dimethyl-1-azirine underoges a smooth thermal

isomerisation to 1-dimethylamino-3-methyl-2-azabutadiene, which is a useful reagent for the synthesis of pyridines or dihydropyridines (A. Demoulin *et. al.*, J. Amer. chem. Soc., 1975, *97*, 4409).

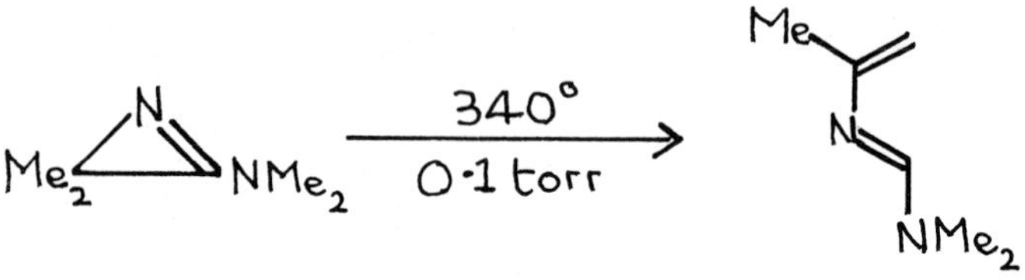

2-Phenyl-1-azirine on heating in carbon disulphide affords a 1:1 adduct with a thiazole ring system. The formation of the adduct appears to proceed *via* a regio-selective cycloaddition of carbon disulphide to the π bond of the 1-azirine. Whether this involves a concerted [π2s + π2a] pathway or a stepwise mechanism involving a dipolar transition state is not known (Nair and Kim, J. org. Chem., 1975, *40*, 1348).

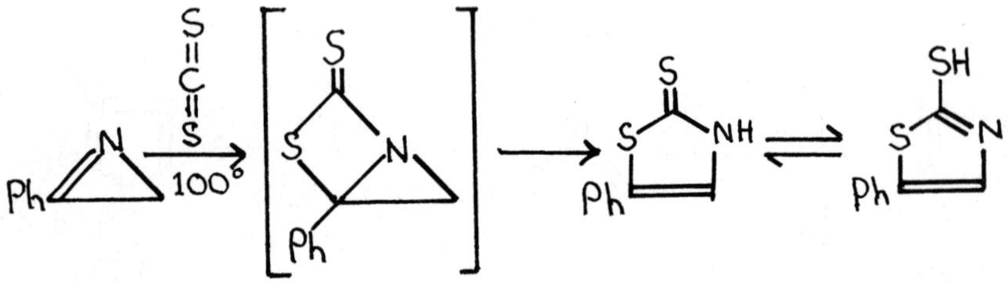

2-Dimethylamino-3,3-dimethyl-1-azirine reacts with carbon disulphide to give a product, which exists as the thiazolinethiolate (12) in the solid state and in an open chain structure (13) in solution (Chaloupka *et. al.*, Helv., 1976, *59*, 2566).

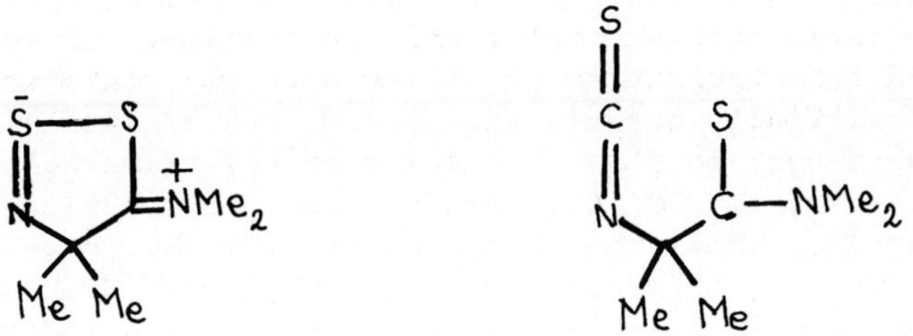

(12)

(13)

When substituents other than methyl are present at C-3, other types of products are sometimes formed (E. Schaumann *et. al.*, Tetrahedron Letters, 1977, 1351). Adducts (14) of similar structure to (12) are obtained from the same 1-azirine and activated isothiocyanates (Schaumann, E. Kausch, and W. Walter, Ber., 1977, 110, 820).

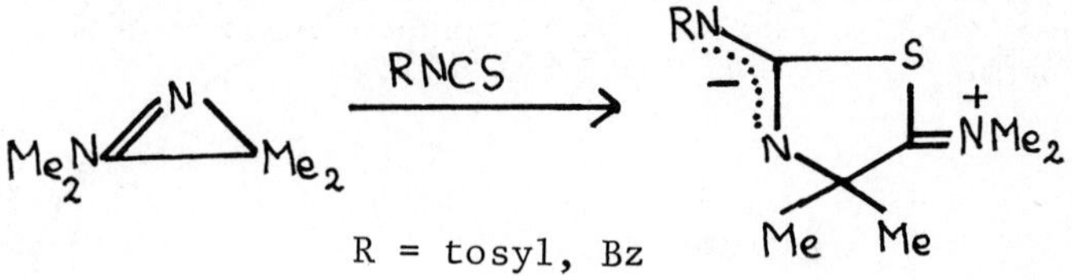

R = tosyl, Bz

(14)

Irradiation of 3-hydroxymethyl-2-phenyl-1-azirine gives 2-phenyl-3-oxazoline.

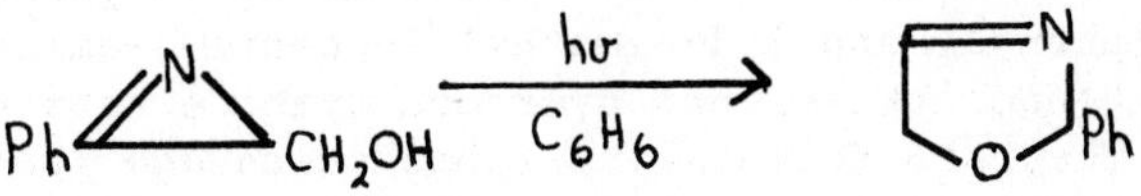

If the OH group is replaced by Cl, Br, $MeCO_2$ or $PhCO_2$
then irradiation of the resulting azirines furnishes an
unexpected rearrangement, which produces azabutadienes
in essentially quantitative yield, *via* a novel 1,4-
-substituent shift. Photolysis of the trimethylsiloxy
derivative (Y = $OSiMe_3$) produces only polymeric
materials, while the methyl ester (Y = OMe) gives a mixture
of *exo-* and *endo-*(2,6-dimethoxymethyl)-4,5-diphenyl-1,3-
-diazabicyclo[3.1.0]hex-3-ene. The results indicate that
the migrating substituent (Y) must be a reasonable good
leaving group in order for the rearrangement to occur
(Padwa, J.J. Rasmussen, and Tremper, Chem. Comm., 1976,
10).

Products obtained on heating 3-methyl-2-phenyl-,
3-ethyl-2-phenyl, and 3,3-dimethyl-2-phenyl-1-azirine are
styrenes, benzonitrile, and hydrogen cyanide or acetonitrile. They are formed by C-C bond cleavage leading
initially to iminocarbene intermediates. Evidence
indicates that the primary mode of product formation from
such an intermediate is a 1,4-hydrogen shift, giving
2-azabutadiene, which then fragments (*via* a small
equilibrium concentration of substituted 1-azetine) leading
to the final products. At high temperatures the

azabutadienes are converted to dihydroquinolines as well
(L.A. Wendling and R.G. Bergmann, J. org., 1976, **41**, 831).

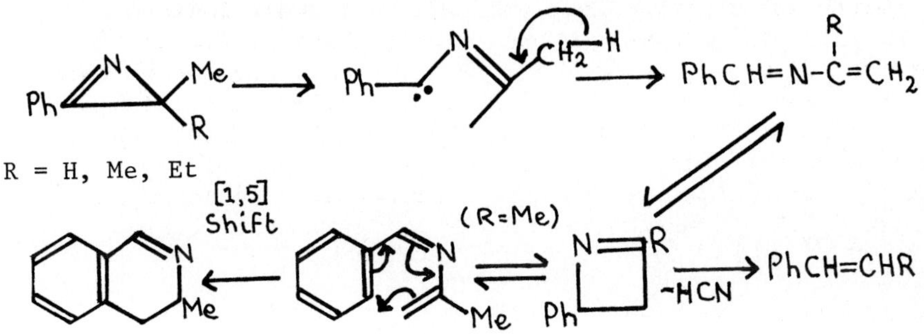

R = H, Me, Et

The regioselectivity of cycloaddition reactions of
photochemically generated benzonitrile isopropylides from
3,3-dimethyl-2-phenyl-1-azirines has been studied
(U. Gerber *et. al.*, Helv., 1977, **60**, 687).

Benzoylazirines are important intermediates formed
during the thermal decomposition of β-benzovinyl-azides
to give heterocycles (K. Isomura *et. al.*, Heterocycles,
1978, **9**, 1207). 2-(Dimethylamino)-3,3-dimethyl-1-azirine
with 2,4-diphenyl-3-methyl-1,3-oxazolium-5-olate yields
a 1:1 adduct, an aminooxazoline (J. Lukác, J.H. Bieri and
H.Heimgartner, Helv., 1977, **60**, 1657). 8-Membered heterocycles
have been synthesised from 2-dimethylamino-3,3-dimethyl-1-
-azirine and saccharin or phthalimide (S. Chaloupka *et.
al.*, *ibid.*, p.2476).

The reaction between 2,3-diaryl-1-azirine-3-
carboxamides and hydrazine is quite rapid at room
temperature and gives 1,2,4-triazin-6-ones (T. Nishiwake
and T. Saito, Chem. Comm., 1970, 1479).

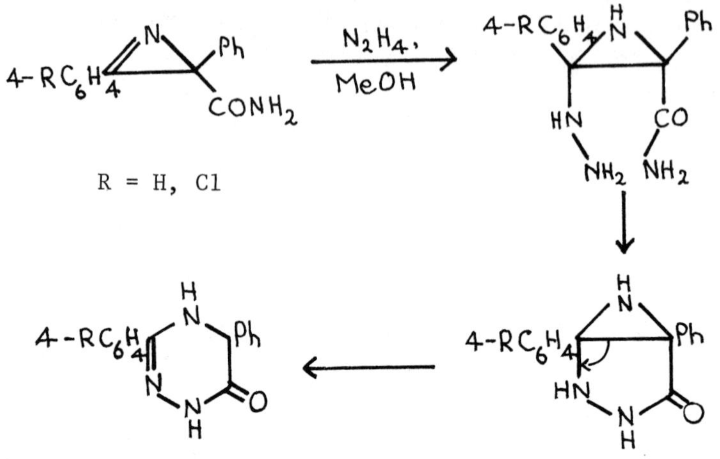

R = H, Cl

With triphenylphosphine and carbon tetrachloride it yields the triphenyliminophosphorane (17) which can be explained in terms of equilibration of the cyano intermediate (15) with the vinylnitrene (16), which probably favours markedly the very reactive nitrene species. Rapid coupling with tervalent phosphorus compounds would assist this shift in the equilibrium (Nishiwaki, *ibid.*, 1972, 565).

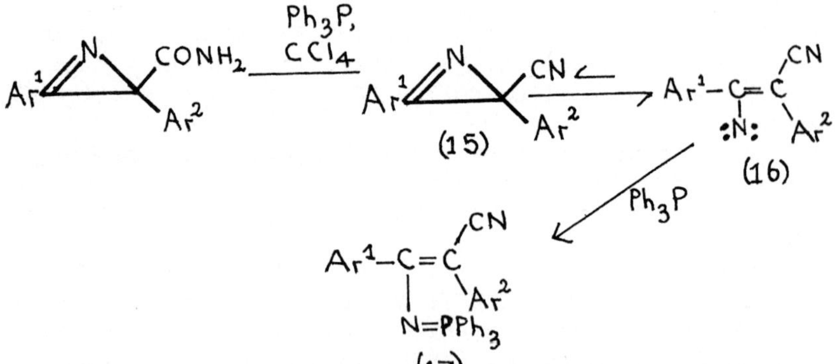

1-Azirines produce bicyclic 1:2 adducts or cyclic 1:1 adducts, depending on the substituents on the azirine ring, for instance, 2-phenyl-1-azirine with 2 or 1 equivalents of diphenylketene gives adduct (18), while 2,3-diphenyl-1-azirine under similar conditions affords (19) (Hassner, A.S. Miller and J.J. Haddadin, Tetrahedron Letters, 1972, 1353).

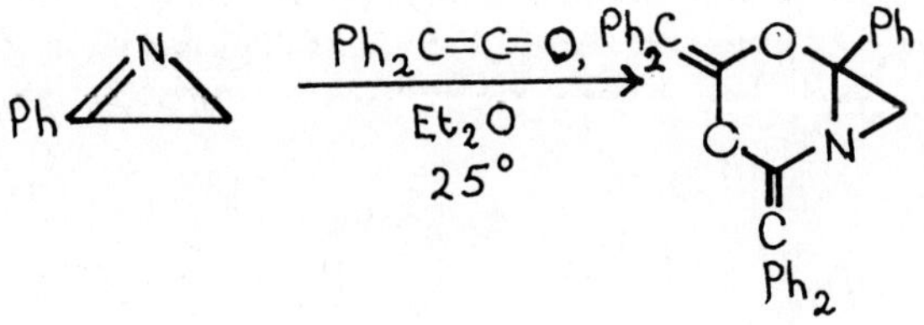

(18)

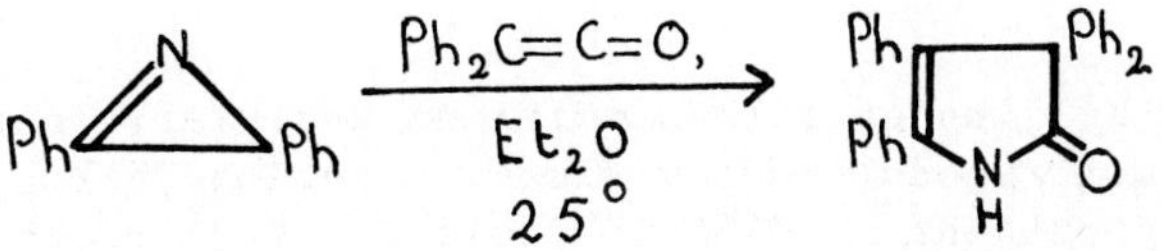

(19)

Reaction of 2-phenyl-1-azirine with an equimolar amount of molybdenum hexacarbonyl results in ring cleavage to give a mixture of 2,5-diphenylpyrazine, 2,5-diphenyl--3,6-dihydro- and 3,6-diphenyl-3,6-dihydro-pyrazine. With chromium hexacarbonyl and tungsten hexacarbonyl only the first and last products are obtained (H. Alper and S. Wollowitz, J. Amer. chem. Soc., 1975, 97, 3541).

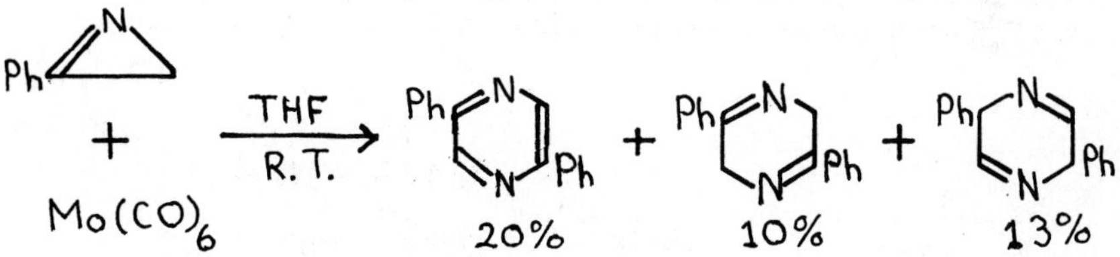

No pyrazines or dihydropyrazines are formed when 2-aryl-1-azirines react with di-iron nonacarbonyl but instead 2,5-diarylpyrroles, and several binuclear iron carbonyl complexes are formed (Alper and J.E. Prickett, Chem. Comm., 1976, 191).

2-Aryl-1-azirines with silver perchlorate at room temperature are converted into 2,5-diarylpyrazines (*ibid.*, p.983).

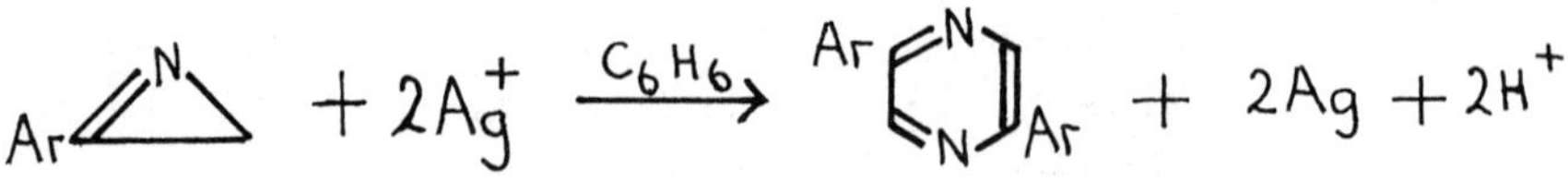

$$Ar \overset{N}{\diamond} + 2Ag^+ \xrightarrow{C_6H_6} Ar \overset{N}{\diamondsuit}_{N} Ar + 2Ag + 2H^+$$

Similarly under mild conditions 2-arylazirines react with chlorodicarbonylrhodium dimer, or chlorocarbonylbis-(triphenylphosphine)rhodium to yield 2-styrylindoles (*ibid.*, p.483).

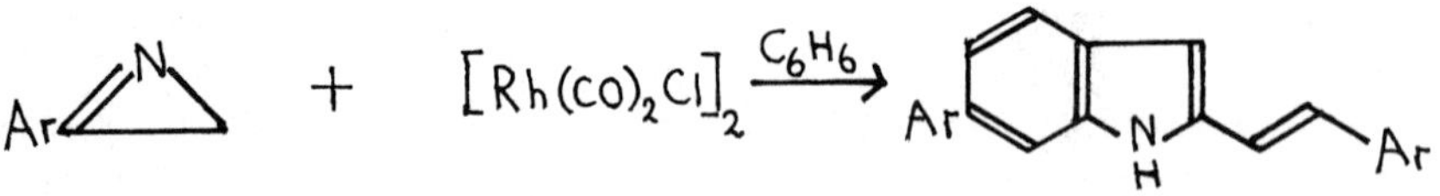

$$Ar \overset{N}{\diamond} + [Rh(CO)_2Cl]_2 \xrightarrow{C_6H_6} Ar \text{—indole—} Ar$$

(iii) 2-Azirines (1H-Azirines): synthesis

Anion (2) obtained by the reactions indicated is particularly stable and does not undergo elimination to the corresponding antiaromatic 2-azirine (G.M. Rubottom *et. al.*, Tetrahedron Letters, 1972, 3591).

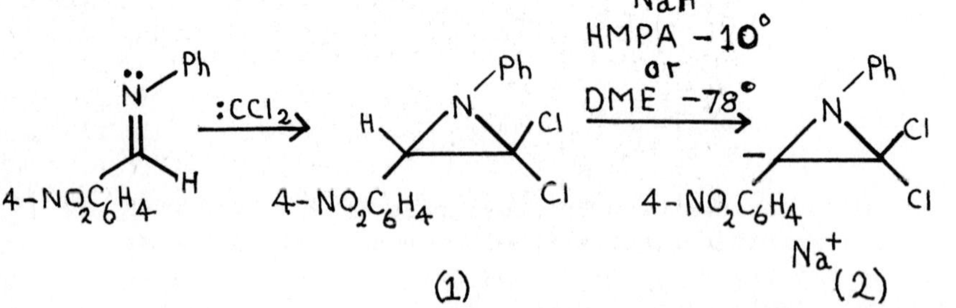

(1)   (2)

HMPA hexamethylphosphoramide
DME   dimethoxyethane

The combined electron withdrawing effect of two chlorines, the ring nitrogen, and the 4-nitrophenyl group make the ring proton of (1) acidic. The 4-nitrophenyl substituent also stabilises the conjugated base (2) by providing for electron release into the aryl ring.

Investigation of the gas-phase pyrolysis of isatin to give aniline and 1-cyanocyclopentadiene, indicates that the iminocyclohexadienylidenes (3) and (5) interconvert quantitatively with the elusive 1H-benzazirine (4) prior to ring contraction to cyanocyclopentadiene (C. Thétaz and C. Wentrup, J. Amer. chem. Soc., 1976, 98, 1258).

Since both 1,4-dimethyl-5-phenyl- and 1,5-dimethyl--4-phenyl-triazole on pyrolysis give 3-methylisoquinoline as a product. The carbenes derived from these triazoles may be interconverted *via* a symmetrical 2-azirine intermediate (T.L. Gilchrist, G.F. Gymer, and C.W. Rees, Chem. Comm., 1973, 835) (see p. 60).

6. *Three-Membered Rings With More Than One Hetero-Atom*

(a) *Dioxirane*

It has been reported that dioxirane (1) has been detected by mass spectrometry (R.I. Martinez, R.E. Huie, and J.T. Herron, Chem. Phys. Letters, 1977, 51, 457) and microwave spectroscopy (F.J. Lovas and R.D. Suenram, *ibid.,*

p.453), as a transient intemediate in the gas-phase
ozonolysis of ethene (W.R. Wadt and W.A. Goddart, J. Amer.
chem. Soc., 1975, 97, 3004; Herron and Huie, *ibid.*, 1977,
99, 5430).

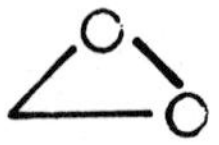

(1)

(b) Oxaziridines (oxaziranes)

The oxidation of *N*-4-nitrobenzylidene-*tert*-butylamine
with 3-chloroperbenzoic acid (MCPBA) in an aprotic solvent
yields an oxaziridine. Kinetic studies suggest a
nucelophilic attack by the Schiff base on a peroxy acid
dimer (V. Madan and L.B. Clapp, *ibid.*, 1970, 92, 4902).

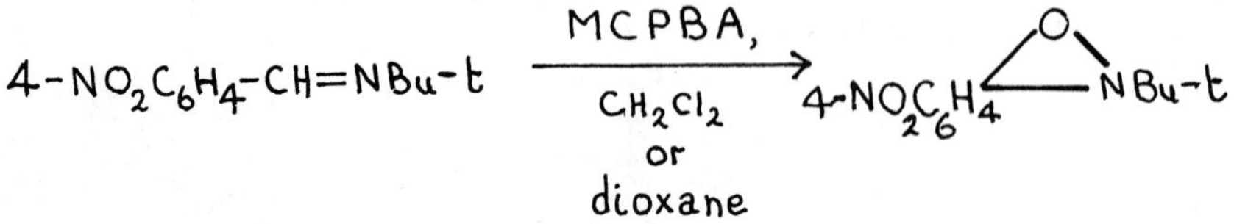

The *tert*-amyl hydroperoxide oxidation of Schiff bases
in the presence of $MoCl_5$ or $MO(CO)_6$ leads to the
formation of oxaziridines. The reaction proceeds rapidly
in benzene and is more convenient than the peracid
oxidation   (G.A. Tolstikov *et. al.*, Tetrahedron Letters,
1971, 2807).

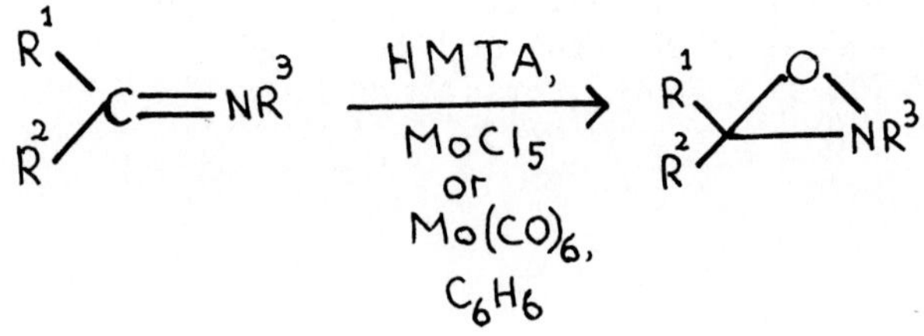

$$\begin{array}{c} R^1 \\ R^2 \end{array}\!\!C\!\!=\!\!NR^3 \quad \xrightarrow[\substack{MoCl_5 \\ or \\ Mo(CO)_6, \\ C_6H_6}]{HMTA,} \quad \begin{array}{c} R^1 \\ R^2 \end{array}\!\!\!\!\bigtriangleup\!\!\!\!^{O}_{}\!\!-\!NR^3$$

It has been shown that 1,3,5-oxadiazepines are obtained by rearrangement of the cycloadducts resulting from the reaction between nitrile oxides and 2-phenyl-
-benzazete, presumably *via* an oxaziridine (see p.148 azetes) (C.W. Rees *et. al.*, Chem. Comm., 1975, 740).

(c) Diaziridines and related compounds

Condensation of benzaldehyde and methylhydrazine in dry tetrahydrofuran containing 2-3 drops of acetic acid, followed by passage of diborane, yields 1-benzyl-2-methyl-
-3-phenyldiaziridine (J.A. Blair and R.J. Gardner, J. chem. Soc., Perkin I, 1972, 485).

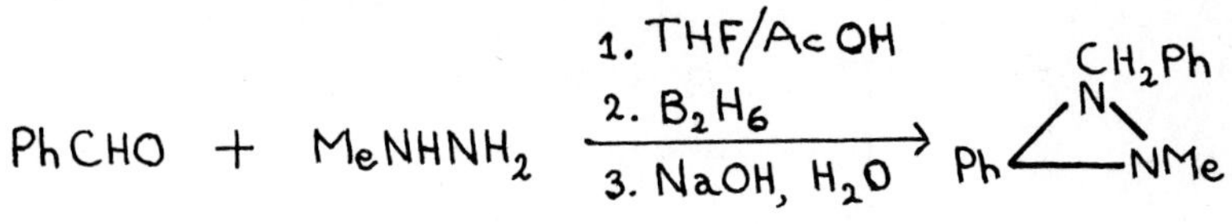

$$PhCHO \;+\; MeNHNH_2 \quad \xrightarrow[\substack{3.\,NaOH,\,H_2O}]{\substack{1.\,THF/AcOH \\ 2.\,B_2H_6}} \quad Ph\!\!\bigtriangleup\!\!\begin{array}{c} N\!-\!CH_2Ph \\ \\ -NMe \end{array}$$

Irradiation of 2-tetrazolines gives diaziridines in reasonable yields (T. Akiyama *et. al.*, Chem. Letters, 1974, 185).

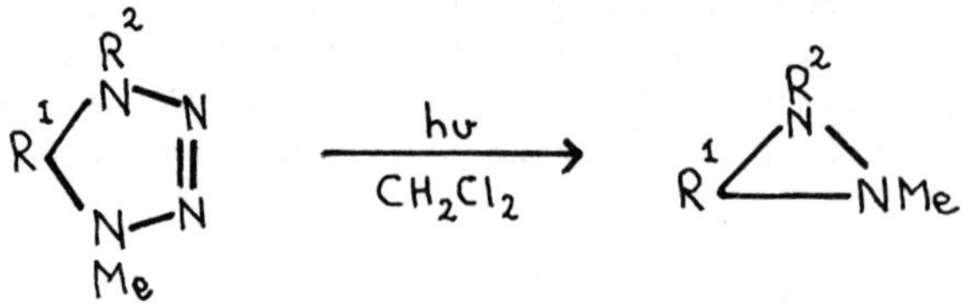

R^1, R^2 = Ph, Me; Me, Ph; Ph, Ph

3,3-Dialkyldiaziridines react with phthaloyl chloride
in the presence of triethylamine to afford 1,1-dialkyl-1H-
-diazirino[1,2-b]phthalazine-3,8-diones (1) which on
heating yield 2-(1-alken-1-yl)-4-hydroxy-1(2H)-phthalza-
inones (2) (H.W. Heine *et. al.*, J. org. Chem., 1974, 39,
3187).

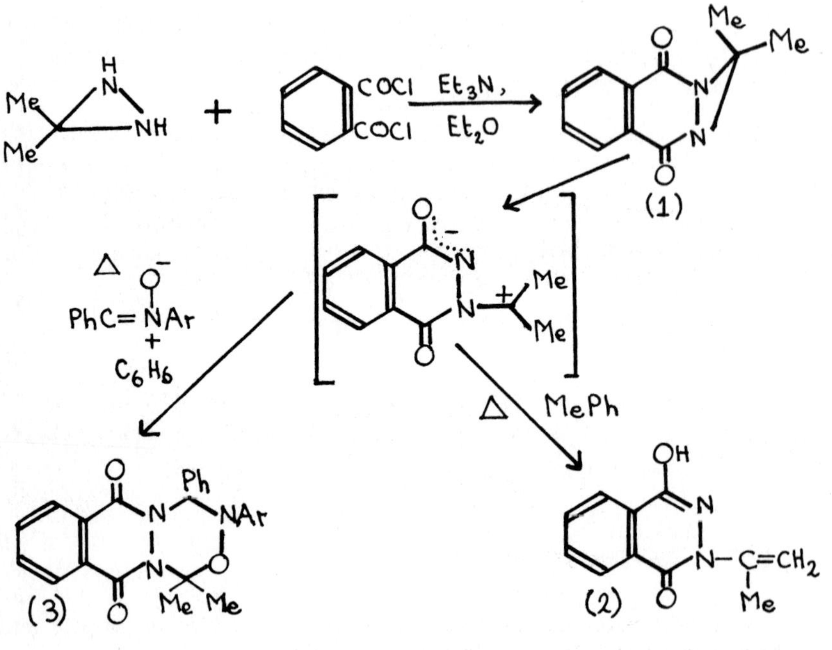

The 3,8-diones react with nitrones to give
1H-[1,2,4,5]oxatriazino[4,5-b]phthalazine-6,11-diones (3).
The tricyclic system is formed in what may be regarded as
the cycloaddition of two different 1,3-dipolar species
(Heine and L. Heitz, *ibid.*, p.3192).

The photolysis of the thermally highly stable
2,5-dimethyltetrazolone yields 1,2-dimethyldiaziridinone
(H. Quast and L. Bieber, Angew. Chem. internat. Edn., 1975,
14, 428).

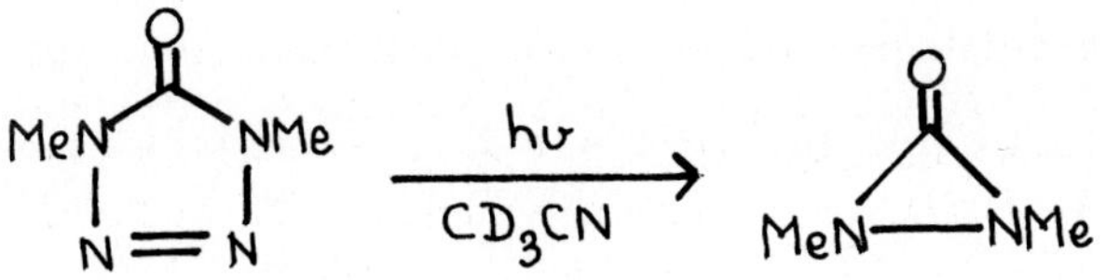

When the photolysis is carried out in either ether or 2-
propanol only secondary products of radical reaction with
solvent are obtained.

Ethyl diazoacetate reacts vigorously in a mild exothermic
manner with 4-phenyl-1,2,4-triazoline-3,4-dione to give a
1:1 adduct, 6-ethoxycarbonyl-3-phenyl-1,3,5-triazabicyclo-
[3.1.0]hexane-2,4-dione. This is the first example of
carbene addition to an azo linkage (R.A. Izydore and
S. McLean, J. Amer. chem. Soc., 1975, **97**, 5611).

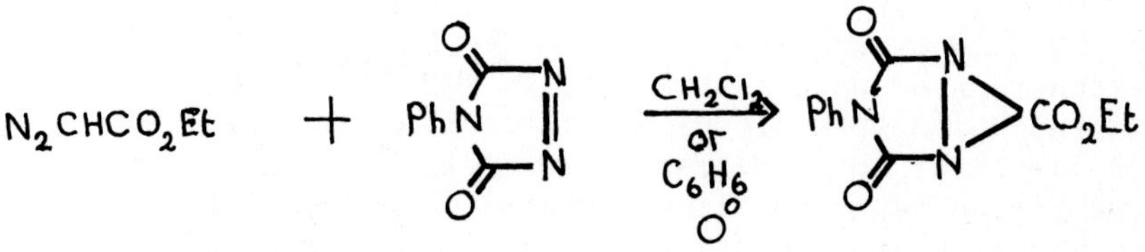

Cycloaddition to hetero analogues of methylencyclo-
propanes are steadily gaining in preparative and
mechanistic importance. The [3+2] cycloaddition of phenyl
isocyanate to triisopropyldiaziridininmine gives a

1.2.4-triazolidine (Quast and E. Spiegel, Angew. Chem.
internat. Edn., 1977, _16_, 109).

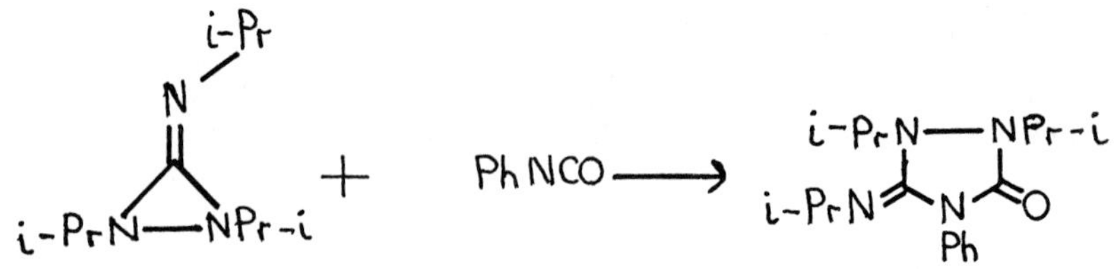

Diaziridinimine (4) reacts with dimethyl acetylenedicarboxylate to give the 1:3 adduct (5), not all alkyl
derivatives of (4) behave in the same way (Quast _et. al._,
ibid., p.177).

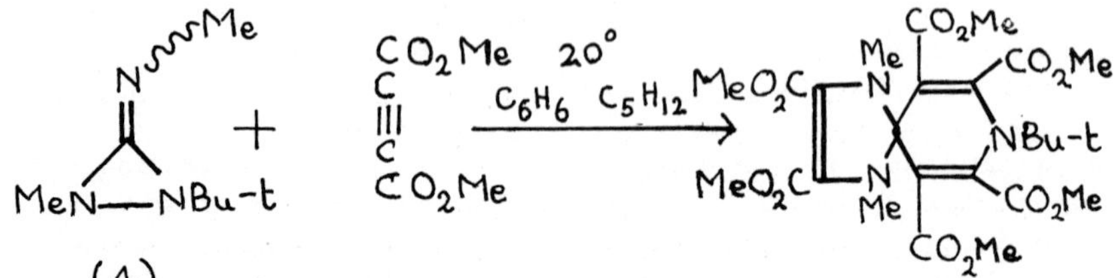

3-Oxidopyridazinium betanes (6, R^1 = Me, R^2 = R^3 = H)
and the related phthalazines (6, R^1 = H, R^2R^3 = C_4H_4)
on irradiation form bicyclic valence isomers (7), which
show little or no amide resonance in their ir spectrum.
In protic solvents the five-membered ring opens;
recyclisation and aromatisation affords the rearranged
pyridazinone (8) (Y. Maki _et. al._, Chem. Letters, 1977,
1005).

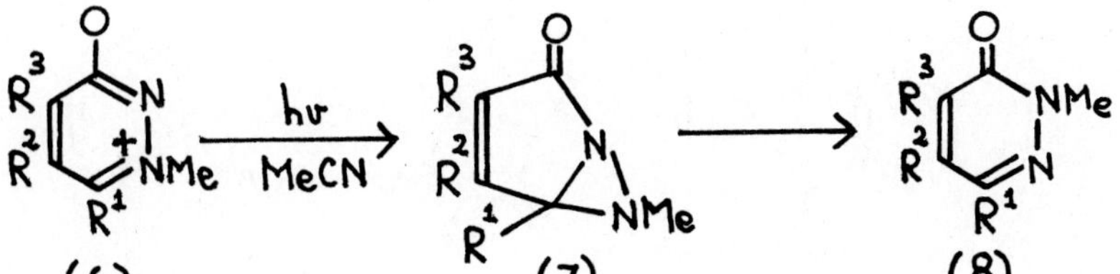

$$(6) \quad (7) \quad (8)$$

Irradiation of solutions of diaziridines in di-*tert*-butyl peroxide has resulted in diaziridinyls detected by esr spectroscopy (A.R. Forrester and J.S. Sadd, Tetrahedron Letters, 1976, 4205).

An extension of the method used for the preparation of 1-azaspiropentane (p. 43) i.e. the reaction of N-isopropylalleninime with organic azides fails to give diazaspiropentanes (J.K. Crandall, L.C. Crawley, and J.B. Komin, J. org. Chem., 1975, **40**, 2045).

Phthalimidonitrene adds to methyl 2-methylbuta-2,3-dienoate to give a 2-methoxycarbonyl-2-methyl-1,5-di-(phthalimido)-1,5-diazaspiro[2,2]pentane. The product is a mixture of invertomers and not diastereoisomers. The nmr spectrum corresponds to that of the major invertomer (better than 90%). Equilibration occurs at ca $-5°$ as indicated by the appearance of signal due to the minor invertomer (R.S. Atkinson and J.R. Malpass, Tetrahedron Letters, 1975, 4305).

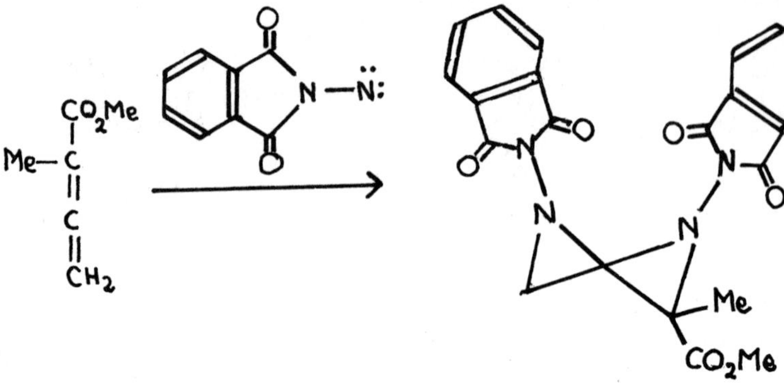

The diazirine isomer (9) has been detected in the thermally reversible thermochromic system 2-oxo-3-diazo-indoline/2-oxospiro[indoline-3,3'-diazirine] (E. Voight and H. Meier, Angew. Chem. internat. Edn., 1975, 14, 103).

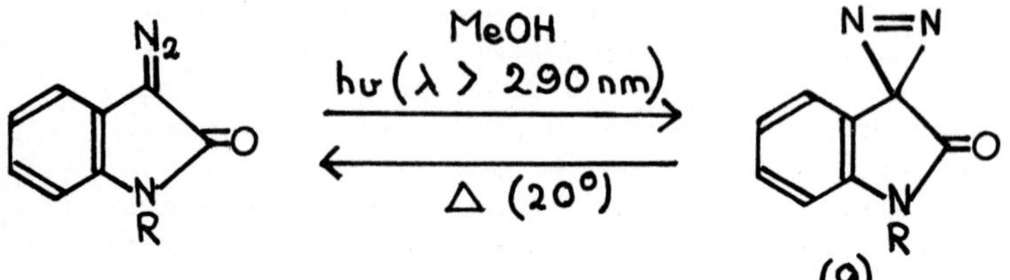

R = H or Me

Although the photoisomeriation of diazo-compounds by visible light to give diazirines appeared to be restricted to α-diazoamides [e.g. diazoacetamide on irradiation gave the diazirinecarboxamide (10); R.A. Franich, G. Lowe, and J. Parker, J. chem. Soc. Perkin I, 1972, 2034], it is now known that the photochemically induced valence isomerisation of the α-diazoketones (11) and (12) yield diazirines (T. Miyashi, T. Nakajo, and

T. Mukai, Chem. Comm., 1978, 442).

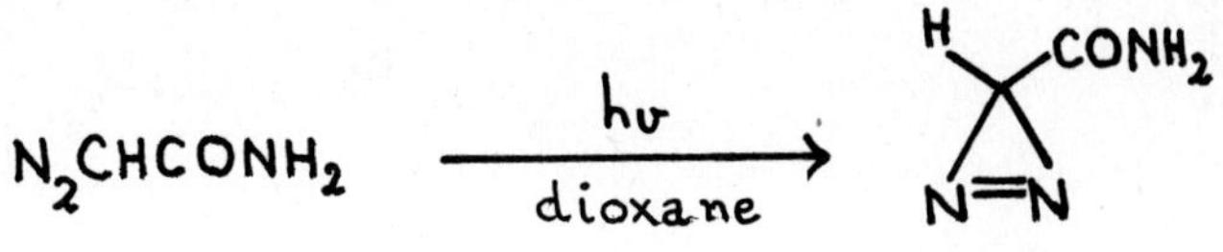

(10)

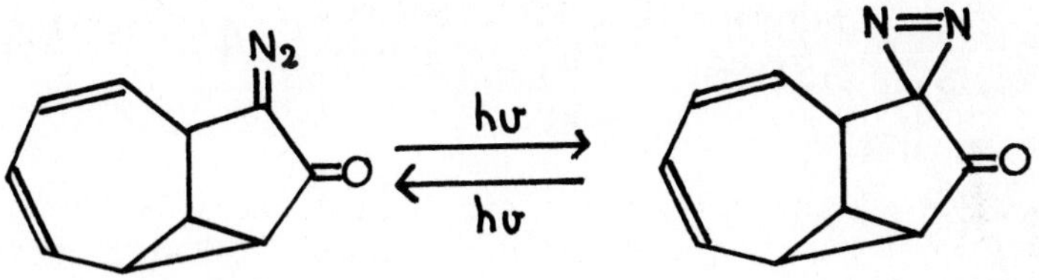

(11

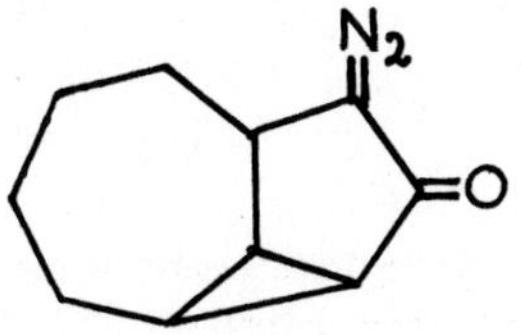

(12)

Irradiation of isopropylidene diazomalonate (13) in a large excess of cyclohexene at 350 nm results in the production of the diazirine (14), a Meldrum's acid derivative, and the cyclopropane (15) as the minor product. Direct irradiation at 243.7 nm yields the cyclopropane (15) in high yield (T. Livinghouse and R.V. Stevens, J. Amer. chem. Soc., 1978, 100, 6479).

Theoretical *ab initio* SCF methods have been used to make calculations on the photochemical diazomethane--diazirine interconversion (B. Bigot *et. al., ibid.,* p.6575).

Evidence has been reported which clearly establishes the intermediacy of diazoalkanes in the thermal decomposition of diazirines (B.M. Jennings and M.T.H. Liu, J. Amer. chem. Soc., 1976, **98**, 6416).

Evidence for the diazomethane intermediate in the
photolysis of diazirine is also available, as in the
preparation of a number of 3-aryl-3H-diazirines (R.A. Smith
and J.R. Knowles, J. chem. Soc. Perkin II, 1975, 686).

(d) Thiaziridines and thiazirines

The addition of *tert*-butylamine sulphonyl chloride
to a mixture of *tert*-butyldiazomethane and triethylamine
in ether gives a thiaziridine dioxide (Quast and F. Kees,
Angew. Chem. internat. Edn., 1974, 13, 742).

$$\text{ClSO}_2\text{NHBu-}t \xrightarrow[\substack{\text{Et}_3\text{N, Et}_2\text{O} \\ -78°}]{t\text{-BuCHN}_2,} \quad t\text{-Bu}\overset{\displaystyle S}{\diagup\!\!\diagdown}\text{NBu-}t$$

The photolysis of phenyl-substituted C-, N-, and
S- containing heterocyclic compounds embedded in
poly(vinylchloride) at 10-15K has been studied. A compound
believed to be phenylthiazirine (I) is formed in each case
and is stable at 10-15K, but on heating it rearranges to
benzonitrile sulphide, which at higher temperatures
decomposes to benzonitrile and sulphur. The thiazirines
belong to the group of compounds designated antiaromatic
by Breslow (A. Holm, N. Harrit, and Ib Trabjerg, J. chem.
Soc. Perkin I, 1978, 746; Holm and N.H. Toubro, *ibid.*,
p.1445).

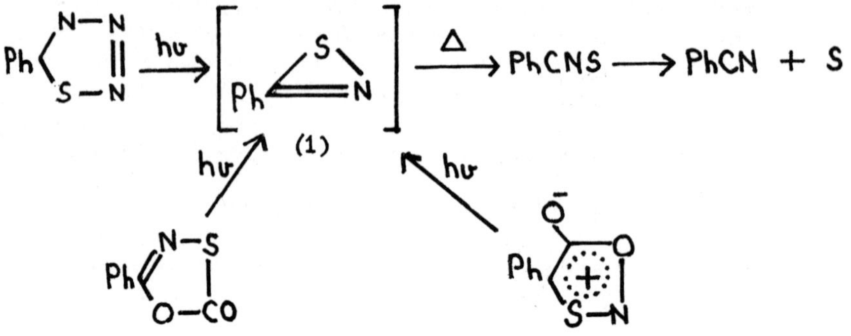

7. Four-Membered Rings With One Oxygen Atom

(a) Oxetanes (trimethylene oxides)

(i) Synthesis

(1) 2-Methyleneoxetane derivatives are obtained by the photochemical cycloadditions of carbonyl compounds with allenes, for example irradiation of benzophenone in the presence of tetramethylallene gives a mixture of products containing some 3,3-dimethyl-2-dimethylmethylene-4,4--diphenyloxetane (D.R. Arnold and A.H. Glick, Chem. Comm., 1966, 813).

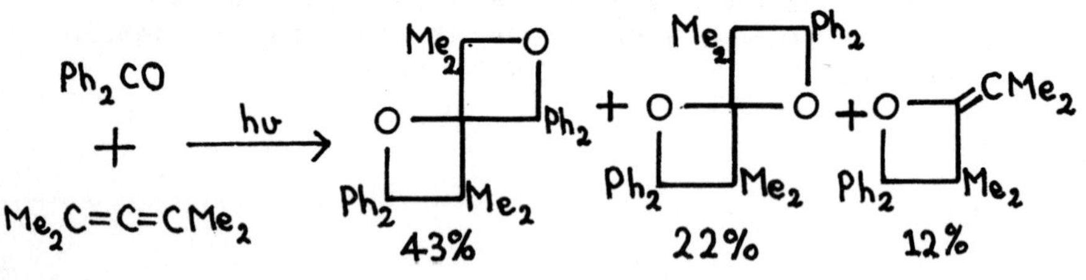

Other examples of the formation of oxetanes from aldehydes or ketones and allenes have been reported (H. Gotthardt,

R. Steinmetz, and G.S. Hammond, J. org. Chem., 1968, <u>33</u>, 2774;   J.K. Crandall and C.F. Mayer, *ibid.*,   1974, <u>39</u>, 2814).

2-Methyleneoxetanes are also formed from the reaction of certain ketenes with olefins (T. DoMinh and ͻ.P Strauss, J. Amer. chem. Soc., 1970, <u>92</u>, 1766;   D.C. England and C.G. Krespan, J. org. Chem., 1970, <u>35</u>, 3312).

2-Methyleneoxetane is prepared by pyrolysis of the anthracene adduct (1).   The oxetane reacts with excess of phenylithium to give 4-phenylbutan-2-one (P.F. Hudrlik and A.M. Hudrlik, Tetrahedron Letters, 1971, 1361).

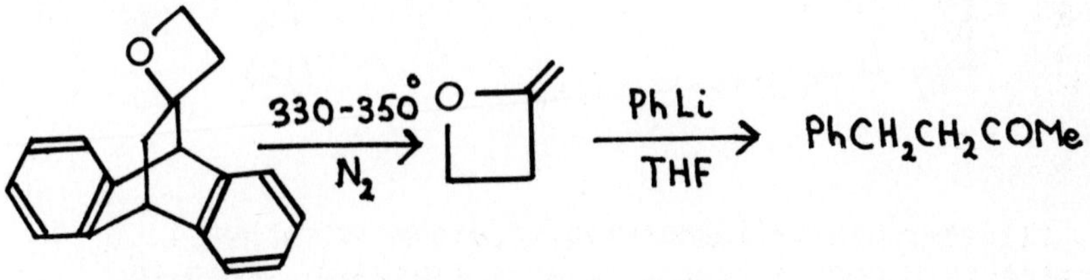

(1)

(2) Uv irradiation of a mixture of acetone and 2,3 -dimethylbut-2-ene gives the oxetane (2) along with the unsaturated ethers (3) and (4).   The latter arising from 1,5-hydrogen atom transfer and the former by ring closure of a biradical intermediate (H.A.J. Carless, Chem. Comm., 1973, 316).

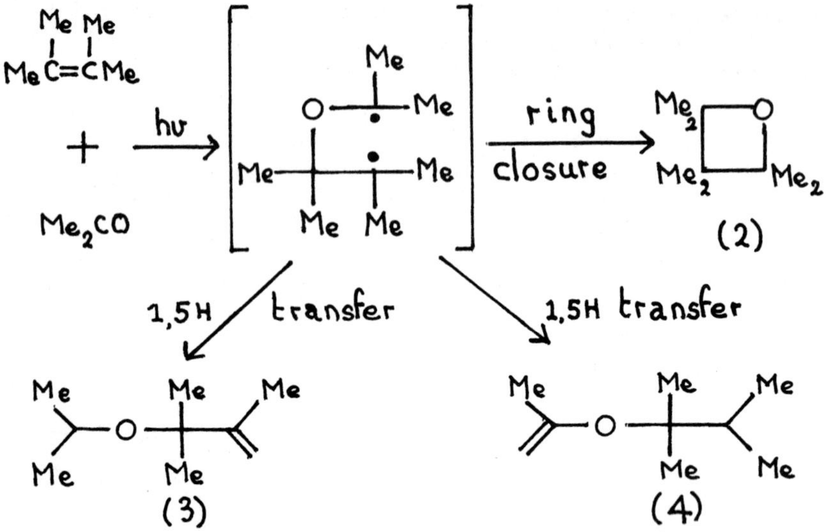

$$(2)$$

(3) The photocycloaddition of aromatic carbonyl compounds to simple alkenes, affords oxetanes. This is now a well established reaction of broad scope (R.A. Caldwell, G.W. Sovocool, and R.P. Gajewski, J. Amer. chem. Soc., 1973, 95, 2549).

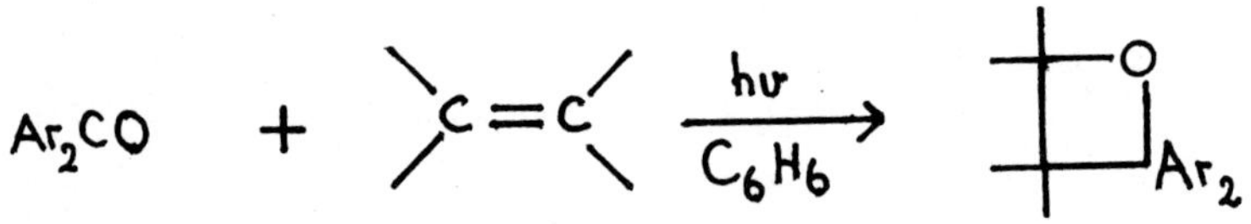

The primary interaction between the benzophenone triplet and simple alkenes has been shown to be the irreversible formation of a complex.

Methyl benzoate and methyl 4-methoxybenzoate on photolysis effect both substitution and [2+2] cycloaddition with alkenes to give oxetanes and ketones. These reactions occur *via* excited singlet states of the esters (T.S. Cantrell, Chem. Comm., 1973, 468).

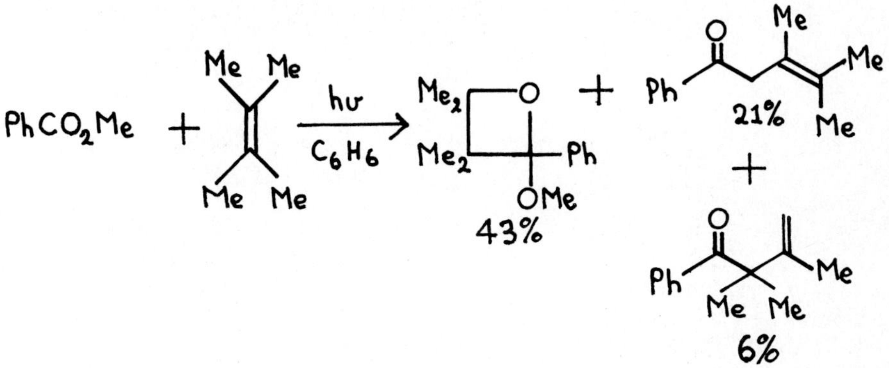

(4) The irradiation of 3-methyl-4-oxa-5-hexen-2-one yields *cis*-1,6-dimethyl-2,5-dioxabicyclo[2.2.0]hexane (J.C. Dalton and S.J. Tremont, Tetrahedron Letters, 1973, 4025).

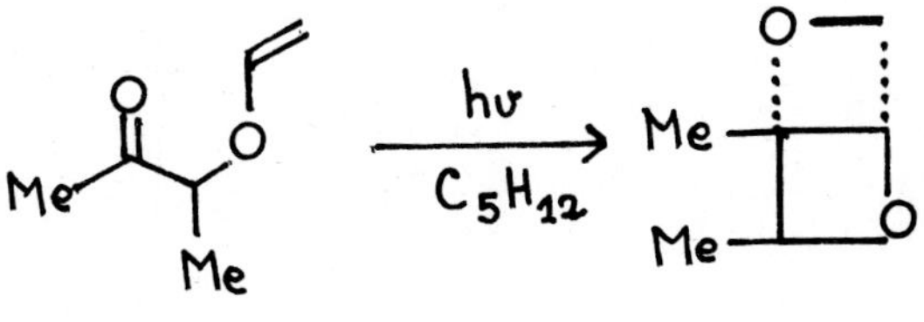

(b) Oxetanones

With bromine the conformationally rigid carboxylate salt (1) gives the oxetan-2-one (β-lactone) (2) resulting from bromolactonization of the βγ-unsaturated acid. No β-lactone is found following the iodination of the unsaturated acid (W.E. Barnett and J.C. McKenna, Chem.

Comm., 1971, 551).

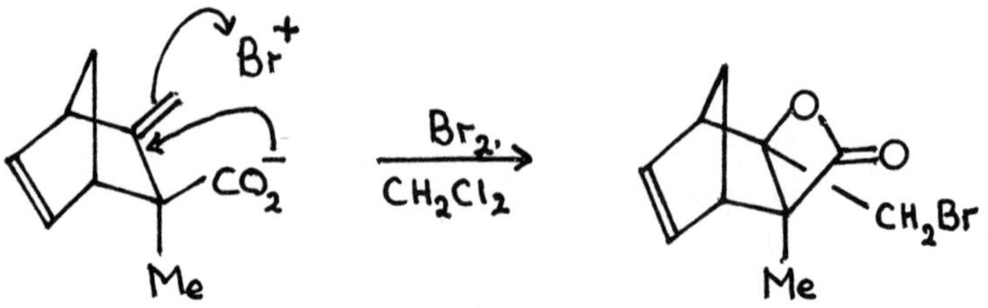

(1)   (2)

    The conformationally less rigid, open chain $\beta\gamma$-
-unsaturated carboxylate salts (3) cyclize readily to
γ-bromo-β-lactones (4) when aqueous solution of the salts
are treated with one equivalent of bromine in carbon
tetrachloride or methylene chloride. Thus affording a
new general method for the synthesis of oxetan-2-ones
(W.E. Barnett, and J.C. McKenna, Tetrahedron Letters, 1971,
2595).

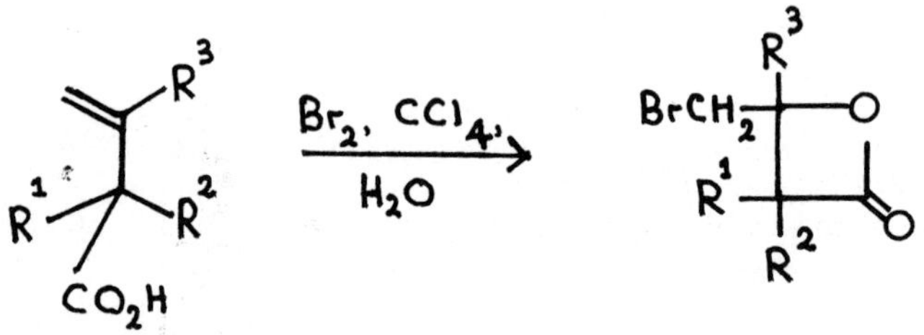

(3)   (4)

    Irradiation of α-pyrone, matrix isolated in argon
at 8°K, equilibrates the α-pyrone and the aldehyde-ketene
and slowly converts the α-pyrone to the β-lactone (bicyclo-
[2.2.0]pyran-2-one). Removal of the Pyrex filter and
continued irradiation (through quartz) at 8°K causes
destruction of the matrix isolated β-lactone and gives

carbon dioxide and cyclobutadiene. The concentration of cyclobutadiene rises to a maximum and then slowly decreases as irradiation is continued (O.L. Chapman, C.L. McIntosh, and J. Pacansky, J. Amer. chem. Soc., 1973, 95, 614).

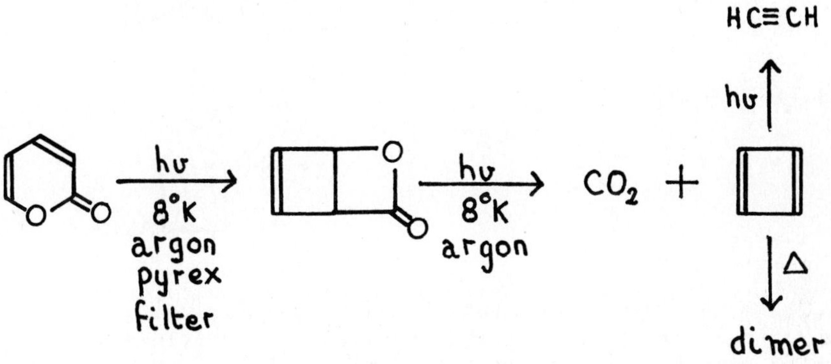

The appropriate deuteriated α-pyrones on photoisomerisation yield the related β-lactones, which on photoelimination of carbon dioxide yields monodeuterio-, 1,2-dideuterio- and 1,3-dideuterio-cyclobutadiene (Chapman *et. al.*, *ibid.*, p.1337).

Irradiation of phthaloyl peroxide matrix isolated in argon at 8°K gives rise to carbon dioxide, a ketoketene, and benzo[1,2-b]oxeten-2-one (benzopropiolactone) (Chapman *et. al. ibid.*, p.4062).

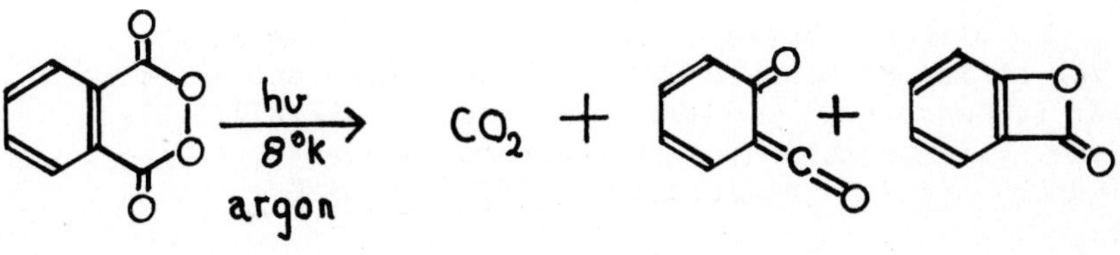

The addition of two equivalents of ozone to 1,1,3--tri-*tert*-butylallene yields 2,4,4-tri-*tert*-butyl-1,5-

-dioxaspiro[2.2]pentane as the only significant product observed by nmr at -78°. It rearranges cleanly on standing to give 2,2,4-tri-*tert*-butyloxetan-3-one (J.K. Crandall and W.W. Conover, Chem. Comm., 1973, 340).

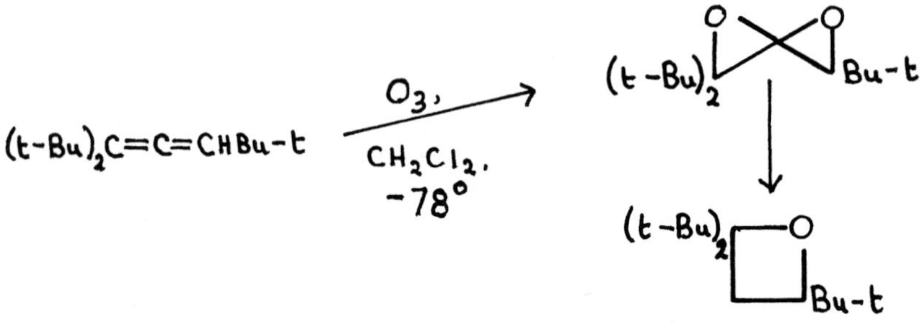

(t-Bu)$_2$C=C=CHBu-t
O_3,
CH_2Cl_2,
-78°
(t-Bu)$_2$
Bu-t
O
O
(t-Bu)$_2$
O
Bu-t

It is claimed that oxetan-2,4-diones (malonic anhydrides) are formed by the ozonlysis of diketenes at -78°. Over the course of an hour at–30° the dione undergoes partial decomposition to carbon dioxide and ketene (C.L. Perrin and T. Arrhenius, J. Amer. chem. Soc., 1978, 100, 5250).

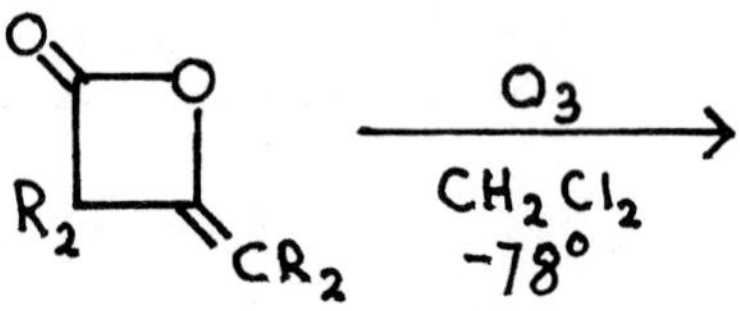

O_3
CH_2Cl_2
-78°
R_2
CR_2
R_2

R = H or Me

(c) Oxetes or oxetenes

An oxetene intermediate has been detected in the photoreaction of benzaldehyde with but-2-yne and it forms a [2+2] cycloaddition product, namely a 2,5-dioxa[2.2.0]-hexane, with benzaldehyde at low temperatures (L.E. Friedrich and J.D. Bower, *ibid.*, 1973, 95, 6869).

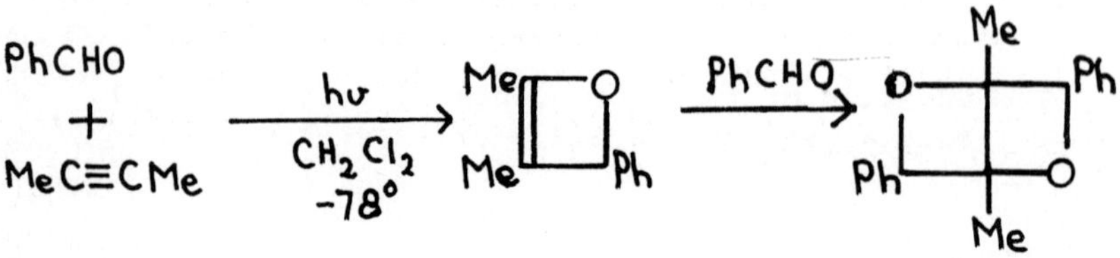

PhCHO
+
MeC≡CMe
hν
CH_2Cl_2
-78°
Me
Me
O
Ph
PhCHO
Me
Ph
Ph
Me

The acid catalysed reaction of 3-chloro-3-dimethyl-amino-propenal-1-^{13}C gives (E)-3-chloro-N,N-dimethylacryl-amide-3-[13]C *via* the intermediate oxetene (1) (M. Neuenschwander and A. Niederhauser, Chimia, 1973, <u>27</u>, 379).

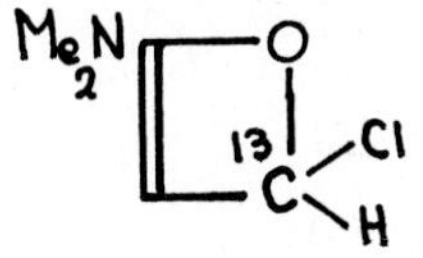

(1)

8. Four-Membered Rings With One Sulphur Atom

(a) Thietanes (trimethyl sulphides)

(i) Synthesis

Irradiation of adamantanethione in the presence of α-methylstyrene, ethyl vinyl ether, or 1,1-diphenylethene affords thietanes (C.C. Liao and P. de Mayo, Chem. Comm., 1971, 1525).

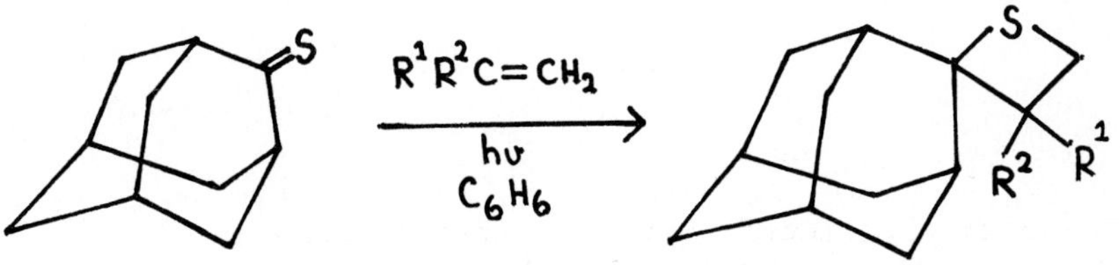

5-Thiabicyclo[2.1.1]hexane, m.p. 151-152° which contains a thietane ring, is obtained on treating a mixture of equal amounts of *cis-* and *trans-*3-chlorocyclopentyl thioacetate with a boiling aqueous ethanol solution of potassium hydroxide (I. Tabushi, Y. Tamaru, and Z. Yoshida, Tetrahedron Letters, 1970, 2931).

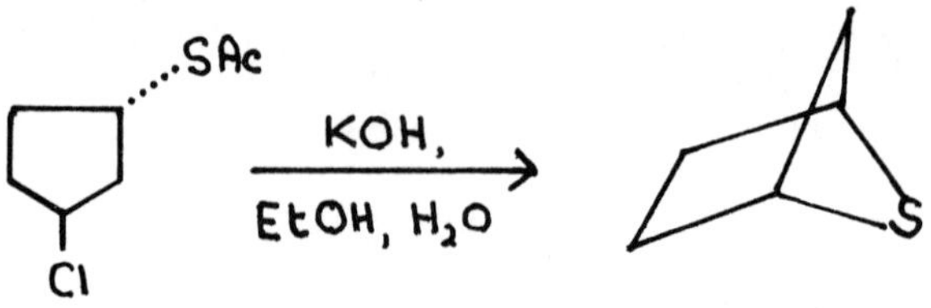

Thiete 1,1-dioxide reacts with 2-methyl-1-dimethyl-
aminoprop-1-ene to give 5,5-dimethyl-6-*exo*-dimethylamino-2-
-thiabicyclo[2.2.0]hexane 2,2-dioxide, m.p.102-103°.
A similar condensation of the dioxide with diethyl 1-propyl-
amine affords the unsaturated 2-thiabicyclo[2.2.0]hexane,
which an acid hydrolysis yields *exo*-5-methyl-2-thiabicyclo-
-[2.2.0]hexan-6-one 2,2-dioxide, m.p.75-78° (L.A. Paquette,
R.W. Houser, and M. Rosen, J. org. Chem., 1970, 35, 905).

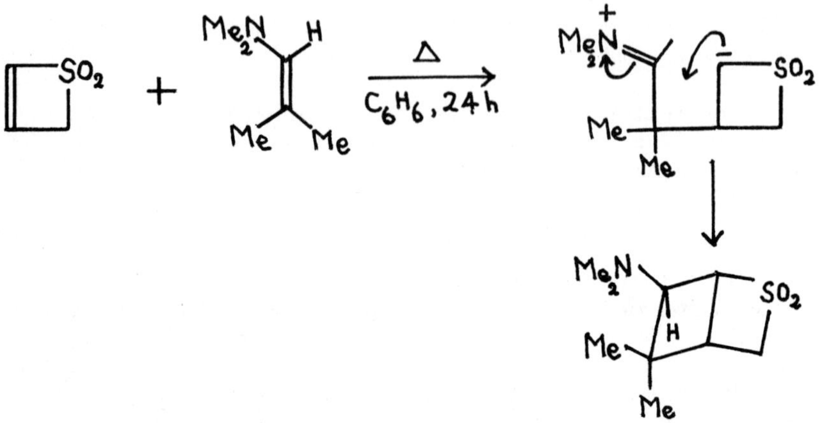

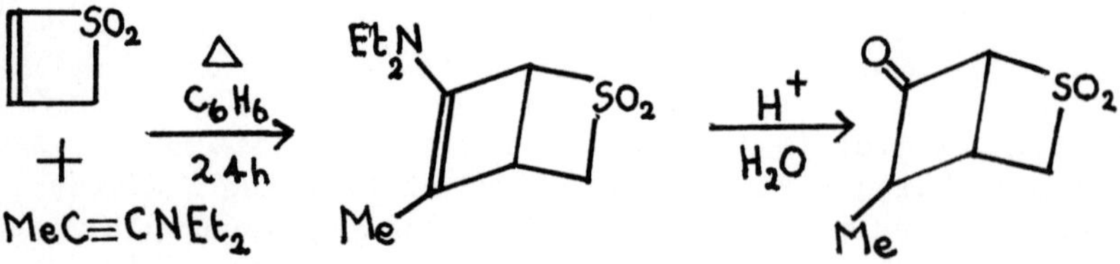

The propynylamine also undergoes a 2+2 cycloaddition to 2,2-dimethylthiete 1,1-dioxide and the corresponding carbonyl compound on diborane reduction gives the *endo* hydroxy sulphone.

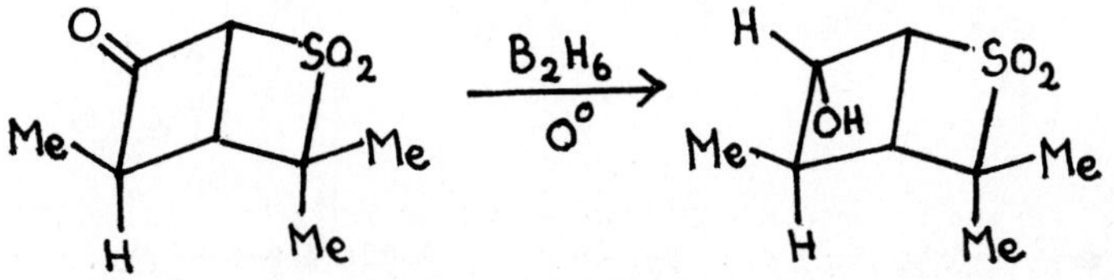

Care must be taken on carrying out the oxidation of 3-hydroxythietane to 3-hydroxythietane 1,1-dioxide, m.p.100°, using hydrogen peroxide. Under no circumstances must the peroxide-containing solution be concentrated in a closed system (D.C. Dittmer *et. al.*, J. org. Chem., 1971, 36, 1324).

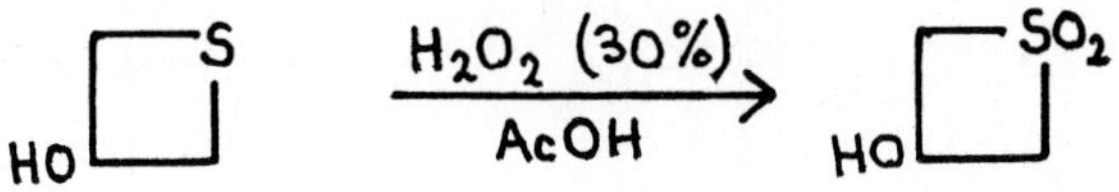

The extent of asymmetric induction arising from cycloaddition of sulphene to the optically active enamines (1) and (2) one with an acyclic tertiary amine residue and the other a cyclic amine, has been found to be 6% and 25%, respectively. Most notably, stereospecificity is seen to increase as the rotational degrees of freedom are reduced.

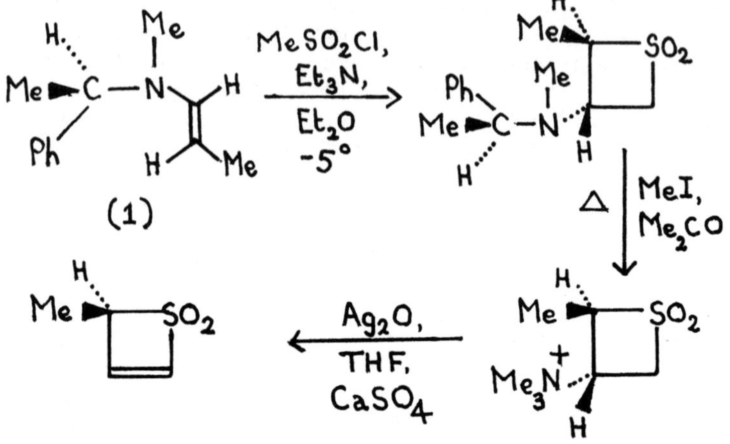

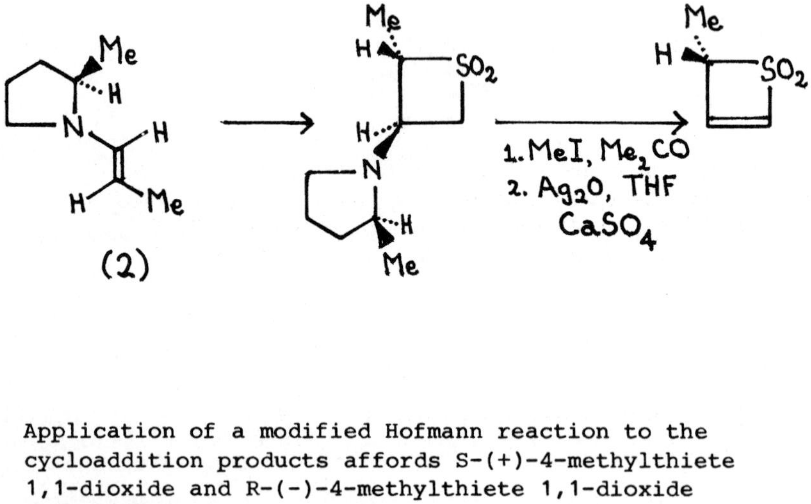

Application of a modified Hofmann reaction to the cycloaddition products affords S-(+)-4-methylthiete 1,1-dioxide and R-(-)-4-methylthiete 1,1-dioxide respectively (L.A. Paquette, J.P. Freeman, and S. Maiorana, Tetrahedron, 1971, 27, 2599).

Diarylthioketones, thiophosgene, and thione

carbonates react with substituted alkenes under
photochemical conditions to give thietanes and in somes
cases, 1,4-dithianes. The uv light-induced cycloaddition
of diphenylthiocarbonate to an alkene gives thietane (3)
(also see p. 113) and the thiocarbonate from pyrocatechol
with 2,3-dimethylbut-2-ene affords the spirothietane (4)
(H. Gotthardt and M. Listl, Tetrahedron Letters, 1973,
2849).

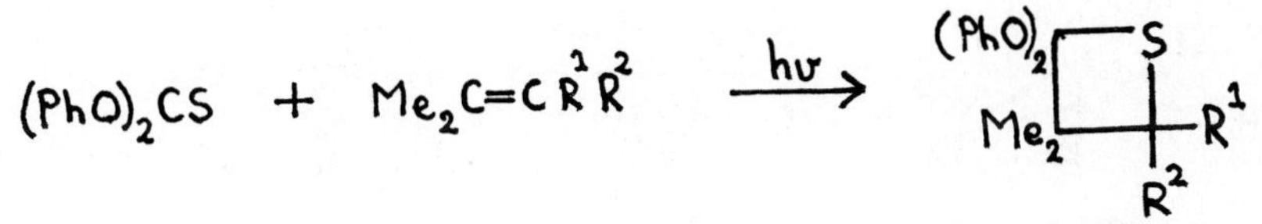

$$(PhO)_2CS \ + \ Me_2C{=}CR^2R^2 \ \xrightarrow{h\nu} \ (3)$$

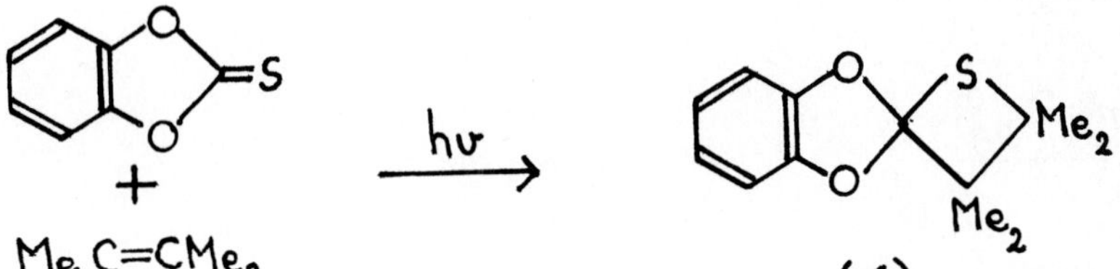

(4)

When a benzene solution of 1,3-diphenyl-2-
-thioparabanate and ethoxyethene is irradiated a spiro
thietane (5) m.p.198.5-200.5° is obtained in 88% yield
(Gotthardt and S. Nieberl, *ibid.*, 1974, 3397).

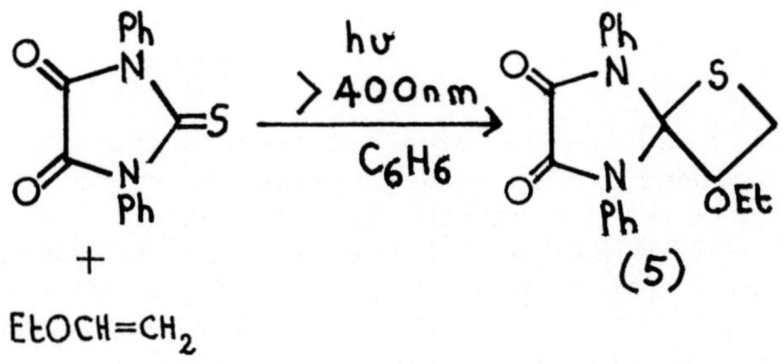

(5)

$EtOCH{=}CH_2$

(ii) Properties and reactions

Both *cis*- and *trans*-2,4-diphenylthietane 1,1-dioxide rearrange on treatment with *tert*-butoxymagnesium bromide to give *cis*-3,5-diphenyl-1,2-oxathiolane *cis*-2-oxide (1) and *trans*-3,5-diphenyl-1,2-oxathiolane (2,3)-*cis*-2-oxide (2) respectively (R.M. Dodson, P.D. Hammen, and R.A. Davis, J. org. Chem., 1971, 36, 2693).

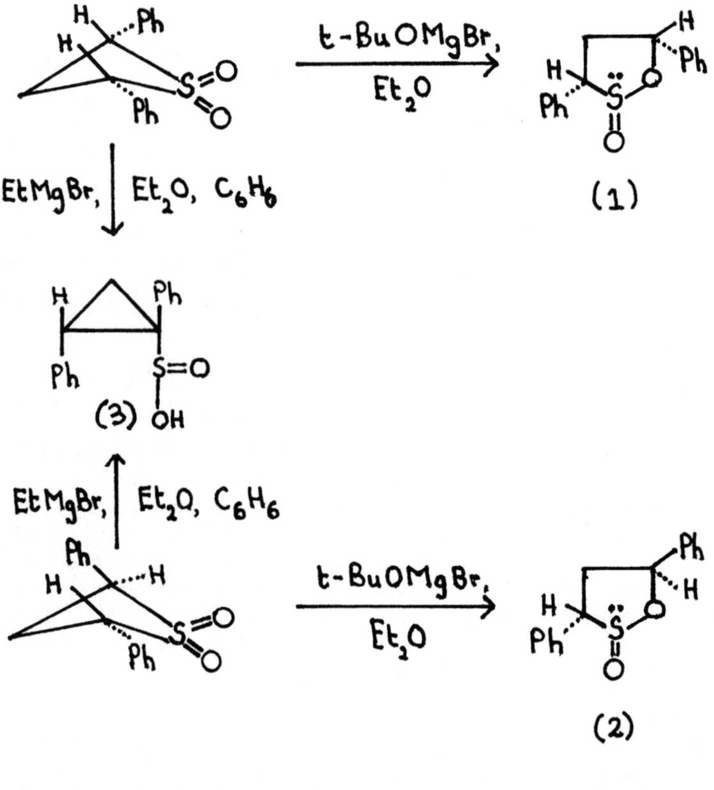

(1)

(3)

(2)

The *cis* and *trans*-2,4-diphenylthietane 1,1-dioxides on treatment with ethylmagnesium bromide rearrange in a highly stereoselective manner into *trans*-1,2-diphenylcyclopropane-sulphinic acid (3) (Dodson *et. al., ibid.,* p.2698).

Reactions of either *cis-* or *trans*-2,4-diphenylthietane 1-oxide with potassium *tert*-butoxide afford a mixture of *cis*-1,2-diphenylcyclopropanethiol (4) and *cis*-1,2-diphenyl-cyclopropanesulphinic acid (5) (Dodson, Hammen, and J. Yu Fan, *ibid.*, p.2703).

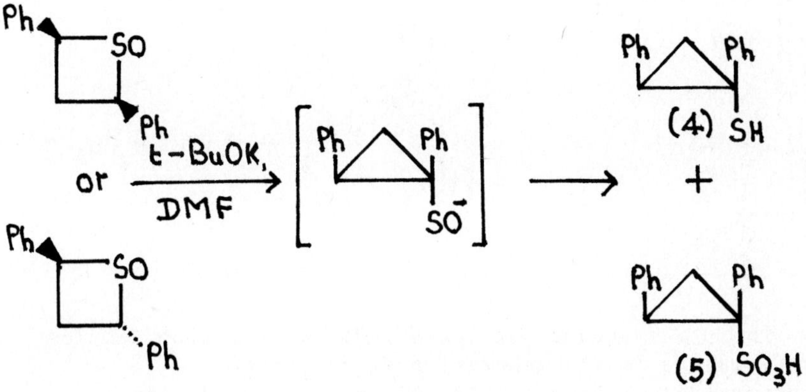

Treatment of *trans*-2,4-diphenylthietane with the same reagent gives a number of products, which could arise from the potassium salt of 1,2-diphenylcyclopropanethiol (6). Some stages involving hydrolysis or air oxidation (Dodson and Yu Fan, *ibid.*, p.2708).

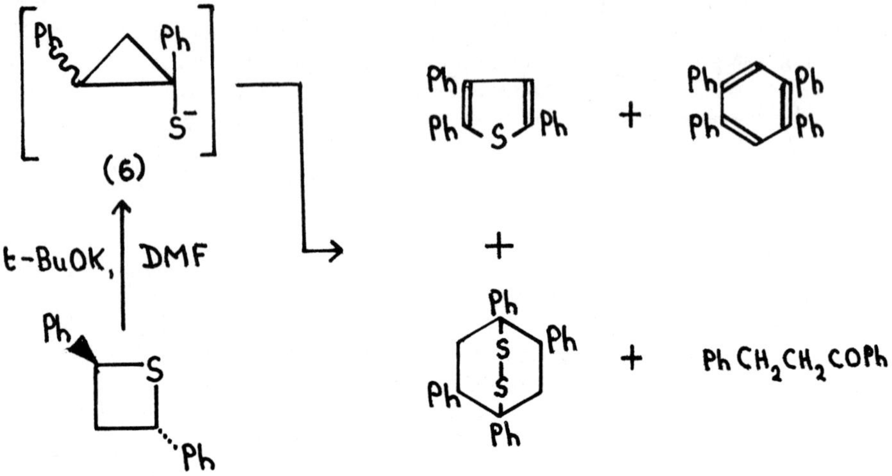

The flash vacuum pyrolysis (FVP) of thietane 1-oxide results in the facile generation of sulphine (thioformaldehyde S-oxide) (E. Block *et. al.*, J. Amer. chem. Soc., 1976, 98, 1264).

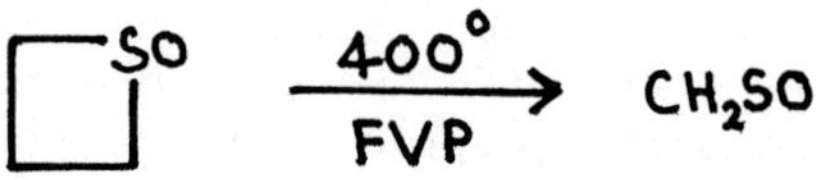

It is reported that the flash vacuum pyrolysis of 3-dimethylmethylene-2,2,4,4-tetramethylthietane 1,1-dioxide gives a mixture of a cyclopropane derivative (7) and a trimethylenemethane derivative (R.J. Bushby and M.D. Pollard, Tetrahedron Letters, 1978, 3855).

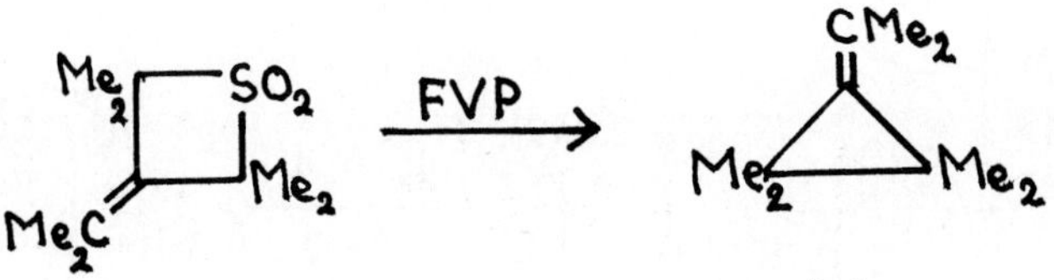

(7)

(b) *Thietes or thietenes and their oxides*

Thietes are of theoretical interest because they have the potential of forming anions and cations isoelectric with the anion and cation of cyclopentadiene.

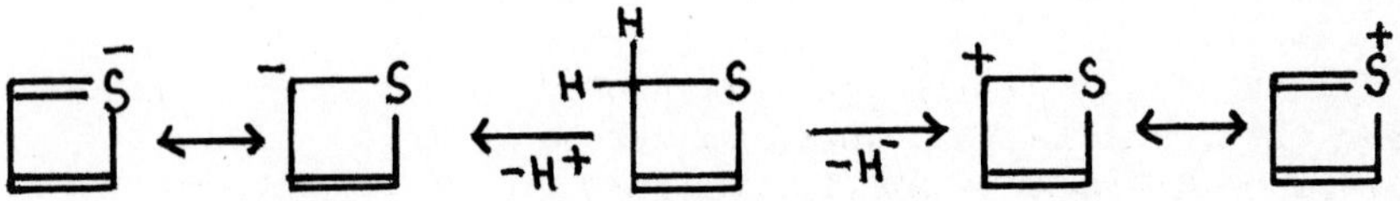

They are obtained by facile Hofmann elimination from 3-aminothietanes, usually derived from the products formed by the addition of sulphene to enamines.

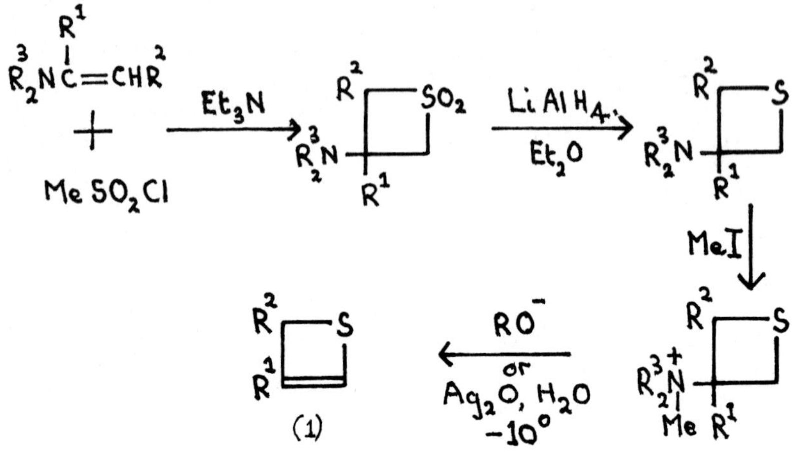

(1)

$R^1 = R^2 = H;$ $R^1R^2 = (CH_2)_4,$ $(CH_2)_5;$
$R^1 = Et,$ $R^2 = Me;$ $R^1 = n\text{-}Pr,$ $R^2 = Et$

Thietes, although relatively stable at low temperatures, show a marked tendency to polymerise or to decompose over a period of time. Thietes [1, $R^1R^2 = (CH_2)_4$ and $R^1R^2 = (CH_2)_5$], in which the heterocyclic ring is fused to a six- or seven-membered carbocyclic ring, are less stable than thiete or thietes substituted with alkyl groups.

Thiete undergoes an acid-catalysed hydration of the double bond on treatment with acidic 2,4-dinitrophenyl-hydrazine, followed by hydrolysis of the thiohemiacetal to β-mercaptopropionaldehyde which gives the 2,4-dinitro-phenylhydrazone.

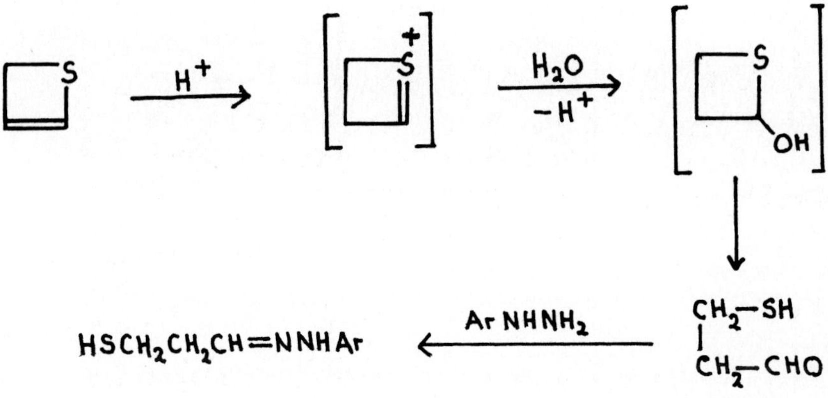
S
H+
S
+
H₂O
-H+
S
OH
Ar NHNH₂
HSCH₂CH₂CH=NNHAr
CH₂-SH
CH₂-CHO

Ar = 2,4$(NO_2)_2C_6H_3$

Thietes [1, $R^1R^3 = (CH_2)_4$ and $R^1R^2 = (CH_2)_5$]
react quite differently yielding the 2,4-dinitrophenylhydrazones of cyclohexene- and cycloheptene-thioaldehyde respectively (D.C. Dittmer *et. al.*, J. org. Chem., 1972, 37, 1111).

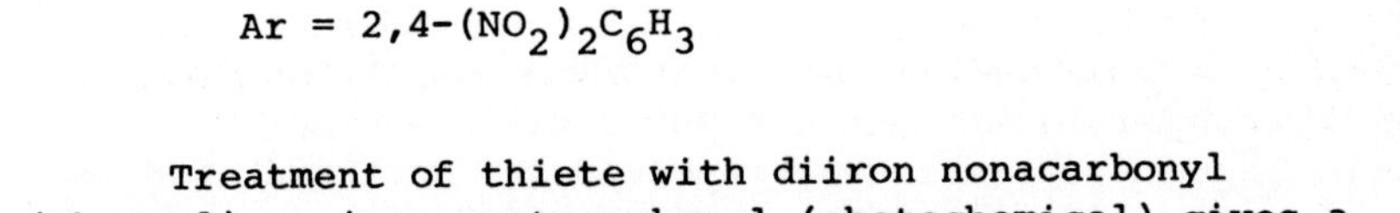
S
H+
CHS
Ar NHNH₂
CH=NNHAr

Ar = 2,4-$(NO_2)_2C_6H_3$

Treatment of thiete with diiron nonacarbonyl (thermal) or iron pentacarbonyl (photochemical) gives a complex, which is converted into the monomeric, red iron dicarbonyl triphenylphosphine complex (2), of thioacrolein (K. Takahashi *et. al.*, J. Amer. chem. Soc., 1973, 95, 6113).

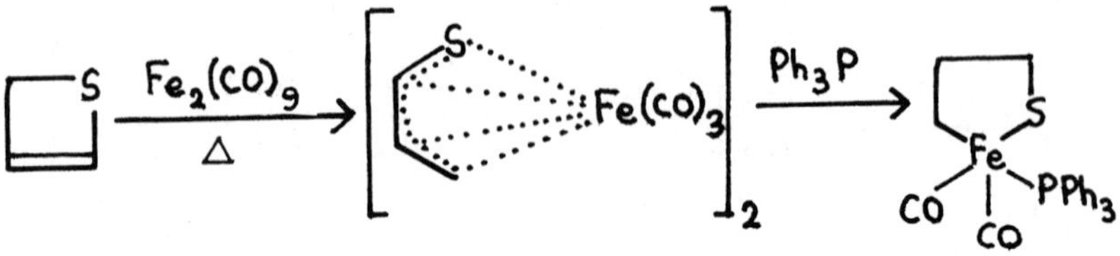

(2)

Thermolysis of thiete 1,1-dioxide in the presence of norbornene gives a cycloadduct of the Diels-Alder type ([4+2] cycloaddition reaction) resulting from the trapping of vinyl sulphene formed by ring opening of the thiete 1,1-dioxide (Dittmer *et. al.*, J. org. Chem., 1977, 42, 1910).

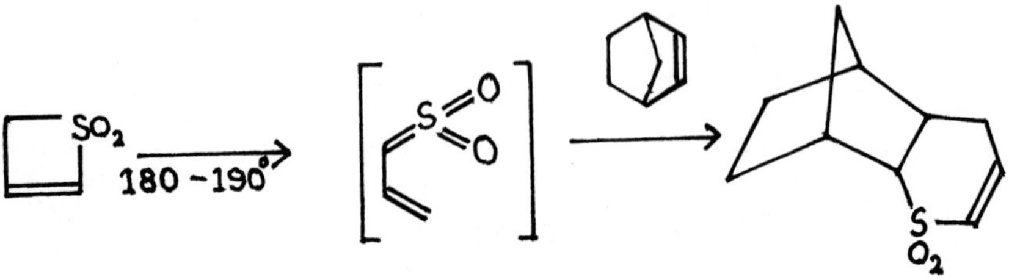

Other cyclic or acyclic alkenes give little or no reaction.

The irradiation of 1,3-dimethyl-2-thioparabanate with diphenylacetylene gives a spirothiete and not a spirothiopyran as previously reported (H. Gotthardt and O.M. Huss, Tetrahedron Letters, 1978, 3617).

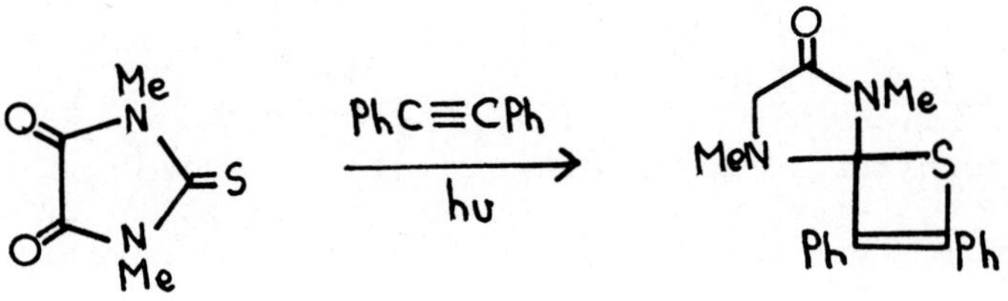

7-Thiabicyclo[4.2.0]-1(6)-octene 7,7-dioxide (3)
on treatment with N-bromosuccinimide followed by triethylamine gives 2H-benzo[b]thiete 1,1-dioxide (4) in low yield
(Dittmer and F.A. Davies, J. org. Chem., 1967, 32, 3872).

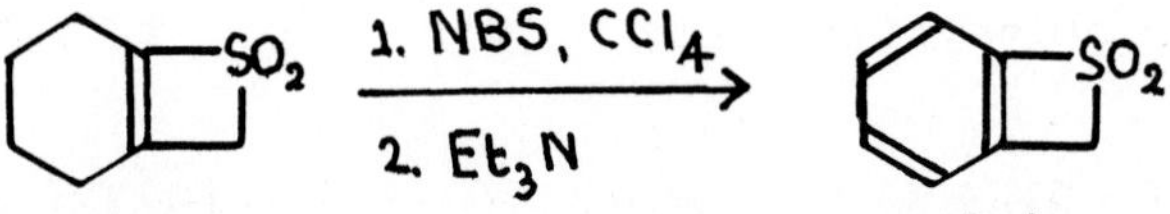

2H-Benzo[b]thiete 1,1-dioxide is synthesised in high
yields by the following route (Dittmer and T.R. Nelsen,
ibid., 1976, 41, 3044).

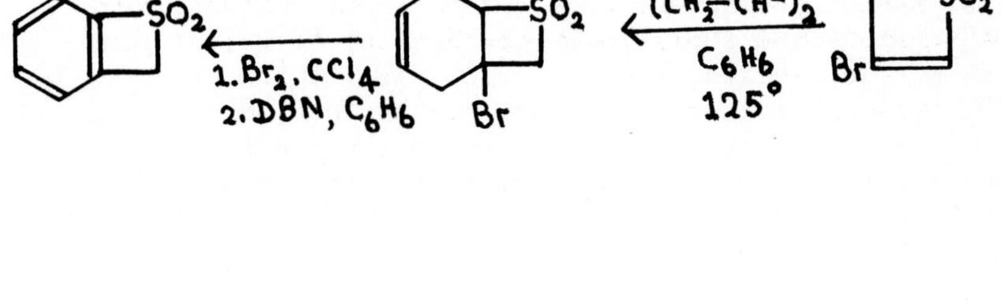

Stabilisation of the thiete ring occurs by incorporation of the double bond of the system in an annelated benzene ring.

Thermolysis of benzothiete 1,1-dioxide gives the sultine, 3H-2,1-benzoxathiole 1-oxide (5) and reduction with lithium tetrahydridoaluminate affords 2-toluenethiol (6).

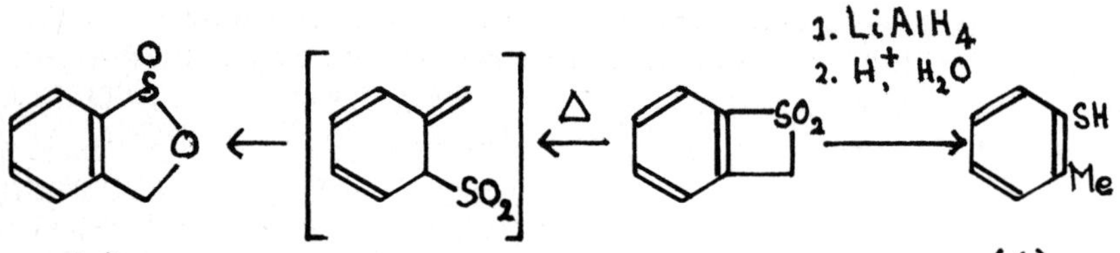

(5)   (6)

This is surprising as the naphthothiete 1,1-dioxide (7) on reduction yields a benzyl mercaptan.

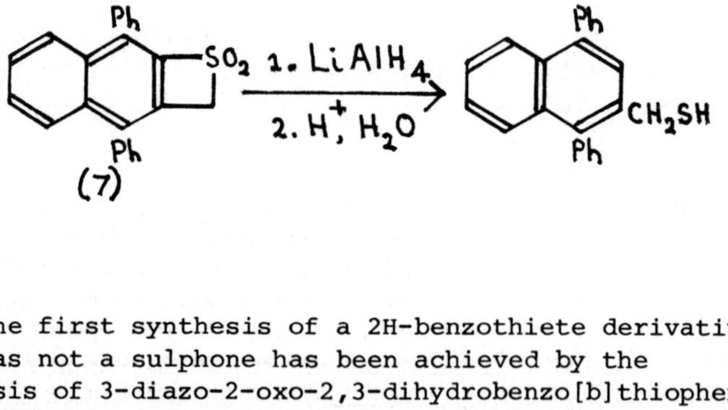

The first synthesis of a 2H-benzothiete derivative, which was not a sulphone has been achieved by the photolysis of 3-diazo-2-oxo-2,3-dihydrobenzo[b]thiophene (8). The carbene formed undergoes a Wolff rearrangement to the ketene and addition of methanol gives methyl benzo-thiete-2-carboxylate, b.p. 73°/0.06 torr, which predictably is more stable than thiete. Ring-opening to its valence tautomer (9), which would involve loss of aromaticity in the benzene ring, is not observed (H.N.C. Wong and F. Sondheimer, Angew. Chem. internat. Edn., 1976, 15, 117).

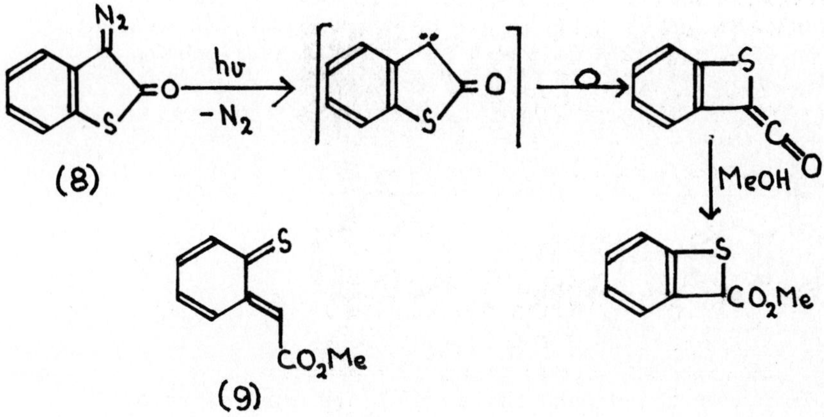

(8)

(9)

Benzothiete is obtained by flash vapour-phase pyrolysis (1000°, 5x10⁻⁴ mm Hg) of benzo[b]thiophene 1,1-dioxide, it is fairly stable at room temperature but dimerises above 100° yielding 6H,-12H-dibenzo[b,f][1,5]-dithiocin. Oxidation of benzothiete with peracid gives the 1,1-dioxide (W.J.M. van Tilborg and R. Plomp, Chem. Comm. 1977, 130).

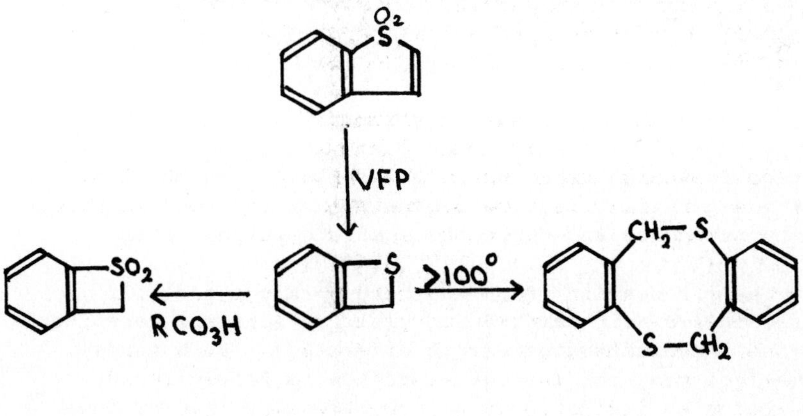

Uv irradiation of naphtho[1,8-cd]1,2-dithiole
1,1-dioxide gives naphtho[1,8-bc]thiete, m.p. 40-42^0, in
up to 97% yield (J. Meinwald *et. al.*, J. Amer. chem. Soc.,
1976, **98**, 6643).

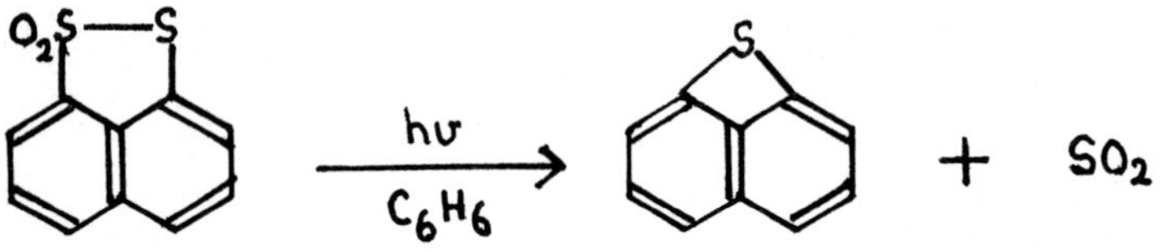

Oxidation of the naphthothiete with one equivalent of
3-chloroperbenzoic acid gives the 1-oxide, m.p.104-105°,
whereas oxidation with excess peracid gives the
1,1-dioxide. Treatment of the naphthothiete with lithium
tetrahydridoaluminate followed by methylation of the base-
-soluble product yields methyl 1-naphthyl sulphide.

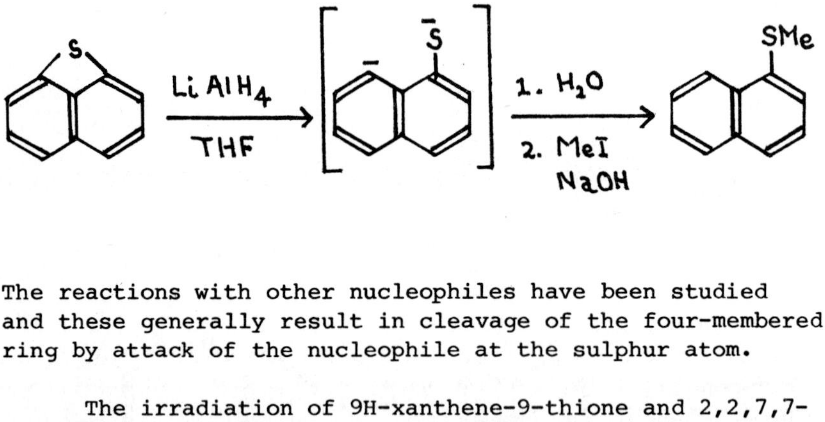

The reactions with other nucleophiles have been studied
and these generally result in cleavage of the four-membered
ring by attack of the nucleophile at the sulphur atom.

The irradiation of 9H-xanthene-9-thione and 2,2,7,7-
-tetramethyl-3,-dithiaoct-4-yne yields a spirothietexanthene
(10) and the unsaturated dithio ester (11). Both forms
can be isolated, but the equilibration is fairly rapid
in solution as indicated by nmr spectroscopy (A.C. Brouwer
et. al., Tetrahedron Letters, 1978, 4839).

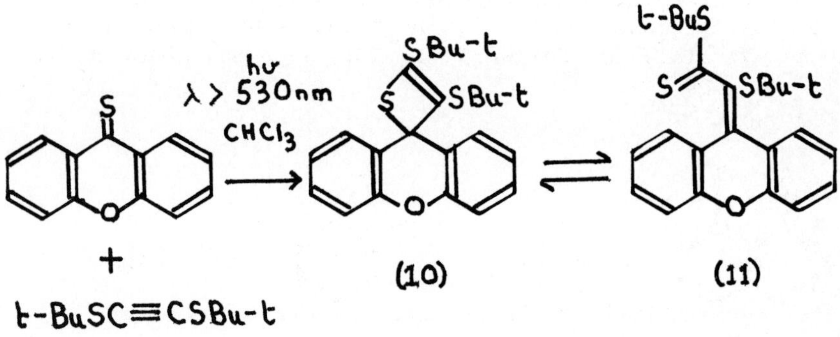

$$t\text{-}BuSC \equiv CSBu\text{-}t$$

(c) Thietanones

Irradiation of thiophosgene and 2,3-dimethylbut-2--ene in benzene gives 2,2-dichloro-2,3,4,4-tetramethyl thietane, which on hydrolysis yields 3,3,4,4-tetramethyl-thietan-2-one (a β-thiolactone) (H. Gotthardt, Tetrahedron Letters, 1973, 1221).

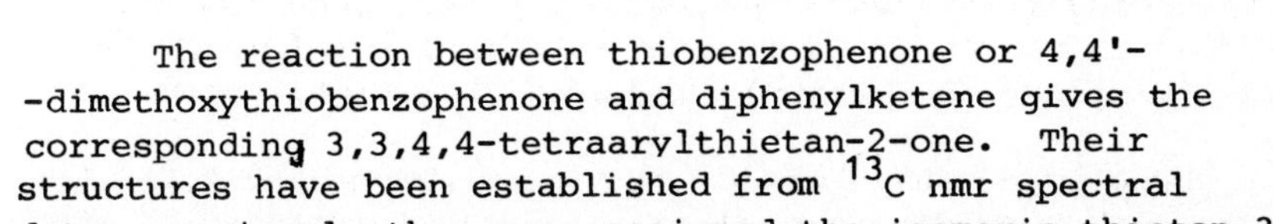

The reaction between thiobenzophenone or 4,4'--dimethoxythiobenzophenone and diphenylketene gives the corresponding 3,3,4,4-tetraarylthietan-2-one. Their structures have been established from ^{13}C nmr spectral data, previously they were assigned the isomeric thietan-3--one structure (H. Kohn, P. Charumilind, and Y. Gopichand, J. org. Chem., 1978, 43, 4961).

$$4\text{-}RC_6H_4\text{-}\overset{\displaystyle 4\text{-}RC_6H_4}{\underset{\displaystyle |}{C}}=S \quad + \quad Ph\,C=C=O \quad \xrightarrow{\hspace{1cm}} \quad (4\text{-}RC_6H_4)_2 \diagup\!\!\!\square\!\!\!\diagdown S \atop Ph_2 \quad O$$

R = H or OMe

    4-Phenylbutan-2-one at room temperature with an excess of thionyl chloride in the presence of catalytic amounts of pyridine gives rise to a vigorous, exothermic reaction which affords 2-benzylidenethietan-3-one, m.p. 113-114°. Base-catalysed condensation of the thietan-3-one with benzaldehyde yields bisbenzylidenethietan-3-one, m.p. 168-169°. Other thietan-3-ones may be similarly synthesised in one step from readily available ketones (A.J. Krubsack, T. Higa, and W.E. Slack, J. Amer. chem. Soc., 1970, 92, 5258).

$$PhCH_2CH_2COMe \xrightarrow[C_5H_5N]{SOCl_2,} \quad \underset{O}{\square}S \xrightarrow[NaOH]{PhCHO,} \quad \underset{O}{\overset{PhCH}{\square}}S \atop CHPh$$

9. *Four-Membered Rings Containing One Nitrogen Atom*

(a) *Azetidines (trimethyleneamines)*

(i) *Synthesis*

    The sequence employed by Vaughan *et. al.,* for the synthesis of azetidine has been modified. The cyclisation of the propylamine derivative is effected by aqueous sodium hydroxide and the reductive detosylation of the 1-tosyl derivative brought about by sodium naphthalenide in

diglyme. The yield of azetidine is approximately 70%
(J. White and G. McGillvray, J. org. Chem., 1974, __39__,
1973).

$$\underset{CH_2-NHTos}{\overset{CH_2-NHTos}{|}} \quad \xrightarrow[H_2O]{NaOH,} \quad \boxed{}-NTos \quad \xrightarrow[diglyme]{C_{10}H_8, Na,} \quad \boxed{}-NH$$

N-Aroylazetidines are prepared in fair yield by use
of the azetidine-diglyme solution and the appropriate aroyl
chloride.

Ar	m.p. (°C)	Ar	m.p. (°C)
Ph	61-62	4-Tolyl	86-86.5
4-NO$_2$Ph	122-123	3,5-DiNO$_2$Ph	155-156
4-ClPh	107	2-Furyl	120.120.5

The photorearrangement of 2-cyano-1-pyrroline 1-
oxides in benzene gives 1-cyanoformylazetidines *via*
oxaziridines (D. St. C. Black, N.A. Blackman, and
A.B. Bosacci, Tetrahedron Letters, 1978, 175).

$$\text{(Me}_2\text{,N-oxide pyrroline with CN, }R^1, R^2_2) \quad \xrightarrow[C_6H_6]{h\nu} \quad \text{(Me}_2\text{ azetidine NCOCN, }R^1, R^2_2)$$

R^1 = H, R^2 = H, Me;
R^1 = Me, Ph, R^2 = H

The photolysis of α - N-alkylamidoacetophenones, especially when R^3 - tosyl, results in very high yields of the corresponding N-substituted azetidin-3-ols, which on hydrolysis afford 3-phenylazetidine-3-ols (E.H. Gold, J. Amer. chem. Soc., 1971, 93, 2793).

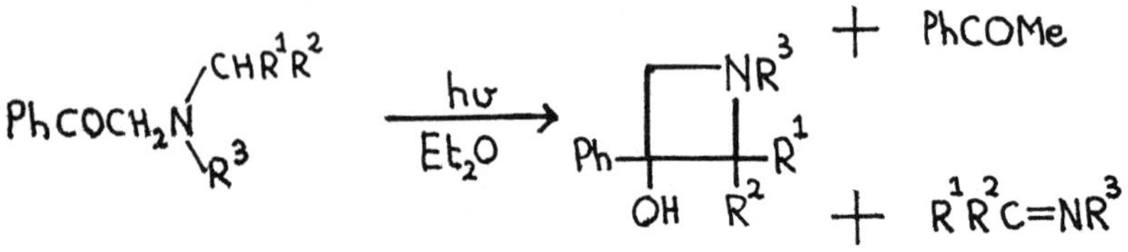

$R^1 = R^2 = H$; $R^1 = R^2 = Me$; $R^1 = H$, $R = Me$;
$R^3 = PhCO$ or Tos

Azetidinols may be obtained from 1-azabicyclo[1.1.0]butanes (pp. 54,61).

The acetone sensitized photolysis of 7-$tert$-butoxy-carbonyl-2,3-benzo-7-azabicyclo[2.2.1]hepta-2,5-diene yields the polycyclic azetidine (1) (P.D. Rosso, J. Oberdier, and J.S. Swenton, Tetrahedron Letters, 1971, 3947).

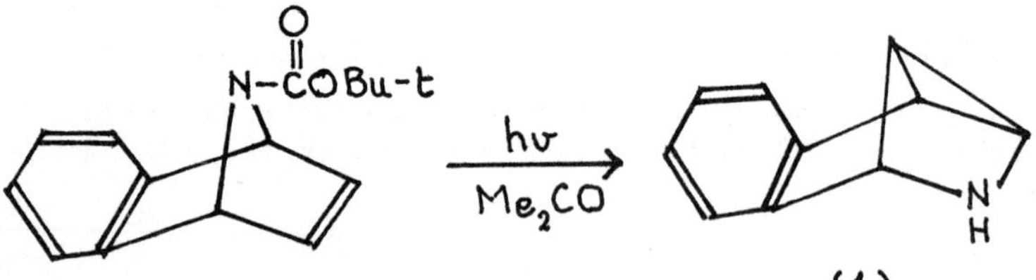

(1)

1-Isopropyl-2-methyleneaziridine (N-isopropylallen-imine) reacts with toluene-p-sulphonyl azide to give a quantitative yield of 1-isopropyl-2-(toluene-4-sulphonyl-imino)azetidine (N-isopropyl-N^1-tolune-p-sulphonyl-β-
-lactamimide), which on basic hydrolysis affords

N-isopropyl-β-aminopropionic acid (J.J. Crandall and J.B. Komin, Chem. Comm., 1975, 436).

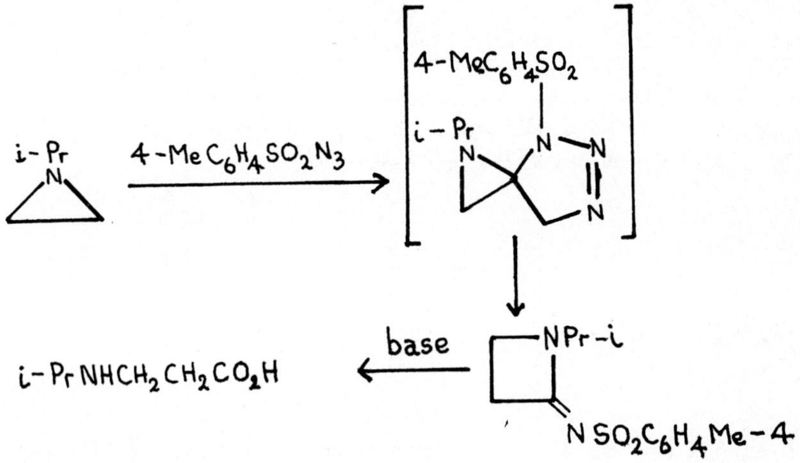

Azetidine-imines are also formed by [3+1] cycloaddition of azomethine ylides to various isocyanides (K. Burger, F. Manz, and A. Braun, Synthesis, 1975, 250).

(b) Azetidinones; oxoazetidines

(i) Synthesis

The ethyl hemiketal of cyclopropanone reacts with sodium azide in buffered solution to form an azetidin-2-one (β-lactam).

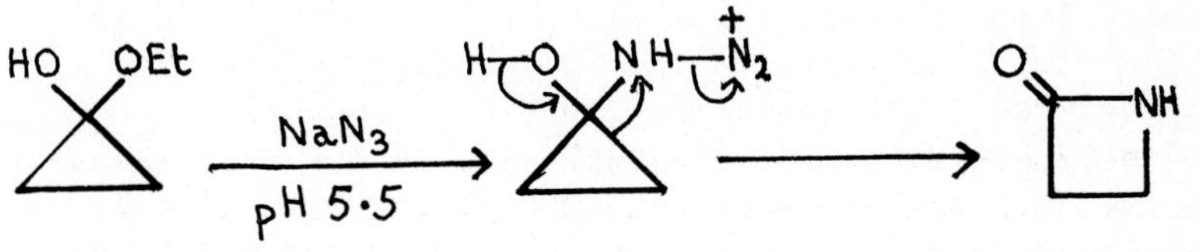

A more general application of this ring enlargement sequence takes place through the electron-deficient (nitrenium ion) species (2a) analogues to the cyclo-propylcarbinyl cation, or by the concerted process (2→3). The reaction involves the formation of carbinol-amines of type 1, which may be converted into *N*-chloro derivatives using *tert*-butyl hypochlorite (Gassman's procedure). Treatment of the *N*-halogeno derivative with silver ion in acetonitrile leads to the β-lactam (H.H. Wasserman, H.W. Adickes, and O. Espejo de Ochoa, J. Amer. chem. Soc., 1971, **93**, 5586).

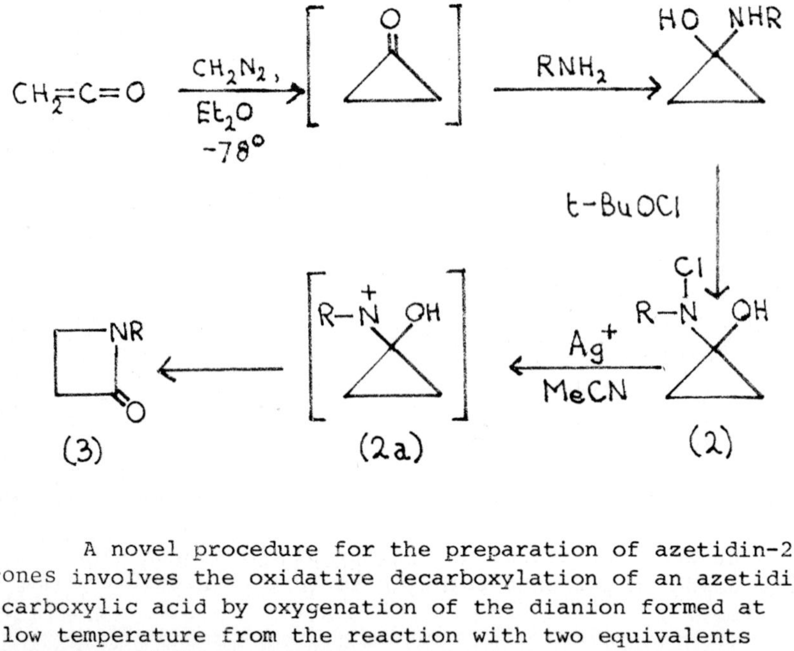

A novel procedure for the preparation of azetidin-2--ones involves the oxidative decarboxylation of an azetidine carboxylic acid by oxygenation of the dianion formed at low temperature from the reaction with two equivalents of lithium diisopropylamide. Uptake of oxygen by the dianion affords the dilithium salt of the hydroperoxy acid which, on acidification, rapidly decomposes to the azetidin-2-one (H.H. Wasserman and B.H. Lipshutz, Tetrahedron Letters, 1976, 4613).

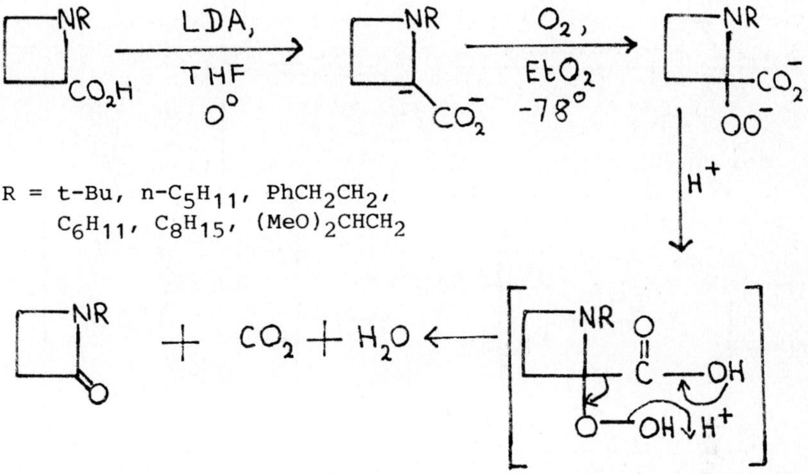

$R = t\text{-}Bu,\ n\text{-}C_5H_{11},\ PhCH_2CH_2,$
$C_6H_{11},\ C_8H_{15},\ (MeO)_2CHCH_2$

The addition of N,N-dimethyl-1-chloro-2-methyl-1-
-propenylamine (4), to solutions of Schiff bases in
methylene chloride yields the 2-azetidinylideneammonium
chloride. Hydrolysis of the salts in 0.5M sodium hydroxide
gives the corresponding 2-azetidinones (M. De Poortere,
J. Marchand-Brynaert, and L. Ghosnez, Angew. Chem. internat.
Edn., 1974, <u>13</u>, 267).

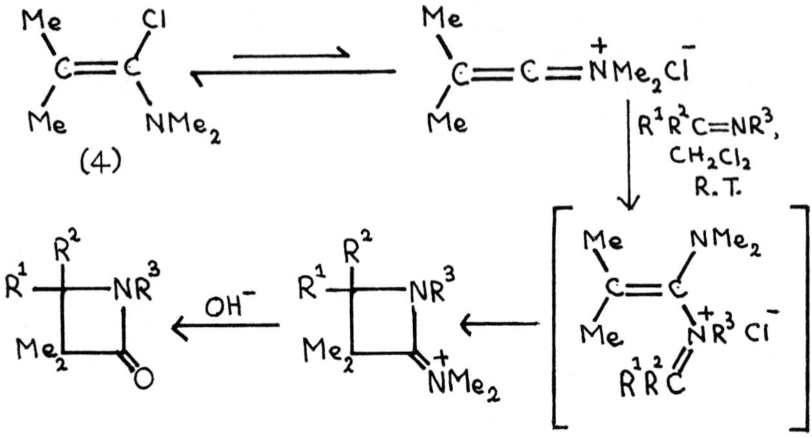

R^1	R^2	R^3
Ph	H	Me
Ph	H	Ph
PhCH$_2$S	H	t-Bu

Treatment of the anion of 1-acyl-5,5-dimethylpyra-zolidine-3-one in glyme with an equivalent of O-mesitylene-sulphonylhydroxylamine in methylene chloride in the presence of three equivalents of yellow mercuric oxide, results in a ring contraction with he formation of 1-acyl-4,4-dimethylazetidine-2-one (P.Y. Johnson, N.R. Schmuff, and J.E. Hatch, Tetrahedron Letters, 1975, 4089).

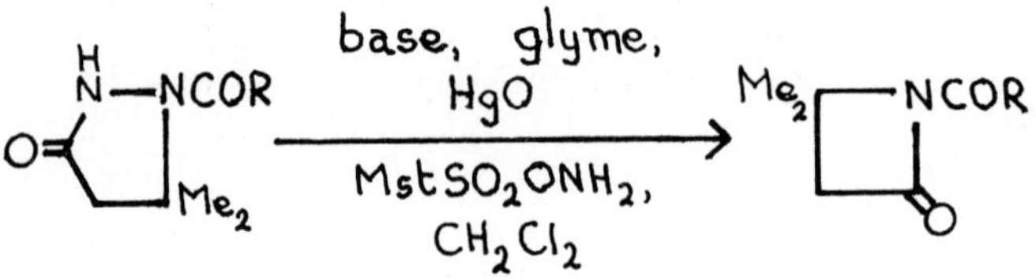

| R = Ph | 72% |
| R = CCl$_3$CH$_2$O | 39% |

The reaction between copper (I) and phenylacetylide and nitrones in dry pyridine followed by hydrolysis of the product yields *cis*-azetidin-2-ones. This is a useful stereoselective reaction for the synthesis of *cis*-azetidin--2-ones, no reaction giving only a *cis*-β-lactam has been previously reported (M. Kinugasa and S. Hashimoto, Chem. Comm., 1972, 466).

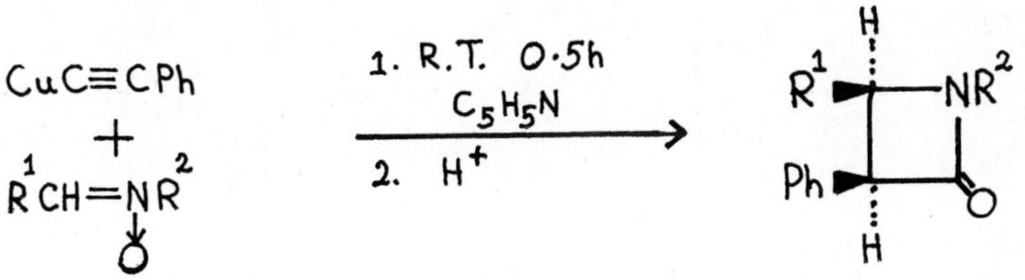

$R^1 = R^2 = Ph$; $R^1 = 2\text{-MeC}_6H_5$, $R^2 = Ph$;
$R^1 = Ph$, $R^2 = 4\text{-ClC}_6H_5$; $R^1 = 2\text{-ClC}_6H_5$, $R^2 = Ph$

Diphenyl-*N*-4- and dimethyl-*N*-4-tolylketenimine react thermally with isocyanates (R^2 = Ph, 4-tolyl, 4-anisyl, 4-tosyl) to give the corresponding 4-iminoazetidin-2-one (Naser-ud-Din, J. Riegl, and L. Skattebøl, Chem. Comm., 1973, 271).

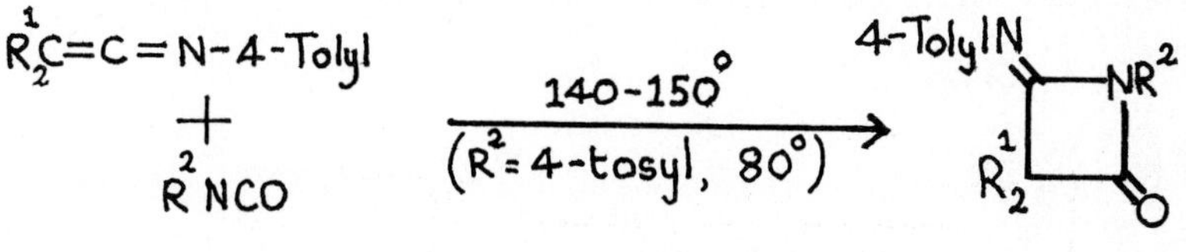

R^1 = Ph or Me

A variety of carboxylate activating groups convert aziridine carboxylates into 3-halogeno-2-azetidinones,

for example, the reaction of sodium 1-*tert*-butylaziridine-
-2-carboxylate with thionyl chloride and sodium hydride
affords 3-chloro-1-*tert*-butylazetidin-2-one instead of
the usual acid chloride.

(5)

The reaction is stereospecific and thought to proceed *via*
a 1-azabicyclo[1.1.0]butan-2-one cation (5). Ring
contraction of the 3-halogenoazetidinone has also been
observed (J.A. Deyrup and S.C. Clough, J. org. Chem., 1974,
39, 902).

 3-Chlorocarbonylazetidin-2-ones may be prepared from
imines or thioimidates and substituted malonyl chlorides.
The β-lactam formation is stereospecific and it is found
that the chlorocarbonyl group is *cis* to the hydrogen or
the alkylthio group at C-4 of the azetidinone. The 3-aryl-
substituted acid chlorides on treatment with 3-chloroper-
benzoic acid and triethylamine at low temperature yield
the *cis*-3-arylazeditidin-2-one. This dechlorocarbonylation
constitutes a short, stereospecific synthesis of *cis*-β-
-lactams under mild conditions (A.K. Bose *et. al., ibid.,*
p.312).

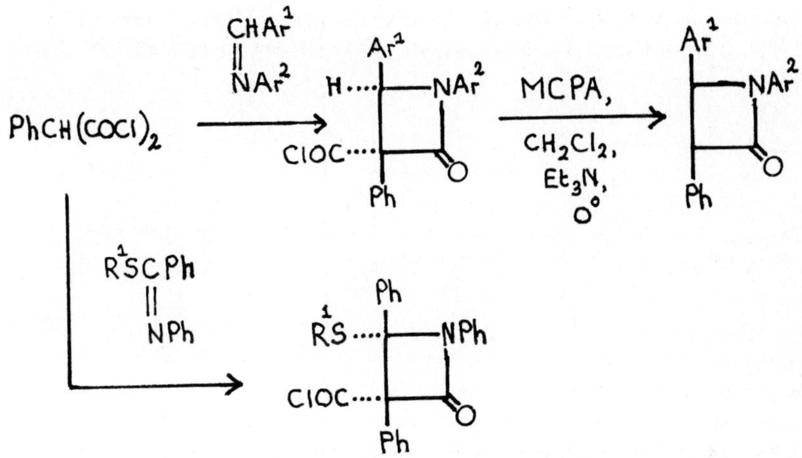

Azetidin-2-ones, unsubstituted in positions 1 and 3 react with two equivalents of n-butyllithium in tetrahydrofuran to form 1,3-dilithio salts, which react with various electrophiles, for example, ketones, alkyl halides, iodine, to yield azetidin-2-ones substituted at C-3 (T. Durst, R. Van Den Elzen, and R. Legault, Canad. J. Chem., 1974, 52, 3206).

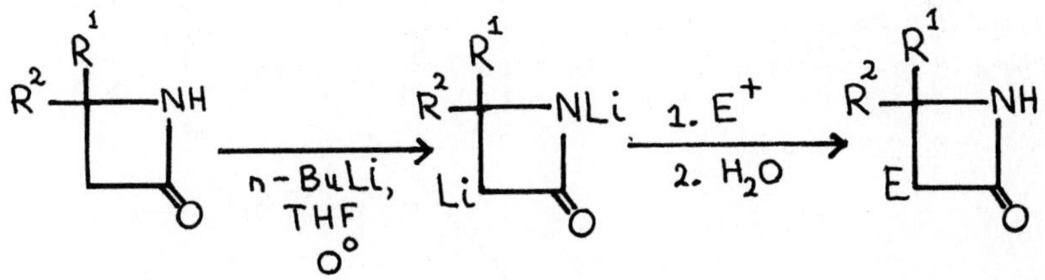

The thermolysis of 4-azido-3-chloro-5-methoxy-1-
-methyl-3-pyrrolin-2-one yields 3-chloro-3-cyano-4-methoxy-
-1-methylazetidin-2-one (58%) (H.W. Moore, L. Hernandez,

and A. Sing, J. Amer. chem. Soc., 1976, <u>98</u>, 3728).

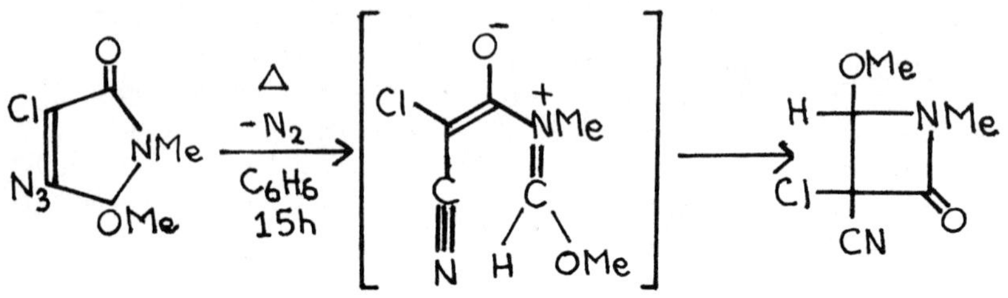

When a benzene solution containing stoichiometric amounts of the lactone (6) and dicyclohexylcarbodiimide (DCC) is boiled 3-chloro-3-cyano-1-cyclohexyl-4-cyclohexylimino-azetidin-2-one (88%), m.p.69-70°, is obtained, but when the thermolysis of the lactone is accomplished in the presence of ethyl *N*-phenylformimidate the product is 2-chloro-3-cyano-4-ethoxy-1-phenylazetidin-2-one (48%), m.p.66-67°.

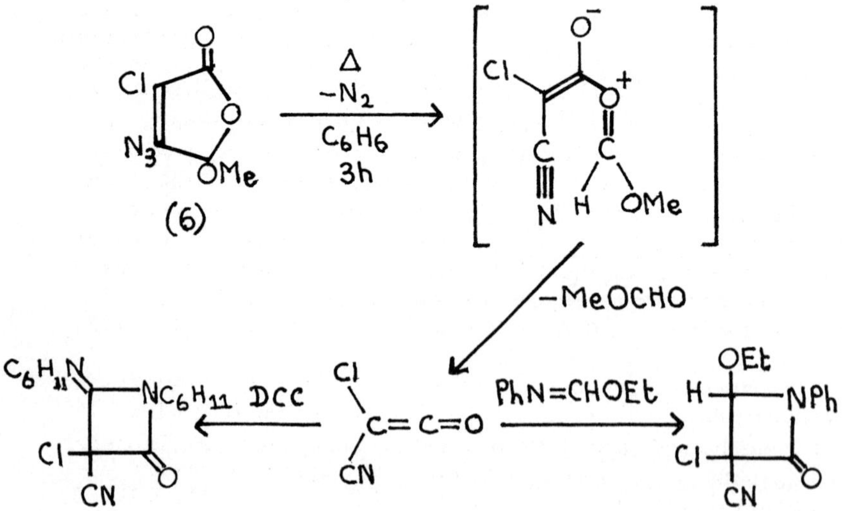

(ii) Reactions

The thermal cleavage of the β-lactams, *cis* and *trans*-3,4-dimethylazetidin-2-one and the analogous diethyl compounds gives an alkene (>90% yield) and cyanuric acid (80-90%). The thermal fragmentations proceed with virtually total retention of stereochemistry. The high stereochemical purity of the olefins could result from the operation of the concerted [σ2s + σ2a] pathway (L.A. Paquette, M.J. Wyvratt, and G.R. Allen, Jr., J. Amer. chem. Soc., 1970, 92, 1763).

cis-1,4-Diphenyl-3-phthalimidoazetidin-2-one is epimerised completely to the corresponding *trans*-β-lactam in the presence of 1,5-diazabicyclo[4.3.0]non-5-ene in benzene solution at 100° in about 22 hours but the *trans*-isomer is unchanged on similar treatment: an equilibrium mixture containing 30% of the *cis*-isomer is obtained when either the *cis*- or the *trans*-3-bromo-1,4-diphenylazetidin--2-one is heated on a steam bath with the same base in benzene solution or treated with a solution of the base in dimethylsulphoxide at room temperature. No reaction occurs if triethylamine is used in either benzene or dimethyl sulphoxide. Epimerisability, therefore depends on the functional group at C-3 and the organic base used while the solvent plays a smaller role (A.K. Bose, C.S. Narayanan, and M.S. Manhas, Chem. Comm., 1970, 975).

Irradiation of the azetidinone, *exo*-aza-4-ketobenzo-tricyclo[4.2.1.0^{2,5}]non-7-ene (1) in methanol yields *exo*-2-methoxy-3-aza-4-keto-7,8-benzobicyclo[4.2.1]nonene (2), a reaction which involves ring expansion of the β-lactam moiety in (1) plus the addition of methanol (H.L. Ammon *et. al.*, J. Amer. chem. Soc., 1973, 95, 1968).

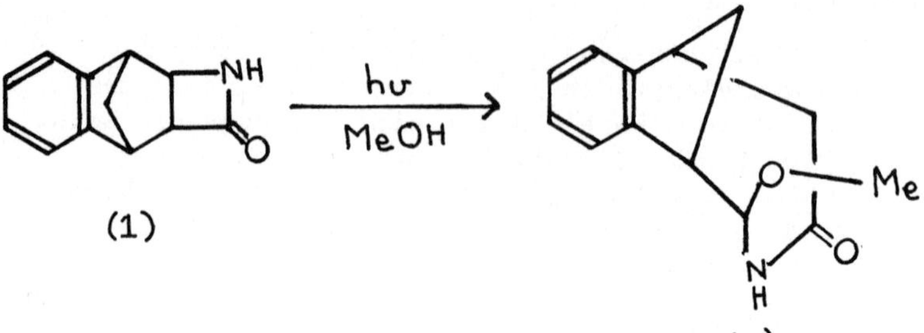

(1) → (2)

The cyanomalondiimide on heating cyclises to give the 2,4-dioxo-3-azetidinecarbonitrile (3), which can be converted into the 2-azetin-4-one (4) (L. Capuano and R. Zander, Ber., 1973, **106**, 3760).

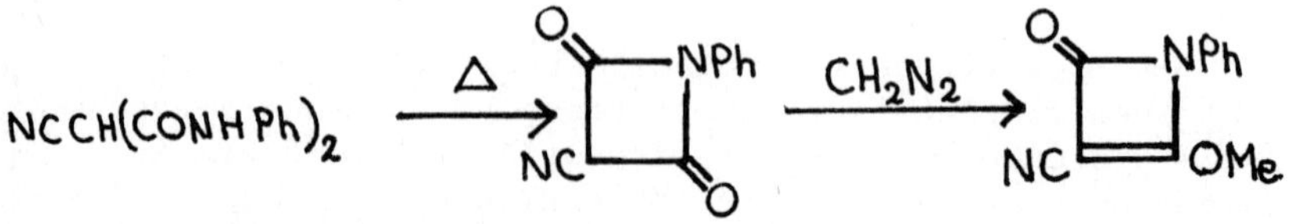

$NCCH(CONHPh)_2$

(3)   (4)

The azetidin-2,4-dione (5) and the related 4-thioxo-azetidin-2-one have been prepared (M.D. Bachi, O. Goldberg and A. Gross, Tetrahedron Letters, 1978, 4167).

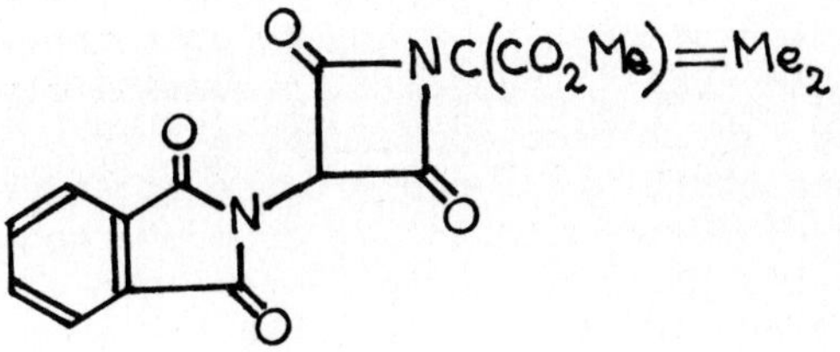

(5)

(c) Azetines

Stepwise nucleophilic addition of trichloromethyl-lithium to 3-methyl-2-phenyl-1-azirine followed by treatment with base gives 2,3-dichloro-*cis*-4-methyl-3--phenyl-1-azetine presumably *via* a azabicyclobutane intermediate (A. Hassner *et. al.*, Angew. Chem. internat. Edn., 1970, **9**, 731).

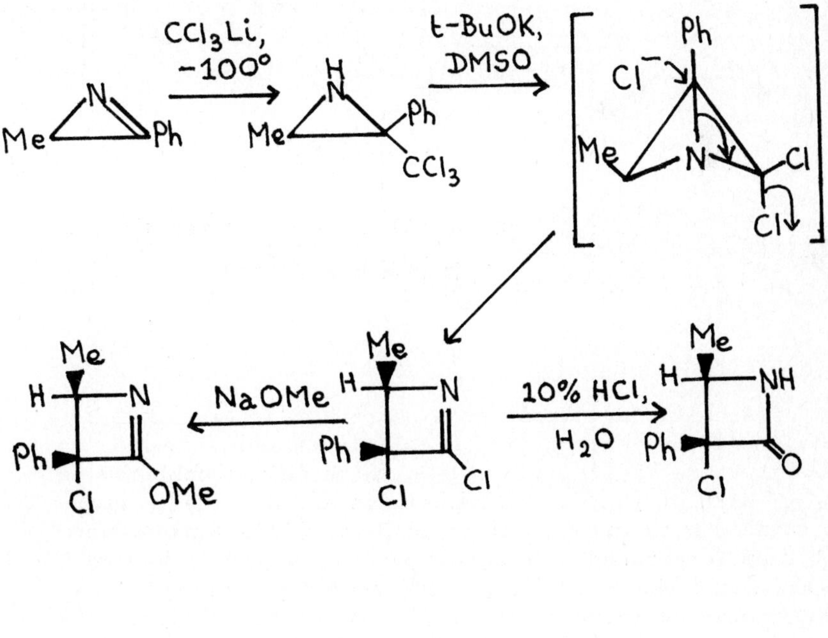

The reactivity of the chloroazetine is significantly different from that of azabicyclobutanes, which are very acid sensitive. Although mild sensitivity to air is observed, it is recovered unchanged on stirring in anhydrous ether saturated with hydrogen chloride. In 10% aqueous hydrogen chloride it is converted to 3-chloro-*cis*-4-methyl-3-phenylazetidin-2-one. The chloroazetine with sodium methoxide is slowly converted to 3-chloro-2-methoxy--4-methyl-3-phenyl-1-azetine.

1-Azido-2,2-dichlorocyclopropanes decompose smoothly between 105 and 125° in a highly regiospecific manner to give the corresponding 3,3-dichloroazetines, which on acid hydrolysis and lithium tetrahydridoaluminate reduction afford β-amino-α,α,dichloroketones and azetidines respectively (Hassner and A.B. Levy, J. Amer. chem. Soc., 1971, **93**, 2051).

$R^1 = Ph,\ R^2 = R^3 = H;$
$R^1 = Ph,\ R^2 = Me,\ R^3 = H;$
$R^1 = R^2 = Me,\ R^3 = H$

Aromatic nitriles react photochemically with 2,3--dimethylbut-2-ene to give stable 1-azetines. These azetines are crystalline solids which may be sublimed *in vacuo* without thermal decomposition and are stable upon treatment with dilute oxalic acid (0.1%) in ethanol and at room temperature. Polymethylation of the 1-azetine ring stabilises the heterocyclic system with respect to decomposition (N.C. Yang *et. al.*, Chem. Comm., 1976, 729).

2-Phenyl-3,3,4,4-tetramethyl-1-azetine, m.p.57-58°;
2-(1-napthyl)-3,3,4,4-tetramethyl-1-azetine, m.p.96-98°.

$ArC{\equiv}N$
$+$
$Me_2C{=}CMe_2$ $\xrightarrow[n-C_6H_{14}]{h\nu}$

[structure: Me₂—N / Me₂—Ar ring] $\xrightarrow{h\nu}$ [structure with Me groups and Ar]

$\Big\downarrow H^+, H_2O$

[structure: Ar–C(=O)–C(Me Me)–C(Me Me)–NH₂]

The reaction between isocyanides and 4,5-dihydro-
-1,3,5-oxazophospholes, having the types of substituents
shown, yields 3-imino-1-azetines (K. Burger, J. Fehn, and
E. Muller, Chem. Ber., 1973, 106, 1).

[structure: $(F_3C)_2$ ... $N{=}R^1$ / P–O / $(OR^3)_3$] $\xrightarrow{CNR^2}$ [structure: $(F_3C)_2$ ... N ring with R^1 and R^2N]

R^1 = t-Bu, Ph, subst-Ph;
R^2 = cyclohexyl, CHMePh, $4{-}NO_2C_6H_4$

The irradiation of the iminoazetine (1) in benzene
solution results in the cycloelimination of isocyanide.
The nitrile ylide (2) formed can be trapped by alcohols
as *N*-(hexafluoroisopropyl)benzimidic esters, by dimethyl
acetylenedicarboxylate as 2H-pyrroles, and by dimethyl
maleate or fumarate as 1-pyrrolines (Burger, W. Thenn,
and Müller, Angew. Chem. internat. Edn., 1973, 12, 155).

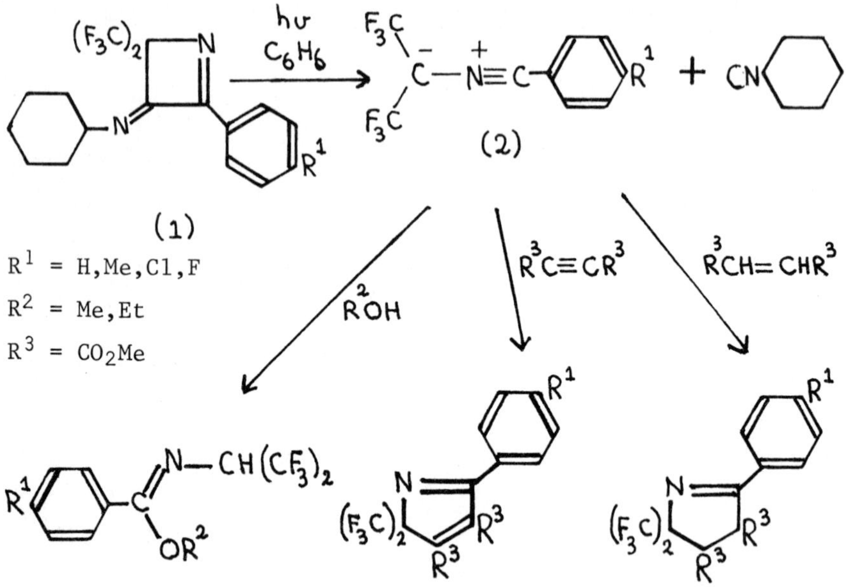

(2)

(1)

R^1 = H,Me,Cl,F

R^2 = Me,Et

R^3 = CO$_2$Me

The *cis*- and *trans*-3,4-diethyl-2-methoxy-1-azetine prepared by treating the related 3,4-diethylazetidin-2-ones with trimethyloxonium fluoroborate, followed by drying over pellets of potassium hydroxide, give on pyrolysis a mixture of products, but no olefin (cf. 3,4-dimethyl-azetidin-2-one p.125). Although direct comparisons of reaction rates in the two different systems could not be made in this instance, it is quite possible that such behaviour is the result of two distinctly different concerted process: [σ2s + σ2a] retrogression in the case on p.125 and conrotatory opening in the case of the 3,4-diethyl-2-methoxy-1-azetines (Paquette, Wyvratt, and Allen, *loc. cit.*, p.125).

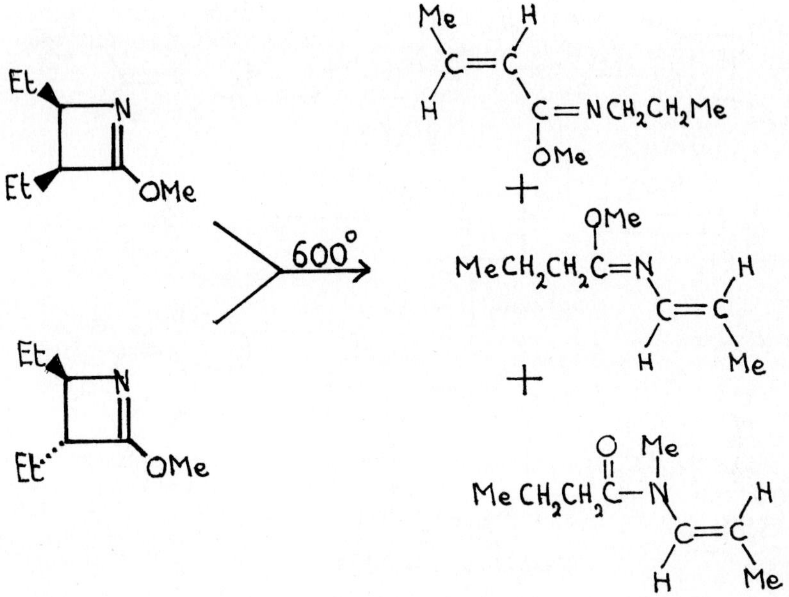

4,4-Dimethyl-2-methoxyazetine and 2-methoxy-3,4,4-
-trimethylazetine react with dimethyl acetylenedicarboxylate to form the respective 1:1 linear fumarate
and maleate adducts (3) and (4) *via* 1,4-dipolar ions.
In nonpolar solvents the 1,4-dipolar ions dimerise to yield
eight-membered ring 2:2 adducts (5) (D.H. Aue and
D. Thomas, J. org. Chem., 1975, **40**, 2360).

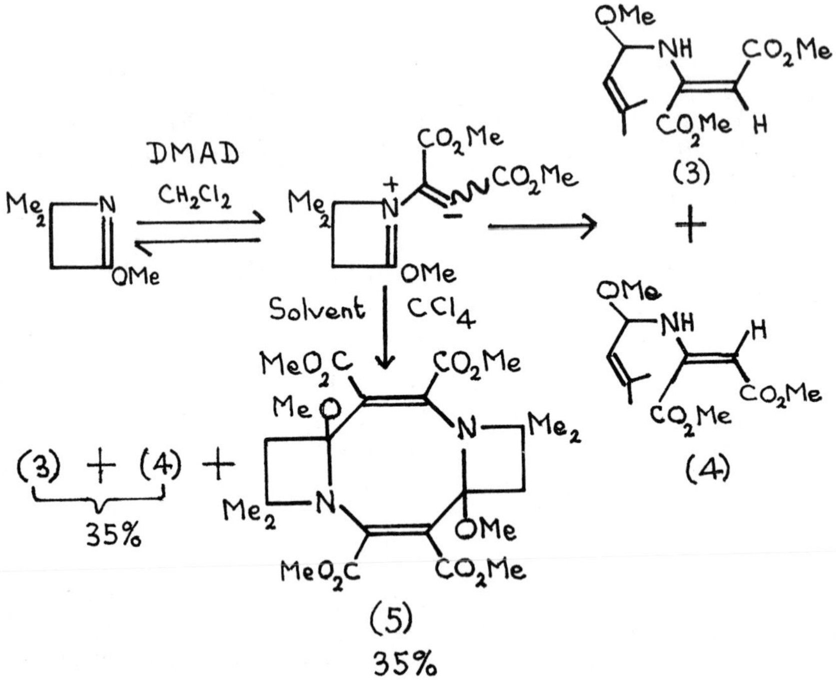

$$(3) + (4) + (5)$$
$$35\% \qquad 35\%$$

2-Alkoxy-1-azetines gives 2:1, cycloadducts, 8,8-dimethyl-3,5-di(toluene-4-sulphonyl)-6-methoxy-1,3,5-–triazabicyclo[4.2.0]octa-2,4-diones, in high yields on treatment with 2 equivalents of toluene-p-sulphonyl isocyanate (*ibid.*, p.2356).

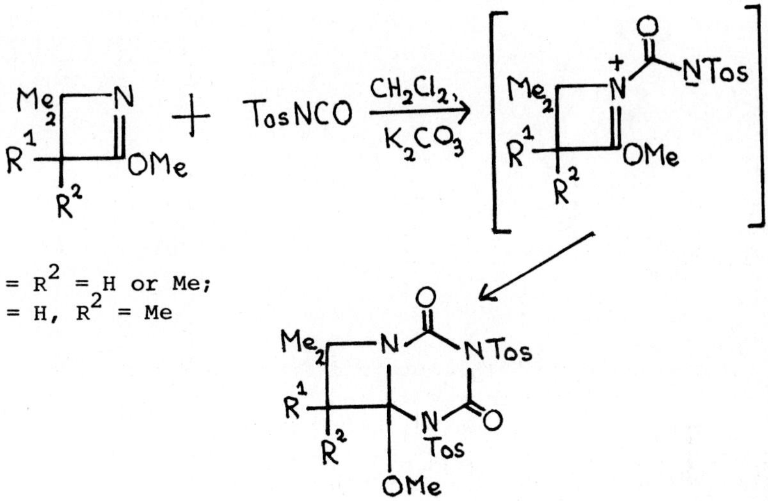

$R^1 = R^2 = H$ or Me;
$R^1 = H$, $R^2 = Me$

The cycloaddition of 3,3-dichloro-4-methyl-2-phenyl-
-1-azetine to diphenylketene leads to the formation of
a 1:2 adduct (Hassner, M.J. Haddadin, and Levy, Tetrahedron
Letters, 1973, 1015).

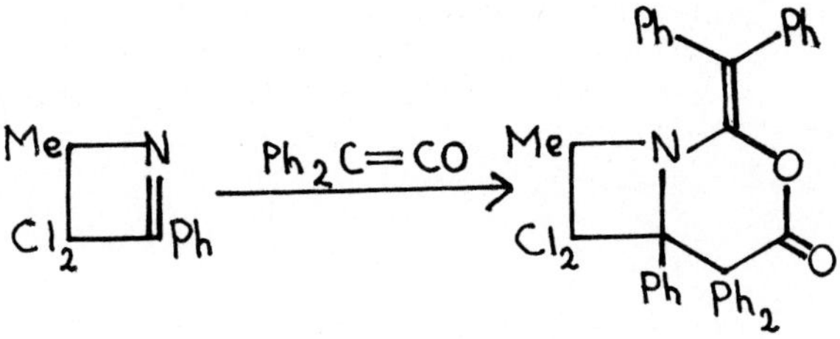

2,3,3-Trimethyl-1-azetine 1-oxide, the first example
of a cyclic four-membered nitrone, is obtained when the
oxime of 2,2-dimethyl-1-tosyloxybutan-3-one, assumed to
be the E isomer, is heated with one equivalent of 1,8-bis-
(dimethylamino)naphthalene in boiling benzene. All attempts

to cyclise the oxime using common bases failed
(D. St. C. Black *et. al.*, *ibid.*, 1974, 4283).

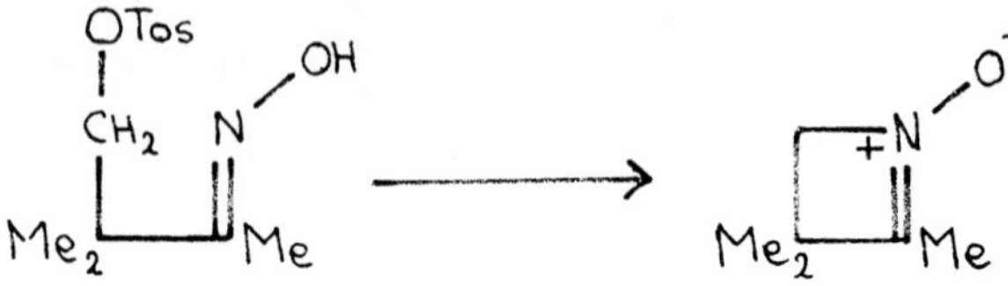

The 1:1 adduct (6) obtained from the appropriate
substituted cyclopentadienone and the photoisomer derived
from *N*-methyl-α-pyridone, gives an irradiation in deuterio-
chloroform the tricyclic diene (7) which on further
irradiation fragments to form the aromatic hydrocarbon
(8) and presumably the thermally labile non-isoable
1-methylazetine-2-ones (9). If the irradiation is carried
out in methanol a mixture of *(Z)*- and *(E)*- methyl
β-methylaminoacrylate is also obtained. These are formed
by the addition of methanol to the iminoketene (10) derived
from the azetin-2-one (9) (G. Kretschmer and R.N. Warrener,
ibid., 1975, 1335).

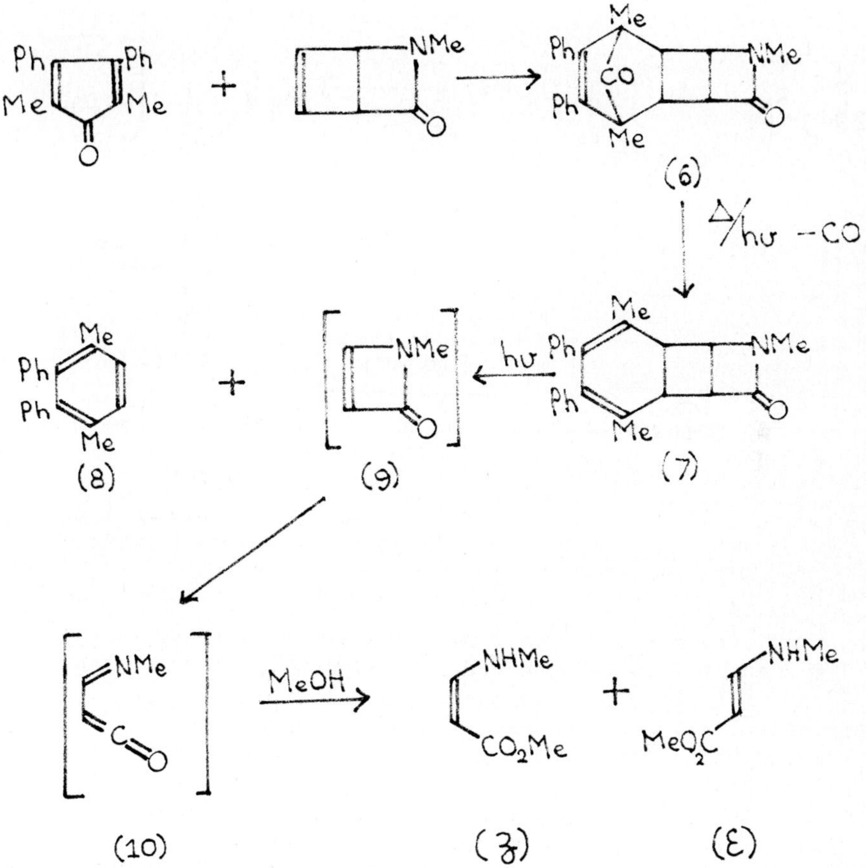

The photoadduct, 7.7-dimethoxy-4-methyl-6-phenyl-1-
-aza-5-oxabicyclo[4.2.0]octa-2-one (11), m.p.103-104°,
obtained by irradiating a methylene chloride solution of
6-methyl-2-phenyl-1,3-oxazin-4-one and 1,1-dimethoxyethene,
on pyrolysis at 225° gives 3,3-dimethoxy-2-phenyl-1-
-azetine (12). The structure is supported by spectral
data. The azetine (12) on stirring at ambient temperature
for 0.25h with 3M hydrochloric acid yields 3-amino-2,2-
-dimethoxy-1-phenylpropanone (T.H. Kock, R.H. Higgins,
and H.F. Schuster, *ibid.*, 1977, 431).

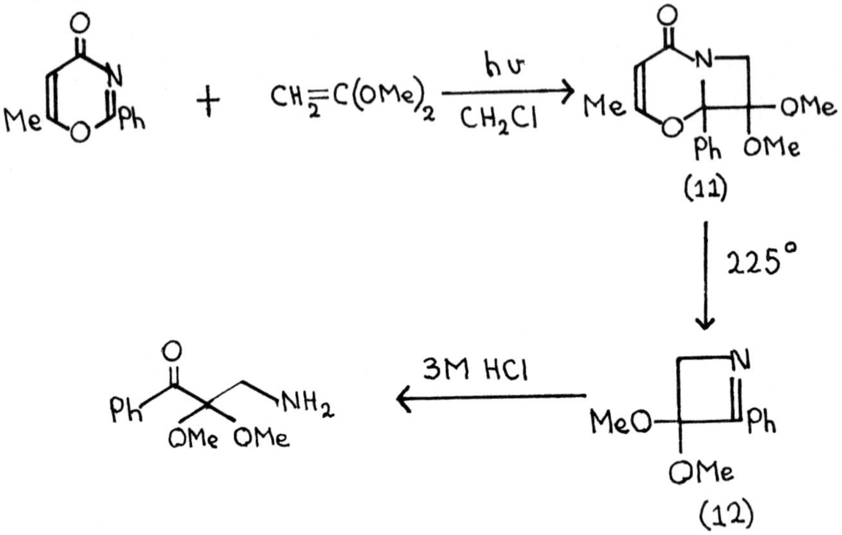

The reaction of 2,5-dimethyl-3,4-diphenylcyclopenta-
-2,4-dienone with *N*-methoxycarbonyl-2-azabicyclo[2.2.0]hex-
-5-ene in boiling benzene gives a 2.5:1 mixture of two
stereoisomeric 1:1 adducts. The major isomer (13) on
irradiation in chloroform at -20° is smoothly
decarbonylated to the diene (14), which fragments to give
1,4-dimethyl-2,3-diphenylbenzene and 1-methoxycarbonyl-2-
-azetine (15), isolated by short path distillation as a
colourless liquid (Warrener, Kretschmer, and M.N. Paddon-
-Row, Chem. Comm., 1977, 806).

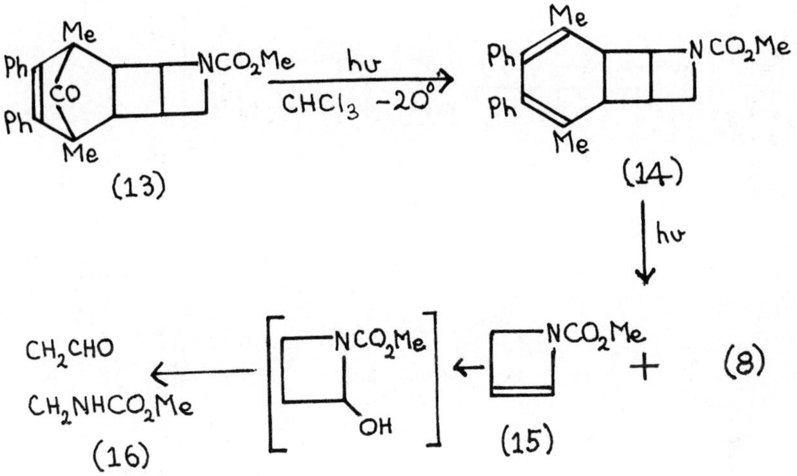

The azetine (15) is thermally stable below 50°, but undergoes rapid polymerisation above this temperature and on treatment with acid slowly gives the ring-opened aldehyde (16) (t½ 18h, room temperature). The reaction is accelerated (t½ 0.5h at 0°) by irradiation. Similar attempts to produce the parent azetine (15 , CO_2Me = H) or its 1-methyl derivative have failed even at -50°. The 1-tosyl compound (15, CO_2Me = $4-MeC_6H_4SO_2$) has been obtained by this method and is a potential precursor to azete, but treatment with *tert*-butoxide has failed to yield products associated with azete.

Irradiation of 1-acetyl-2-cyano-4-methyl-1,2--dihydroquinoline (17a) gives a mixture of the 1-acetyl-benzoazetine (18a) and the indoline (19a) whose relative yields depend upon the period of irradiation and the solvent. Irradiation of (18) gives (19) (M. Ikeda *et. al.*, *ibid.*, 1975, 575).

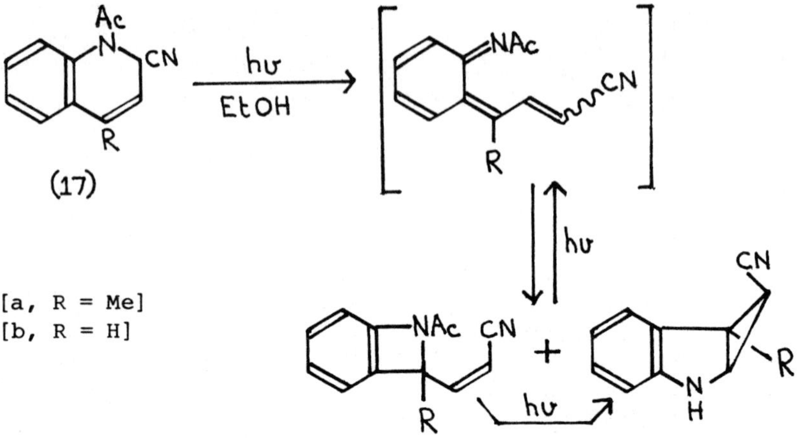

[a, R = Me]
[b, R = H]

Similarly, irradiation of dihydroquinoline (17b) in ether yields 1-acetylbenzoazetine (18b) in quantitative yield as an unstable solid which rapidly decomposes on exposure to air.

Photolysis of the 3-substituted benzotriazinones (20, R = aryl or benzoyl) leads to the formation of the 1H-benzoazetin-2-ones (21). The benzoazetinones cannot. be isolated but their presence in solution is shown by spectroscopic means and their interception with added nucleophiles (E.M. Burgess and G. Milne, Tetrahedron Letters, 1966, 93; G. Ege, Ber., 1968, 101, 3079; Ege and F. Pasedach, *ibid.*, p.3089).

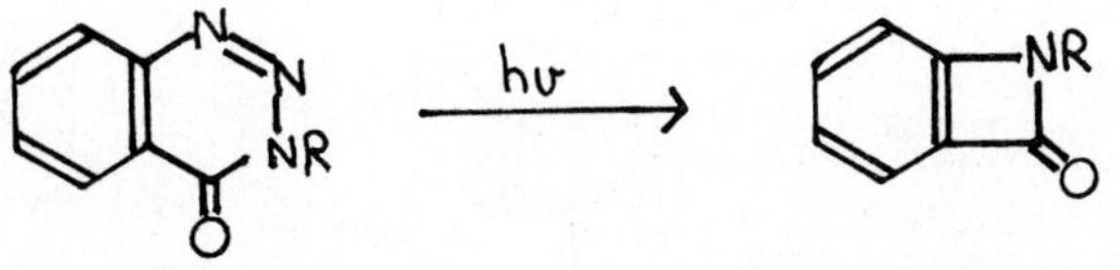

(20)   (21)

The photolysis of 3-amino-3H-naphtho[2,3-d]-v-
-triazin-4-one affords 1-amino-1H-naphtho[2,3-b]azetin-2-
-one as a labile crystalline solid. On heating or when
treated with acetic acid it is converted into 2H-benz[f]-
indazol-3-one (22) and on oxidation by lead tetraacetate
it gives 2-naphthoic acid. 1-Phenyl-naphtho[2,3-b]azetin-
-2-one is obtained as a labile crystalline solid from
related naphthotriazinone (N. Bashir and T.L. Gilchrist,
J. chem. Soc., Perkin I, 1973, 868).

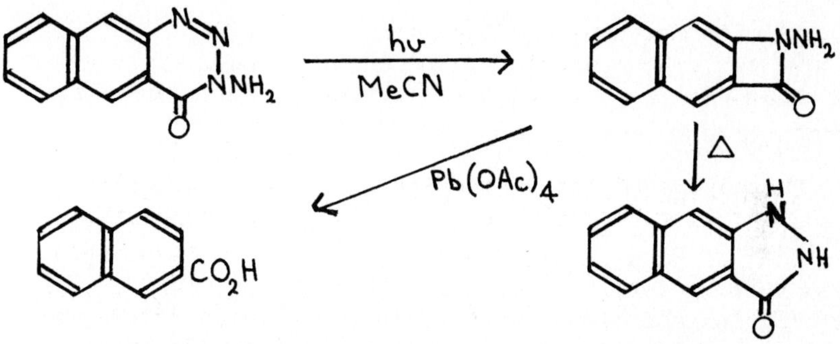

(22)

Ketenimines (23), valence tautomers of benzazetinones
(24) have been postulated as intermediates in the
thermolysis of 1,2,3,-3H-benzotriazin-4-one, and its
3-phenyl- and 3-hydroxy- (25) dervatives. Thermolysis of
the latter affords excellent yields of 3-(2-aminobenzoyl-
oxy)-1,2,3-benzotriazin-4-one (26) (P Ahern, T Navratil,

and K Vaughan, Tetrahedron Letters, 1973, 4547).

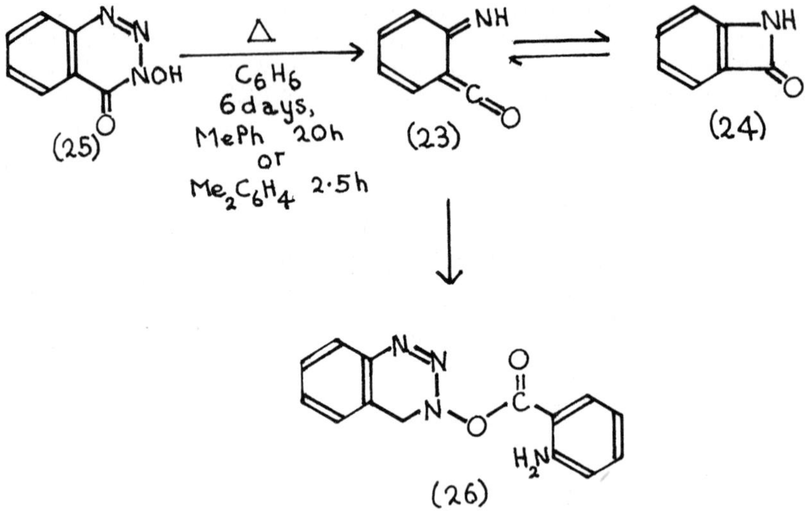

(d) Azetes

Vapour phase pyrolysis of 4-phenylbenzo-1,2,3-
-triazine affords 2-phenylbenzazete (an azacyclobutadiene)
(*ca.* 60%) along with biphenylene (*ca.* 20%) and 9-phenyl-
acridine (*ca.* 15%). It dimerises and reacts with
nucleophiles and dienes very rapidly but is surprisingly
stable up to *ca.* -40°. 2-Phenylnaphtho[2,3-b]azete is
appreciably stable at room temperature. (B.M. Adger *et.
al.*, Chem. Comm., 1973, 19; B.M. Adger, C.W. Rees, and
R.C. Storr, J. chem. Soc., 1975, 45).

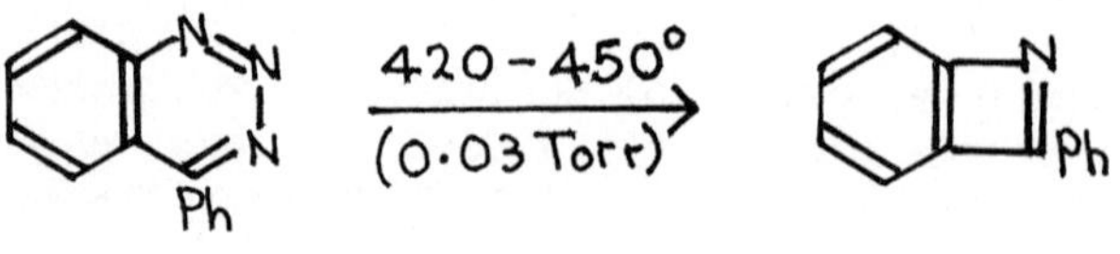

The pyrolysis of 4-methyl- and 4-phenyl-1H-2,3-
-benzoxazin-1-one yields biphenylene, probably by a stepwise
process, which presumably involves the extrusion of carbon
dioxide to give 2-phenylbenzazete and then the loss of
acetonitrile and benzonitrile, respctively, to give benzyne
which dimerises to biphenylene (M.P. David and
J.F.W. McOmie, Tetrahedron Letters, 1973, 1361).

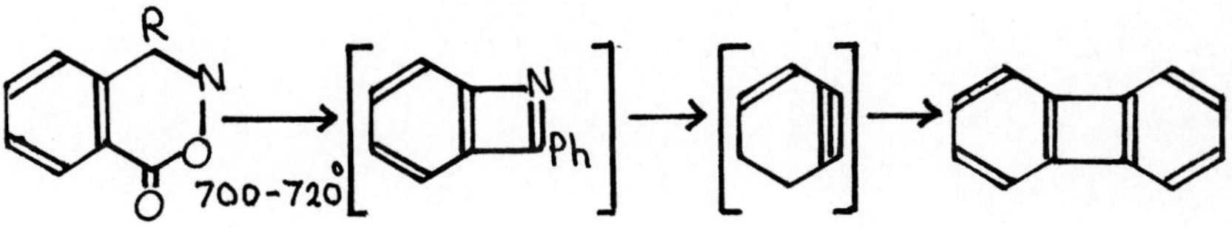

R = Me or Ph

2-Arylbenzazetes are also produced by photolysis
of 4-arylbenzotriazines and at -80° can be intercepted
in cycloaddition reactions, for example, with either
4-nitrobenzonitrile oxide, or cyclopentadiene. The aryl-
benzazetes dimerise thermally to give the angular dimers
and, in the presence of Lewis acids, the linear dimers
(Rees, Storr, and P.J. Whittle, Chem. Comm., 1976, 411).

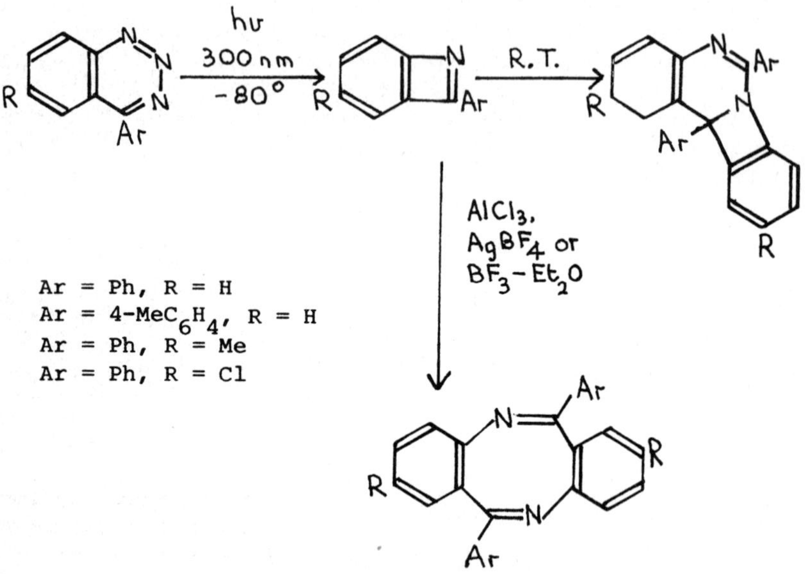
hν
300 nm
-80°
R.T.
AlCl3,
AgBF4 or
BF3-Et2O
Ar = Ph, R = H
Ar = 4-MeC6H4, R = H
Ar = Ph, R = Me
Ar = Ph, R = Cl

No evidence has been obtained for the formation of 2-methylbenzazete in significant amounts, but there is evidence for the formation of *tert*-butylbenzazete from 4-*tert*-butyl-1,2,3-benzotriazine. The chemistry of 2-arylbenzazetes cannot be usefully extrapolated to alkylbenzazetes (*idem.*, Tetrahedron Letters, 1976, 4647).

Nitrile oxides react with 2-phenylbenzazete to give initially labile 1,3-dipolar cycloaddition products, which rearrange very readily to give 1,3,5-oxadiazepines (2) presumably *via* an oxaziridine (1) (Rees *et. al.*, Chem. Comm., 1975, 740).

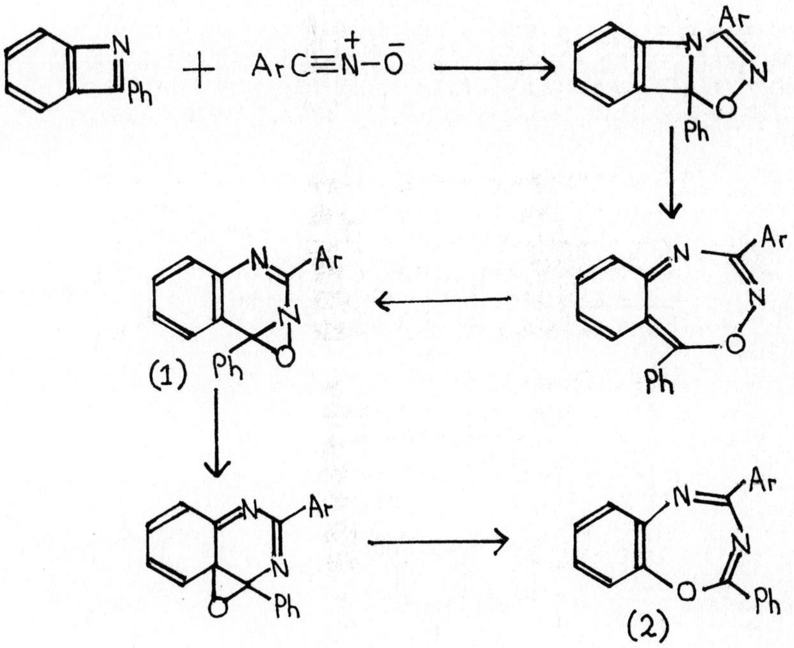

Ar = Ph, 4-MeC$_6$H$_4$, 4-ClC$_6$H$_4$, 4-NO$_2$C$_6$H$_4$

The primary cycloadducts from 2-phenylbenzazete and diarylnitrile imines rearrange spontaneously to 1,2,4--triarylbenzo[f]-1,3,5-triazepines, when Ar = Ph, 4-MeC$_6$H$_4$, 4-ClC$_6$H$_4$, or 4-NO$_2$C$_6$H$_4$.

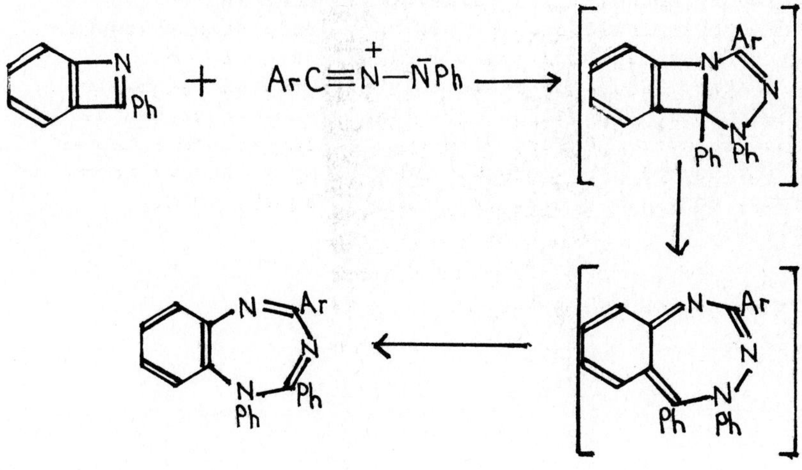

The adduct from 2-phenylbenzazete and diazomethane is isolable but labile and on heating in benzene at 50° is transformed quantitatively into the azidostyrene, and with acid yields a mixture of 2- and 3-phenylindole and an amino-ketone (P.W. Manley *et. al., ibid.,* 1978, 396).

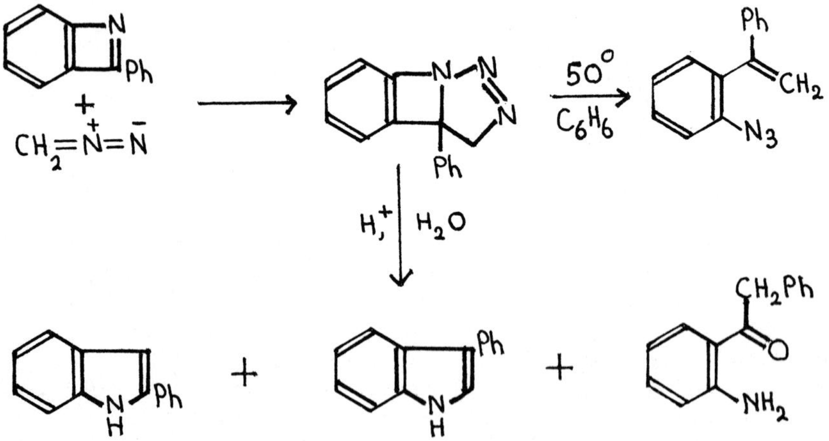

According to the push-pull principle, introducing amino groups, should provide sufficient stabilization for monocyclic aza-(azetes) and 1,3-diaza-cyclobutadienes to exist. It has been concluded that the pyrolyzate "frozen" at -196° from the flash pyrolysis at 527° of tris(dimethylamino)-1,2,3-triazine (3) contains tris(dimethylamino)azete [tris(dimethylamino)azacyclobuta-diene]. [1]H-nmr spectral data indicates a yield of approximately 30%. The relatively high thermal stability and the nmr data suggests that it should be described as a resonance hybrid 4a-4d, with special importance attached to 4d (G. Seybold, U. Jersak, and R. Gompper Angew. Chem. internat. Edn., 1973, 12, 847; H.-U. Wagner, *ibid.,* p.848).

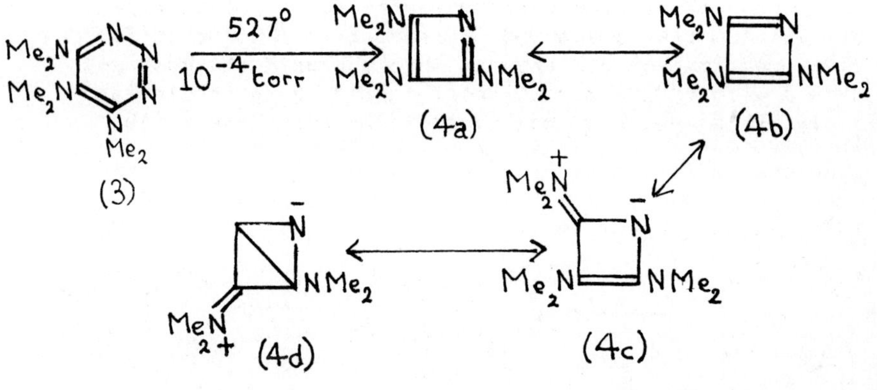

(3) → (4a) ↔ (4b), (4c), (4d)

T. Kurihara and M. Mori (Tetrahedron Letters, 1976,
1825) reported that boiling an acetic acid solution of
3-benzylideneacylpyruvic acid with two equivalents of
hydroxylamine hydrochloride yielded a solid, m.p.70-71°,
believed to be 3-methyl-4-(2-chloropheny)azeto[3,2-d]-
isoxazoline (5), and a small amount of 3-methyl-4-(α-amino-
-2-chlorophenyl)-5-isoxazoline (6). Evidence has been
provided, however to show that the solid, m.p.70-71°,
is in fact 5-(2-chlorophenyl)-4-cyano-3-methylisoxazole
(7) (Rees, Starr, and Whittle, *ibid.*, 1976, 3931).

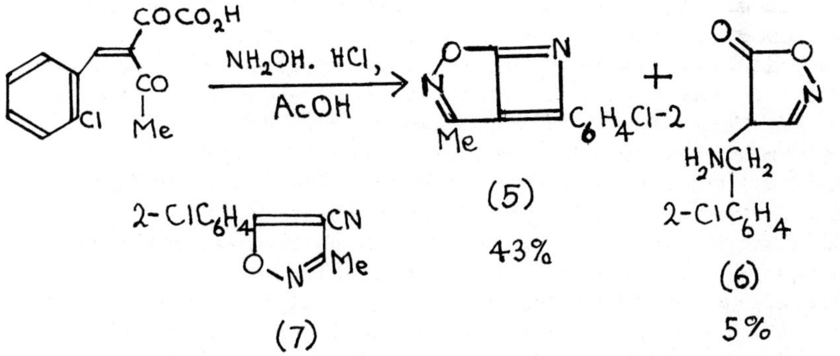

Irradiation (>270 nm) of oxazinone (8) at $-70°$ gives the bicyclic compound (9), which after approximately 12h eliminates carbon dioxide to afford the azetes (10) and (11) in photoequilibrium. Although (10) and (11) are not detected owing to the formation of a dimer, their intermediacy is suggested by the photofragmentation products at 7K (Ar matrix) (A. Krantz and B. Hoppe, J. Amer. chem. Soc., 1975, 97, 6590; G. Maier and U. Schäfer, Tetrahedron Letters, 1977, 1053).

10. Three- and Four-Membered Rings Containing Other Elements

Three- and four-membered rings containing unusual heteroatoms are becoming more common and the pursuit of related potential aromatic and antiaromatic compounds has resulted in the preparation of a number of novel compounds.

(a) Phosphiranes (phosphacyclopropanes) and phosphirenes

Phosphirane, b.p. $36.5°$ is obtained by the reaction between 1,2-dichloroethane and sodium phosphinide in anhydrous ammonia, and by the reduction of 2-bromoethyl-phosphonous dibromide with lithium tetrahydridoaluminate in diethylene glycol diethyl ether. The former reaction does not follow the nucleophilic substitution pathway

normally observed with simple alkyl halides, but gives
a mixture of phosphirane, phosphine, and ethene. A
mechanism has been suggested (R.I. Wagner *et. al.*, J. Amer.
chem. Soc., 1967, <u>89</u>, 1102).

$$ClCH_2CH_2Cl \xrightarrow[NH_3]{2NaPH_2,} \overset{H}{\underset{}{\triangle P}} + PH_3 + CH_2{=}CH_2$$

$$BrCH_2CH_2PBr_2 \xrightarrow[(EtOCH_2CH_2)_2O]{LiAlH_4,} \overset{H}{\underset{}{\triangle P}} + EtPH_2 + PH_3 + CH_2{=}CH_2$$

   Phosphirane is thermally unstable in the liquid phase
and decomposes at 25° to give a viscous non-volatile
liquid, ethylphosphine and ethene. Hydrogen chloride
decomposes it rapidly below room temperature, but it is
not rapidly decomposed by methanol at room temperature.
There is evidence that like aziridine it is an unusually
weak base.

   Tri-*tert*-butylphosphirane 1-oxide is obtained by
treating the phosphane oxide (1) with *sec*-butyllithium
in a mixture of tetrahydrofuran and 2-methylbutane at
-78°, followed by chlorination using carbon tetrachloride
at -100° and ring closing the product with lithium
diethylamide in ether at -80° (H. Quast and M. Heuschmann,
Angew. Chem. internat. Edn., 1978, <u>17</u>, 867).

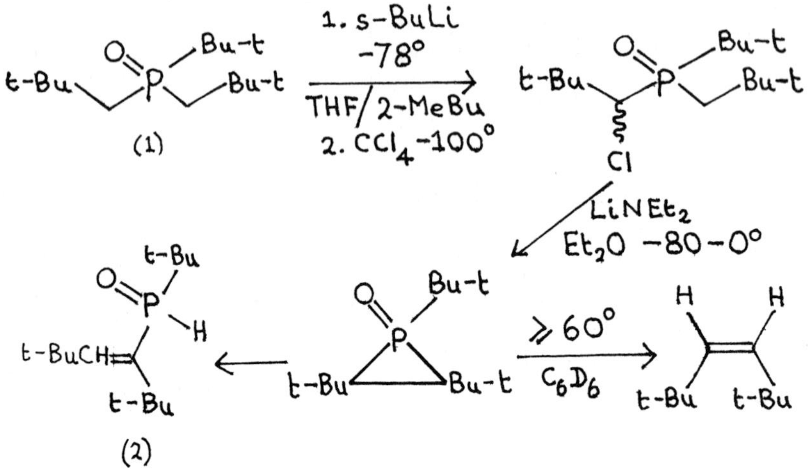

(1)

(2)

Tri-*tert*-butylphosphirane is remarkably stable towards concentrated hydrochloric acid, but it slowly isomerises to (2) with strong bases, such as lithium piperidide or lithiumdiethylamide. In [D_6]-benzene at $\geqslant 60°$ it slowly decomposes quantitatively to *(Z)*-2,2,5,5-tetramethylhex-3--ene.

1,2,3-Triphenylphosphirene 1-oxide a member of a new class of potentially aromatic phosphacyclopropenes is obtained by the addition of 1,5-diazabicyclo[4.3.0]non--5-ene to a solution of bis(α-bromobenzyl)phenylphosphine oxide in benzene. It is fairly stable but on pyrolysis yields diphenylacetylene and on treatment with aqueous sodium hydroxide gives 1,2-diphenylvinylphosphinic acid, which on pyrolysis affords *cis*- and *trans*- stilbene (3.1) (E.W. Koos *et. al.*, Chem. Comm., 1972, 1085).

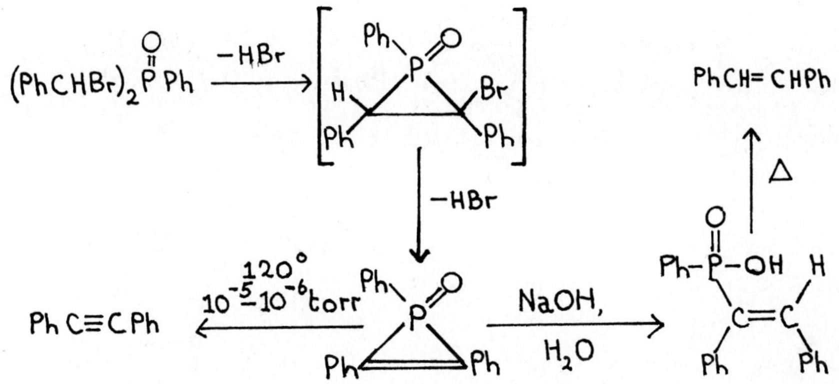

(b) Phosphetanes

A phosphetane derivative, 2,2,3,4,4-pentamethyl-1-
-phenylphosphetane 1-oxide (1) is formed on treating
2,4,4-trimethylpent-2-ene with the complex from phenylphos-
phonous dichloride and aluminium chloride. Reduction of
phosphetane (1) using lithium tetrahydridoaluminate,
followed by quaternization with methylene di-iodide gives a
phosphetanium salt, (S.E. Fishwick *et. al., ibid.*, 1967,
1113).

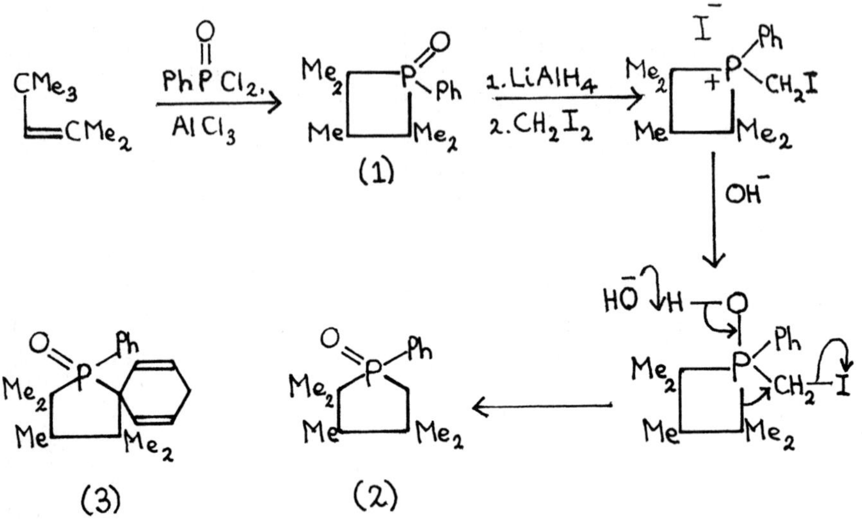

(1)

(2)

(3)

Similar treatment of 1,2,2,3,4,4-hexamethyl-1-phenyl-
-phosphetanium iodide affords the phospholane 1-oxide (3).
Since the phenyl group cannot be lost, migration of the
apical CMe_2 to the phenyl group occurs. The reaction
between 2,2,3,4,4-pentamethyl-1-phenylphosphetane 1-oxide
and phenyllithium gives a cyclohexiadienyl anion which
cannot be protonated, and consequently the five-membered
ring opens and the resulting anion when quenched with
methyl iodide gives an open chain phosphine oxide
(S.E. Cremer and R.J. Chorvat, Tetrahedron Letters, 1968,
413).

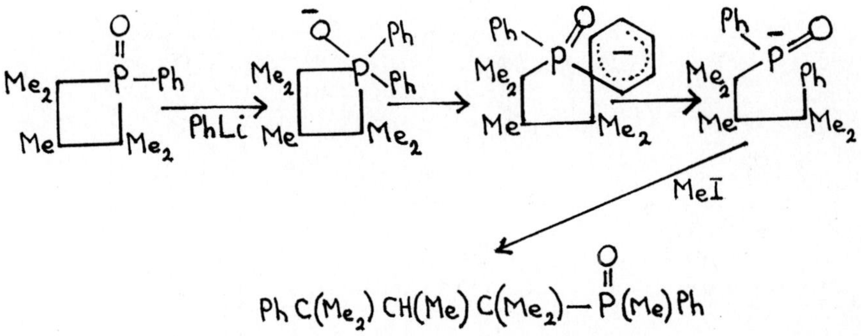

$$Ph\,C(Me_2)\,CH(Me)\,C(Me_2) - \overset{\overset{O}{\|}}{P}(Me)\,Ph$$

Treatment of 1-chloro-2,2,3,4,4-pentamethylphosphe-
tane 1-oxide with *tert*-butyllithium at 0° affords
2,2,3,4,4-pentamethyl-1-*tert*-butylphosphetane 1-oxide,
which on reduction with trichlorosilane gives 2,2,3,4,4-
-pentamethyl-1-*tert*-butylphosphetane (Cremer *et. al.*,
ibid., p.5799).

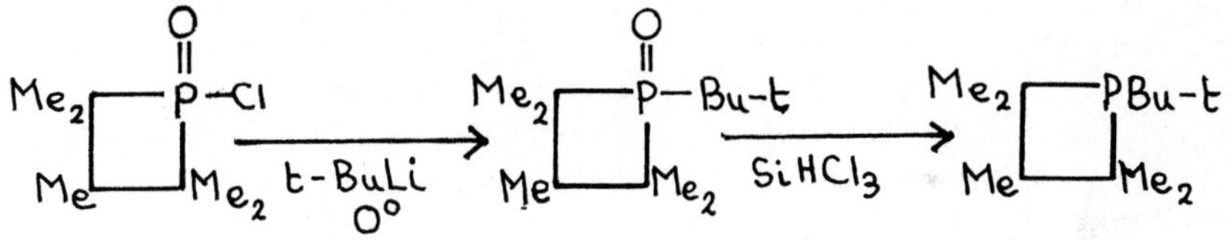

(c) Siliranes (silacyclopropanes) and silirenes

A stable silirane (3), m.p.72-74° is obtained by
treating silane (2), prepared by the reaction between the
lithium reagent from *gem*-dibromocyclopropane (1) and
dichlorodimethylsilane. It is air sensitive and has to
be handled under an inert atmosphere.

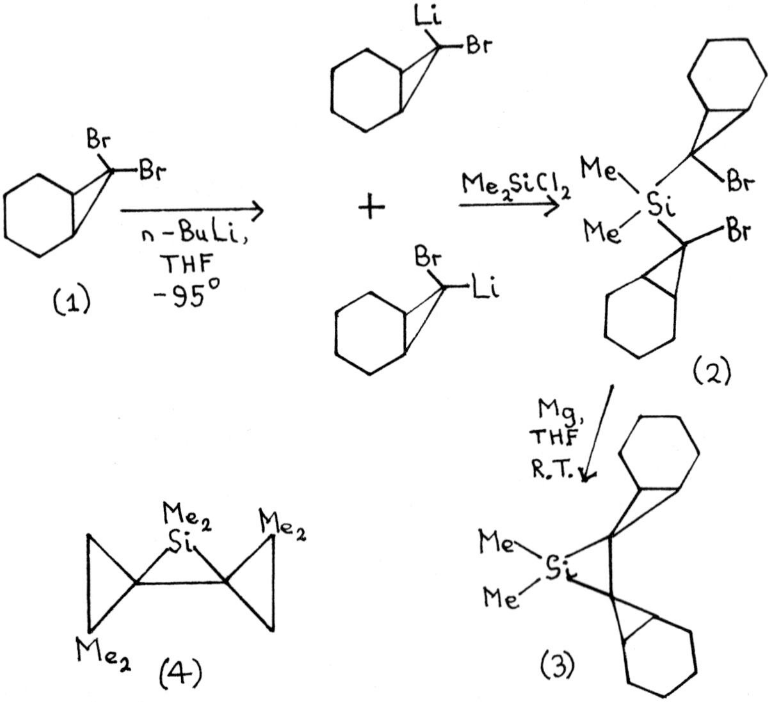

1,1-Dimethyl-*trans*-bis-2,3-(2,2-dimethylcyclopropyl-idene)silirane (4), b.p.50-51°/3.5mm, is prepared in a similar manner from appropriate starting materials (R.L. Lambert, Jr., and D. Seyferth, J. Amer. chem. Soc., 1972, 94, 9246; Seyferth, C.K. Haas, and D.C. Annarelli, J. organometal. Chem., 1973, 56, C7). 1,1,2,2,3,3-Hexa-methylsilirane is obtained in admixture with other compounds on reacting dimethylbis(α-bromoisopropyl)silane with magnesium in tetrahydrofuran. It does not survive attempted fractional distillation or purification by gas chromatography (Seyferth and Annarelli, J. Amer. chem. Soc., 1975, 97, 2273).

It is claimed that the addition of dimethylsilene, obtained by gas-phase flow-pyrolysis of 1,2-dimethoxy-1,1,-

2,2-tetramethyldisilane, to dimethylacetylene at 600°
yields 1,1,2,3-tetramethylsilirene.

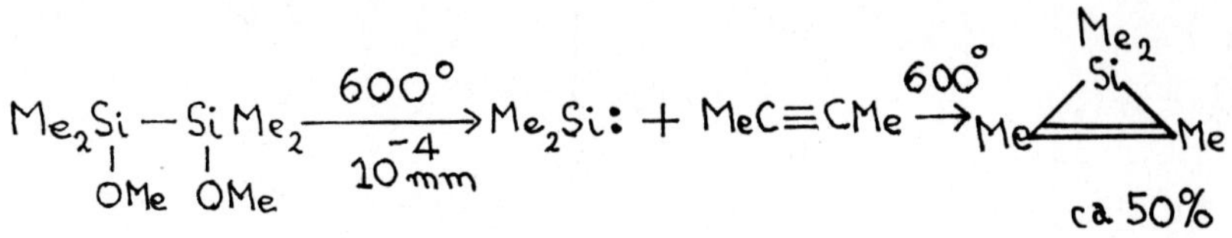

$$Me_2Si-SiMe_2 \xrightarrow[10^{-4}mm]{600°} Me_2Si: + MeC\equiv CMe \xrightarrow{600°} Me\underset{\overset{Si}{Me_2}}{\triangle}Me$$

ca 50%

The reported preparation of 1,1-dimethyl-2,3-diphenyl-
silirene gives the dimer (5) (R.T. Conlin and P.P. Gaspar,
ibid., 1976, 98, 3715).

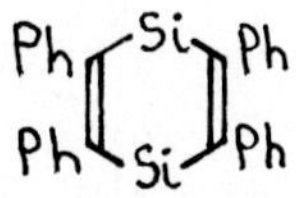

$$Ph\underset{Ph}{\overset{Si}{\prod}}\underset{Ph}{\overset{Ph}{Si}}$$

(5)

Hexamethylsilirane reacts with bis(trimethylsilyl)-
acetylene to give 2,3-bis(trimethylsilyl)-1,1-dimethyl-
silirene (Seyferth, Annarelli, and S.C. Vick, *ibid.*,
p.6382). Treatment of the silirene with water or an
alcohol results in ring-opening and formation of product
(6).

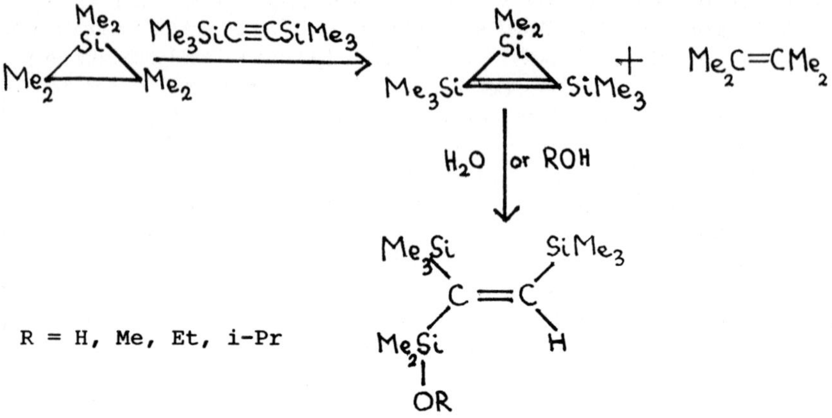

$$Me_2Si\text{(silirene)} \xrightarrow{Me_3SiC\equiv CSiMe_3} Me_3Si\text{(silirene)}SiMe_3 + Me_2C=CMe_2$$

$$\xrightarrow{H_2O \text{ or } ROH}$$

R = H, Me, Et, i-Pr

$$Me_3Si\text{—}C=C\text{—}SiMe_3$$

(6)

Irradiation of a benzene solution of (pentamethyl-disilanyl)phenylacetylene affords 1,1-dimethyl-2-phenyl-3--trimethylsilylsilirene, which is fairly stable in solution. Addition of methanol to the solution causes ring-opening within a minute to yield the ethene derivatives.

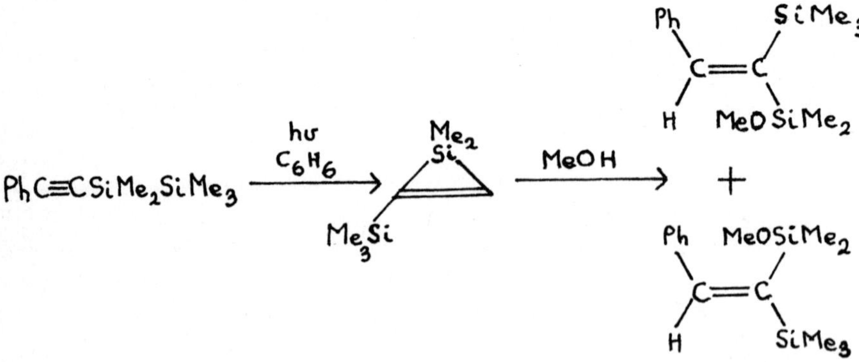

$$PhC\equiv CSiMe_2SiMe_3 \xrightarrow[C_6H_6]{h\nu} \text{(silirene)} \xrightarrow{MeOH}$$

The reaction between 1,1-dimethyl-2-phenyl-3-trimethyl-
silylsilirene and various ketones gives 1-sila-2-oxacyclopent-
-4-enes, and the palladium complex-catalysed cycloaddition
with some acetylenes yields silacyclopentadienes
(H. Sakurai, Y. Kamiyama, and Y. Nadaira, *ibid.*, 1977,
99, 3879).

Irradiation of 1,1-dimethyl-2-phenyl-3-trimethylsilyl-
silirene in the presence of a nitrile results in the
formation of aza-2,8-disilabicyclo[3.2.1]octa-2,6-diene
derivatives, whereas with acrylonitrile besides the related
derivative another new ring system, 1-aza-2,8-disilabicyclo-
[3.3.o]octa-3,6-diene is formed (Sakurai, Kamiyama, and
Nakadaira, Chem. Comm., 1978, 80).

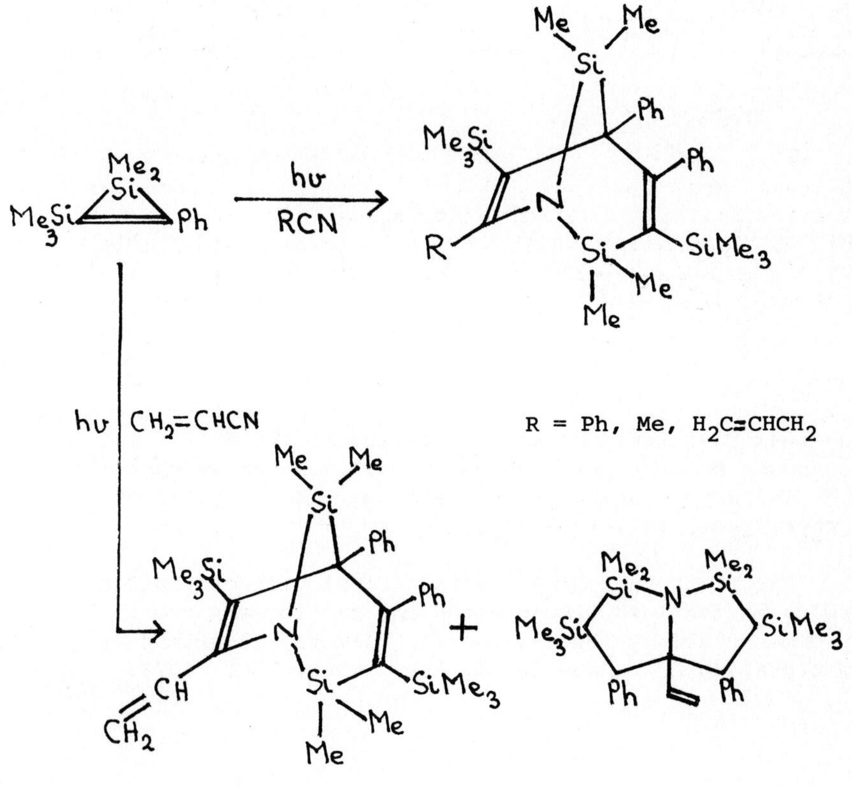

(d) Silacyclobutanes (siletanes), siletenes, and siletes

The name silacyclobutanes is more commonly used for this class of compounds.

Boiling 3-chloropropylmethyldichlorosilane in ether with magnesium activated by iodine affords 1-chloro-1--methylsilacyclobutane, which on treatment with methyl-magnesium iodide gives 1,1-dimethylsilacyclobutane (1,1-dimethylsiletane) and on reduction with lithium tetrahydridoaluminate yields 1-methylsilacyclobutane (N.S. Nametkin, V.M. Vdovin, and P.L. Grinberg, Doklady Akad. Nauk SSSR, 1963, 150, 799).

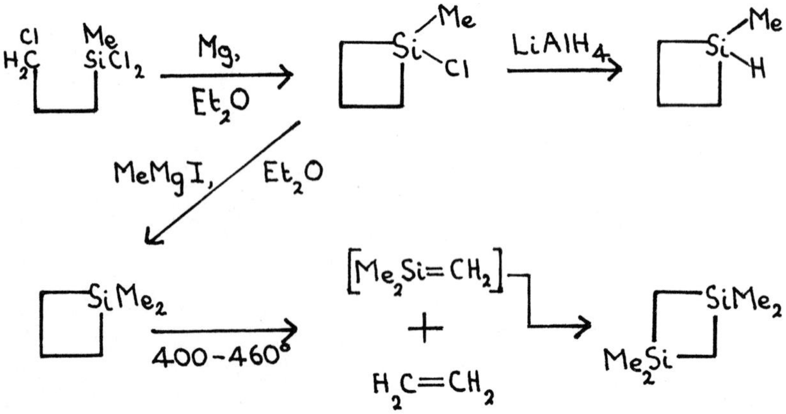

1,1-dimethylsilacyclobutane on heating to 400-460° decomposes to give only ethene and 1,1,3,3-tetramethyl--1,3-disilacyclobutane (M. Flowers and L.E. Gusel'nikov, J. chem. Soc., (B), 1968, 419).

The flash vacuum pyrolysis of diallyldimethylsilane results in the loss of propylene and the formation of 1,1-dimethylsilacyclobut-2-ene (1,1-dimethyl-2-siletene) (E. Block and L.R. Revelle, J. Amer. chem. Soc., 1978, 100, 1630).

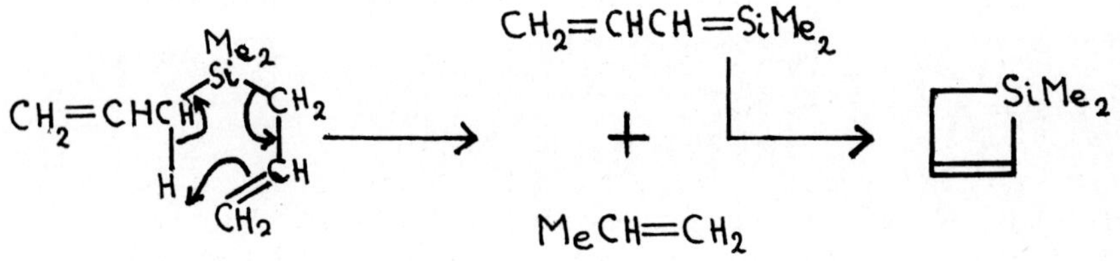

Me2
Si
CH2=CHCH
CH2
H
CH
CH2
CH2=CHCH=SiMe2
+
MeCH=CH2
SiMe2

The reaction between 1,1-dialkyldichlorosilanes and 1,8-dilithionaphthalene gives 1,1-dialkylnaphtho[1,8-bc]-siletes, 1,1-diethylnaphtho[1,8-bc]silete, b.p.80°/0.02mm, 1,1-dimethylnaphtho[1,8-bc]silete, b.p.65°/0.04 mm (L.S. Yang and H. Shechter, Chem. Comm., 1976, 775).

Li
Li
R2SiCl2,
Et2O
0°
R2
Si

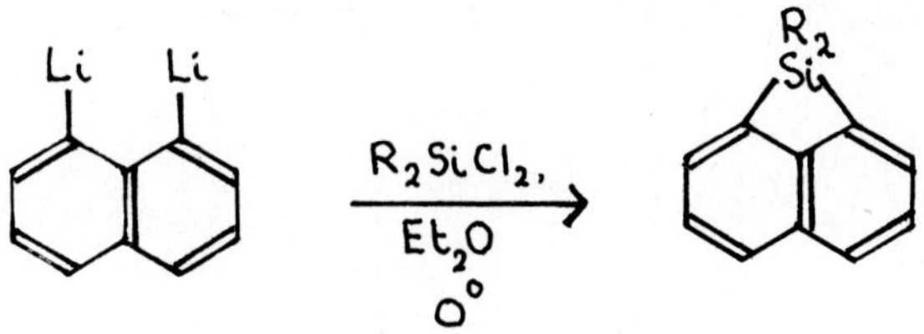

R = Me or Et

(e) Germacyclobutanes

Derivatives of the four-membered ring containing one germanium atom, namely germacyclobutane, have been obtained by reacting di-n-butyl- or diethyl-chloro(3--chloropropyl)germane with sodium or liquid sodium-potassium eutectic alloy respectively in toluene (P. Mazerolles, M. Lesbre, and J. Dubac, Compt. rend., 1965, 260, 2255).

$$\begin{matrix} CH_2-GeClR_2 \\ | \\ CH_2-CH_2Cl \end{matrix} \quad \xrightarrow{\begin{bmatrix} (R=n\text{-}Bu),\ Na,\ MePh \\ \hline (R=Et),\ Na/K,\ MePh \end{bmatrix}} \quad \square\text{-}GeR_2$$

The ring in germacyclobutanes is highly strained and consequently it is opened by a large variety of reagents, for example, bromine, hydrogen halides and other protonic acids, germanium tetrachloride, lithium tetrahydridoaluminate, alcoholic silver nitrate, mercuric chloride and sulphuryl chloride. Ring expansion is effected by the insertion of dichlorocarbene into geranium-carbon bond of 1,1-diethylgermacyclobutane (Seyferth *et. al.*, J. organometal. Chem., 1969, **16**, 503).

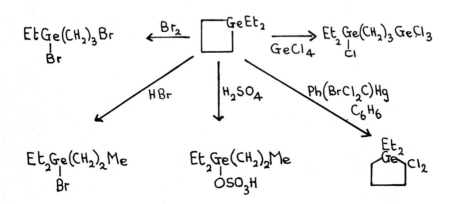

11. *Four-Membered Rings With Two Hetero-Atoms*

(a) *Dioxetanes*

Irradiation of a solution of *cis*-diethoxyethene in fluorotrichloromethane (Freon 11) containing tetraphenylporphin at -78° in a stream of oxygen, results in the

formation of *cis*-3,4-diethoxy-1,2-dioxetane as a
crystalline solid. Similar results are obtained when the
reaction is carried out in deuterioacetone containing Rose
Bengal. Photooxidation of *trans*-diethoxyethene under the
latter condition affords *trans*-3,4-diethoxy-1,2-dioxetane
(P.D. Bartlett and A.P. Schaap, J. Amer. chem. Soc., 1970,
92, 3225).

Each dioxetane on warming decomposes quantitatively to
ethyl formate. A small sample of the crystalline *cis*-
compound on warming to room temperature, melted and
exploded. It has been shown that during the first-order
decomposition of *cis*-diethoxydioxetane to excited ethyl
formate; energy transfer to fluorescers causes
chemiluminescence (T. Wilson and Schaap, *ibid.*, 1971, **93**,
4126). A very narrow band chemiluminescence is observed
when 3,3,4-trimethyl-1,2-dioxetane is thermally decomposed
in the presence of certain lanthanide chelates, for
example, europium tris(thenoyltrifluoroacetonate)-1,10-
-phenanthroline (P.D. Wildes and E.H. White, *ibid.*,
p.6286).

Photooxygenation of tetramethoxyethene in ether,
at -70°, sensitized by either zinc tetraphenylporphine
dinaphthalene-thiophene with visible light gives 3,3,4,4-
-tetramethoxy-1,2-dioxetane (S. Mazur and C.S. Foote,
ibid., p.3225). Stable bicyclic 1,2-dioxetanes are formed
by the addition of singlet oxygen to 1,4-dioxene and
1,3-dioxole (Schaap, Tetrahedron Letters, 1971, 1757).

Dioxetane intermediates have been isolated from the ozonolysis of several olefins (P.R. Story, E.A. Whited, and J.A. Alford, J. Amer. chem. Soc., 1972, 94, 2143), but other results have been reported which support the formation of a peroxy-epoxide intermediate (P.G. Gassman and X. Creary, Tetrahedron Letters, 1972, 4411). Other studies on the ozonolysis of olefins have been reported (L.A. Hull, I.C. Hisatsune, and J. Heicklen, J. Amer. chem. Soc., 1972, 94, 4856; C.W. Gillies, and R.L. Kuczkowski, *ibid.*, p.7610).

Triphenylphosphine reacts with 3,3,4,4-tetramethyl--1,2-dioxetane in benzene at ~6° to give the phosphorane (1). Decomposition of the phosphorane (1) at 55° yields quantitative amounts of triphenylphosphine oxide and 2,2,3,3-tetramethyloxirane (P.D. Bartlett, A.L. Baumstark, and M.E. Landis, *ibid.*, 1973, 95, 6486).

(1)

Dioxetanes are isolated in good yield by low-temperature photooxygenation of indenes in methanol or methanol-acetone in the presence of Rose Bengal sensitizer. After vacuum evaporation (<0°), and purification by chromatographing on silica gel with chloroform they are obtained as pale yellow crystals from cold ether-pentane. The dioxetanes cleave smoothly to dicarbonyl products, but decompose explosively above their melting points on rapid heating (P.A. Burns and C.S. Foote, J. Amer. chem. Soc., 1974, 96, 4339).

$R^1 = R^2 = Ph, R^3 = H$ $R^1 = i-Pr, R^2 = R^3 = H$
$R^1 = Me, R^2 = Ph, R^3 = H$ $R^1 = Me, R^2 = H, R^3 = t-Bu$

3,4-Diadamantyl-1,2-dioxetane is a stable 1,2-
-dioxetane (J.J. Wieringa *et. al.*, Tetrahedron Letters,
1972, 169). The synthesis and spectral data of a stable,
optically active 1,2-dioxetane, a ketal derivative of 3,4-
-diadamanyl-1,2-dioxetane, has also been presented
(H. Wynberg and H. Numan, J. Amer. chem. Soc., 1977, 99,
603, 3870). X-Ray data for dispiro(adamantane-2,3'-
[1,2]dioxetan-4',2"-admantane) has been presented and
supports suggestions regarding the reasons for its unusual
stability, including that it may be due to a high energy
barrier arising from the large inertial masses of the rigid
adamantane units (Numan *et. al.*, Chem. Comm., 1977, 591).

Endoperoxide (3) obtained on photooxygenation of
2-(2'-anthryl)-1,4-dioxene (2) at -78°, rearranges
quantitatively to 1,2-dioxetane (4) at ambient temperature
upon treatment with silica gel. The decomposition of (4)
at 80° is attended by chemiluminescence and gives only
the expected cleavage product (5) (Schaap, Burns, and
K.A. Zaklika, J. Amer. chem. Soc., 1977, 99, 1270).

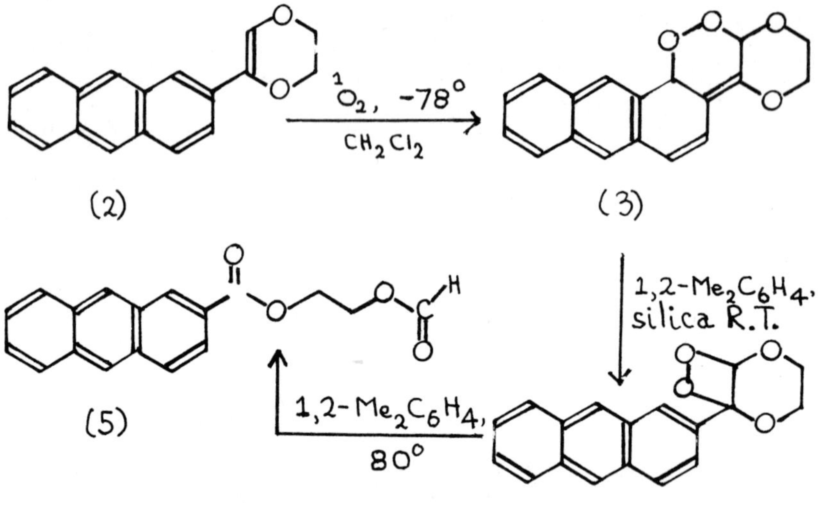

It has been found that silica gel can be used to advantage as a heterogeneous catalyst for the rearrangement of an endoperoxide to a 1,2-dioxetane. The relatively unstable 1,2-dioxetane (4) can be "stored" as (3) and generated when needed. This system therefore has the potential for a practical chemical light source.

It has been sugested that the tetraoxaspirocyclo-heptane (6) is an intermediate in the formation of aldehydes and ketones by the photooxygenation of allenes in carbon disulphide in the presence of eosin (T. Greibrokk, Tetrahedron Letters, 1973, 1663).

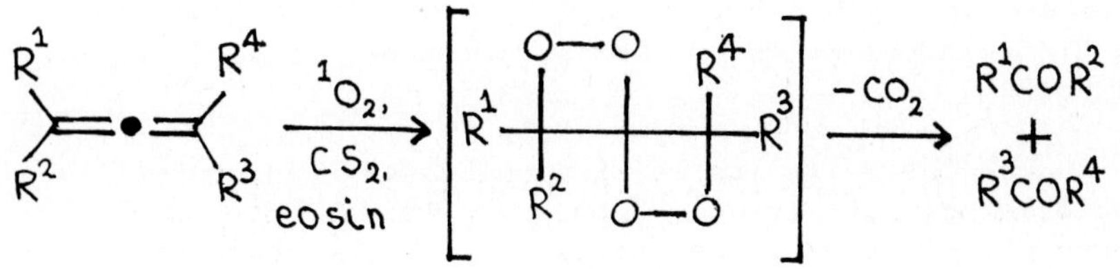

(6)

$$R^1 = Ph, \quad R^2 = R^3 = R^4 = H \qquad R^1 = R^2 = R^3 = Ph, \quad R^4 = H$$
$$R^1 = R^3 = Ph, \quad R^2 = R^4 = H \qquad R^1 = R^2 = R^3 = R^4 = Ph$$

The cyclisation of the α-hydroperoxy acid (7) in the presence of an equimolar quantity of dicyclohexyl-carbodiimide (DCCD) in carbon tetrachloride below -10° affords 4-*tert*-butyl-1,2-dioxetan-3-one (an α-peroxy lactone), which can be isolated by flash distillation of the reaction mixture at -10°. Its approximate lifetime is 5-8 minutes at room temperature and it gives pivalaldehyde with evolution of carbon dioxide. A sample warmed to room temperature in the dark clearly displays the expected luminescence. The suggestion of dioxetan-3--ones as intermediates in bioluminescence and chemilumin-escence has been well documented in recent years (W. Adam and J.-C. Liu, J. Amer. chem. Soc., 1972, **94**, 2894).

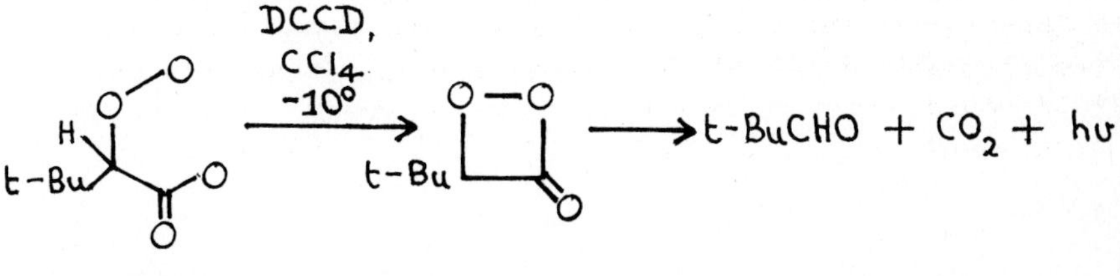

(7)

The instability of the highly water-sensitive 4-adamantyl-1,2-dioxetan-3-one has so far hindered its isolation in pure form. It decomposes to give

adamantane-1-carboxaldehyde and carbon dioxide (Adam and
H.-C. Steinmetzer, Angew. Chem., internat. Edn., 1972,
11, 540).

The thermally labile di-*tert*-butyl-, *tert*-butyl,
and dimethyl-1,2-dioxetan-3-ones are readily prepared by
dicyclohexylcarbodiimide cyclisation of the respective
α-hydroperoxy acids in methylene chloride at sub-ambient
temperatures. They are isolated and purified by low-
temperature flash distillation and exhibit a characteristic
carbonyl frequency at $1860 \pm 10 cm^{-1}$ (Adam *et. al.*, J.
Amer. chem. Soc., 1977, 99, 5768; Adam and O. Cueto, J.
org. Chem., 1977, 42, 38).

The thermochemical values and activation parameters
have been estimated for the thermal decomposition of
4-*tert*-butyl-1,2-dioxetan-3-one, 4,4-dimethyl-1,2-dioxetan-
-3-one, and 1,2-dioxetan-3,4-dione. The observed and the
calculated half-life for the first compound are in good
agreement and the calculated value given for the last
compound is only 0.54 seconds (W.H. Richardson and
H.E. O'Neal, J. Amer. chem. Soc., 1972, 94, 8665).

(b) Dithietanes

1,3-Dithietane m.p.105-106° is obtained by reducing
1,3-dithietane 1-oxide with excess 1M tetrahydrofuran-
-borane. The 1-oxide, m.p.71-73.5°, is prepared from
bis(chloromethyl)sulphoxide, which in the presence of the
phase transfer catalyst "tricaprylmethylammonium chloride
reacts rapidly and exothermically at room temperature with
aqueous sodium sulphide (E. Block *et. al.*, *ibid.*, 1976,
98, 5715).

Treatment of the 1-oxide at -20° with potassium

permanganate-magnesium sulphate in acetone gives
1,3-dithietane 1,1-dioxide (96%), m.p.141-143°, and
oxidation with iodobenzene dichloride in aqueous pyridine
at -30° or with 3-chloroperbenzoic acid in methylene
chloride at 0° yields, respectively, a 3:1 or 2:3 mixture
of the *cis*- and the *trans*-1,3-dithietane 1,3-dioxide
[m.ps.260° (decomp.) and 203-205° (decomp).]. The
1,1-dioxide on treatment in chloroform with peracetic
acid at 0° yields 1,3-dithietane 1,1,3-trioxide (90%),
m.p.231-234°. Oxidation of the 1,1-dioxide, the *cis*-
or the *trans*-1,3-dioxide, or the 1,1,3-trioxide with excess
peracetic acid at 100° gives the 1,1,3,3-tetraoxide.

Sulphur dichloride reacts with active methylene
compounds under mild conditions to give either 1,3-dithie-
tanes or sulphurated products depending upon the reactivity
of the active methylene moiety and the stoichiometry of the
reaction. 2,4-Bis(acetyl)-2,4-bis(methoxycarbonyl)-1,3-
-dithietane, m.p.125-126°, is obtained by the reaction
of equimolar quantitites of methyl acetoacetate and sulphur
dichloride (S.K. Gupta, J. org. Chem., 1974, $\underline{39}$ 1944).

$$2MeCOCH_2CO_2Me + 2SCl_2 \xrightarrow[25-28°]{CHCl_3}$$

Photodimerisation of adamantanethione in n-pentane
yields the 1,3-dithietane (1) (C.C. Liao and P. de Mayo,
Chem. Comm., 1971, 1525).

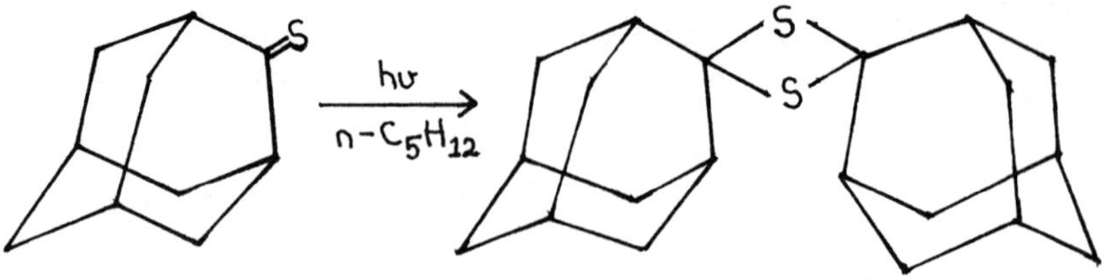

The reaction of (dimethylsulphuranylidene)aceto-
phenone with carbon disulphide in chloroform at room
temperature gives 2-benzoyl-4-benzoylmethylene-1,3-
-dithietane (78%), m.p.145-147° (Y. Hayashi *et. al.*,
Tetrahedron Letters, 1971, 1781).

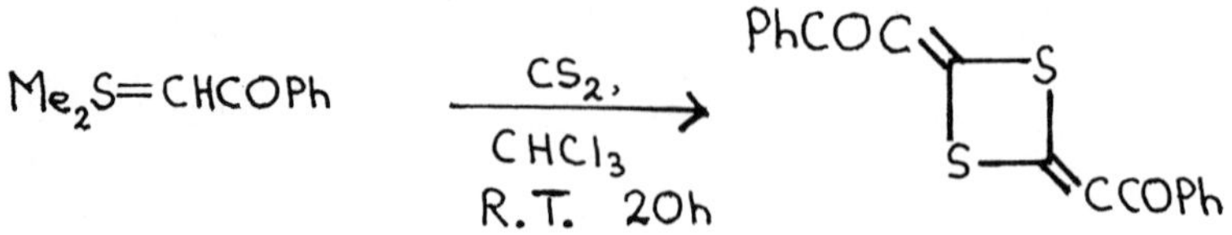

When the reaction is carried out in ethanol 3,5-diphenacy-
lidene-1,2,4-trithiole is obtained. Desulphurization of
2-benzoyl-4-benzoylmethylene-1,3-dithietane using zinc-
-acetic acid affords acetophenone (45%) and propiophenone
(15%).

(c) Oxathietanes

β-Hydroxy sulphoxides react with *N*-bromo- or
N-chloro- succinimide or sulphuryl chloride in methylene
chloride or carbon tetrachloride at room temperature to
give initially 1,2-oxathietane 2-oxides (β-sultines) which
have only a limited thermal stability. In most cases they
readily lose sulphur dioxide within a few minutes to afford
olefins. The 1,2-oxathietane 2-oxides are probably formed
via intramolecular cyclisation of the initially formed
β-hydroxychlorooxosulphonium chloride to the alkoxyoxo-

sulphonium salt, which undergoes fragmentation to the oxathietane and *tert*-butyl chloride (F. Yung, N.K. Sharma, and T. Durst, J. Amer. chem. Soc., 1973, 95, 3420).

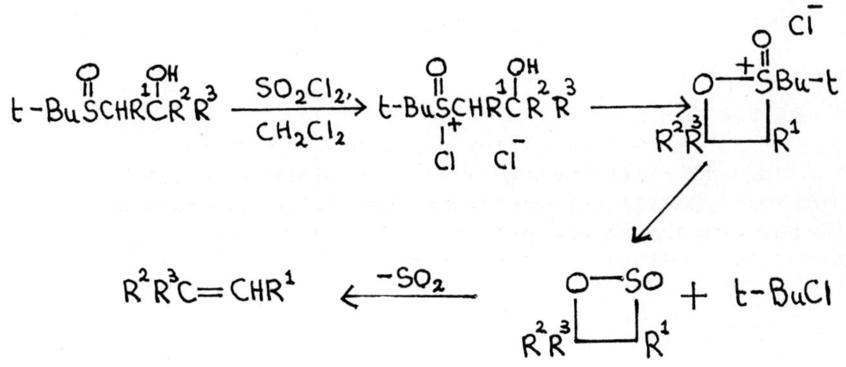

R^1, R^2, R^3 = alkyl or aryl

3,3-Dimethyl-2,2-diphenyl-1,2-oxathietane 2-oxide is stable at room temperature for several days but decomposes quantitatively into 1,1-dimethyl-2,2-diphenylethene and sulphur dioxide when warmed to 30° in methylene chloride (Durst and B.P. Gimbarzevsky, Chem. Comm., 1975, 724).

The uv irradiation of mixtures of sulphur dioxide and ketene in argon or nitrogen matrices at 10-20K give an adduct, 1,2-oxathietane-4-one 2-oxide, identified on the basis of its ir spectrum and photodecomposition to carbon dioxide. The other expected product sulphine has not been identified, this is probably because it is decomposed photochemically as rapidly as it is formed (I.R. Dunkin and J.G. MacDonald, *ibid.*, 1978, 1020).

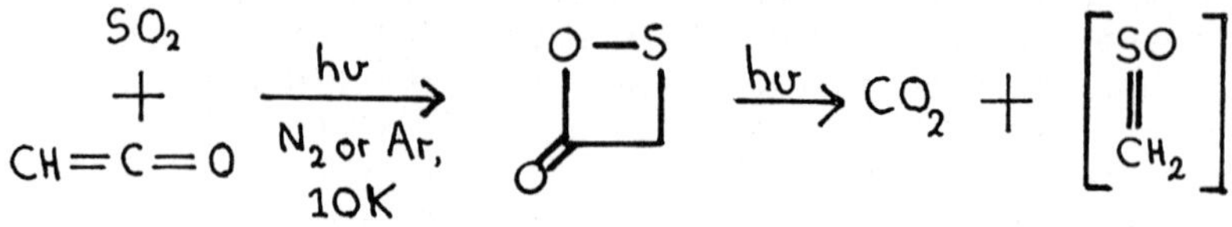

(d) Dithietenes

3,4-Bis-(trifluoromethyl)-1,2-dithietene reacts with 1-phenyl-2,2,4,4-tetramethylphosphetane to give a stable sulphur containing phosphorane, m.p. 109-112°, in high yield (N.J. De'Ath and D.B. Denney, *ibid.*, 1972, 395).

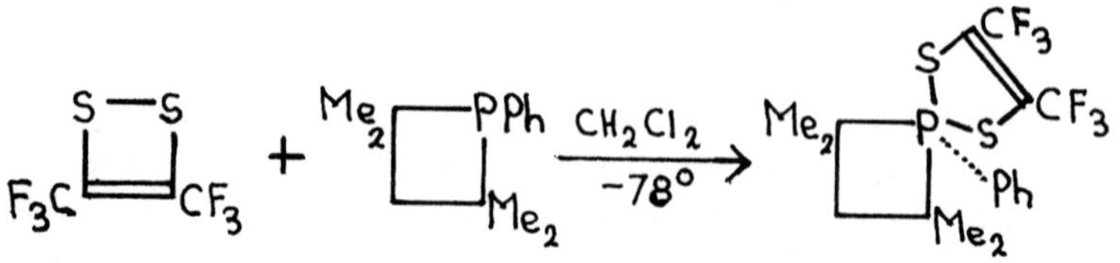

It has been reported that α-hydroxyketones, or less effectively α-diketones, are converted to 1,2-dithietene radical cations (1), by treatment with a mixture of sodium sulphide (or alternatively sodium thiosulphate or sodium dithionite) and sulphuric acid. The radical cations are identified by esr spectroscopy (G.A. Russell, R. Tanikaga, and E.R. Talaty, J. Amer. chem. Soc., 1972, 94, 6125).

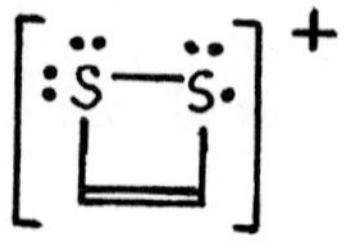

(1)

Irradiation of a benzene solution of the vinylene
dithiocarbonate (2) affords the dithione (3), which in
solution gives an equilibrium mixture of the α-dithione
form and the valence tautomeric form (a 3,4-diaryl-1,2-
-dithietene derivative). The position of the equilibrium
is solvent-, light-, and temperature-dependent (W. Küsters
and P. de Mayo, *ibid.*, 1973, **95**, 2383).

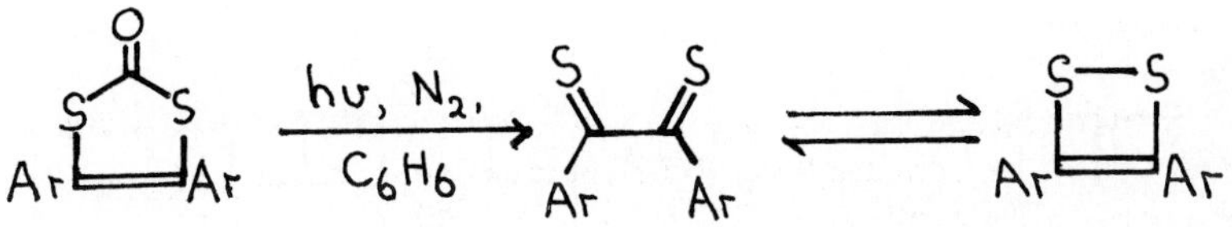

(e) Diazetidines

Cyclohexanone enamines react with dibenzoyldiimide
at room temperature to give a 1:1 addition product, a
1,2-diazetidine, in almost quantitative yield. The
diazetine on mild acidic hydrolysis affords 2-(*N,N*'-diben-
zoyl)hydrazincyclohexanone (L. Marchetti and G. Tosi,
Tetrahedron Letters, 1971, 3071).

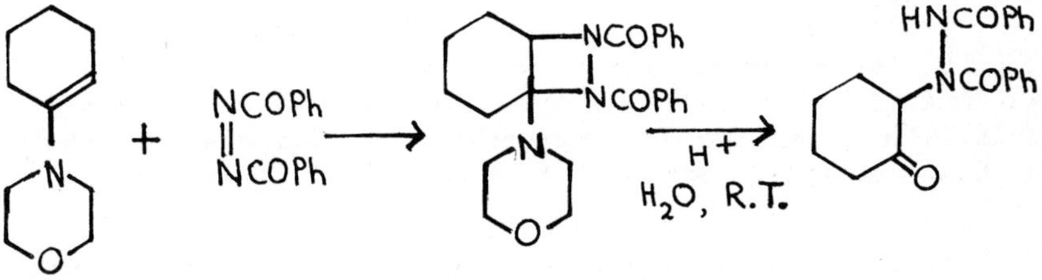

(f) Diazetines (dihydrodiazetes)

Dimethyl 1,2-diazet-3-ine-1,2-carboxylate (1) is
prepared by the photolysis of the 1:1 adduct of
2,5-dimethyl-3,4-diphenylcyclopenta-2,4-dienone and
2,3-dimethoxycarbonyl-2,3-diazabicyclo[2.2.0]hex-5-ene.

It is thermally unstable even at ambient temperatures and isomerises to the ring-opened product (2). Hydrogenation using palladium-carbon catalyst gives the 1,2-diazetidine. The ring system of (1) formally satisfies the Hückel (4π+2) rule, but no pronounced stablity is observed, and the absence of a ring current suggests that it is not aromatic by this criterion (E.E. Nunn and R.N. Warrener, Chem. Comm., 1972, 818).

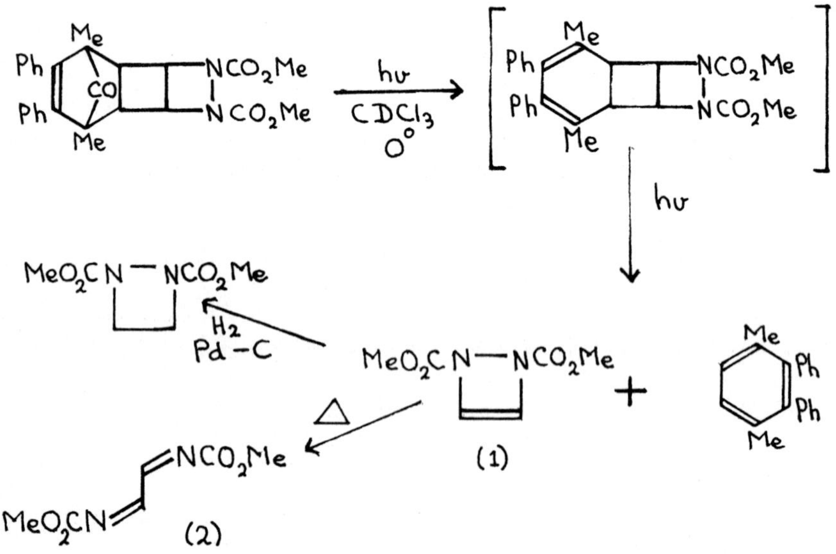

(g) *Rings containing one nitrogen and one oxygen or sulphur atom*

(i) *4H-1,2-Oxazete*

4,4-*tert*-Butyl-1,2-oxazete (2, R = H) generally known as a reactive intermediate is condensed on a cold-finger trap at -196° following the vacuum pyrolysis of the nitrosoalkene (1). On heating it decomposes to give only di-*tert*-butylketone and hydrogen cyanide (K. Wieser and A. Berndt, Angew. Chem., 1975, 87, 73).

$$(t\text{-Bu})_2 C{=}CRNO \xrightarrow[\substack{10 \text{ torr}}]{\substack{220^\circ \\ -4}} \underset{(t\text{-Bu})_2}{\overset{O-N}{\square}}R \xrightarrow{\triangle} (t\text{-Bu})_2 CO \; + \; HCN$$

$$(1) \qquad\qquad\qquad\qquad (2)$$

R = H or Me

Treatment of $(t\text{-Bu})_2C(NO_2)C({=}NOH)Me$ with methanolic sodium methoxide affords 3-methyl-4,4-*tert*-butyl-1,2-oxazete (2 , R = Me).

Reaction of 1,1-di-*tert*-butylallenes with nitrogen tetroxide in ether at 0° allows the isolation of the thermally labile 4H-1,2-oxazete 2-oxides, which on standing are converted into the more stable carbonyl derivatives (*idem.*, *ibid.*, p.72).

$$(t\text{-Bu})_2 C{=}C{=}CHR \xrightarrow[\substack{0^\circ}]{\substack{N_2O_4, \\ Et_2O}} \underset{(t\text{-Bu})_2}{\overset{O-NO}{\square}}CHRNO_2 \longrightarrow \underset{(t\text{-Bu})_2}{\overset{O-NO}{\square}}COR$$

R = H, Br, or Cl

(ii) 2H-1,3-Thiazetes

2,2,6,6-Tetrakis(trifluoromethyl)-6H-1,3,5-oxathiazines obtained from thiocarboxamides and two equivalents of hexafluoroacetone, on thermolysis afford 2H-1,3-thiazetes as the result of a retro-Diels-Alder reaction followed by an electrocyclic ring closure (K. Burger,

J. Albanbauer, and M. Eggersdorfer, Angew. Chem. internat. Edn., 1975, 14, 766).

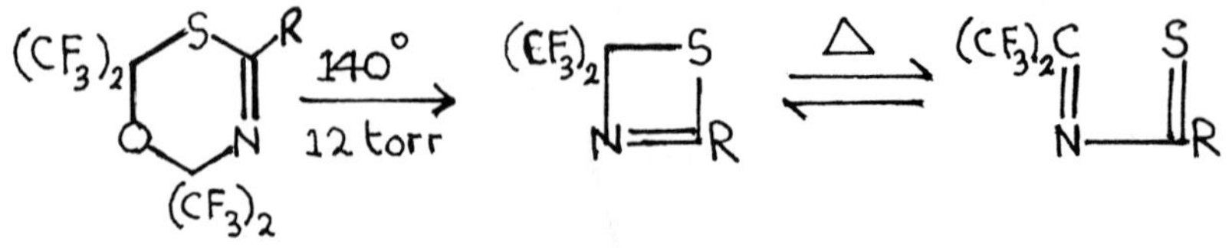

The 2H-1,3-thiazetes exist exclusively in cyclic form at room temperature but are in valence-tautomeric equilibirum with the open chain N-(perfluoroisopropylidene)-thiocarboxamides at elevated temperature, and react with isocyanides and phosphorus pentasulphide to give 5-imino-4,-4-bis(trifluoromethyl)-1,3-thiazol-2-ines and 3,3-bis(tri-fluoromethyl)-3H-1,2,4-dithioazoles, respectively (Burger, Albanbauer, and W. Foag, *ibid.*, p.767).

The irradiation of thioketene in acetonitrile yield N-thioacetylketimines *via* a thiazete intermediate (D.S.L. Blackwell, P. de Mayo, and R. Suau, Tetrahedron Letters, 1974, 91).

Chapter 2

COMPOUNDS CONTAINING A FIVE-MEMBERED RING WITH ONE HETERO
ATOM OF GROUP VI;   OXYGEN

R. LIVINGSTONE

1. *The Furan Group*

Since the publication of Volume 1VA most of the
additional information available is in areas already well
described.   There are a number of new syntheses of known
furan derivatives and extra physical data, especially
relating to ^{1}H and ^{13}C-nmr spectra has been reported
for a number of groups of well known derivatives.   Various
spectral data and other physical properties have been
related to the conformations of some derivatives, including
furan-2-carboxaldehyde.

Hetero[5]annulenes fused to four-membered rings have
become molecules of considerable interest, since the strain
in such systems is increased when compared with the
corresponding benzene analogues and the chemical and
physical properties of the aromatic ring are expected to be
unusual.   Some simple furans have been elaborated into a
variety of useful 'synthons', for example, 4-substituted
5-hydroxy-3-oxocyclopenetene by molecular rearrangement of
2-furyl carbinols.

(a) Furan and its substitution products

(i) Synthetic methods

(1) 2-Allyl- and 2-benzyl-furans may be prepared by
boiling the tosyl derivatives of the allyl or benzyl
alcohol with furan in acetonitrile in the presence of a
catalyst, lithium perchlorate.   Some alcohols can react
directly with furan under acidic conditions (P.H. Boyle,
J.H. Coy, and H.N. Dobbs, J. chem. Soc., Perkin I, 1972,
1617).

$$\text{furan} \xrightarrow[\text{MeCN, LiClO}_4]{\text{PhCH}_2\text{OTos,}} \text{2-benzylfuran (CH}_2\text{Ph)}$$

(2) 2-Alkylfurans are obtained by the iodine induced decomposition of the complex formed by treating lithiated furan with trialkylboranes (I. Akimoto and A. Suzuki, Synth., 1979, 146). Propionate radical adds slowly to furan yielding methyl β-(2-furyl)propionate (20%) (S.E. Schaafsma et. al., Tetrahedron Letters, 1973, 827).

(3) The reaction of diazoaminobenzenes with furan and isopentyl nitrite offers a convenient new method for the arylation of furan. 2-Arylfurans of the type 2-$(RC_6H_4)C_4H_3O$ (R=H, 4-Me, 4-Br, 4-Cl, 4-NO_2, 4-CO_2Me, 4-CO_2H), 2-$(R^1R^2C_6H_3)C_4H_3O$ (R^1R^2=3,4-diCl, R^1R^2=4-Cl-3-CF_3) and 2-(3-pyridyl)furan have been prepared. The diazoamino-benzene decomposes and the resulting aryl radical arylates the furan.

$$\text{furan} \xrightarrow[\substack{C_5H_{11}ONO, \\ 30^\circ, \ 24h}]{\text{ArNHN=NAr,}} \text{2-Ar-furan} \ + \ N_2 \ + \ ArN_2OH$$

Methyl 4-aminobenzoate reacts with furan and isopentyl nitrite to give 2-(4-methoxycarbonylphenyl)furan and 4,4'--di(methoxycarbonyl)diazoaminobenzene. The latter compound is unstable in the reaction medium and with isopentyl nitrile and furan at 30° affords 2(4-methoxycarbonyl-phenyl)furan (L. Fisera, J. Kovác and E. Komanova, Tetrahedron, 1974, 30, 4123; Fisera, Kovác and B. Hasova, Chem. Zvesti, 1976, 30, 480). Reactions between furan and other radical precursors are known and arylation only occurs at position 2, whereas thiophene is arylated at positions 2 and 3 (L. Benati et. al., J. heterocyclic Chem., 1972, 9, 919).

(4) Many naturally occurring furans have a substituent in position 3 and an uncomplicated route to 3-substituted furans involves initially the photochemical addition of a carbonyl compound to furan. This occurs regioselectively to furnish derivatives of 2,7-dioxa-bicyclo[3.2.0]hept-3-ene(oxetanes). The outcome of the photochemical cycloaddition is strongly dependent on the carbonyl compound used; yields vary from acetophenone (1%) to benzaldehyde (35%). The oxetanes are isomerised in the presence of acids, for example, 4-toluenesulphonic acid, in aprotic solvents at room temperature, to 3-furylmethanols (A. Zamojski and T. Koźluk, J. org. Chem., 1977, $\underline{42}$, 1089).

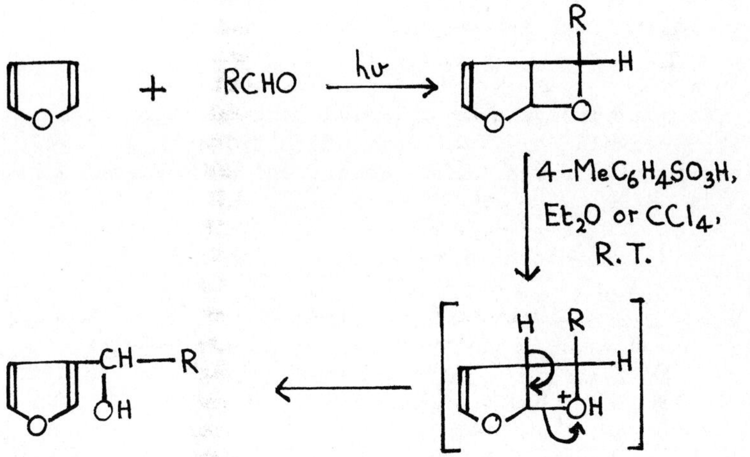

The 3-furylmethanol (R=CO_2Bu) on reduction with lithium tetrahydridoaluminate yields diol (1), which with lead tetraacetate gives furan-3-carboxaldehyde (65%).

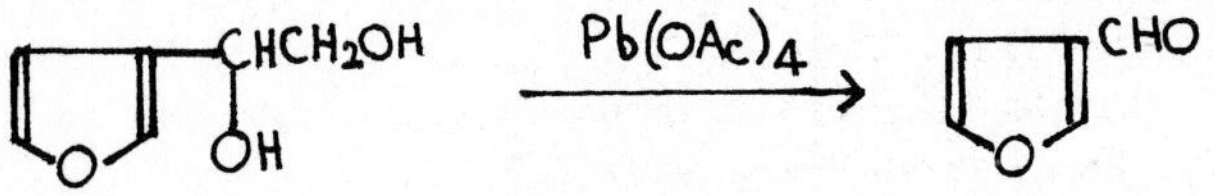

(1)

(5) Furans are prepared by treating the appropriate singlet oxygen adducts (1:2 dioxenes) (2) of dienes with base, followed by acidification (K. Kondo and M. Matsumoto, Chem. Letters, 1974, 701).

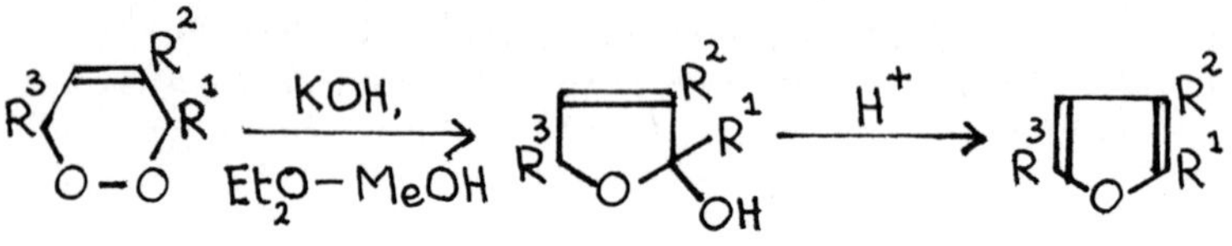

(2)

R^1, R^2, R^3 = H, Me, Ph

(6) The pyrolysis of 1,3-dioxolan-4-yl benzoates yield 1,3-dioxolan-4-ylium ions, which fragment and cyclo-condense to give furans (H.D. Scharf and E. Wolters, Ber., 1978, 111, 639).

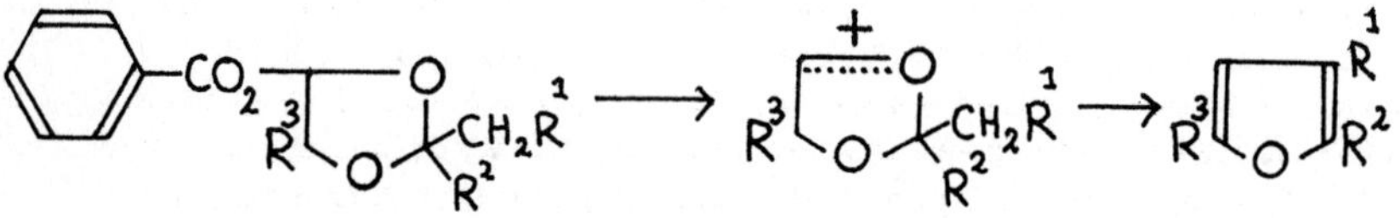

R^1, R^2, R^3 = H, alkyl

Information on "Furans, Synthesis and Application" has been given by A. Williams (Chemical Technology Review, 1973, 18, 303pp.) and on their synthesis by M.E. Garst (Diss. Abs. Int. B, 1974, 35, 1201).

(ii) General properties and reactions

Properties. The ground state aromaticities of furan, thiophene, selenophene, and tellurophene have been compared using different criteria: the nmr dilution shift, the difference in chemical shifts of the β- and α-protons, the effect of a 2-methyl substituent on the ring proton chemical shifts, the diamagnetic susceptibility exaltation, the sum of the bond orders, the Julg parameter, and the mesomeric dipole movement. The agreement among the results obtained by using the different methods is remarkably good. The following order of decreasing aromaticity has been established, benzene>thiophene>selenophene>tellurophene> furan (F. Fringuelli *et. al.*, J. chem. Soc., Perkin II, 1974, 332).

Calculation of molecular orbitals by Hückel's method for furan and some of its derivatives, indicate that C-5 is the most favoured position for electrophilic attack in 2-substituted furans (Y. Rodriguez Gutierrez, C. Aguiar Punal and L.A. Montero Cabrera, Centro, Ser.: Quim. Tecnol. Quim., 1977, 5, 3). A study has been made of the reactivity towards bromination of furan, pyrrole, thiophene, and related 2-carboxylic acids and benzo derivatives using a Hückel MO delocalisation model (C. Decoret and B. Tinland, Austral. J. Chem., 1971, 24, 2679).

^{13}C-nmr spectra of furan (F.J. Weigert and J.D. Roberts, J. Amer. chem. Soc., 1972, 94, 6021), of 2- and 3- monosubstituted furans, including methyl and halogeno, and a comparison with thiophenes and selenophenes (S. Gronowitz, I. Johnson, and A.B. Hornfeldt, Chem. Scr., 1975, 7, 211), and of 2-furyl systems (D.A. Forsyth and G.A. Olah, J. Amer. chem. Soc., 1979, 101, 5309); vapour phase uv spectra of 2-methyl-, 2-vinyl-, and 2-acetyl-furan and furan-2-carboxaldehyde and its 5-substituted derivatives (Montero Cabrera, Rev. Cenic, Ciene. Fis., 1974, 5, 195); ir and Raman spectra of some 2-substituted and 2,5-disubstituted derivatives of furan (J.H.S. Green and D.J. Harrison, Spectrochim. Acta, Part A, 1977, 33A, 843), and ir intensities of ring stretching bands for 2-substituted furans (J.M. Angelelli *et. al.*, Tetrahedron, 1972, 28, 2037) have been reported.

Reactions. The reaction between furans and 2,4--dinitrobenzenediazonium sulphate is very dependent on the reaction conditions. Furan, 2-, and 3-methylfuran in an aqueous acetic acid medium undergo ring opening to give a *N*-(2,4-dinitroanilino)-5-hydroxypyrrol-2(5H)-one, but in acetic anhydride-sodium acetate medium undergo arylation.

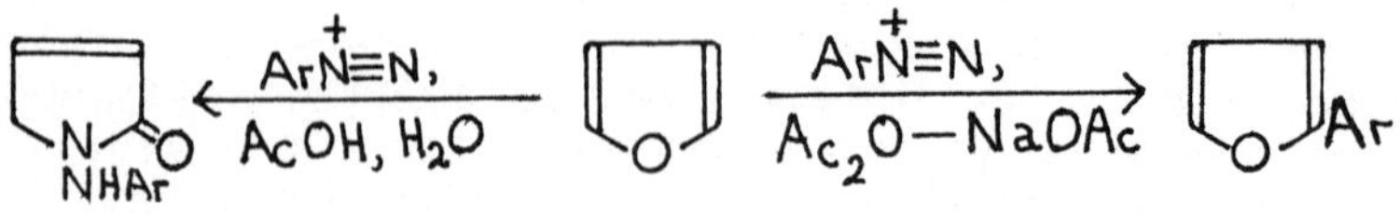

$$Ar = 2,4-(NO_2)_2C_6H_3$$

2,5-Dimethylfuran undergoes coupling at the 3-position in aqueous acetic acid, coupling at the 2-methyl group in acetic anhydride-sodium acetate, and ring opening in ethanol (M.G. Bartle *et. al.*, J. chem. Soc., Perkin I, 1978, 401).

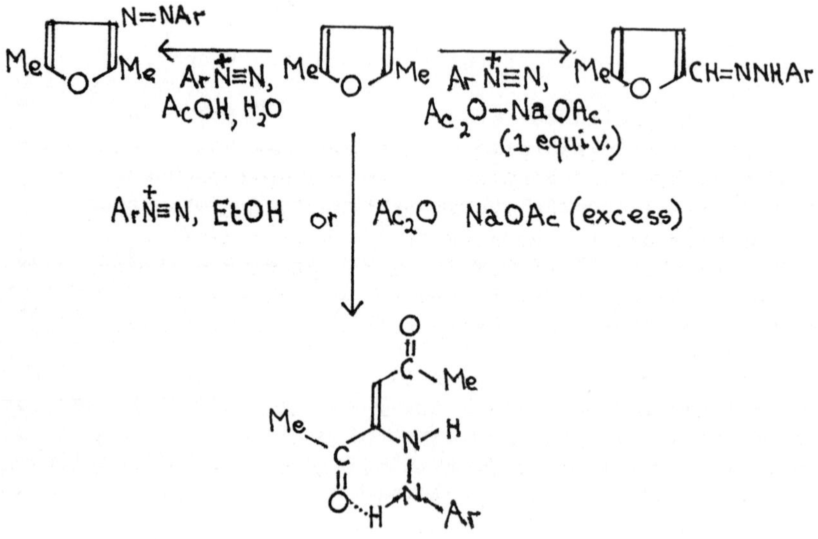

The anodic oxidation of 2,5-dimethylfuran in a methanolic solution of sodium cyanide affords a 2:1 mixture

of *cis*- and *trans*-2-cyano-2,5-dimethyl-5-methoxy-2,5-
-dihydrofurans. Also formed as by-products are small
amounts of *cis*- and *trans*-2,5-dimethoxy-2,5-dimethyl-
-2,5-dihydrofurans and traces of 2-methoxymethyl-5-
-methylfuran and 2,5-bis(methoxymethyl)furan. It is
concluded that the overall reaction proceeds *via* a polar
mechanism involving the initial oxidation of 2,5-dimethyl-
furan to give an intermediate cation radical and is non-
stereospecific (K. Yoshida and T. Fueno, J. org. Chem.,
1971, **36**, 1523).

It has been found that a platinum electrode controls the
stereochemistry of the anodic mixed 1,4-addition of cyano-
and methoxy- groups to 2,5-dimethylfurans; the isomer
ratio of products varies significantly with the type of
aromatic additive (*tert*-butylbenzene, pentamethylbenzene,
anisole, naphthalene, pyridine) as well as the initial
concentration of the substrate (Yoshida, Chem. Comm., 1978,
20).

3-Nitrosobut-3-en-2-one adds to furan and 2,5-
-dimethylfuran to give cycloadducts (1), but the furan
adduct is unstable and completely isomerizes to the cxime (2)
within a few hours at room temperature (T.L. Gilchrist and
T.G. Roberts, *ibid.*, 1978, 847).

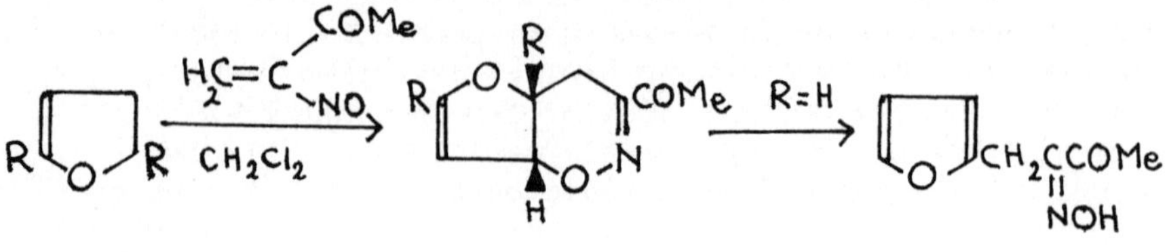

$$(1) \qquad\qquad (2)$$

Adduct (3) is obtained from α-nitrosostyrene and furan (R. Faragher and Gilchrist, J. chem. Soc., Perkin I, 1979, 249).

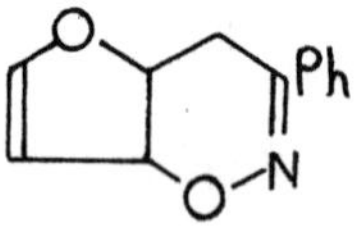

$$(3)$$

Ring expansion occurs when furans are heated with trichlorosilane or hexachlorodisilane to give compound (4) (E.A. Chernyshev *et. al.*, Zh. obshch. Khim., 1978, **48**, 2798).

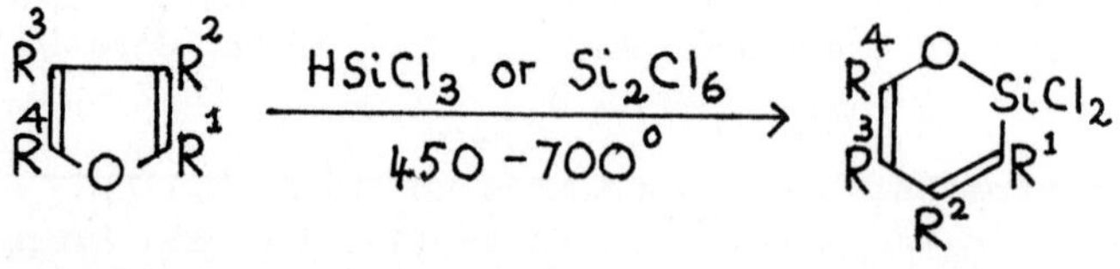

R^1, R^2, R^3 = H, R^4 = H, Me; (4)
R^1 = R^4 = Me, R^2 = R^3 = H;
R^1, R^2 = H, CH=CH–CH=CH, R^3R^4 = CH=CH–CH=CH

Comparative experiments of the cyclo addition of tetracyanoethene oxide to furan and other related hetero- cycles show the following order of reactivity furan> 2-methylthiophene~benzo[b]furan>benzo[b]thiophene>seleno- phene>thiophene (Gronowitz and B. Uppstrom, Acta Chem. Scand., 1975, B29, 441).

1,3-Diethoxycarbonylallene undergoes cycloaddition with furan and 2,5-dimethylfuran with remarkable ease to give an adduct. The reactions are characterised by low reaction temperatures and good yields (A.P. Kozikowski, W.C. Floyd, and M.P. Kuniak, Chem. Comm., 1977, 582).

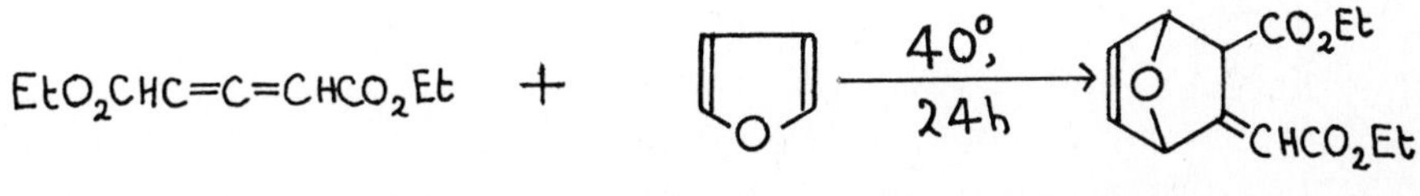

The cycloaddition reaction between aryl- and aryloxy- -furans and acetylenedicarboxylate esters affords the benzene derivatives (6) *via* the adduct (5) (A.F. Oleinik *et. al.*, Khim. Geterotsikl. Soedin., 1979, 17).

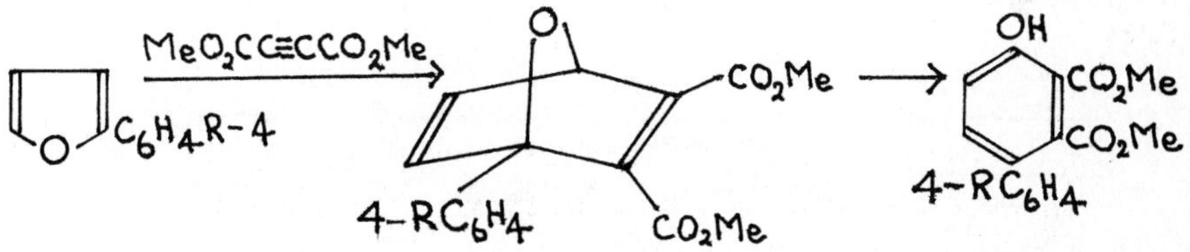

(5)   (6)

The structures of the Diels-Alder reaction products of furan derivatives with maleic anhydride as shown by ^{13}C-nmr data have been published (T. Suzuki *et. al.*, Heterocycl., 1978, 9, 1759). Cycloaddition reaction involving furan and 3-bromo-3-methyl-2-(trimethylsiloxy)-

but-1-ene leads to a bicyclic system (H. Sakurai,
A. Shirahata, and A. Hosomi, Angew. Chem. internat. Edn.,
1979, 18, 163).

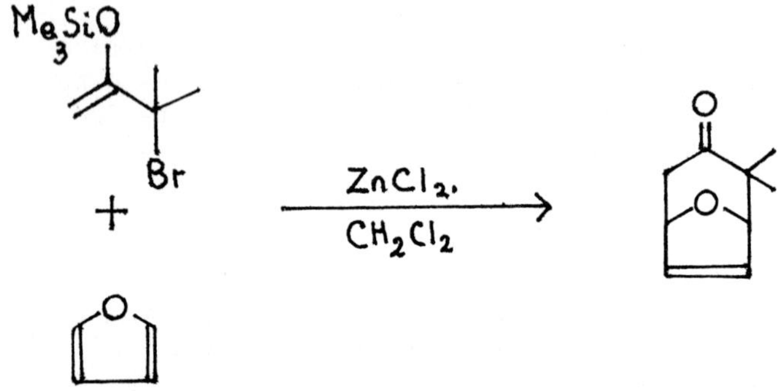

Furan undergoes a [4+4] photochemical cycloaddition
to benzene and the product readily undergoes reversible
thermal, or irreversible photochemical Cope rearrangement
to a 2,3-, 1',2'-isomer (J. Berridge, D. Bryce-Smith, and
A. Gilbert, Chem. Comm., 1974, 964).

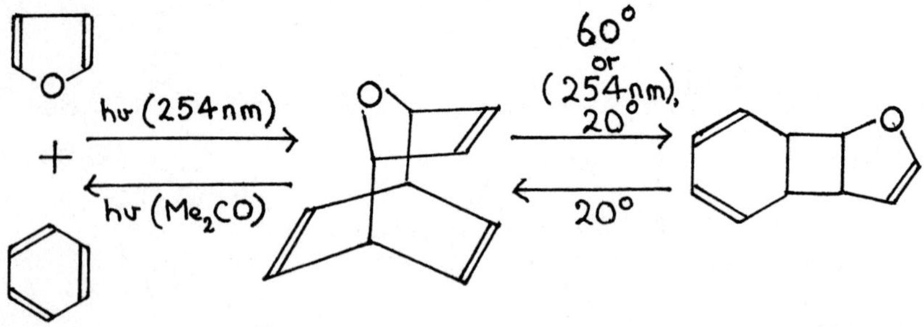

[4+4] Cycloadducts are also obtained by the photoreaction
of 9-cyanoanthracene with furan, 2- and 3-methyl-, and
2,5-dimethyl-furan (K. Mizuno, C. Pac, and Sakurai, J.
chem. Soc., Perkin I, 1974, 2360).

Oxyallyl cations generated from α,α'-dibromoketones

by copper-sodium iodide, undergo cycloaddition reactions
with furan (A.P. Cowling and J. Mann, *ibid.*, p.1564).

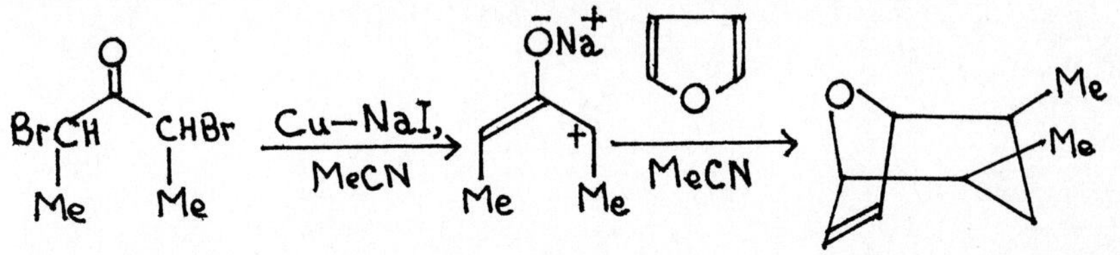

When a mixture of 2,4-dibromo-2,4-dimethylpentan-3-one
($R^1=R^2=Me$) and diiron nonacarbonyl in furan ($R^3=H$) is
heated the cyclic adduct (7) is obtained (R. Noyori *et.
al.*, Tetrahedron Letters, 1973, 1741).

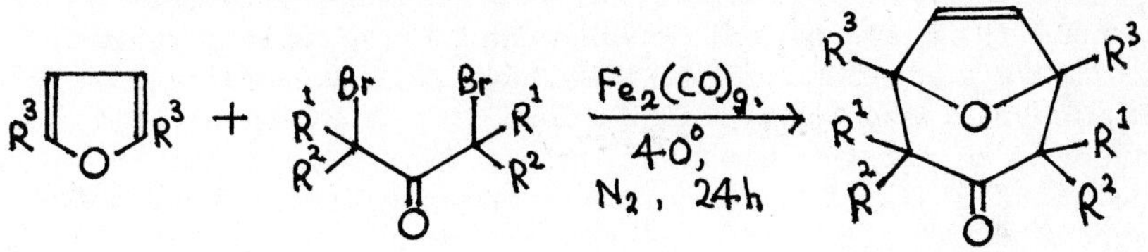

(7)

    Chloromethylenecyclopropane undergoes [2+4] cyclo-
addition with furan (A.T. Bottini and L.J. Cabral,
Tetrahedron, 1978, <u>34</u>, 3187). The Diels-Alder reaction of
cyclopentene-1,3-dione with furan gives adduct (8)
(M. Oda and S. Kawanishi, Jap. Pat. 78, 127,439/1978;
127,495/1978).

(8)

(iii) Furan

Furan and maleic anhydride are simultaneously obtained by heating a mixture of but-1-or-2-ene, steam, and oxygen at 350-450° in the presence of a catalyst, prepared from ammonium paramolybdate (H. Inoue, K. Mizutani, and H. Ito, Jap. Pat. 72 38,423/1972). Using an ammonium paramolybdate-phosphoric acid catalyst, furan is obtained from buta-1,3-diene, steam and air at 400° *(idem, ibid..* 43,545/1972). A number of other preparations of furan by catalytic oxidation of butadiene have been reported (H. Hasegawa, T. Hayashi and A. Takahasi, Chem. Abs., 1972, 76, 99427y; J.B. Bertus, D.C. Tabler, and M.M. Johnson, U.S. Pat. 3,984,056/1975; F.E. Farha, Jr., Johnson, and Tabler, U.S. Pat. 3,894,055; 3,912,763; 3,928,389/1975). The oxidative decarbonylation of furan-2- -carboxaldehyde to furan is effected by mixing it with water and air, and passing the mixture through a lead melt at 730° (L. Meszaros, Acta phys. Chem., 1972, 18, 99). The aldehyde is converted into furan on passing it over nickel-carbon, nickel-chromic oxide, or Raney nickel at 180-300° (V.N. Sokolova *et. al.,* U.S.S.R. 502,890/1976). The mechanism of the vapour-phase catalytic decarbonylation of furan aldehydes (A. Ya Karmil'chik and S. Hillers, Chem. Abs., 1972, 76, 126685q; G. Gardos *et. al.,* Hung. J. Ind. Chem., 1976, 4, 139) and decarbonylation of furan-2- -carboxaldehyde using different catalysts have been investigated (*idem, ibid.,* 1975, 3, 577, 589; 1976, 4, 125; Sokolova *et. al.,* V sb., Ispol'z Pentozansoderzh. Syr'ya, 1976, 72; Hillers *et. al., ibid.,* p.71, p.75; Z. Dudzik, R. Sawala, Z. Szuba, Pol. 95,215/1978). Furan is formed by the gas-solid reaction between furan-2- -carboxaldehyde and soda lime (I. Fujiyoshi and C. Tanaka,

Chem. Abs., 1977, <u>86</u>, 170528w). Furan is obtained by the catalytic gas-phase oxidation of crotonaldehyde with oxygen in the presence of steam using a catalyst composed of oxides of Mo, Bi, and Zn(Sb or Te) (S. Maeda *et. al.*, Japan 72 08,817/1972). The yield of furan from the oxidation of crotonaldehyde in solutions of transition metal salts is increased by the addition of Cu^{2+} or Pd^{2+} salts and by increasing the temperature (L. Elefteriu *et. al.*, Zh. Org. Khim., 1978, <u>14</u>, 2490).

The pK_a value of furan determined spectrophotometrically using Hammet Ho indicator method is -0.11. An extrapolation method is used to overcome the problem of polymerisation of furan in solutions of high acid concentration (S.I. Vohra and N.A. Naqvi, Pak. J. Sci. Ind. Res., 1971, <u>14</u>, 470).

^{1}H- and ^{13}C-nmr spectra of furan and thiophene; C--C, C-H, and H-H internuclear distance ratios have been compared with earlier nmr and microwave results (P. Diehl and H. Boesiger, J. mol. Struct., 1976, <u>33</u>, 249).

MINDO/3 calculations of molecular vibration frequencies are reported for furan (M.J.S. Dewar and G.P. Ford, J. Amer. chem. Soc., 1977, <u>99</u>, 1685).

Both 2- and 3-furyllithium react with butyl borate to give 2- and 3- furanboronic acids (1 and 2), m.pp.110° (decomp.) and 126-128°, respectively (B.P. Roques, D. Florentin, and M. Callanquin, J. heterocyclic Chem., 1975, <u>12</u>, 195). Nmr spectral data is reported for 2- and 3-furanboronic acids and their formyl derivatives (Florentin and Roques, Bull. Soc. chim. Fr., 1976, 1999).

(1.)   (2)

Furan, and 2-methylfuran are metallated in the

α-position on treatment with n-butylcesium and n-
-butylpotassium, prepared by the reaction of cesium and
potassium respectively, with dibutyl mercury (P. Benoit and
N. Collignon, *ibid.*, 1975, 1302). Boiling furan with
n-butyllithium and *NNN'N'*-tetramethylethenediamine in
hexane gives almost a quantitative yield of 2,5-dilithio-
furan (D.J. Chadwick and C. Willbe, J. chem. Soc., Perkin
I, 1977, 887). Also reported is the lithiation of 2- and
3- methylfuran, 2,5-dimethylfuran, and benzo[b]furan.

(iv) Alkyl- and aryl-furans

(1) *Preparation of alkyl- and aryl-furans.* 2-Methyl-
furan may be prepared by the catalytic hydrogenation of
furan-2-carboxaldehyde (V.A. Poteryakhin *et. al.*, Chem. Abs.,
1978, 88, 152327s). 2-Alkylfurans are obtained by
lithiating furan and then treating with the appropriate
trialkylborane and decomposing the resulting complex with
iodine (I. Akimoto and A. Suzuki, Synth., 1979, 146).

$$R^1 \diagdown\!\!\!\square\!\!\!\diagup_O Li \xrightarrow[\text{2. } I_2]{\text{1. } R_3^2 B} R^1 \diagdown\!\!\!\square\!\!\!\diagup_O R^2$$

R^1 = H, Me R^2 = Pr, Bu, CH_2CHMe_2, CHMeEt

2,3-Dimethylfuran may be prepared by the lithium
tetrahydridoaluminate reduction of 3-chloromethyl-2-methyl-
furan obtained from ethyl 2-methyl-3-furoate (A.Craveiro
and E.L. Sanchez, Chem. Abs., 1977, 86, 171158n). 2,4-
And 2,5-dimethylfuran are obtained by method 6 (p. 4).

The reaction of n-butylthiomethylene derivatives of
ketones with dimethylsulphonium methylide gives good yields
of 3- and 3,4-substituted furans (M.E. Garst and
T.A. Spencer, J. Amer. chem. Soc., 1973, 95, 250).

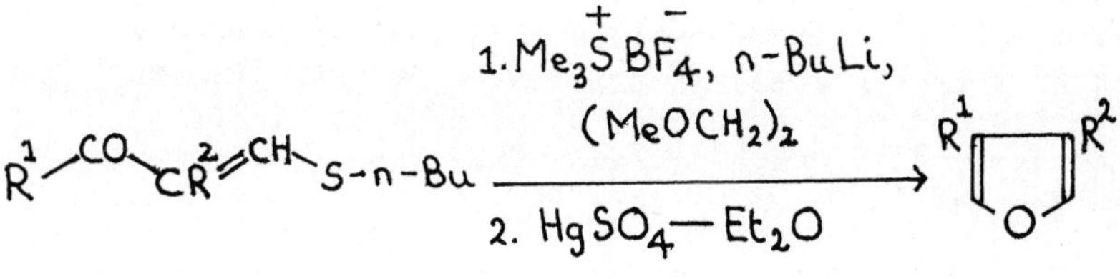

$$R^1\text{-CO-}CR^2\text{=CH-S-}n\text{-Bu} \xrightarrow[\text{2. }HgSO_4\text{—}Et_2O]{\substack{1.\,Me_3\overset{+}{S}\,\overset{-}{BF_4},\ n\text{-BuLi,}\\ (MeOCH_2)_2}} R^1\overset{\displaystyle\frown}{\underset{O}{}}R^2$$

This method permits the facile synthesis of the natural products perillene (4) and dendrolasin (6) from ketones (3) and (5), respectively. Previously they were synthesized by considerably more complicated pathways starting from preformed furans.

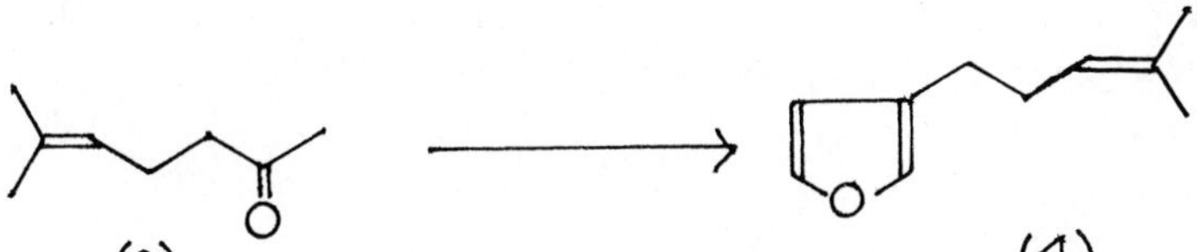

(3) $\longrightarrow$ (4)

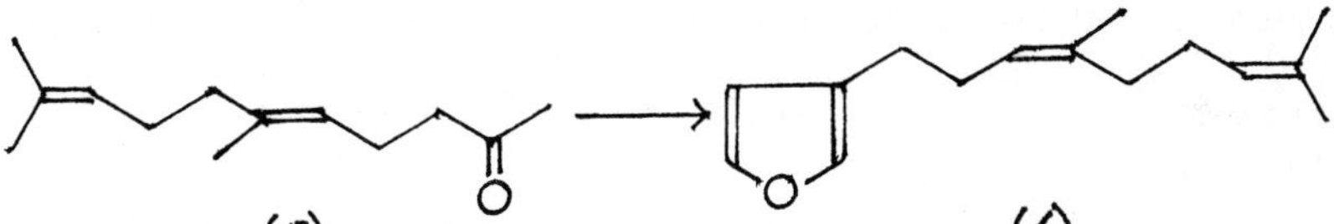

(5) $\longrightarrow$ (6)

Tetraphenylfuran is formed along with other products when 3,4,5,6-tetraphenylpyridazine 1-oxide is irradiated in acetone or methylene chloride (T. Tsuchiya, H. Arai, and H. Igeta, Tetrahedron Letters, 1971, 2579). (Z)-4-Bromo--1,3-diphenylbut-2-en-1-one on boiling in 95% ethanol yields 2,4-diphenylfuran, m.p.110-111° (R. Faragher and T.L. Gilchrist, J. chem. Soc., Perkin I, 1976, 336).

(2) *Properties and reactions of alkyl- and aryl-furans.*

2,3-Dimethylfuran undergoes a Diels-Alder reaction with dimethyl acetylenedicarboxylate to give adduct (1), which on photolysis yields compound (2). On treatment with aqueous methanolic hydrogen chloride (2) affords (3) and (4) (E. Wenkert, A.A. Craveiro, and E.L. Sanchez, Synth. Comm., 1977, 7, 85).

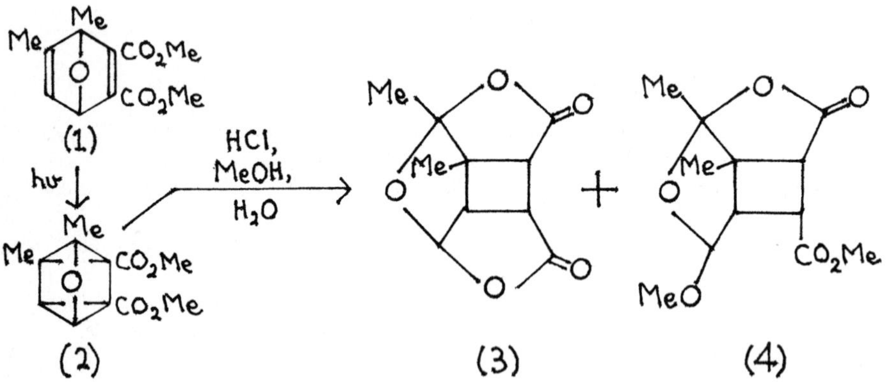

2,5-Di-*tert*-butylfuran on treatment with *tert*-butyl chloride under Friedel-Craft conditions gives 2,3,5-tri--*tert*-butylfuran (85%) (H, Wynberg and U.E. Wiersum, Tetrahedron Letters, 1975, 3619).

Nitrosocarbonyl compounds (5), the intermediates in hydroxamic acid oxidations react, with 2,5-dimethylfuran to give 3-substituted-5-methyl-5-(*cis*-3-oxobutenyl)-1,4,2--dioxazoles (6) (C.J.B. Dobbin *et. al.*, Chem. Comm., 1977, 703).

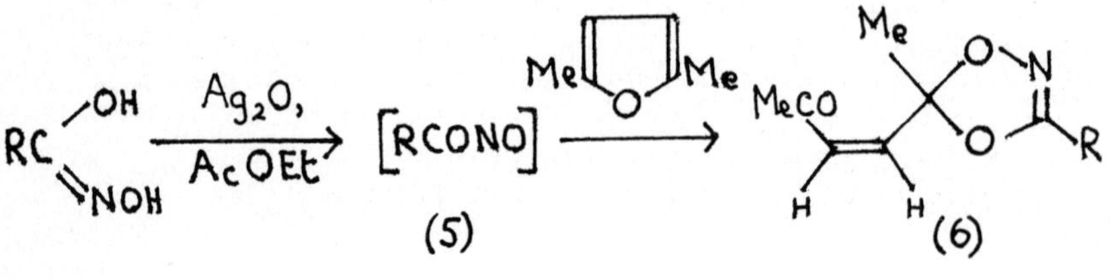

R = Ph, t-Bu, 4-BrC$_6$H$_4$, 4-NO$_2$C$_6$H$_4$

The reaction is reversible and probably occurs by way of an initially formed Diels-Alder adduct. The product regenerates the original Diels-Alder participants on heating.

[13]C nmr spectra of 2,5-diphenylfuran (A. Caspar, S. Altenburger-Combrisson, and F. Gobert, Org. mag. Res., 1978, <u>11</u>, 603); [1]H-nmrs spectra of 2,5-disubstituted furans (S. Andreae, Z. Chem., 1979, <u>19</u>, 110).

(v) Halogenofurans

(1) *Preparation.* The 2- and 3-chlorofurans may be obtained from furan and 3-bromofuran, respectively, by treatment with either ethyllithium or butyllithium followed by treatment with hexachloroethane (S. Gronowitz, A.B. Hornfeldt, and K. Pettersson, Synth. Comm., 1973, <u>3</u>, 213). 3-Iodofuran is prepared from the rection of 3-furyllithium, obtained from the 3-bromo derivative, with iodine (Z.N. Nazarova, B.A. Tertov, and T.G. Stepanosova, U.S.S.R. 427,932/1974).

Nmr spectral data indicates the formation of the covalent adducts, *cis*- and *trans*-2,5-dibromo-2,5-dihydro-furans and *trans*-2,3-dibromo-2,3-dihydrofuran as intermediates during the bromination of furan in carbon disulphide (E. Baciocchi, S. Clementi, and G.V. Sebastiani, Chem. Comm., 1975, 875).

(2) *Properties and reactions.* 3-Bromofuran with phenyllithium gives a mixture of the 2- and the 3-lithio derivatives and 2,3-dibromofuran affords 3-bromo-2--lithiofuran with butyllithium. Treatment of 3-bromofuran with lithium di-isopropylamide, followed by trimethylsilyl chloride yields 3-bromo-2-trimethylsilylfuran. A similar reaction sequence with 2,3-dibromofuran gives 2,3-dibromo--5-trimethylsilylfuran. Attention is drawn to the regio-specific metallation of these compounds by lithium di--isopropylamide and the synthetic potential of such metallated compounds bearing halogen substituents (G.M. Davies and P.S. Davies, Tetrahedron Letters, 1972, 3507).

Grignard reagents derived from 2-bromofuran and bromo derivatives of pyridine, thiophene, and selenophene may be synthesized *via* an exchange reaction between the heterocyclic bromides and isopropylmagnesium chloride (M.B. Mechin *et. al.*, J. organometal. Chem., 1974, <u>67</u>, 327).

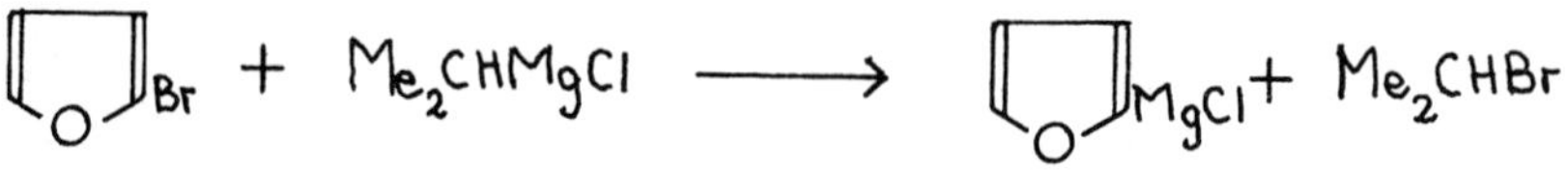

    2-Furylmagnesium bromide and 3-methyl-2-furyl--magnesium bromide are obtained from the corresponding bromide using 90% magnesium-copper alloy. Reaction between the latter Grignard reagent and 1-bromo-3-methylbut-2-ene gives rosefuran, 3-methyl-2-(3-methylbut-2-enyl)furan (1) (A. Takeda, K. Shinhama, and S. Tsuboi, Bull. chem. Soc. Japan, 1977, <u>50</u>, 1903). Other syntheses of rosefuran and related compounds are reported (N.D. Ly and M. Schlosser, Helv., 1977, <u>60</u>, 2085; S.A. Firmenich, Japan, Pat. 79 12,368/1979).

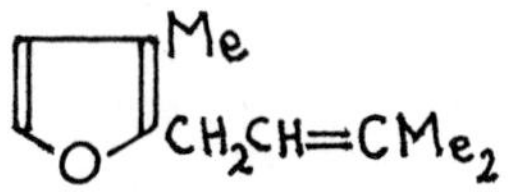

(1)

2,5-Diaryl-3-bromofurans react with butyllithium to give the corresponding 2,5-diarylfurans and the allenes (2) and/or the acetylenes (3). If the reaction is carried out in hexane at 65-70° the allenes and acetylenes can be obtained in good yields (T.L. Gilchrist and D.P.J. Pearson, J. chem. Soc., Perkin I, 1976, 989).

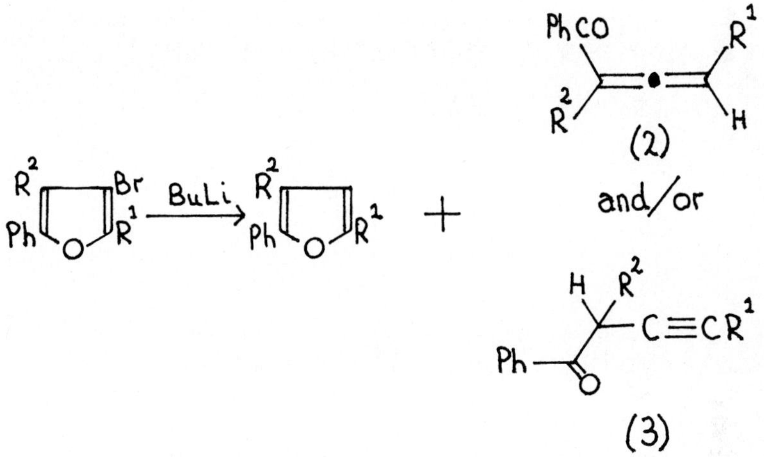

If the reaction mixture from 3-bromo-2,4,5-triphenylfuran is quenched with acetic acid a 1:1 mixture of the allene and the isomeric acetylene is otained, but attempts to separate them results in isomerisation of the acetylene to the allene.

Photooxidation of 3-chloro- and 3-bromo-furan in methanol affords a mixture of 3-halogeno-3-formyl-acrylic acid and the pseudo ester, 3-halogeno-5-methoxy-2-oxodihydrofuran (4). On heating the acrylic acid derivative is converted into the pseudo ester. It is believed that the initially formed cyclic peroxide (5) undergoes a prototropic rearrangement to yield the acrylic acid which is then esterified (F. Farina *et. al.*, Chem. Abs., 1978, **89**, 197237z).

(4)

R = Cl or Br

(5)

It has been shown that the first step in the electrochemical reduction of 2-halogeno-5-nitrofurans in dimethylformamide is dehalogenation (I.M. Sosonkin *et. al.*, Khim. Geterotsikl. Soedin., 1977, 23). The anion-radical mechanism of the exchange of bromine for iodine in 2-bromofuran derivatives has been studied (V.N. Novikov, *ibid.*, 1976, 1601). The reactivity differences for the reaction between 5-halogeno-2-nitrofurans (halogeno=Cl,Br,I) and dimethylamine correspond to differences in the ir, uv, and nmr spectra of the compounds (Novikov and S.V. Borodaev, *ibid.*, p.1316).
The Diels-Alder addition reaction between 2-bromofuran and dimethyl acetylenedicarboxylate has been decribed (H. Brunner and S. Loskot, J. organometal. Chem., 1973, 61, 401).

For the ^{13}C, ^{1}H nmr spectra of 2- and 3-halogeno-furans and other mono-substituted furans see Gronowitz, I. Johnson, and Hornfeldt, Chem. Scr., 1975, 7, 211. The ir spectra of some related derivatives are reported (M. Senechal and P. Saumagne, Compt. rend., 1973, 276B, 79). Bond lengths determined for C-halogen in 2-chloro- and 2-bromo-furan by electron diffraction methods are 1.707(6) and 1.840(8) Å respectively (G.A. Shcherbak *et. al.*, Zh. struckt. Khim., 1979, 20, 532). The C-Br bond length in 3-bromofuran is 1.853(7) $\overset{o}{A}$ and the C=C-Br angle is 125.4° (*idem. ibid.*, p.530).

(vi) Furansulphonic acids

The kinetics of the reaction between furan-2- -sulphonyl chloride and some 4-substituted anilines and between furan-3-sulphonyl chloride in methanol solution give, on comparison with related compounds the following reactivity order: benzene>thiophene-3->furan-3->furan- -2≏thiophene-2-sulphonyl chloride (A. Arcoria *et. al.*, J. org. Chem., 1974, **39**, 3595). Furan-2-sulphonyl chloride, b.p.74-76°/6 mm, is prepared by treating sodium furan-2-sulphonate with phosphoryl chloride. Furan-3- -sulphonyl chloride, b.p.54-56°/0.2 mm is obtained from ammonium furan-3-sulphonate and phosphoryl chloride. Furan-2-sulphonic acid on irradiating with an ionizing ray in an organic solvent or water gives a semiconductor (Y. Nagai *et. al.*, Jap. Pat. 73 39,921/1973).

(vii) Nitrofurans

The 2- and 3-nitrofuran may be prepared by treating 2- and 3-lithiated furan with *trans*-chlorovinyliodoso dichloride and reacting the resulting di(furan-2- and -3-yl)iodonium chlorides with sodium nitrite (S. Gronowitz and B. Holm, Synth. Comm., 1974, **4**, 63).

2-Bromo-5-nitrofuran on nitration with a mixture of nitric acid and sulphuric acids gives 2-bromo-3,5-dinitro- furan, m.p.78-80°. The yield is low presumably due to oxidation of the furan nucleus by the nitrating medium. Treatment of 2-bromo-3,5-dinitrofuran in acetic acid with potassium iodide affords 2-iodo-3,5-dinitrofuran m.p.111- 112°.

Nitration of 5-methyl-2-furoic acid and 5-methyl-2-
-nitrofuran occurs at the position adjacent to the methyl
substituent (M.E. Sitzmann, J. heterocyclic. Chem., 1979,
16, 477).

   In the radical methylation of 2-nitrofuran using
dimethyl sulphoxide and hydrogen peroxide, an unusual
displacement of the nitro group by a methyl group occurs
(U. Rudgvist and K. Torssell, Acta Chem. Scand., 1971, 25,
2183). 2-Nitrofuran undergoes smooth photosubstitution of
the nitro group by a nucleophile, for example, CN^-, CNO^-,
OMe^-, H_2O (M.B. Groen and E. Havinga, Mol. Photochem.,
1974, 6, 9) and along with its 5-methyl and 5-formyl
derivatives it exhibits phosphorescence in ethanol. The
excited state involved in its photolysis is a triplet
(W. Kemula and J. Zawadowska, Bull. Acad. Pol. Sci., Ser.
Sci. Chim., 1976, 24, 155). The esr spectra of anion
radicals of some 5-nitrofuran derivatives in water-dimethyl-
formamide have been studied (R. Gavars *et. al.*, Khim
Geterotsikl. Soedin., 1972, 435) and polargraphic and esr
data are reported for 2-acyl-5-nitrofurans (L. Baumane,
Chem. Abs., 1978, 89, 75149b).

(viii) Amino- and cyano-furans

   A number of 2-amino-3-cyanofurans have been prepared
by the interaction of acyloins with malononitrile in aqueous
base.

However, it is reported that the reaction between hydroxy-
propan-2-one and malononitrile under basic conditions does
not give 2-amino-3-cyano-4-methylfuran but 2,4-diamino-3,5-
-dicyano-3a,6-dimethyl-3a,4,7,7a-tetrahydro-*endo*-4,7-
-epoxybenzofuran (J.L. Isidor, M.S. Brookhart, and
R.L. McKee, J. org. Chem., 1973, 38, 612).

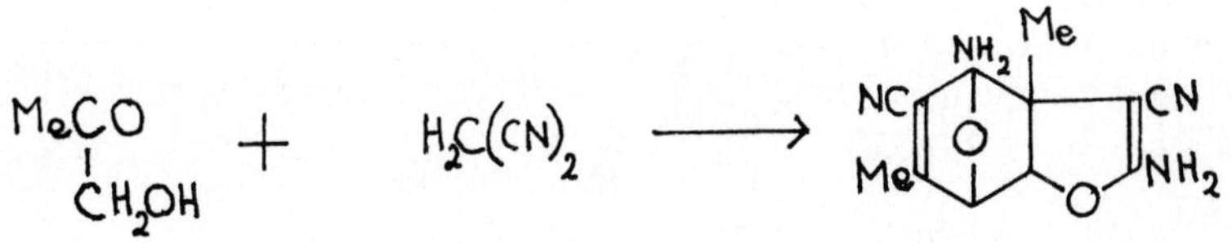

$MeCO$ — CH_2OH + $H_2C(CN)_2$ ⟶ [structure]

N-3-Furylbenzamide (1) gives a conventional Diels-Alder adduct (2) with maleic anhydride but with dimethyl maleate and methyl acrylate further reactions take place at the enamide grouping of the initially formed adducts (J.N. Bridson, Canad. J. Chem., 1979, **57**, 314).

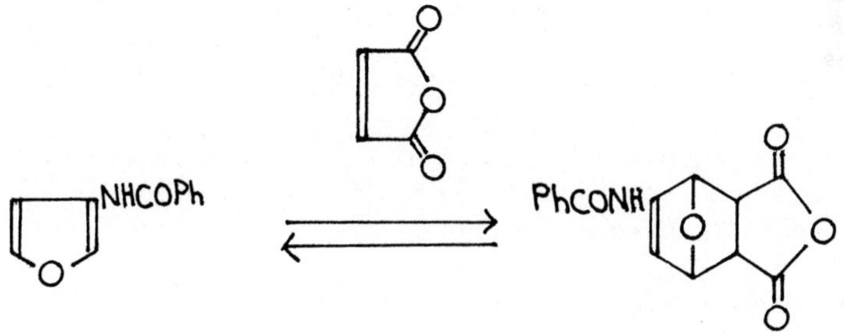

(1)   (2)

(ix) Oxodihydrofurans; butenolides (hydroxyfurans)

3,5,5-Trisubstituted 2-oxodihydrofurans [2(5H)-
-furanones, 2,5-dihydrofuran-2-ones] are prepared in one step by cyclisation on further heating of the aldehydo ester obtained, by treating potassium phenylacetate with an α-bromo substituted aldehyde in the presence of 18-crown-6 ether (A. Padwa and D. Dehm, J. org. Chem., 1975, **40**, 3139). The intermediate may be isolated and cyclised.

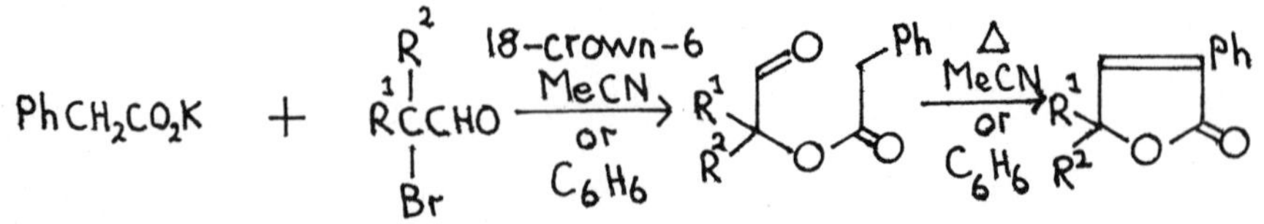

3-Halogeno-5-methoxy-2-oxodihydrofurans are obtained from 3-halogenofurans (p. 19).

The acetylenic oxazole (1) on heating gives a furan by intramolecular cycloaddition of the acetylene to the ring, followed by loss of acetonitrile. Acid hydrolysis converts the furan into the butenolide (2) (P.A. Jacobi and T. Craig, J. Amer. chem. Soc., 1978, 100, 7748).

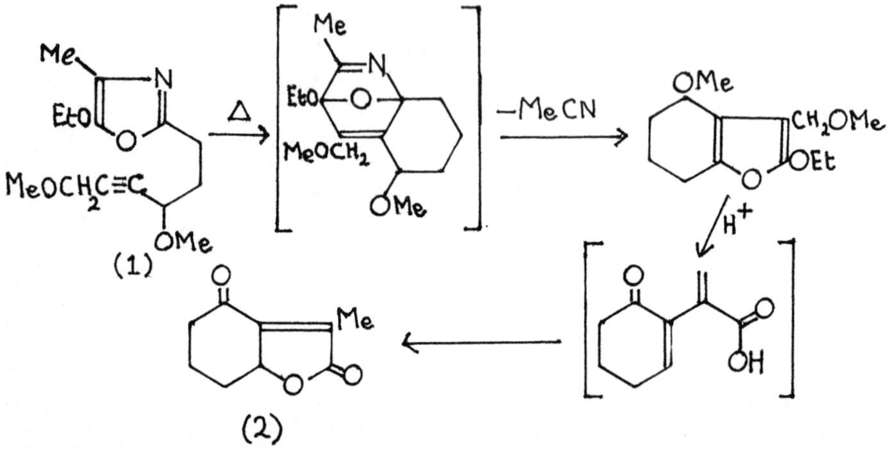

Narthogenin (3) the aglycon of nartheside obtained from *Narthecium ossifragum* has been synthesized by the bromination of 4-methoxy-2-oxo-dihydrofuran followed by hydrolysis (T. Reffstrup and P.M. Boll, Phytochem., 1979, 18, 325).

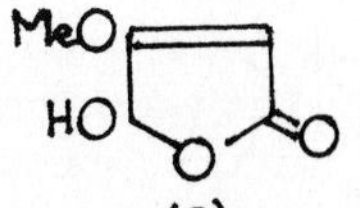

(3)

The absolute configuration of carlosic acid
(4- butyryl-3-hydroxy-5-oxodihydrofuran-2-acetic acid) from
Penicillin charlesii NRRL 1887 has been established as *S*
(J.L. Bloomer and F.E. Kappler, J. chem. Soc., Perkin I,
1976, 1485).

3-Oxodihydrofurans [3(2H)-furanones, 2,3-dihydrofuran-
-3-ones] have been used as building blocks for the
synthesis of muscarins and because of their pleasant and
varied odour are valued by perfumers. A new and efficient
route to these heterocycles uses 2-dimethylamino-4-methyl-
ene-1,3-dioxolanes, obtained by the debromination of $\alpha,\alpha'-$
-dibromo ketones with a zinc-copper couple in dimethylformamide and dimethylacetamide, as the key intermediates.
On heating in a 1M solution in dimethylformamide they form
the α-amino ethers (4). As β-ketoamines which are rendered
even more labile by an α-ether grouping, they rapidly lose
dimethylamine, affording 3-oxodihydrofurans (B.K. Carpenter
et. al., J. Amer. chem. Soc., 1972, 94, 6213).

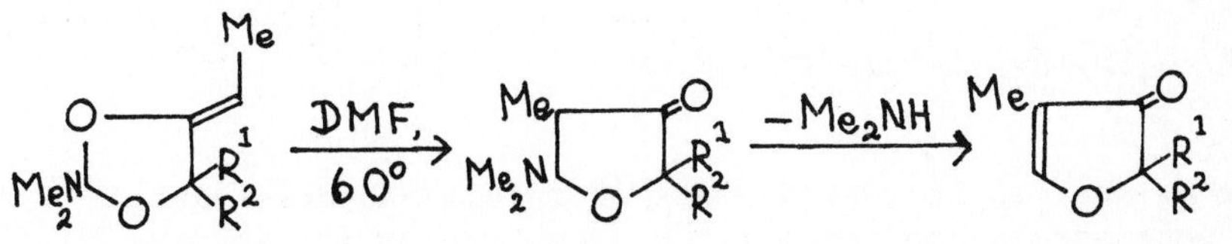

(4)

$R^1 = R^2 = Me;$ $R^1 = Me, R^2 = H$

2-Benzylidene-4-ethoxycarbonyl-5-methyl-3-oxodi-
hydrofurans are obtained by Knoevenagel condensation
between 4-ethoxycarbonyl-5-methyl-3-oxodihydrofuran and
substituted benzaldehydes in the presence of toluene-4-
-sulphonic acid (I. da. R. Pitta, M.do S. Lacerda, and C.L.
Duc, J. heterocyclic Chem., 1979, 16, 821).

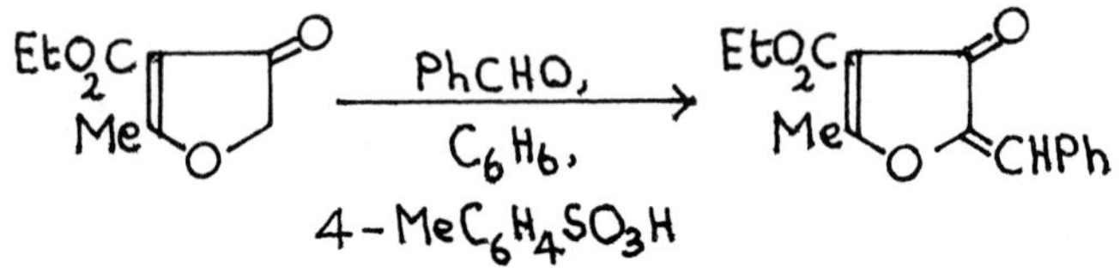

2,3-Dioxodihydrofurans (2,3-dihydrofuran-2,3-diones)
are prepared in good yield by the addition of oxalyl
chloride to a solution of alkenyloxysilanes.

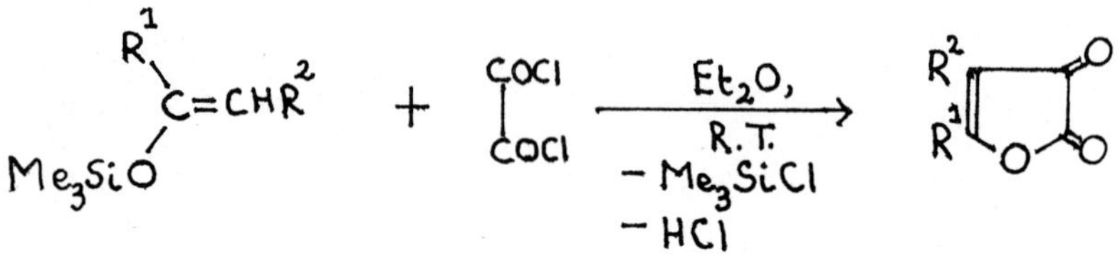

R^1 = Ph, Ar, t-Bu; R^2 = H, Me

They are topological isomers of maleic anhydride and
decompose on heating *via* the α-oxoketenes (5) to give 3H-
-pyran-2,4-diones (S. Murai, K. Hasegawa, and N. Sonoda,
Angew. Chem. internat. Edn., 1975, 14, 636). The
α-oxoketenes may be trapped with chloral.

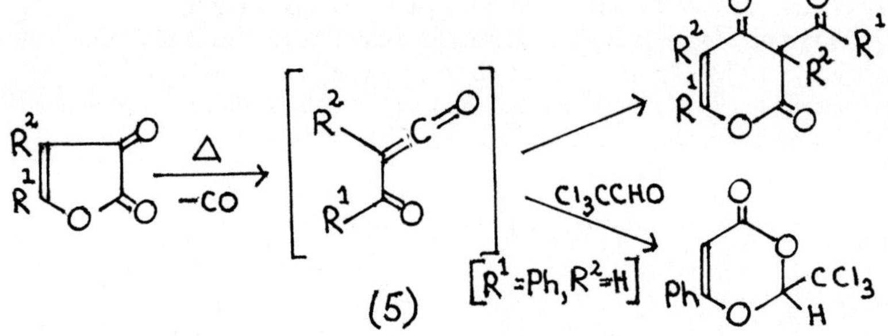

(5)

$[R^1=Ph, R^2=H]$

(x) Furfuryl alcohol and related compounds

Furfuryl alcohol may be obtained by the reduction of furan-2-carboxaldehyde with dibutylstannane (R. Knocke and W.P. Neumann, Ann., 1974, 1486) or thiourea dioxide in alkaline-ethanol [S.-L. Huang and T.-Y Chen, J. Chin. chem. Soc. (Taipei), 1974, 21, 235].

2-Furyl carbinols, obtained by the reaction between furan-2-carboxaldehyde and the appropriate Grignard reagent, undergo a molecular rearrangement, on acid-catalyzed hydrolysis in acetone-water giving directly 4-substituted 5-hydroxy-3-oxocyclopentene. The rearrangement proceeds stereospecifically. The 3-oxocyclo-pentene molecule and its 5-hydroxy derivatives are present in several biologically active natural products as major structural features (G. Piancatelli, A. Scettri, and S.Barbadoro, Tetrahedron Letters, 1976, 3555).

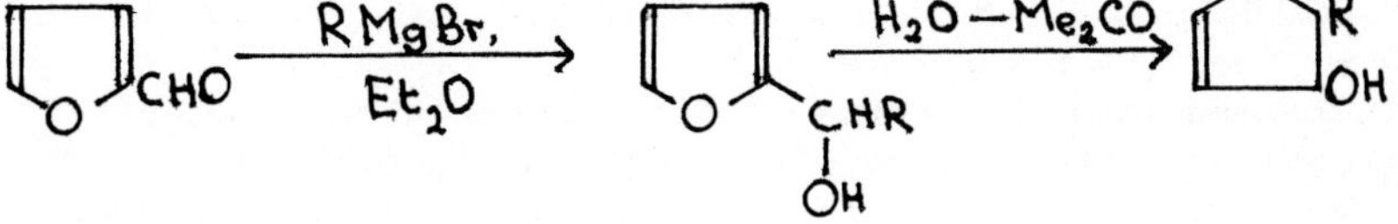

$R = Ph, Me, Me(CH_2)_4CH_2$

tert-Butyl 8-(2-furyl)-8-hydroxyoctanoate is prepared by the reaction of *tert*-butyl 8-formyloctanoate with 2-furyllithium (Piancatelli and Scettri *ibid.*, 1977, 1131).

tert-Butyl 3-hydroxy-5-oxo-1-cyclopenteneheptanoate and its 3-deoxy derivative, important postaglandin intermediates can be obtained from *tert*-butyl 8-(2-furyl)-8-hydroxy-octanoate.

(xi) Aldehydes

   (1) Preparations.   Information is available on the preparation of furan-2-carboxaldehyde (furfural) from various sources including: optimum conditions of corncob hydrolysis for the extraction of furan-2-carboxaldehyde [A.M. Maur, V.P. Repka, and V.G. Panasuyk, Khim. Teknol. (Kharkov), 1971, 24, 72]; the production of furan-2- -carboxaldehyde and acetic acid from cellulosic materials such as corncobs, bagasse, oat husks by hydrolysis (W. Jaeggle, Chem. Age India, 1976, 27, 521); reviews on aspects of furan-2-carboxaldehyde manufacture (J.P. Gupta, Indian Sugar, 1973, 22, 915; P.K.N. Panicker, Chem. Age India, 1974, 25, 793; 1975, 26, 101). Furan-2- and -3-carboxaldehyde are prepared by oxidation of penta-1,3- -diene and 2-methylbuta-1,3-diene, respectively, with oxygen using a Mo-Cu-As-Te-SiC catalyst (T. Vrbaski and T.D. Sheehan, U.S. Pat. 3,630,964/1971).

   5-Arylfuran-2-carboxaldehydes and related carbonyl compounds including carboxylates are obtained by the Meerwein condensation of the appropriate carbonyl compound with either a 4-nitro- or 2,4-dinitro-benzene diazonium salt (S. Farinas, C.R. Rodriguez, and I. Ramso Raimundo, Chem. Abs., 1979, 91, 175100p).

$$\text{\includegraphics{furan}} \quad \xrightarrow[\substack{4-NO_2C_6H_4N_2^+ \\ \text{or} \\ 2,4-(NO)_2C_6H_3N_2^+}]{CuCl_2, \ Me_2CO,} \quad \text{\includegraphics{nitrophenylfuran}}$$

R^1 = H, Me, Et, Pr R^2 = H, NO$_2$

Furan-3-carboxaldehyde is prepared by the oxidation of 3-hydroxymethylfuran with pyridinium chlorochromate (W.C. Still, J. Amer. chem. Soc., 1978, 100, 1481).

(2) Properties and reactions.   *(a) Furan-2-carbox-aldehyde.*   The Grignard reaction between furan-2-carbox-aldehyde and alkyl- or phenyl-magnesium bromide gives exclusively a 1,2-addition product, whereas the corresponding dialkyl- or diphenyl- cadium affords both 1,2- and 1,4-addition products (M. Gocmen, G. Soussan, and P. Freon, Bull. Soc. chim. Fr., 1973, 1310).   The reaction between the aldehyde and benzylmagnesium chloride results in 1,2-, 1,4-, and abnormal additions.   When the reagent is in great excess 1,4-addition predominates and abnormal addition is favoured by a low ratio of reagent to substrate.   Two other compounds also isolated are 3-benzylfuran-2-carboxaldehyde and 2,5-dihydrofuran-2-one, formed by air or oxygen oxidation of 1,4-addition product (R. Sjoholm, Acta Chem. Scand., 1978, B32, 105).

Furan-2-carboxaldehyde reacts with bromine in dichloroethane in the presence of dihydroquinone to give 5-bromofuran-2-carboxaldehyde.   Treatment of the bromo derivative with lithium chloride in dimethylformamide affords 5-chlorofuran-2-carboxaldehyde (74%), and potassium iodide in acetic acid yields 5-iodofuran-2-carboxaldehyde (80%).   All the above derivatives may be interconverted by related methods (R. Mocelo and V. Pustovarov, Chem. Abs., 1977, 86, 72309f).

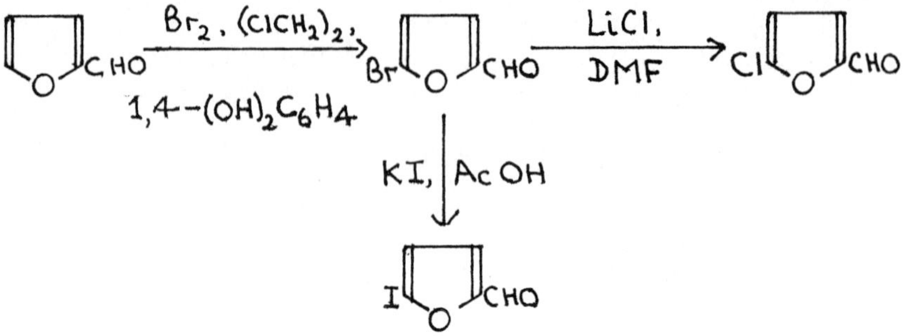

The bromination of furan-2-carboxaldehyde in aqueous solution using sodium bromate-hydrogen bromide results in oxidation and formation of *cis*- and *trans*-isomers of compounds (1) and (2) (H. Greuter and T. Winkler, Helv., 1978, **61**, 3103).

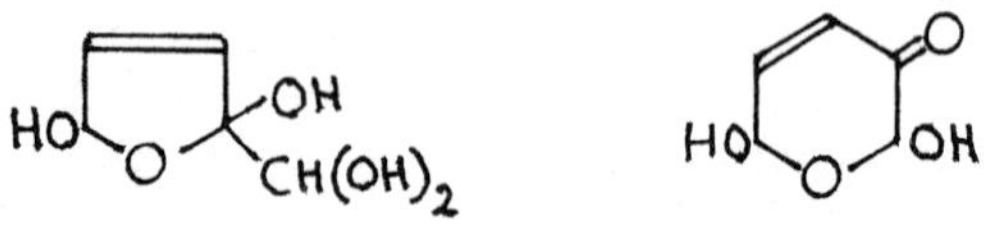

(1)   (2)

Nitration of furan-2-carboxaldehyde with a mixture of fuming nitric acid and acetic anhydride at -40 to -20° yields 5-nitrofuran-2-carboxaldehyde; results are better at this temperature than at 0° (K. Venters and M. Trusule, Chem. Abs., 1977, **87**, 68045v). 5-Nitrofuran-2-carboxalde- hyde diacetate is obtained in 80% yield by adding furan-2- -carboxaldehyde and a nitric-sulphuric acid mixture simultaneously to acetic anhydride at -5-0° (S. Hillers *et. al., ibid.,* 1978, **88**, 89510d). Nitration of furan-2- -carboxaldehyde followed by conversion to the nitrofurfuryl bromide and subsequent treatment with trialkyl phosphite affords dialkyl 5-nitrofurfurylphosphonates (H. Seeboth and

S. Andrae, Z. Chem., 1976, 16, 399).

The reaction of furan-2-carboxaldeyde with acetylene at $\leqslant 0°$ in the presence of potassium hydroxide gives ethynyl furyl carbinol (L.P. Kirillova, A.V. Rechkina, and L.I. Vereshchagin, Zh. Org. Khim., 1971, 7, 2469); with acetaldehyde in 0.1M sodium hydroxide at $-8°$, furyl-acrolein (65%) (H. Thielemann, Z. Chem., 1974, 14, 436); and with acetic anhydride in the presence of methane-sulphonic acid, phosphoric acid, or sulphuric acid, furylidene diacetate (F. Freeman and E.M. Karchefski, J. chem. Eng. Data, 1977, 22, 355). The Claisen-Schmidt condensation of furan-2-carboxaldehyde with acetone in aqueous sodium hydroxide, ethanolic sodium hydroxide, or sodium hydroxide-phenol to give (3) and (4) involves the formation of intermediate (5) (D.A. Isacescu and F. Avramescu, Chem. Abs., 1978, 89, 146061e). Presentation of experimental and literature data on the condensation of furan-2-carboxaldehyde with acetone (idem, ibid., p.661). Furan-2-carboxaldehyde adds smoothly to acrylonitriles to give furyloxobutyronitriles (H. Stetter and H. Kuhlmann, Tetrahedron, 1977, 33, 353).

CH=CHAc

(3)

CH=CHCOCH=CH

(4)

CH(OR)O⁻ Na⁺   R = H, Et, Ph

(5)

The cycloaddition reaction between furan-2-carbox-aldehyde and dichloroketene results only in the isolation of the decarboxylated product, 1,1-dichloro-2-(2-furyl)ethene (H.O. Krabbenhoft, J. org. Chem., 1978, 43, 1305).

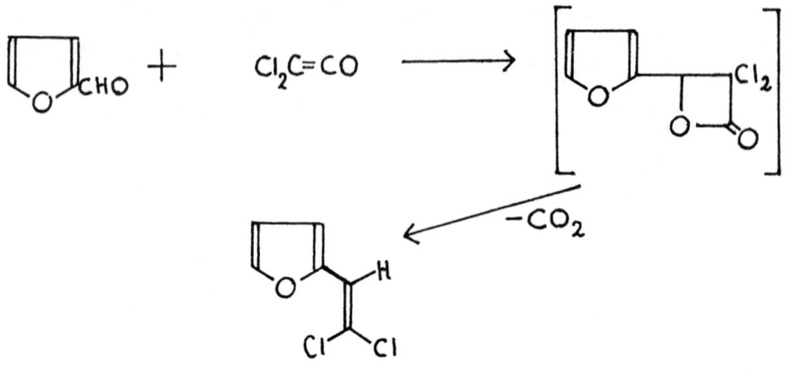

Oxidation of furan-2-carboxaldehyde with sodium hypochlorite followed by acidification affords 2-furoic acid (B.P. Pant and P.K. Ramachandra, Indian J. Pharm., 1977, 39, 117) and vapour-phase ammoxidation yields 2-furonitrile (P. Singh, Y. Miwa, and J. Okada, Chem. pharm. Bull., 1978, 26, 2838).

Furan-2-carboxaldehyde tosylhydrazone, m.p.125-126°. Pyrolysis of diazo(2-furyl)methane and its substituted analogues, as generated from the corresponding tosyl-hydrazone sodium salts give γ, δ-acetylenic α,β-olefinic carbonyl products resulting from opening of the furan ring (R.V. Hoffman, G.G. Orphanides, and H. Shechtes. J. Amer. chem. Soc., 1978, 100, 7927).

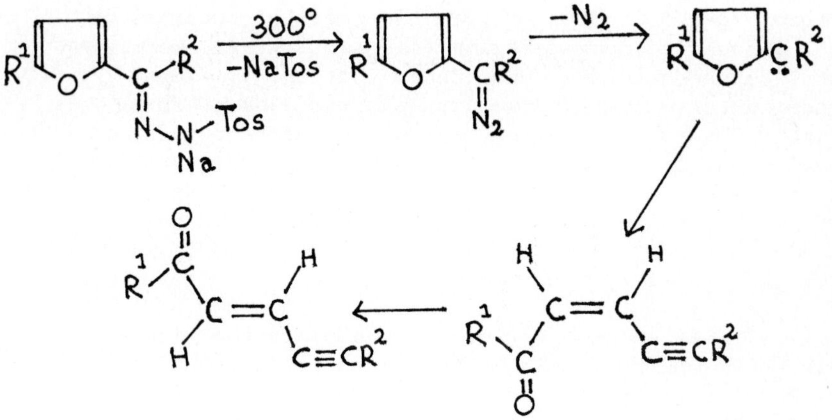

The question of the conformational preference of furan-2-carboxaldehyde has been studied using a wide variety of methods. Unfortunately much confusion has arisen because of the apparently contradictory results obtained by different techniques (C.L. Cheng *et. al.*, J. chem. Soc. Perkin II, 1975, 744). It is now recognised, however, that the position of equilibrium between alternative rotational isomers of a particular molecule is markedly medium-dependent. For example, in the case of furan-2-carboxaldehyde, the *trans*-form is preponderant in the vapour phase, but the *cis*-isomer is undoubtedly the more abundant in dipolar solvents.

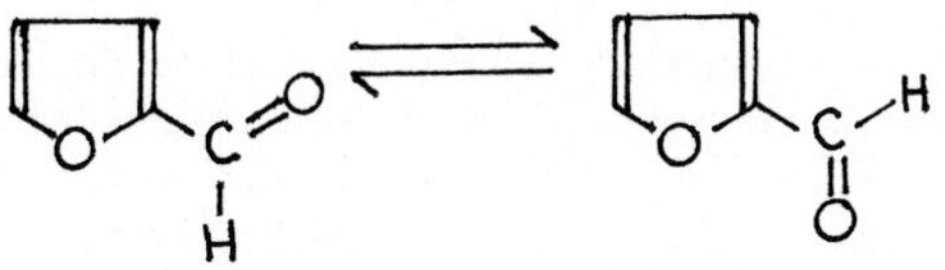

trans-form   *cis*-form

^{13}C-nmr spectral data below $-60°$ of furan-2-
-carboxaldehyde indicates that the ratio of rotamers is
5:1. Shielding considerations suggest, in agreement with
but independently of ^{1}H-nmr data, that the *cis*-form
predominates. The 220 MHz nmr spectral data of furan-2-
-carboxaldehyde and 2-acetylfuran, including benzene
solvent shifts, induced chemical shifts by the lanthanide
shift reagent Eu(FOD)$_3$, and the Nuclear Overhauser
effect show that the furan derivatives exist in the ratio
ca. 1:1.05∿1:1.18, *cis:trans* (S. Nagata *et. al.*, Tetrahedron,
1973, 29, 2545).

^{13}C-nmr spectra of furan-2-carboxaldeyhde (also
related thiophenes, selenophenes, and tellurophenes)
(F. Fringuelli *et. al.*, Acta Chem. Scand., Ser. B, 1974,
28, 175); of furan-2-carboxaldehyde (S. Gronowitz,
I. Johnson, A.B. Hornfeldt, Chem. Scr., 1975, 7, 211;
M.T.W. Hearn, Austral. J. Chem. 1976, 29, 107); of furan-
2-carboxaldehyde and 2-acetylfuran and determination of
conformations (T.N. Huckerby, J. mol. Struct., 1976 31,
161; B.P. Roques, S. Combrisson, and F. Wehrli,
Tetrahedron Letters, 1975, 1047); of furan-2-carbox-
aldehyde and 2-acetylfuran and their 1:1 and 1:2 complexes
with aluminium chloride (L. Belen'kii, I.B Karmanova, and
S.V. Rykov, Chem. Scr., 1976, 10, 201). ^{1}H-nmr spectra of
derivatives of furan-, thiophene-, and pyrrole-2-carbox-
aldehydes referring to conformational dependence of the ^{4}J
coupling between side chain and ring protons in the formyl-
heterocycles (Roques and Combrisson, Canad. J. Chem., 1973,
51, 573). Ir spectra of furan-2-carboxaldehyde
(C.G. Andrieu *et. al.*, Compt. rend., 1972, 275C, 559;
Andrieu, C. Chatain-Cathaud, and M.C. Fournie-Zaluski, J.
mol. Struct., 1974, 22, 433; K. Volka *et. al.*, J.
Radioanal. Chem., 1976, 30, 205). A study of the
5-deuterio-analogues shows that the multiple ir carbonyl
absorption of furan-2-carboxaldehyde, alkyl furan-2-
-carboxylates and 2-furoyl chloride are caused by Fermi
resonance and/or rotational isomerisation (D.J. Chadwick
et. al., Chem. Comm., 1972, 742). Mass spectra of
5-substituted furan-2-carboxaldehyde (5-substituents are
Me, Cl, Br, I, NO$_2$, CO$_2$Me, CO$_2$H) (N. Lauzardo *et. al.*,
Ciencias, Ser. 3, 1972, 13, 17).

An *ab initio* molecular orbital study of the
conformational preferences of furan-2-, furan-3-, pyrrole-2-

and pyrole-3- carboxaldehyde has been reported (I.G. John,
R.L.D. Ritchie, and L. Radom, J. chem. Soc. Perkin II,
1977, 1601; C. Liegeois, J.M. Barker, and H. Lumbroso,
Bull. Soc. chim. Fr., 1978, 329).

(b) *Furan-3-carboxaldehyde*. Furan-3-carboxaldehyde
with tributylstannyllithium followed by a α-chloroethyl
ether gives the alkoxystannane (6), which on treatment with
butyllithum can be lithiated and consequently methylated in
the side chain. Attempted deprotonation of O-ethoxyethyl-
furan-3-methanol leads largely to ring metallation
(W.C. Still, J. Amer. chem. Soc., 1978, 100, 1481).

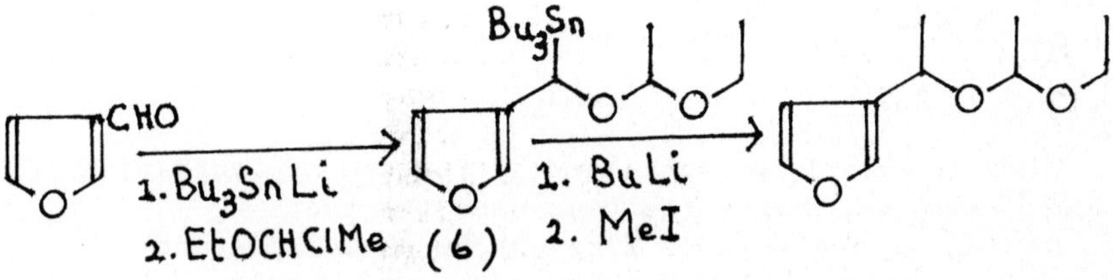

Furan-3-carboxaldehyde tosylhydrazone, m.p.116-
119°. Pyrolysis of diazo(3-furyl)methane, generated from
furan-3-carboxaldehyde tosylhydrazone sodium salt, because
ring opening as in the case of the related 2-furyl
derivative (p.32) is impossible, affords *cis*- and *trans*-
-1,2-di(3-furyl)ethenes (Hoffman, Orphanides, and
Sketchtes, *loc. cit.*).

Nmr spectral data obtained by the use of lanthanide induced chemical shifts has been used for the conformational analysis of furan-3-carboxaldehyde and 3-acetylfurans and related thiophenes and pyrroles (G. Gacel, Fournies-Zaluski, and Roques, Org. mag. Reson., 1976, $\underline{8}$, 525). ^{13}C-nmr spectra of furan-3-carboxaldehyde (Gronowitz, Johnson and Hornfeldt, *loc. cit.*, Hearn, *loc. cit.*). [1]H-nmr spectra of derivatives of furan-, thiophene-, and pyrrole-3-carboxaldehydes (Roques and Combrisson, *loc. cit.*). Ir spectra of furan-3-carboxaldehyde (Andrieu *et. al.*, *loc. cit.*; Chatain-Cathaud, and Fournie-Zaluski, *loc. cit.*, K. Volka *et. al.*, *loc. cit.*; P. Adam *et. al.*, Sb. Vys. Sk. Chem.-Technol. Praze, Anal. Chem., 1977, $\underline{H12}$, 193.

(xii) Ketones

Acylfurans (1) (R = 2-MeOC$_6$H$_4$ or 2,4,6-Me$_3$C$_6$H$_2$) and (2) (R = 2-furyl or 2-thienyl) are otained by treating furan with the appropriate acid chloride in the presence of stannic chloride. Derivatives of type (3) (R = Ph, 4-MeC$_6$H$_4$, 4-MeOC$_6$H$_4$, 4-BrC$_6$H$_4$) and (4) (R = Ph, 2-furyl, 2-thienyl) are prepared by lithiating 3-bromofuran-2-carboxaldehyde diethyl acetal and 2-(3-furyl)-1,3-dioxolane, respectively and treating the product with the appropriate

dimethylamide (RCONMe$_2$) (Fournie-Zaluski and Chatain-
Cathaud, Bull. Soc. chim. Fr., 1974, Pt.2, 1571).

Alkyl 2-furyl ketones have been obtained by acylation
of furan with carboxylic acid anhydrides in the presence of
trifluoromethane sulphonic acid (Yu. A. Nikolyukin *et. al.*,
U.S.S.R., 447,403/1974). High yields of 2-acetyl- and
2-acetyl-3-methyl-furan may be obtained by acetylation of
the appropriate furan using acetyl toluene-4-sulphonate in
acetonitrile or benzene (S.I. Pennanen, Heterocycl. 1976,
4, 1021).

Previous estimations of the α:β ratio for the
acetylation of furan showed that the percentage of the
3-substituted isomer was smaller than 0.1%. The value has
now been determined more accurately under three different
experimental conditions and varies from 0.0147 to 0.125%
(G. Ciranni and S. Clementi, Tetrahedron Letters, 1971,
3833).

Treating 2-furyldimethylaminoacetonitrile with sodium
methoxide in 1,2-dimethoxyethane and then with acrylonitrile gives 1-(2-furyl)-3-cyanopropan-1-one on hydrolysis.
Since the initial furan derivative is prepared from furan-2-
-carboxaldehyde, this procedure affords a method of
obtaining the γ-keto-nitrile from the aldehyde (V. Reutrakul
et. al., Chem. Letters, 1979, 339).

$$\text{(furan)CHCN-NMe}_2 \xrightarrow[\substack{\text{2. } H_2C{=}CHCN \\ \text{3. 2M HCl or } CuSO_4 \\ \text{in EtOH}}]{\text{1. NaOMe, } (MeOCH_2)_2} \text{(furan)COCH}_2CH_2CN$$

(xiii) Carboxylic acids

(1) *Preparation,*   2-Furoic acid may be obtained by oxidising furan-2-carboxaldehyde with sodium hypochlorite followed by acidification (p. 32), by the electrooxidation of furfuryl alcohol (J. Kaulen and H.J. Schaefer, Synth., 1979, 513), and by the carboxylation of furan by sodium palladium (II) malonate in the presence of silver acetate (T. Sakakibara and Y. Odaira, J. org. Chem., 1976, 41, 2049).   Esters of some derivatives of 2-furoic acid may be obtained by the isomerisation of 4-alkyl-2-(5-substituted- -2-furyl)-1,3-dioxolanes on heating.   The dioxolanes are prepared by condensing the 5-substituted furan-2- -carboxaldehyde with the appropriately substituted glycol (V.G. Kul'nevich, Z.I. Zelikman, and S.E. Tkachenko, Chem. Abs., 1977, 86, 89489n).

R^1 = H, Me; R^2 = H, Me, NO_2

    For the preparation of 5-arylfuran-2-carboxylates see p. 28.

    2-Furylacetic acid is obtained from 2-furoic acid chloride and diazomethane followed by treatment of the resulting 2-furyl diazomethyl ketone with silver oxide

and water, or by converting 2-hydroxymethylfuran (2-furyl-
-methanol) into 2-chloromethylfuran using thionyl chloride
and followed by reaction with sodium cyanide and hydrolysis
of the resulting 2-furylacetonitrile. On reduction with
lithium tetrahydridoaluminate 2-furyl-acetic acid yields
β-(2-furyl)ethanol (D. Satoh, T. Hasimoto, and M. Shimada,
Chem. Abs., 1978, 89, 42947g).

3-Furoic acid is prepared in 97% yield by lithiating
3-bromofuran in ether with butyllithium at -70° and
treating the resulting 3-furyllithium with solid carbon
dioxide and water (Y. Fukuyama *et. al.*, Synth., 1974,
443).

(2) *Properties and reactions*. Methyl 2-furoate on
nitration with a mixture of fuming nitric acid and acetic
anhydride at -40 to -20° yields methyl 5-nitrofuroate,
the results are better at this temperature than at 0°
(K. Venters and M. Trusule, Chem. Abs., 1977, 87, 68045v).

2-Furoic acid is reduced by lithium in liquid ammonia
and methanol to 2,5-dihydrofuran-2-carboxylic acid
(H.R. Divanfard and M.M. Joullie, Org. Prep. Proced. Int.,
1978, 10, 94).

The lithium liquid ammonia reduction of 3-furoic
acid, in the absence of a proton source proceeds with
β-elimination and ring opening, accompanied with the
consumption of four equivalents of metal to give a mixture
of two tetrahydrofuran derivatives on adding ethanol at the
end of the reduction (J. Slobbe, Austral. J. Chem., 1976,
29, 2553).

$$\text{(furoic acid, } CO_2H) \quad \xrightarrow[\text{2. EtOH}]{\text{1. Li, liq·NH}_3, \; -33^\circ} \quad \text{(HO} \cdots \text{Me)} \; + \; \text{(EtO} \cdots \text{Me)}$$

Furoic acid undergoes efficient and regioselective ring metallation on reaction with lithium diisopropylamide (LDA) to give bis-anionic species, which react well with a number of electrophiles. Thus 3-furoic acid affords the bis-anion (1), which reacts rapidly with carbonyl compounds and methyl iodide at -78° to give high yields of products. Other alkylating agents do not react with (1) at -78° but require temperatures of -20 to -10° and allylic and benzylic halides give only traces of alkylated products.

$$\text{(} CO_2H \text{)} \quad \xrightarrow[\substack{\text{THF} \\ -78^\circ}]{\text{LDA,}} \quad \text{(1)} \quad \xrightarrow{D_2O} \quad \text{(} CO_2H, \; D \text{)} \quad 95\%$$

Me$_2$CO / -78° → CO_2H, C(Me)$_2$OH — 93%

PhCHO / -78° → CO_2H, CHPh / OH — 96%

MeI / -78° → CO_2H, Me — 90%

EtI / -20 to -10° → CO_2H, Et — 42%

2-Furoic acid on similar treatment gives the bis--anion (2) which reacts very rapidly and cleanly at -78° with carbonyl compounds, while coupling with alkylating reagents only occurs at temperatures $>-30^\circ$ but yields are significantly higher than with 3-furoic acid. Again only traces of alkylated products are obtained with allylic and benzylic halides (D.W. Knight, Tetrahedron Letters, 1979, 469).

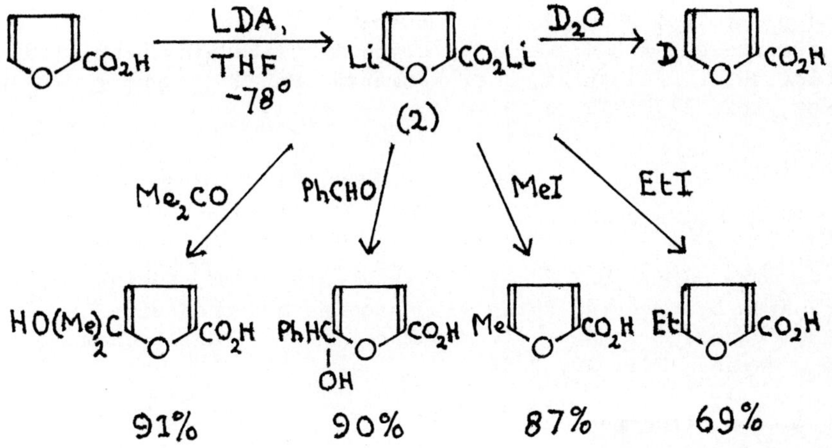

Ethyl 2-furoate or 5-methyl-2-furoate on treatment with sodium in the presence of trimethylchlorosilane gives the trimethylsilyl enol ether of 3,5-bis(trimethylsilyl)-pent-4-yn-1-al or 4,6-bis(trimethylsilyl)hex-5-yn-2-one, respectively, which on hydrolysis yields the corresponding aldehyde or ketone (I. Kuwajima, K. Atsumi, and I. Azegami, Chem. Comm., 1977, 76).

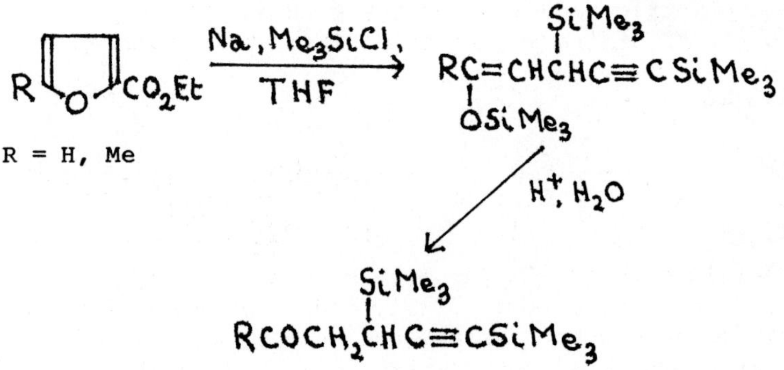

3-Furoic acid is converted into 3-hydroxymethylfuran by borane-dimethylsulphide [R.B. Franklin, C.N. Statham,

and M.R. Boyd, J. Labelled Compd. Radiopharm., 1978, 15
(Suppl. Vol.), 569]. ^{13}C-nmr spectra of methyl
2-furoate and its 1:1 and 1:2 complexes with aluminium
chloride (L. Belenk'ii, I.B. Karmanova and S.V. Rykov,
Chem. Scr. 1976, 10, 201).

(b) Dihydrofurans

(i) 2,3-Dihydrofurans

The *trans*- and *cis*-2,5-diethyl-2,5-dihydrofurans are
prepared by treating furan with bromine and then with
ethylmagnesium bromide (T. Masamune *et. al.*, Bull. chem.
Soc. Japan, 1975, 48, 2294).

The Birch reduction of 2- and 3-furoic acid gives the
2,5- and 2,3-dihydrofuroic acid, respectively. The use of
1,2:5,6-di-O-isopropylidene-α-D-glucofuranose (G) as the
proton source in the Birch reduction causes asymmetric
reduction. Methyl 2,3-dihydro-3-furoate, b.p.83-84°/30mm,
$[\alpha]_D$ -8.7°, is obtained by reducing 3-furoic acid with
sodium and (G) in liquid ammonia and esterifying the
resulting optically active acid with diazomethane. If the
reduction mixture containing the 2,3-dihydro-3-furoic acid
is acidified and left for several hours it is assumed to
give 5-hydroxytetrahydro-3-furoic acid, which on oxidation
with silver oxide yields paraconic acid (1), b.p.150-160°/
0.05mm, $[\alpha]_D$ +1.9° (R configuration, $[\alpha]_D$ - 60.4°),
thus the asymmetric reduction occurs with an optical yield
of *ca.* 3% (T. Kinoshita and T. Miwa, Chem. Comm., 1974,
181).

Reaction of 2,3-dibenzoyl-2,3-dihydro-5-phenylfuran
(R=H), m.p.116-119°, with hydrazine in boiling ethanol
results in a novel ring transformation to give 1,2,4a,8a-

-tetrahydro-3,5,8-triphenylpyridazino[4,5-c]pyridazine
(B.E. Landberg and J.W. Lown, J. chem. Soc. Perkin I, 1975,
1326).

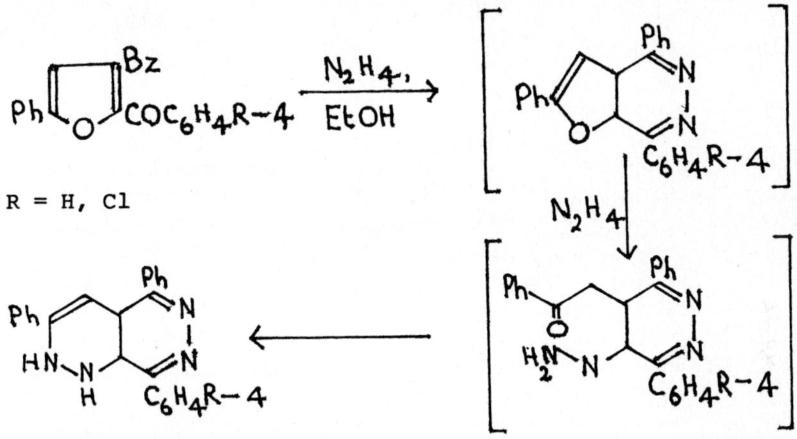

R = H, Cl

Two moles of phenylcyclopropenone react with the
enamine (2) to yield the spirolactone (4), which contains a
2-oxo-dihydrofuran ring. The reaction probably goes *via*
the betaine (3) (T. Eicher and S. Bohm, Ber., 1974, 107,
2215).

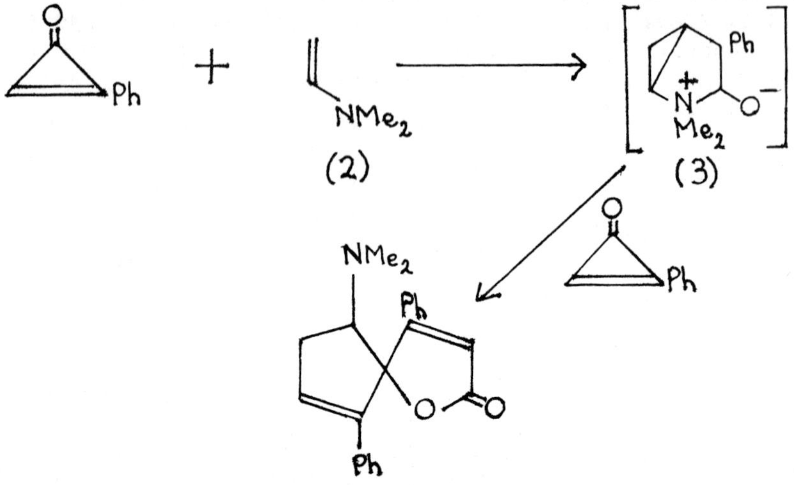

Treatment of phenylcyclopropenone with Cu^{2+} ions affords the tetramer (5), a derivative of 5-methylene-3-
-phenyl-2-oxodihydrofuran (Eicher and N. Pelz, Tetrahedron Letters, 1974, 1631).

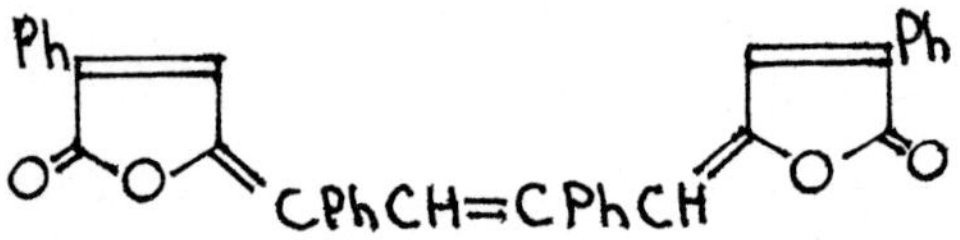

(ii) *2,5-Dihydrofuran*

2,5-Dihydro-2,3-di-*tert*-butyl-4-methy-5-methylene-
furan is obtained when 2,3-dimethyl-5-*tert*-butylfuran is treated with *tert*-butyl chloride and a molar equivalent of

aluminium chloride in carbon disulphide at $0°$ (H. Wynberg and U.E. Wiersum, *ibid.*, 1975, 3619). The reaction appears to be general, applying also to cycloalkyl[b]furans, but normal ortho-*tert*-butylation occurs when the exomethylene group cannot be formed.

2,5-Dimethoxy-2,5-dihydrofuran is obtained by the electrolytic methoxylation of furan in methanol in the presence of sodium bromide (H. Nohe and H. Hannebaum, Ger. 2,710,420/1977; V. Krishnan and A. Muthukumaran, Chem. Abs., 1979, 91, 183932b). Electrolytic methoxylation of 3-acetylfuran at approximately $-50°$ in the presence of ammonium bromide affords 2,5-dihydro-2,5-dimethoxy-3-(1,1--dimethoxyethyl)furan, which on cis hydroxylation followed by acetylation gives 3,4-diacetoxy-2,5--dimethoxy-3-(1,1-dimethoxyethyl)-tetrahydrofuran. The diacetoxy derivative is converted to the related hydroxyl derivative by treatment with sodium methoxide in methanol (J. Srogl, M. Janda, and I. Stibor, Coll. Czech. chem. Comm., 1973, 38, 3666).

2,5-Dihydro-2-furoic acid is obtained from 2-furoic acid (p. 38). The *cis*- and *trans*-2-cyano-2,5-dihydro-2,5--dimethyl-5-methoxyfurans are obtained by the cyanomethoxylation of 2,5-dimethylfuran (p. 6).

Methyl 2,5-dihydro-2-furoate, b.p.$94°$/35mm, $[\alpha]_D$ $+3.5°$ is prepared by Birch asymmetric reduction of 2-furoic acid followed by esterification (Kinoshita and Miwa, *loc. cit.*).

Gas phase oxidation of furan over a 10% MoO_3-TiO_2 catalyst proceeds with 88% selectivity to give 2,5-dioxo-2,5--dihydrofuran (maleic anhydride) (M. Blanchard and J. Goichon, Bull. Soc. chim. Fr., 1975, 289).

(c) Tetrahydrofuran

Tetrahydrofuran is obtained by the dehydration and cyclisation of butan-1,4-diol in the presence of polyphosphoric acid (K. Akagane and G.G. Allan, Shikizai Kyokaishi, 1972, 45, 293), by catalytic dehydration of the diol over an Al, Ni, Mo oxide catalyst at 320° (M. Khasanova and M.F. Abidova, Chem. Abs., 1976, 85, 46276c), and by heating the diol in the presence of pyridine hydrochloride. The latter method also yields 2,5-dihydrofuran from but-2-ene--1,4-diol (J. Egyed, P. Demerseman, and R. Royer, Bull. Soc chim. Fr., 1973, 3014). The behaviour of products with the passage of time during the hydrogenation of furan and 2-methylfuran to the tetrahydrofurans in an alcohol, in the presence of palladium black or palladium oxide has been investigated (K. Shinozaki, M. Abe, and M. Uchiyama, Chem. Abs., 1979, 90, 87156k). Vapour-phase hydrogenation of furan using a nickel catalyst occurs with high conversion to tetrahydrofuran (Shinozaki, *ibid.*, 71978u). The catalytic hydrogenation of furan and its alkyl derivatives to tetrahydrofurans, with reference to the concurrence of hydrogenolysis and isomerisation, has been reviewed (Z. Dudzik and M. Gasiorek, Przem. Chem., 1975, 54, 637).

5-Alkoxytetrahydrofuran-2-ones are synthesized by first irradiating furan-2-carboxaldehyde in an alkyl alcohol in the presence of eosin-Y, while oxygen is bubbled through the mixture and then hydrogenating the product using a palladium-carbon catalyst (G. Fagan, R.E. Kepner, and A. Webb, Synth. Comm., 1979, 9, 683).

It is believed that 5-hydroxytetrahydro-3-furoic acid is formed on leaving 2,3-dihydro-3-furoic acid to stand overnight in acid solution. On oxidation it is converted to paraconic acid (p.42). 5-Hydroxy- and 5-ethoxy-3--methyltetrahydrofuran-2-one (p.39). (±) Nonactic acid (1) obtained by the hydrolysis of nonactin, produced by *Streptomyces*, has been synthesized (M.J. Arco,

M.H. Trammell, and J.D. White, J. org. Chem., 1976, 41, 2075).

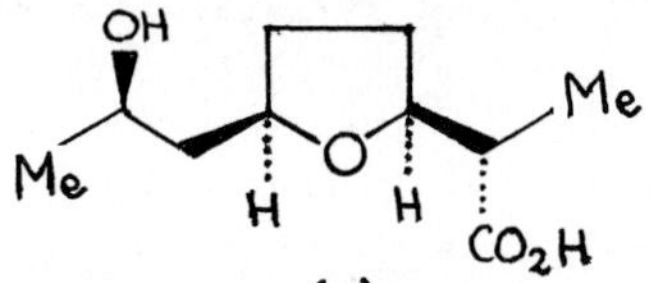

(1)

Tetrahydrofuran, 3,4-dialkyltetrahydrofuran, and 3,3,4,4-tetraalkyltetrahydrofurans are cleaved smoothly by n-butyllithium in hexane affording alkenes and lithium enolates of aldehydes.

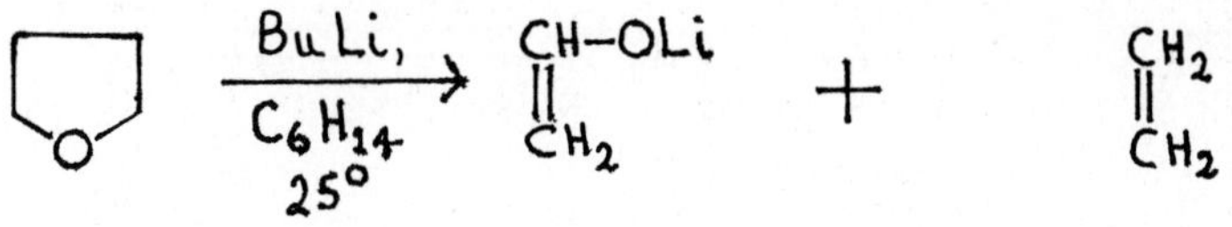

The reactions appear to proceed by abstraction of an α-hydrogen followed by -[π4s + π2s] cycloreversion (R.B. Bates, L.M. Kroposki, and D.E. Potter, J. org. Chem., 1972, 37, 560; P. Tomboulian *et. al., ibid.,* 1973, 38, 322; M.E. Jung and R.B. Blum, Tetrahedron Letters, 1977, 3791). The rate of cleavage of tetrahydrofuran by butyl-lithium at -20 to 20° increases with decreasing temperature (N. Yu. Baryshnikov, N.N. Kaloshina, and G.I. Vesnovskaya, Zh. obshch. Khim., 1977, 47, 2790).

Ring opening of tetrahydrofuran occurs when it reacts with benzyne, resulting in the formation of carbonium ion (2), which reacts with phenol to form 1,4-diphenoxybutane (R.S. Pal and M.M. Bokadia, Pol. J. Chem., 1978, 52, 1473).

$$(2)$$

Ring cleavage also occurs when diphenylethylphosphine reacts with lithium and tetrahydrofuran to give butane derivative (3) which cyclizes in boiling benzene in the presence of hydrogen bromide. Further treatment of the product with sodium chlorate yields 1-ethyl-1-phenylphospholanium perchlorate (4) (W.R. Purdum, S.D. Venkataramu, and K.D. Berlin, Inorg., Synth., 1978, 18, 189).

$$(3) \qquad (4)$$

The γ-ray induced reactions between tetrahydrofuran and carbon tetrachloride, chloroform, and methylene chloride give as the main products 4-chlorobutan-1-ol and 5-chlorooctahydro-2,2'bifuryl (T. Yumoto and K. Iseda, Chem. Abs., 1978, 88, 152325q).

The ring conformations of tetrahydrofuran have been
calculated (A.A. Lugovskoi and V.G. Dashevskii, Mol. Biol.,
1972, 6, 440) and its ^{13}C-nmr spectral data reported
(C. Konno and H. Hikino, Tetrahedron, 1976, 32, 325;
F.J. Weigert and J.D. Roberts, J. Amer. chem. Soc., 1972,
94, 6021). Also reported is the nmr spectral data of boron
trifluoride adducts of tetrahydrofurans (P. Stilbs and
S. Forsén, Tetrahedron Letters, 1974, 3185).

2. Benzo[b]furans and their Hydrogenated Products

Although a few new preparations of benzofurans and
their derivatives have been reported a number of well known
reactions have been investigated in more detail, using
physiochemical and spectroscopic methods. Thus some
intermediates have been identified and reaction mechanisms
proposed.

(a) Benzofurans

(i) General synthetic methods

(1) 2-Substituted benzofurans can be obtained by the
cyclization of allylphenols. The reaction of the sodium
salt of allylphenols with dichlorobis(benzonitrile)palladium
goes cleanly to give the 2-substituted benzofurans
(T. Hosokawa et. al., Tetrahedron Letters, 1973, 739).
2-Methylnaphto[1,2-b]furan and 2-methylnaphtho[2,1-b]furan
may be prepared by this method.

R = H, Ph

R = H (31%)
R = Ph (53%)

(2) The Claisen rearrangement of aryl 2-chloroprop-2-
-enyl ethers (1) to 2-(2-chloroprop-2-enyl)-phenols (2)
proceeds almost quantitatively, when the ether is heated in
NN-diethylaniline or 1,4-di-isopropylbenzene. The phenols
are readily cyclized to the corresponding benzofurans on
treatment with concentrated hydrochloric acid at 85-89°.
When base is excluded from the work up of these reactions,

the unstable intermediate (3) can be observed by nmr
spectroscopy (W.K.A. Anderson and E.J. LaVoie, Chem. Comm.,
1974, 174; Anderson, LaVoie, and J.C. Bottaro, J. chem.
Soc., Perkin I, 1976, 1).

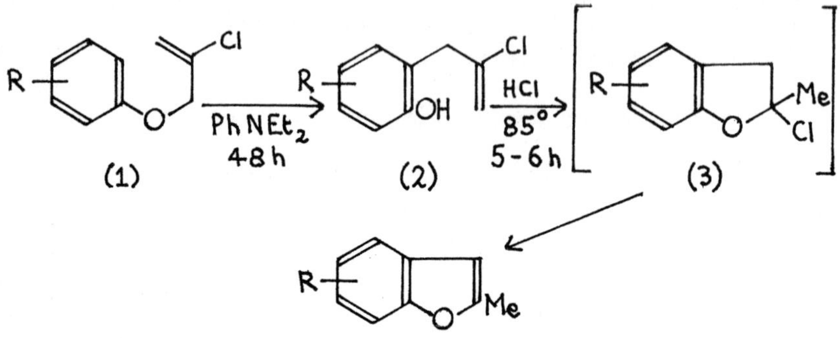

(3) The reaction between phenol and the product from
N-chlorosuccinimide and methylthiomethyl methyl ketone, an
extension of Gassman's indole synthesis, yields 2-methyl-3-
-(methylthio)benzofuran (12%) which on desulphurisation
with Raney nickel gives 2-methylbenzofuran (P.G. Gassman
and D.R. Amick, Synth. Comm., 1975, 5, 325).

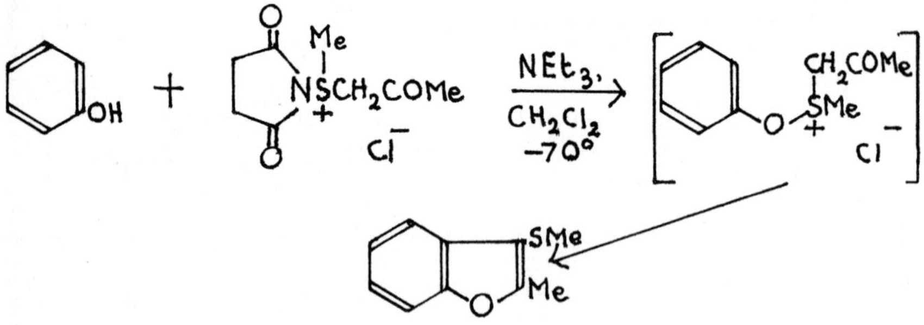

(ii) General properties and reactions

Benzofuran on oxidation with managanese (III) acetate in acetic acid-acetic anhydride affords 2-(acetoxymethyl)-benzofuran, 3a,8b-dihydrofuro[3,2-b]benzofuran-2(3H)-one (1), bis(2-benzofuryl)methane, and 2-benzofurylacetic acid (A. Kasahara *et. al.*, Bull. chem. Soc. Japan, 1976, <u>49</u>, 3711).

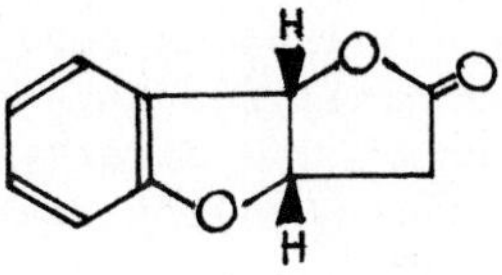

(1)

Sensitized photooxidations of benzofurans show that benzofuran and 2-methylbenzofuran are stable towards singlet oxygen, but that the vinylbenzofurans (2) and (3) are easily photooxygenated to yield the 1,4-*endo*-peroxides (4) and (5) (M. Matsumoto, S. Dobashi, and K. Kondo, *ibid.*, 1977, <u>50</u>, 3026).

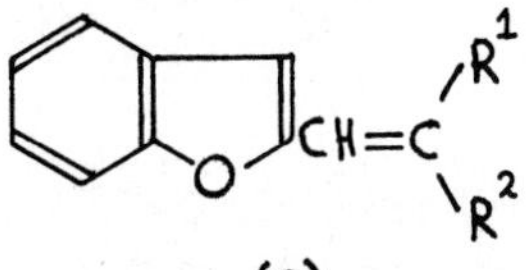

(2)

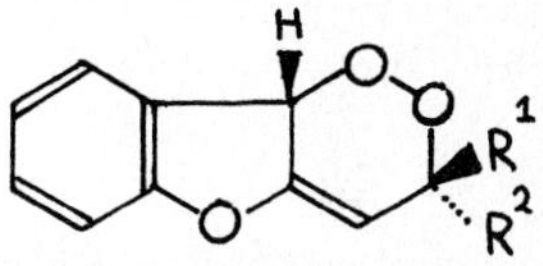

R^1, R^2 = H, Me, Ph

(4)

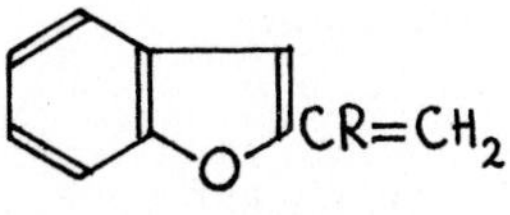

(3)

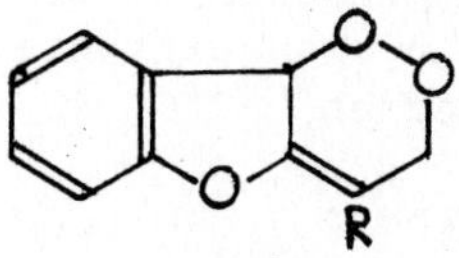

(5)

R = Me, Ph

The hydrogenation of benzofuran in acetic acid over
5% palladium/carbon gives 2,3-dihydrobenzofuran
(R.A. Ellison and F.N. Kotsonis, J. org. Chem., 1973, 38,
4192). The corresponding 2,3-dihydro derivatives are
obtained by the ionic hydrogenation of 2-methyl-,
3-methyl-benzofuran, and benzofuran using trifluoroacetic
acid-triethylsilane. Reduction of 2-methylbenzofuran and
benzofuran with trifluoroacetic acid - triethyldeuterio-
silane gives the corresponding 2-deuteriodihydro
derivatives, whereas 3-methylbenzofuran affords the
3-deuterio derivative. Analogous reduction of 2,3-dimethyl-
benzofuran with the original reagent gives a 18:80 mixture
of the *cis-* and the *trans-* 2,3-dimethyl-2,3-dihydrobenzo-
furan (Karakhanov *et. al.,* Doklady Akad. Nauk SSSR, 1974,
214, 584; Khim. Geterotsikl. Soedin., 1975, 1479).

Benzofuran reacts with iodine azide to give a 42:58
mixture (93%) of the *cis-* and the *trans-* 2,3-diazido-2,3-
dihydrobenzofuran, which on boiling with dimethyl acetylenedicarboxylate yields the corresponding 1:2 cycloadducts
(S. Kwon *et. al.,* Heterocycl., 1977, 6, 33).

Tetracyanoethene oxide adds to a number of heterocycles
including benzofuran and the order of reactivity is furan>
2-methylthiophene~benzofuran>benzothiophene>selenophene>
thiophene>2-chlorothiophene (S. Gronowitz and B. Uppstrom,
Acta Chem. Scand., 1975, B29, 441). The desulphurisation
of epipolythio-2,5-piperazinediones by triphenylphosphine
in the presence of benzofuran, indole, or 3-methyindole
yields cycloaddition products (T. Sato and T. Hino,
Tetrahedron, 1976, 32, 507).

The 1,3-cycloaddition of benzonitrile oxide and
mesitonitrile oxide to benzofuran gives the two
regioisomeric cycloadducts (6) and (7) in the ratio 70:30
and 26:74, respectively. Adduct (7) on treatment with acid
yields 3-aryl-4-(2-hydroxyphenyl)isoxazole (P. Caramella

et. al., J. org. Chem., 1978, <u>43</u>, 3006).

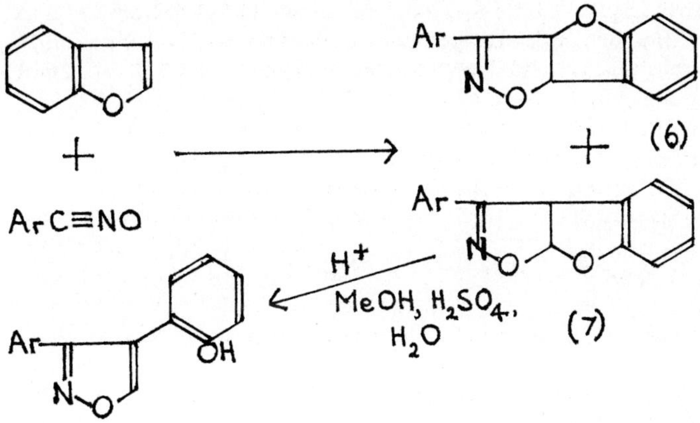

$$Ar C \equiv NO$$

(6)

(7)

3,5-Dichloro-2,4,6-trimethylbenzonitrile oxide adds to furan and thiophene to give a single regioisomer, whilst the analogous cycloaddition to benzothiophene leads to both possible adducts (P.L. Beltrame *et. al.*, J. chem. Soc., Perkin II, 1977, 706).

A concerted mechanism has been proposed for the cycloaddition of $4\text{-}RC_6H_4C(Cl){=}NNHPh$ to benzofuran to give benzofuropyrazoles (Le Quoc Khanh and B. Laude, Compt. rend. 1973, <u>276C</u>, 109).

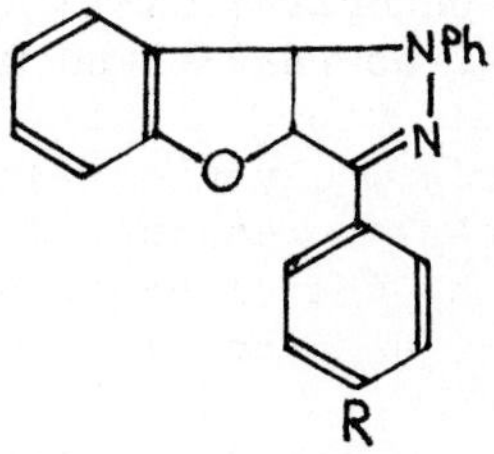

$$4\text{-}RC_6H_4C(Cl){=}NNHPh$$

R = H, Me, MeO, Cl, NO$_2$

The photochemical reaction between benzofuran (8) $[R^1{=}R^2{=}Me; \quad R^1{=}Me, CH_2OH, R^2{=}H; \quad R^1R^2{=}(CH_2)_4]$ and

benzophenone gives the oxetanes (9) (R^3=Ph). Oxetanes
(9; R^3=H) are formed by the reaction of the benzofuran
with benzaldehyde, but yields are not so high because of
their instability. In the cases of 2-methoxycarbonyl- and
2-cyano-benzofuran, dimers of the benzofuran are obtained
(K. Kawase *et. al.*, Bull. chem. Soc. Japan, 1974, *47*, 2660).

The photoreaction between propiophenone or acetophenone and
benzofuran leads both to dimerisation of benzofuran and
oxetane formation (S. Farid, S.E. Hartman, and C.D. De Boer,
J. Amer. chem. Soc., 1975, *97*, 808). The irradiation of
benzofuran with 2-(3-pyridyl)benzofuran or 2-phenylbenzo-
furan results in formation of the head-to-tail *syn* and *anti*
cyclobutane codimers as the main products; an excited
singlet of benzofuran is implicated as the reactive
excitive state (K. Takamatsu, H.-S. Ryang, and H. Sakurai,
Chem. Comm., 1973, 903; J. org. Chem., 1976, *41*, 541).
Cyclobutanes and oxetanes are formed by the irradiation of
charge-transfer complexes of dimethylmaleic anhydride with
benzofuran (S. Farid and S.E. Shealer, Chem. Comm., 1973,
296). The photoreaction of benzofuran in the presence of a
2-thioparabanate (10) gives a spirothietane (11)
(H. Gotthardt and S. Nieberl, Ber., 1978, *111*, 1471).

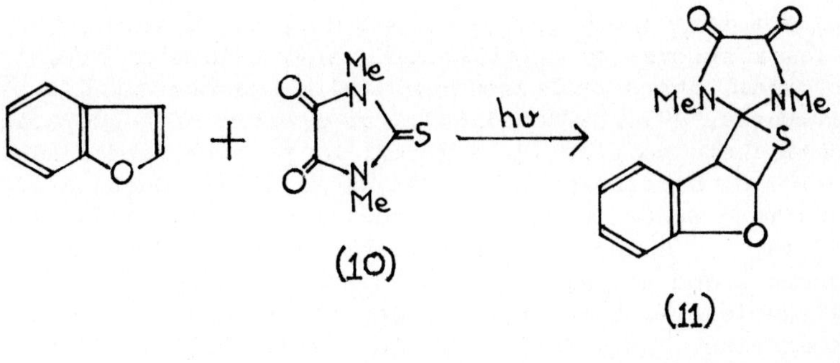

(10)   (11)

The reaction between benzofuran and trichlorosilane
in the vapour phase under the effect of accelerated
electrons gives a mixture of compounds (12) and (13).
Dehydrosilylation of (13) at ~450° affords (12)
(B.T. Vainshtein *et. al.*, Khim. Geterotsikl. Soedin., 1977,
554).

(12)   (13)

Benzofuran on treatment with n-butyllithium and an
equimolar proportion of *NNN'N'*-tetramethylethylenediamine
in hexane gives 93% of the 2-lithio derivative
(D.J. Chadwick and C. Willbe, J. chem. Soc., Perkin I,
1977, 887).

The following spectra have been reported and in some
instances a full analysis is given; ir and Raman spectra
of benzofuran and benzothiophene (G. Mille, G. Davidovics,
and J. Chouteau, J. chim. physiochim. Biol., 1972, 69,
1656); ir and ^{1}H-nmr spectra of methyl substituted

benzofurans and other 5-membered heterocycles
(N.N. Zatsepina *et. al.*, Khim., Geterotsikl. Soedin., 1977,
1110); ir spectra of acetylbenzofurans, oximes of acyl-
benzofurans, and nitrobenzofurans (M.L. Desvoye *et. al.*,
Compt. rend., 1971, 273C, 1284); uv spectra of benzofuran
and other heterocycles [H. Guesten, L. Klasine, and
B. Rusic, Z. Naturforsch. A, 1976, 31A, 1051; Guesten *et.
al.*, Excited States Biol. Mol. Proc. Int. Conf., 1974 (Pub.
1976), 45]; ^{13}C-nmr spectra of benzofuran and methyl-
benzofurans and comparison with thiophene and some
methylindoles (N. Platzer, J.J. Basselier, and
P. Demerseman, Bull. Soc. chim. Fr., 1974, 905); ^{1}H- and
^{13}C-nmr spectra of benzofuran and its methyl derivatives
(N. Platzer, Org. mag. Res., 1978, 11, 350; P.D. Clark,
D.F. Ewing, and R.M. Scrowston, *ibid.*, 1976, 8, 252;
T. Okuyama and T. Fueno, Bull. chem. Soc., Japan, 1974, 47,
1263); ^{13}C-nmr spectra of fluorobenzofurans
(R.J. Abrahams *et. al.*, J. chem. Soc., Perkin II, 1972,
1733); mass spectra of a number of arylbenzofurans
(J. Vebrel, M. Roche, and J. Gore, Org. mass Spec., 1977,
12, 751). The results obtained by Hartree-Fock
calculations of ring currents and proton chemical shifts
for five membered heterocycles, including benzofuran,
indicate that the commonly accepted scale of aromatic
character, based on chemical criteria, is less consistent
than that obtained by spectroscopic methods (E. Corradi,
P. Lazzeretti, and F. Taddei, Mol. Phys., 1973, 26, 41).

Ab initio LCAO-SCF-MO calculations have been
performed for benzofuran, isobenzofuran, indole, and
isoindole [J. Koller, A. Azman, and N. Trinajstic,
Z. Naturforsch., Teill A., 1974, 29, 624; Experimentia,
Suppl., 1976, 23 (Quant. Struct.-Act. Relat.), 205].
Ab-initio LCGO calculations have been reported for
benzofuran, benzothiophene, and indole (M.H.Palmer and
S.M.F. Kennedy, J. chem. Soc., Perkin II, 1974, 1893).
Also reported calculations using the Pariser-Parr approach
(L. Klasine, E. Pop, and Trinajstic, Tetrahedron, 1972, 28,
3465) and MO-LCAO calculations [J. Fabian, Z. physik. Chem.
(Leipzig) 1979, 260, 81].

For basicity of benzofuran and other heterocycles see
M.P. Carmody *et. al.* (Tetrahedron, 1976, 32, 1767). Recent
advances in the chemistry of benzofuran and its derivatives
(P. Cagniant and D. Cagniant Adv. Heterocycl. Chem.,

Vol. 18, Academic Press, New York, 1975, p.337), and related naturally occurring compounds have been reviewed (R.D.H. Murray, Aromat. heteroaromat. Chem., 1977, 5, 472).

(iii) Alkyl- and aryl- benzofurans

The alkylation of benzofuran with isobutene in presence of zinc chloride on alumina in an autocalve gives 2- and 3-*tert*-butylbenzofuran (1:1.2) and small amounts of other products (E.A. Karakhanov *et. al.*, Khim. Geterotsikl. Soedin., 1971, 7, 1020). The same main products are obtained using *tert*-butyl chloride in the vapour phase at 185-260° in the presence of zinc chloride-alumina (Karakhanov, G.V. Drovyannikova, and E.A. Viktorova, *ibid.*, 1974, 172). The kinetics of this reaction have been discussed (Karakhanov *et. al.*, *ibid.*, 1971, 7, 1017). Benzofuran on allylation with allyl iodide in the presence of silver trichloroacetate in methylene chloride, benzene, or nitromethane affords approximately equimolar amounts of 2- and 3-allylbenzofuran (Karakhanov *et. al.*, Vestn., Mosk. Univ. Khim., 1975, 16, 731). Similarly alkenylation with β-methylallyl iodide yields a mixture of 2- and 3-(β-methyl-allyl)benzofuran (A.V. Anisimov *et. al.*, Zh. Org. Khim., 1979, 15, 172). It is reported that treatment of benzofuran with γ,γ-dimethylallyl chloride in ether in the presence of zinc chloride and zinc acetate at 10-16° gives the 2-allyl derivative (Yu. I. Tarnopol'skii and L.I. Denisovich, Khim. Geterotsikl. Soedin., 1970, 29). Benzofuran with cyclopentene at 100-130° in the presence of zinc chloride-alumina gives 2- and 3-cyclopentylbenzo-furan and other products (Kharakhanov *et. al.*, Vestn., Mosk. Univ., Khim., 1974, 15, 225).

2-Acylphenoxyacetic acids on treatment with acetic anhydride and sodium acetate in benzene afford 3-alkyl-benzofurans [G. Rosseels *et. al.*, Ing. Chim. (Brussels), 1971, 53, 37].

R = alkyl

Benzofurans (1: R^1=H,Me; R^2=H; R^3=7-Me, 6-Me, 6-OMe, 5-OMe, 5-SMe, 5-F, 5-I) are prepared by alkylating the salicyclic ester (2) with the appropriate ethyl bromoacetate, hydrolysing the diester, cyclizing the diacid with sodium acetate-acetic acid-acetic anhydride, hydrolysing the product I(R^2=OAc), and reducing the resulting 3-oxo--2,3-dihydrobenzofuran derivative, which on dehydration affords the required benzofuran (P. Cagniant and G. Kirsch, Compt. rend., 1976, 282C, 993).

(1)   (2)

2-Hydroxystilbenes (3) (R^1 = H, OMe; R^2 = H, OMe) are oxidised with lead tetraacetate to give the corresponding 2-arylbenzofuran in moderate yields; poor yields are obtained with manganese (III) acetate (K. Nogami and K. Kurosawa, Bull. chem. Soc. Japan, 1974, 47, 505).

(3)

2-Arylbenzofurans (5: R^1, R^2 = H or Me; R^3 = H, Me, or t-Bu; or R^2R^3 = benzo; R^4 = H, Me, or MeO; R^5 = aryl) are also obtained by the cyclization of ether (4) (R^6 = CHO, COMe, or CR≡NPh) in dimethylformamide, $(Me_2N)_3PO$, or MeCN in the presence of NaOH, KOH, or t-BuOK as condensing agents (W. Sahm, E. Schinzel, and P. Juerges, Ann., 1974, 523).

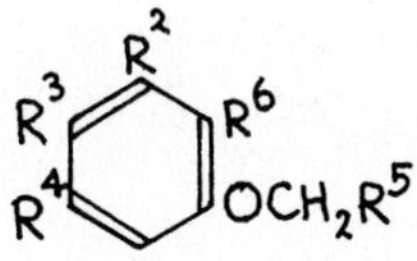

(4)

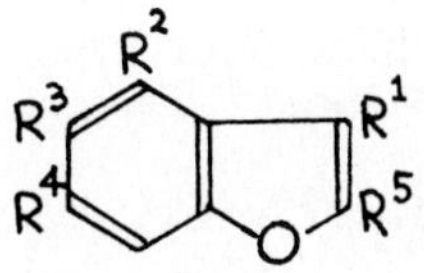

(5)

The Heck reaction between benzofuran and aryl-palladium chloride results in the formation of 2-aryl-benzofuran derivatives. In benzene with palladium, benzofuran gives 2,2'-bibenzofuryl and 2-arylbenzofuran. Benzofuran in the presence of palladium acetate also reacts with olefins to produce benzofuryl-substituted olefins and a small amount of 2,2'-bibenzofuryl (A. Kasahara *et. al.*, Bull. chem. Soc. Japan, 1973, 46, 1220). A number of 2-phenyl-, 2-(hydroxyphenyl)- and 2-(methoxyphenyl)-benzo-furans have been prepared and tested for fungicidal activity (G.A. Carter, K. Chamberlain, and R.L. Wain, Ann. Appl. Biol., 1978, 88, 57).

1,3,4-Oxadiazin-6-one 4-oxides condense with benzyne, generated from benzenediazonium 2-carboxylate to give mixtures of substituted benzofurans and acylbenzofuranones. A side product, a diaryl homophthalic anhydride (6), apparently results from trapping of the benzyne precursor by an intermediate from the primary reaction (J.P. Freeman and R.C. Grabiak, J. org. Chem., 1976, 41, 2531).

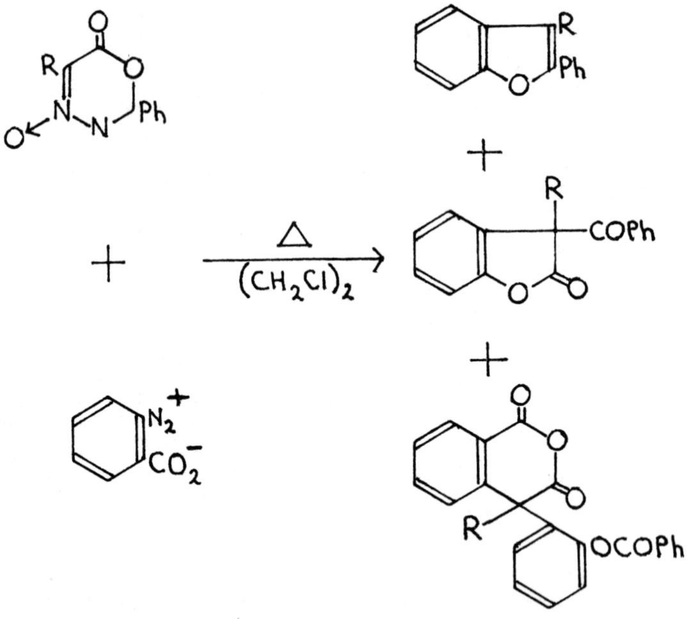

R = Me, Et, Ph

(6)

3-Phenylbenzofuran is obtained together with 2-methoxydeoxybenzoin (7) and 2'-methoxydeoxybenzoin (8) by the deamination of 2-amino-1-(2-methoxyphenyl)-1-phenyl-ethanol with sodium nitrite in aqueous acetic acid (C.E. Spivak and F.L. Harris, *ibid.*, 1972, **37**, 2494).

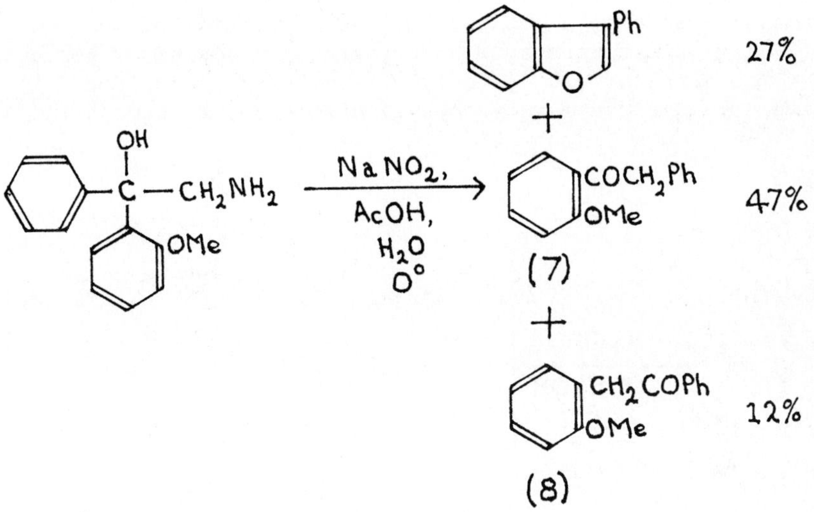

$$27\%$$

$$47\% \quad (7)$$

$$12\% \quad (8)$$

The 5- and the 7-phenylbenzofuran are prepared by the cylclisation of 4- and 2-(2,2-dimethoxyethoxy)biphenyl, respectively with phosphorus pentoxide. Similar treatment of the 3-isomer gives a mixture of 4- and 6-phenylbenzo-furan, which may be separated by column chromatography. The latter compound is also obtained by the reaction of 2-hydroxy-4-phenylbenzaldehyde with ethyl bromomalonate followed by decarboxylating the resulting acid (P. Spangnolo *et. al.*, J. chem. Soc., Perkin I, 1972, 556).

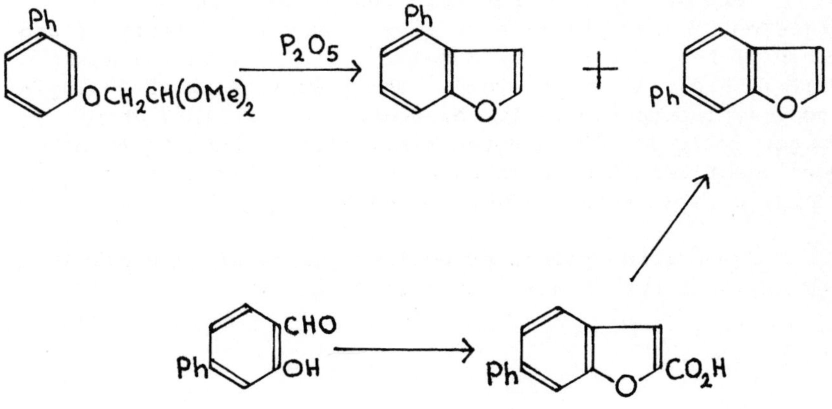

Vinyl bromides with three aryl substituents and containing an oxygen atom at the *ortho* position of one of the β-aryl substituents, when treated with sodium hydroxide in 80% ethanol at 120° give 2,3-diarylbenzofurans, for example, 2,3-bis(4-methoxyphenyl)benzofuran (T. Sonoda *et. al.*, Chem. Comm., 1976, 612).

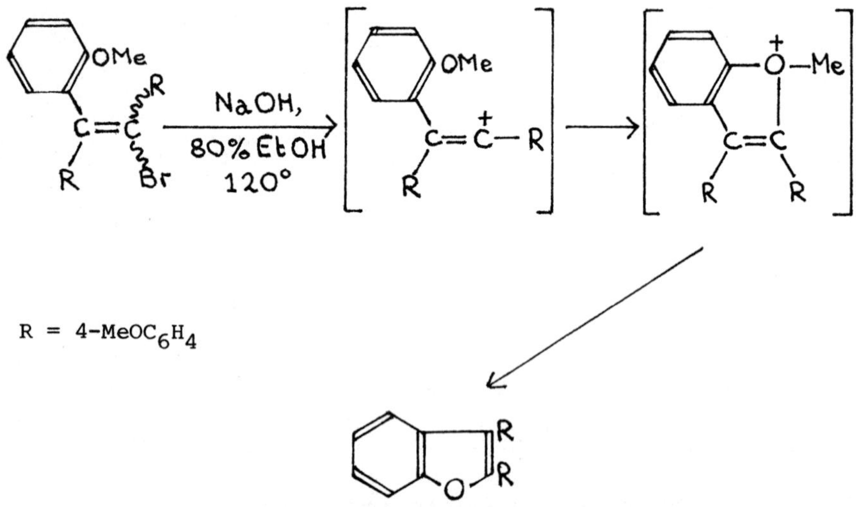

$R = 4\text{-MeOC}_6\text{H}_4$

Benzofuran condenses with benzyl chloride in the presence of zinc chloride to give 2-benzyl-, 3-benzyl-, and dibenzyl-benzofuran (S.K. Ermolaeva *et. al.*, Vestn. Mosk. Univ., Khim., 1974, 15, 236). The preparations of 2-alkyl- and 2-aryl-benzofurans (A. Areschka *et. al.*, Ind. Chim. Belge, 1972, 37, 89) and the alkylation of benzofuran have been described (E.A. Karakhanova *et. al.*, Vestn. Mosk. Univ., Ser. 2, Khim., 1977, 18, 610).

The melting points or boiling points of some alkyl- and arylbenzofurans are listed in Table 1.

TABLE 1

ALKYL- AND ARYL-BENZOFURANS

Substituent	b.p. (^{0}C)	Ref.
4-Phenyl	130-132/1mm	1
5-Phenyl	m.p.64-66	1
6-Phenyl	98-100/0.3mm	1
7-Phenyl	120/0.5mm	1
3-Methyl-2-phenyl	76-79(bath)/0.3mm	2
3-Ethyl-2-phenyl	90-92(bath)/0.3mm	2
5-Chloro-2-methyl	78/1mm	3
	128-133/25mm	4
7-Chloro-2-methyl	55/0.4mm	3
5-Methoxy-2-methyl	92.5/2.7mm	3
	123-125/12mm	5
7-Methoxy-2-methyl	80/2mm	3
	125-127/15mm	6

References

[1] P. Spagnolo *et. al.*, J. chem. Soc., Perkin I, 1972, 556.

[2] J.P. Freeman and R.C. Grabiak, J. org. Chem., 1976, 41, 2531.

[3] W.K.A. Anderson, E.J. LaVoie, and J.C. Bottaro, J. chem. Soc., Perkin I, 1976, 1.

[4] G.H. Coleman and R.H. Rigterink, Chem. Abs., 1952, 46, 3084.

[5] T. Abe and T. Shimizu, Nippon Kagaku Zasshi, 1970, 91, 753.

[6] P.N. Giraldi *et. al.*, Arzneim-Forsch., 1970, 20, 676.

The acetylation of 2,3-dimethylbenzofuran using acetic anhydride-tin (IV) chloride in 1,2-dichloroethane affords, besides the main product 6-acetyl-2,3-dimethylbenzofuran (9), small amounts of the 4-acetyl derivative (10) and 2-acetyl-3-ethylbenzofuran (11). The overall yield of acetylated products is *ca.* 30% and the ratio of 9:10:11 is 89:2:9 (E. Baciocchi *et. al.*, Chem. Comm., 1978, 597).

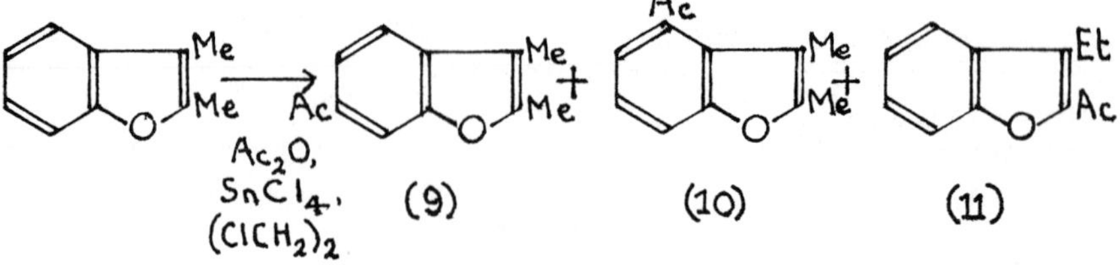

(9)   (10)   (11)

Compound (11) is probably formed *via* attack of the
electrophile at the 2-position of the benzofuran and
subsequent migration of the 2-methyl group.

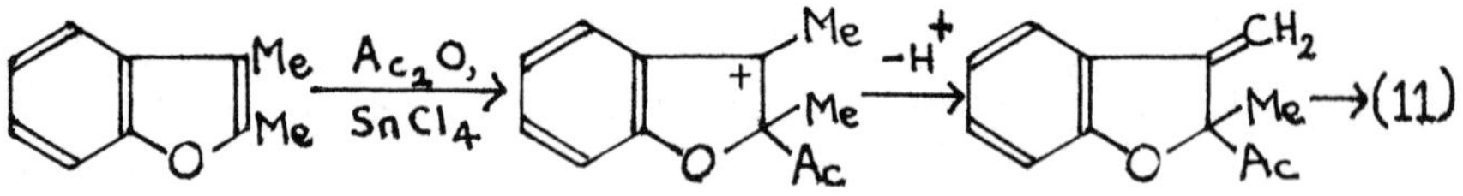

2-Phenylbenzofuran, possessing the highest value of
spin density in its radical anion in the 4-position,
photodimerizes in n-propylamine. On the other hand,
3-phenyl- and 3-phenyl-2,4,7-trimethyl-benzofuran possessing
the highest spin density in their respective radical anions
in the 2-position, are photoreduced to the corresponding
2,3-dihydrobenzofuran derivatives. 2,3-Diphenylbenzofuran
undergoes photocyclization, in n-propylamine and other
solvents, when irradiated under these conditions
(J. Párkányl *et. al.*, J. org. Chem., 1976, **41**, 151).

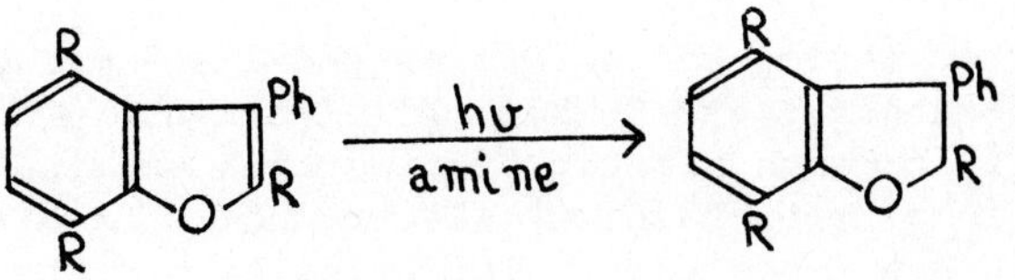

R = H, Me

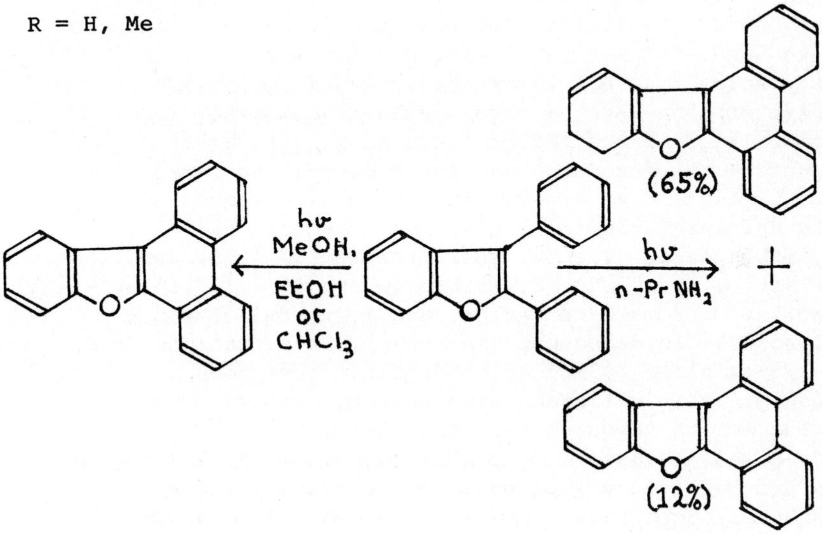

Prostaglandin analogues have been prepared by reacting lithiated benzofuran or benzothiophene with appropriate aldehydes (M. Hayashi, S. Kori, and T. Tanouchi, Ger. Offen. 2,548,006/1976).

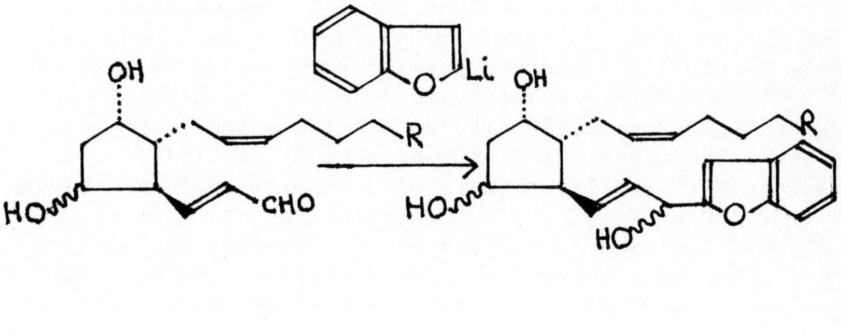

(iv) Halogenobenzofurans

The bromination of benzofuran and its 5-methyl, 6-methyl, 5-methoxy, and 5-chloro derivatives in acetic acid reveals, through kinetic studies that bromine adds to benzofuran in a *trans* fashion, electrophilic bromine attacking at the 2-position. The transition state of the reaction resembles a cyclic bromonium ion intermediate (T. Okuyama, K. Kungiza, and T. Fueno, Bull. chem. Soc. Japan, 1974, <u>47</u>, 1267). The reaction in carbon disulphide at -40° and -50° has been studied using nmr spectroscopy. In all cases the formation of an adduct between bromine and the heterocycle is observed. Decomposition of the adduct leads to ring brominated products with benzofuran and its 2-methyl and 3-methyl derivatives, and to side-chain bromination with 2,3--dimethylbenzofuran (Baciocchi, S. Clementi, and G.V. Sebastiani, J. chem. Soc., Perkin II, 1976, 266). *Syn* eliminating from *trans*-2,3-dibromo-, *trans*-2,3-dichloro-, *trans*-2,3-dibromo-3-deuterio-, and *trans*-2,3-dibromo-5--chloro-2,3-dihydrobenzofuran and *anti* eliminations from *cis*-2,3-dichloro-2,3-dihydrobenzofuran have been investigated in different base-solvent systems. *Anti* elimination is favoured over the *syn* by a factor ~33,000 in EtONa-EtOH, ~10,000 in *t*-BuOK—*t*-BuOH, and >10,000 in *t*-BuOK—*t*-BuOH, in the presence of 18-crown-6 ether (Baciocchi, Sebastiani, and R. Ruzziconi, J. org. Chem., 1979, <u>44</u>, 28).

Chlorination of benzofuran with chlorine in ether at -5° to 0° gives the *trans*- and the *cis*-2,3-dichloro--2,3-dihydrobenzofuran, both of which on treatment with sodium ethoxide in ethanol afford 3-chlorobenzofuran. The rate of the elimination reaction is much greater for the *cis*-dihalogeno derivative than for the *trans*-isomer (Baciocchi, Clementi, and Sebastiani, J. heterocycl. Chem., 1977, <u>14</u>, 359). The acetolysis of *trans*-2,3-dichloro-2,3--dihydrobenzofuran in acetic acid at 102° results in the formation of 2-chlorobenzofuran (1) (65%) and 2-oxo-2,3--dihydrobenzofuran [2(3H)-benzofuranone] (2) (31%), together with smaller amounts of 3-chlorobenzofuran (3) and *trans*-2,3-diacetoxy-2,3-dihydrobenzofuran (4). It is suggested that the acetolysis of the dichloro compound takes place by a reversibly formed intimate ion pair (5) from which elimination and substitution products are

formed in rate-determining steps. A route is indicated
below for the formation of products (2) and (4) (Baciocchi
et. al., J. org. Chem., 1979, __44__, 32).

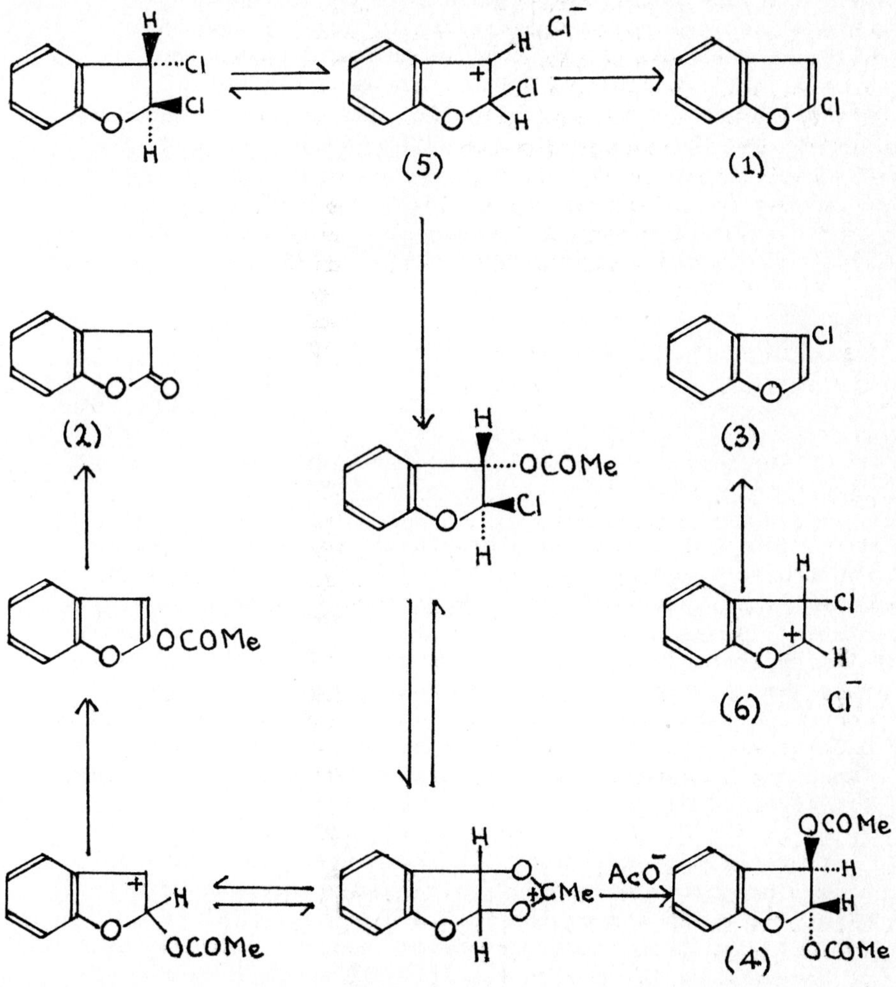

The presence of 3-chlorobenzofuran (3) in the reaction
products indicates that ion pair (6) is also formed.
Similar treatment of *cis*-2,3-dichloro-2,3-dihydrobenzo-
furan at 80° induces isomerization to the *trans*-isomer

in agreement with the greater stability of the *trans*-adduct (Baciocchi, Clementi, and Sebastiani, *loc. cit.*).

Treatment of *trans*-2,3-dibromo-2,3-dihydrobenzofuran with chloride ions results mainly in the formation of *trans*-2,3-dichloro- 2,3-dihydrobenzofuran by indirect halogen exchange. Since a nucleophilic displacement of bromide anion by chloride anion can be excluded, a mechanism involving the following equilibrium is suggested (Baciocchi *et. al.*, J. heterocyclic Chem., 1977, **14**, 949).

$$2Cl^- + Br_2 \rightleftharpoons 2Br^- + Cl_2$$

The chlorination and bromination of 2,3-dimethyl-benzofuran leads predominantly to the formation of side-chain substituted products. When the reaction occurs by a heterolytic mechanism the main product is 3-halogenomethyl--2-methylbenzofuran and 2-halogenomethyl-3-methylbenzofuran when the reaction occurs by a free radical mechanism. Competition between the two mechanisms is observed in conditions usually suitable for heterolytic halogenation (Baciocchi, Clementi, and Sebastiani, J. chem. Soc., Perkin II, 1974, 1882).

Labelling experiments indicate that 2,6-dichloro-phenyl propargyl ether (7) rearranges on heating to give 6-allenyl-2,6-dichlorocyclohexa-2,4-dien-1-one (8) in a [3s,3s]-sigmatropic rearrangement. Radical decomposition of intemediate (8) yields mainly 7-chloro-2-chloromethyl-benzofuran (9) and 3,8-dichloro-2H-benzo[b]pyran (10) and small amounts of other products (N. Sarcevic, J. Zsindely, and H. Schmid, Helv., 1973, **56**, 1457).

5-Chlorobenzofuran-2-carboxylic acid, m·p.266-267°, on decarboxylation with copper powder in quinoline gives 5-chlorobenzofuran, which on treatment with bromine in carbon disulphide at -10° affords 5-chloro-2,3-dibromo--2,3-dihydrobenzofuran. Subsequent addition to potassium hydroxide in 95% ethanol yields 3-bromo-5-chlorobenzofuran, m.p.75-76.5°. Similarly 6-chlorobenzofuran-2-carboxylic acid, m.p.244-246° gives 6-chlorobenzofuran and hence 6-chloro-2,3-dibromo-2,3-dihydrobenzofuran, m.p.63-65°, and 3-bromo-6-chlorobenzofuran m.p.36-37°. [1]H nmr spectra have been reported for the above compounds. 4-Bromo--5-methoxybenzofuran, m.p.88-89.5°; 3,4-dibromo-5--methoxybenzofuran, m.p.84-85°; 5-methoxy-2,3,4-tribromo--2,3-dihydrobenzofuran, m.p.81-83°; 1-(5-chloro-3-benzo-furyl)ethanol, m.p.84-85°, 4-nitrobenzoate, m·p·118-119° 1-(6-chloro-3-benzofuryl)ethanol, 4-nitrobenzoate, m.p.108.5-109.5°. The rates of solvolysis of the above 4-nitrobenzoates have been measured (D·S· Noyce and R.W. Nichols, J. org. Chem. 1972, 37 4311). 5- And 6--chloro-2-phenyl-, and 5- and 6-chloro-3-phenyl-benzofuran have been synthesized and their nmr spectra reported (M. Roche and E. Cerutti, Compt. rend., 1974, 279C, 663). Bromination of 2,3-diphenyl-6-methoxybenzofuran and 2,3-diphenyl-5-methoxybenzofuran with bromine in carbon tetrachloride gives 5-bromo-2,3-diphenyl-6-methoxybenzo-

furan and 4,6-dibromo-2,3-diphenyl-5-methoxybenzofuran,
respectively. Bromination of 2,3-diphenyl-5-methoxy-6-
-nitrobenzofuran affords 4-bromo-2,3-diphenyl-5-methoxy-6-
-nitrobenzofuran, which on reduction yields the amine
(O.H. Hishmat and A.H. Abd el Rahman, J. pr. Chem., 1973,
315, 227); 7-bromo-2-methyl-3-nitrobenzofuran (E.C. Ressmer
et. al., J. med. Chem., 1976, 19, 174).

Tetrachlorobenzyne reacts with acetone in the
presence of butan-2,3-dione to yield an adduct, 4-acetyl-
-2,2,4-trimethylbenzo-1,3-dioxan (11; R^1 = R^2 = Me).
Similar adducts (11; R^1 = R^2 = Et and R^1 = Me, R^2 = Et) are
obtained when acetone is replaced by diethyl ketone and
ethyl methyl ketone. On cleavage with sulphuric acid in
acetic anhydride, each adduct gives 3-acetoxymethyl-2-
-methyl-4,5,6,7-tetrachlorobenzofuran (12). Cleavage with
hydrobromic acid in acetic acid gives the 3-bromomethyl
derivative (13) (H. Heaney and C.T. McCarty, Chem. Comm.,
1970, 123).

3,5,6,7,8-Pentabromocoumarin on warming with alkali
yields 4,5,6,7-tetrabromobenzofuran-2-carboxylic acid,
which decarboxylates on heating with charcoal to give
4,5,6,7-tetrabromobenzofuran (G.P. Ellis and I.L. Thomas,
J. chem. Soc., Perkin I, 1973, 2781).

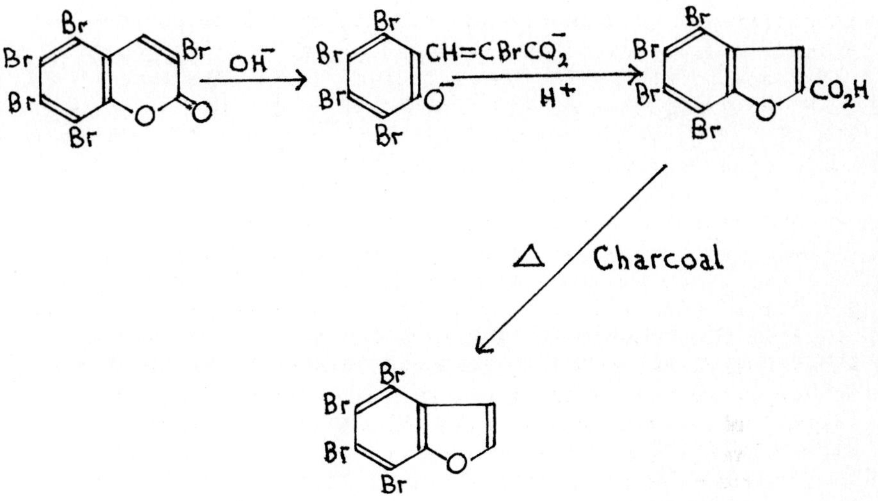

Fluorination of benzofuran with cesium tetrafluorocobaltate at 380° affords octafluorodihydrobenzofuran and other products. Pyrolytic defluorination of the octafluoro compound at 580° over nickel gives hexafluorobenzofuran, b.p.148° (J. Bailey, R.G. Plevey, and J.C. Tatlow, Tetrahedron Letters, 1975, 869).

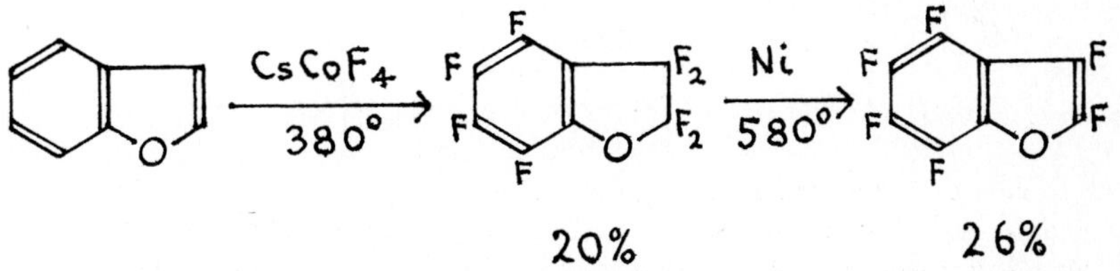

20%   26%

Treatment of benzofuran with trifluoro-oxymethane in freon gives a mixture of adducts (14; R[1] = H, R[2] = OCF_3, OMe; R[1] = OCF_3, F, R[2] = H), which on treatment with base yield benzofuran derivatives, for example (14; R[1] = OCF_3, R[2] =H) affords (15) (D.H.R. Barton et. al., J. chem. Soc., Perkin I, 1977, 2604).

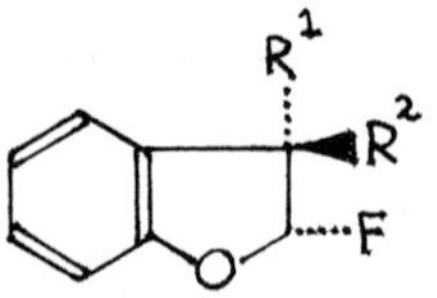

(14) 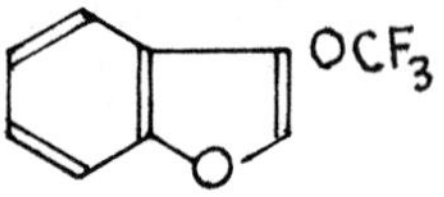   (15)

2- And 3-trifluoromethylbenzofuran are prepared by
the reaction of 2- and 3-bromobenzofuran with trifluoromethyl iodide in the presence of copper powder
(Y. Kobayashi, I. Kumadaki, and Y. Hanzawam, Chem. pharm.
Bull., 1977, 25, 3009).

(v) Nitro-, amino- and cyano-benzofurans

2-Nitrobenzofurans used as bactericides, fungicides,
protozoacides, and *Trichomonas* inhibitors, are prepared by
reaction of salicylaldehydes with nitromethyl bromide, for
instance, boiling salicylaldehyde with nitromethyl bromide
in acetone in the presence of potassium carbonate gives
2-nitrobenzofuran (R. Royer, P. Demerseman, and R. Cavier,
Ger. Offen. 2,131,927/1972).

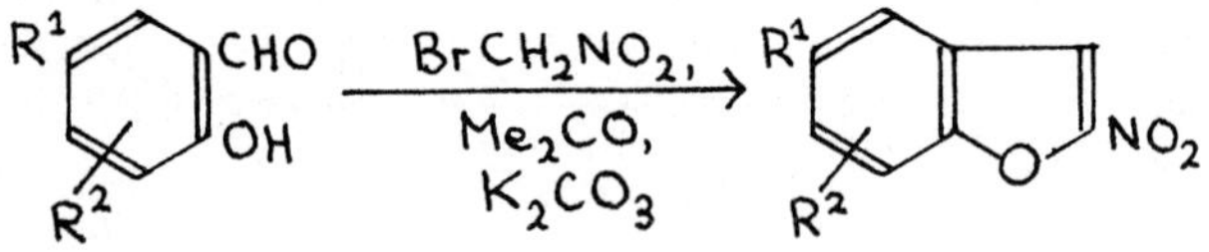

R^1, R^3 = H, Cl, Br, Me, or MeO

Benzofurans containing an electrophilic group in the
2-position, for example, 2-acetyl, 2-ethoxycarbonyl, and
2-cyano, on treatment with nitric acid in acetic anhydride
nitrate mostly in the 5- and 6-positions, and in some
instances in the 4-position and possibly in the 7-position.
In the case of the first two examples the substituent at

the 2-position is replaced by a nitro group. The amounts
of different isomers obtained on nitration are shown in
Table II (G. Lamotte, Demerseman, and Royer, J. heterocyclic
Chem., 1978, 15, 1343).

TABLE II

NITRATION OF SOME WITH BENZOFURANS WITH HNO_3-Ac_2O

Substituent	Amounts of isomers (%)				Amount of Substitution product (%)
	$4-NO_2$	$5-NO_2$	$6-NO_2$	$7-NO_2$	$2-NO_2$
2-Ac	12	31	20	3	18
2-CO_2Et	7	31	37		2(2,5-$diNO_2$)
2-CN	10	33	19	2	

Nitration of 5- and 7-hydroxy-2,3-dimethylbenzofuran gives
the 4-nitro, 6-nitro, and 4,6-dinitro derivatives;
4-methoxy- and 4,7- and 5,7-dimethoxy-2,3-dimethylbenzo-
furan yield the corresponding 6-nitro derivatives; and 4-
nitro derivatives are obtained from 7-methoxy- and 6,7-
-dimethoxy-2,3-dimethylbenzofuran (R. Royer *et. al.*, Bull.
Soc. chim. Fr., 1972, 163). 3,5,6-Trimethylbenzofuran on
nitration with nitric acid in acetic acid at 20° affords
2-nitro-3,5,6-trimethylbenzofuran (*idem*, Ger. Offen. 2,113,
489/1971). A number of 2-nitro-, and 2,4-, 2,5-, 2,6-, and
2,7-dinitro-benzofurans have been synthesized and their
activity against microorganisms investigated (*idem*, Eur.,
J. med. Chem.-Chim. Ther., 1978, 13, 411).

Methods of reducing 5-nitrobenzofuran to 5-amino-
benzofuran (R. Albrecht, Ann., 1972, 762, 55;
P.I. Abramenko, V.G. Zhiryakov, and T.K. Ponomareva, Khim.
Geterotsikl. Soedin., 1975, 1603) and 6-nitro-2,3-dihydro-
benzofuran to 6-amino-2,3-dihydrobenzofuran have been
described (Albrecht, H.J. Kessler, and E. Schroeder, Ger.
Offen. 2,416,519/1975).

Heating 4-(isopropylideneaminoxy)benzonitrile with
hydrogen chloride — acetic acid gives 5-cyano-2-methyl-
benzofuran as the major product. It is proposed that the
rearrangement of the *O*-aryloximes to benzofurans occurs in

a manner analogous to that for the Fischer indole synthesis
(A. Mooradian and P.E. Dupont, Tetrahedron Letters, 1967,
2867).

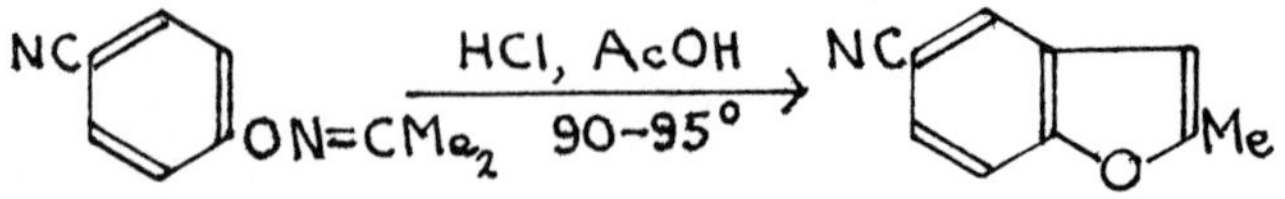

(vi) Hydroxybenzofurans

The 4-, 5-, 6-, and 7-hydroxybenzofuran have been
prepared by the reaction of the appropriate methoxysali-
cylaldehyde with ethyl chloroacetate, when a mixture of
ethyl methoxycoumarilate, methoxycoumarilic acid, and
methoxybenzofuran is obtained. Decarboxylation of the
coumarilic acid derivatives with copper in quinoline and
demethylation of the methoxybenzofurans using pyridine
hydrochloride gives the required hydroxybenzofuran (L. Rene
and Royer, Bull. Soc. chim. Fr., 1973, 2355).

6,7-Dimethyl-5-methoxybenzofuran a minor tobacco
constituent is prepared by the reaction of 2,3-dimethyl-
-4-methoxyphenol with bromoacetaldehyde diethyl acetal and
cyclization of the resulting product (A.J. Aasen,
B. Kimland, and C.R. Enzell, Acta Chem. Scand., 1971, 25,
3537).

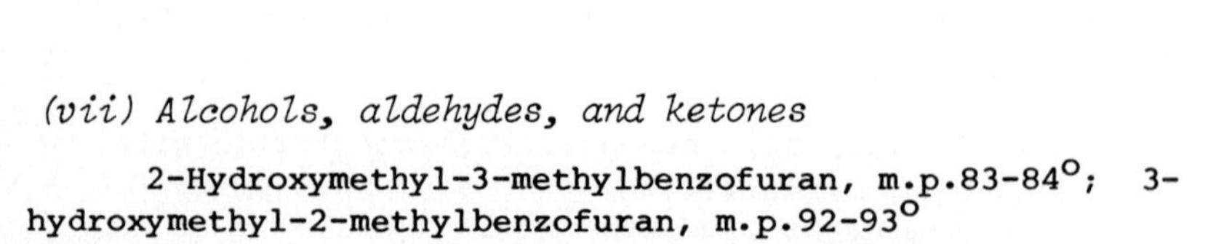

(vii) Alcohols, aldehydes, and ketones

2-Hydroxymethyl-3-methylbenzofuran, m.p.83-84°; 3-
hydroxymethyl-2-methylbenzofuran, m.p.92-93°
(E. Baciocchi, S. Clementi, and G.V. Sebastiani, J. chem.

Soc., Perkin II, 1974, 1882); 1-(5-methoxy-3-benzofuryl)-
ethanol, 4-nitrobenzoate, m.p. 126-127°; 3-acetyl-6-
-methoxybenzofuran, m.p 72.5-73°; 1-(6-methoxy-3-benzo-
furyl)ethanol, 4-nitrobenzoate, m.p. 106.5-108°
(D.S. Noyce and R.W. Nichols, J. org. Chem., 1972, 37,
4311).

The Rosenmund reduction of the acid chlorides of
3-methyl-, 3,5- and 3,7-dimethyl-, and 5- and 7-ethyl-3-
-methyl-benzofuran-2-carboxylic acid yields mainly the
corresponding 2-carboxaldehydes, with small amounts of the
related substituted 2-methylbenzofuran and substituted 1,2-
bis(2-benzofuryl)ethene (J. Wojtanis, B. Sila, and
T. Lesiak, Khim. Geterotsikl. Soedin., 1977, 744). The
aldehydes with potassium cyanide undergo the benzoin
reaction.

Benzofuran-3-carboxaldehyde may be synthesized in
high yield from readily available ethyl 3-methylbenzo-
furan-2-carboxylate (ethyl 3-methylcoumarilate), by
conversion into ethyl 3-hydroxymethylbenzofuran-2-
-carboxylate, followed by oxidation with manganese dioxide,
hydrolysis and decarboxylation (A. Shafiee and
M. Mohamadpour, J. heterocyclic Chem., 1978, 15, 481).

The 5- and 7-hydroxybenzofuran-4-carboxaldehyde (1
and 2, R^1 = R^2 = H), 6-hydroxybenzofuran-5-carbox-
aldehydes (3, R^1 = R^2 = H, R^3 = H, Me), 6-hydroxy-
-7-methylbenzofuran-5-carboxaldehyde (4, R^1 = R^2 = H),
and 4- and 6-hydroxybenzofuran-7-carboxaldehyde (5 and 6,
R^1 = R^2 = H) are prepared by demethylating the related
methoxy derivatives (1)-(6) (R^2 = Me) with pyridine
hydrochloride in quinoline. The methoxy compounds are
obtained by formylating the related coumarilic acid
esters (1)-(6) (R^1 = CO_2Me, CO_2Et, R^2 = Me), then
hydrolysing and decarboxylating the resulting acids (1)-(6)
(R^1 = CO_2H) (Rene, J.P. Buisson, and Royer, Bull. Soc.
chim. Fr., 1975, 2763).

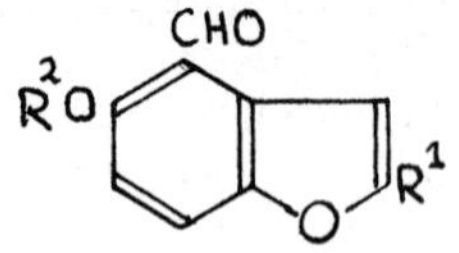

(1)

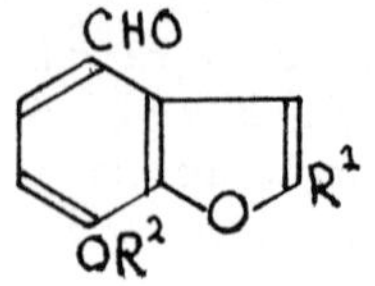

(2)

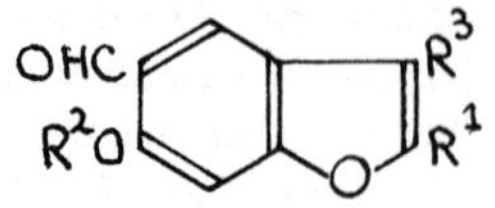

(3)

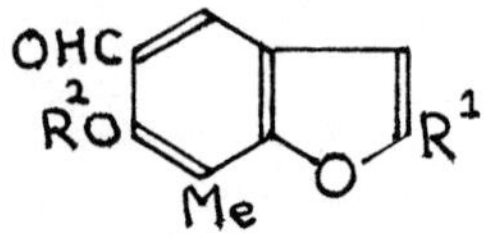

(4)

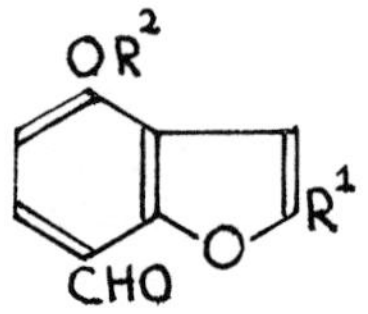

(5)

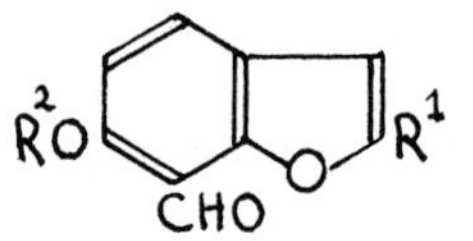

(6)

4-Hydroxybenzofuran-5-carboxaldehyde, 7-hydroxy-benzofuran-6-carboxaldehyde, 5-hydroxybenzofuran-4--carboxaldehyde, 6-hydroxybenzofuran-7-carbaldehyde (Royer *et. al.*, Eur. J. med. Chem. -Chim. Ther., 1978, <u>13</u>, 213). Appropriate benzofuran-2-carboxaldehydes or 2-acetylbenzo-furans are starting materials for a number of benzofurans (7, R^1 = CO_2H, CH_2CO_2H, $CH(OH)CO_2H$, $COCO_2H$, $CH=CR^4CO_2H$, R^4 = H, Me, Et, $4-ClC_6H_4$; R^2 = H, Me, Et, Ph, Cl, $4-ClC_6H_4$; R^3 = H, 5-Cl, 5-OMe, 5-Br, 5-I, 5-F, 6-OMe) (A. Aurozo *et. al.*, *ibid.*, 1975, <u>10</u>, 182).

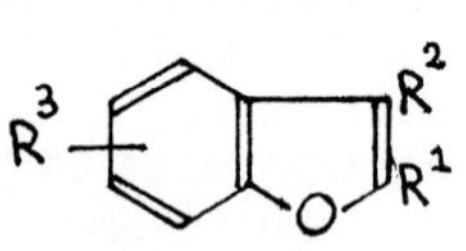

(7)

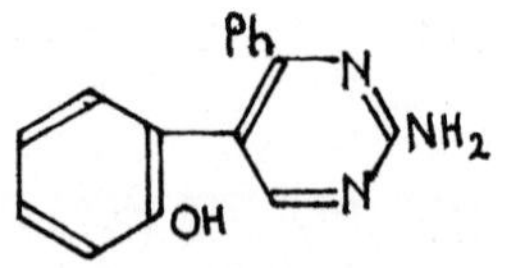

(8)

2-Phenylbenzofuran-3-carboxaldehyde undergoes nucleo-philic ring cleavage when treated with sodium hydroxide or

guanidine-sodium ethoxide to yield 2-hydroxybenzyl phenyl
ketone or 2-amino-5-(2-hydroxyphenyl)-4-phenylpyrimidine
(8) (K. Takagi and T. Ueda, Chem. Pharm. Bull., 1972, 20,
2053). Benzofuran analogues of flavone (9, R = H, 6-, and
7-Me and OMe, 6-Cl, 6-Br) have been obtained from benzo-
furan-2-carboxaldehyde and the appropriate 2-hydroxyaceto-
phenone (L.G. Grishko *et. al.*, Dopov. Akad. Nauk. Ukr. RSR,
Ser. B: Geol., Khim. Biol. Nauki, 1978, 426).

(9)   (10)

The acetylation of 2,3-dimethylbenzofuran provides an
example of a 'non-conventional' Friedel-Crafts reaction
(p.63). The reduction of 2-acetyl-5-halogenobenzofurans
furnishes 2-ethyl-5-halogenobenzofurans, which on Friedel-
Crafts acylation afford 3-acyl-2-ethyl-5-halogenobenzo-
furans (Takagi and Ueda, Chem. Pharm. Bull., 1975, 23,
2427). 2-Benzofuryl vinyl ketones (10, R^1 = H, Me, Et,
R^2 = Me; R^3 = H, OMe, Cl) are prepared from the
appropriate 2-acylbenzofurans *via* the Mannich reaction and
hydrolysis of the resulting Mannich base. When R^2 = H
the vinyl ketones are too unstable to be isolated before
they polymerise (R. Faure and G. Mattioda, Bull. Soc. chim.
Fr., 1973, 3059).

(viii) Benzofuran carboxylic acids

4-Chlorocoumarin on boiling with sodium hydroxide in
aqueous dioxane yields benzofuran-2-carboxylic acid (80%)
(V.A. Zagorevskii, V.L. Savel'ev, and N.V. Dudykina, Khim.
Geterotsikl. Soedin., Sb. 2: Kislorodsoderzhashchie
Geterotsikly. 1970, 155). The pk_a values and C = O
stretching frequencies of benzofuran-2-carboxylic acid and
the 2-carboxylic acids of benzothiophene, benzoselenophene,
benzotellurophene and their parent heterocycles have been

reported (F. Frinquelli and A. Taticchi, J. heterocyclic Chem., 1973, 10, 89). The Rosenmund reduction of 3-methyl-benzofuran-2-carboxylic acid chloride and some of its derivatives gives mainly the related 2-carboxaldehyde (p. 75).

6-Methoxybenzofuran-3-carboxylic acid, m.p. 179-180°, ethyl ester, m.p. 55-56°; ethyl 4-methoxybenzo-furan-2-carboxaldehyde-3-carboxylate, m.p. 153.5-154.5°; ethyl 6-methoxybenzofuran-2-carbaldehyde-3-carboxylate, m.p. 164-165° (D.S. Noyce and R.W. Nichols, J. org. Chem., 1972, 37, 4311). Derivatives (1) of alkyl 5-alkoxybenzo-furan-3-carboxylate with a CH_2SR^1 (R^1 = alkyl) group in the 2-position are prepared by treating alkyl 5-alkoxy--2-methylbenzofuran-3-carboxylate with N-bromosuccinimide and then reacting the resulting alkyl 5-alkoxy-2-bromo-methylbenzofuran-3-carboxylate with the appropriate mercaptan (F.A. Trofimov *et. al.*, U.S.S.R. 349,686/1972).

(1)

R^1, R^2 R^3 = alkyl

(2)

Dimethyl acetylenedicarboxylate and methyl propiolate react generally with 2-vinylfurans to yield the corresponding dimethyl benzofuran-4,5-dicarboxylates and methyl benzofuran-4-carboxylates, respectively. In several cases the dimethyl acetylenedicarboxylate does not react exclusively with the exo-cyclic diene system of the 2-vinylfuran and the corresponding dimethyl 3,6-epoxy-3--vinyl-3,6-dihydrophthates are formed by Diels-Alder addition to the furanoid diene system. At 80° 2-vinyl-furan and dimethyl acetylenedicarboxylate form the 1:2 adduct (2) (W.J. Davidson and J.A. Elix, Austral. J. Chem., 1973, 26, 1059).

5-Acetoxybenzofuran-4-carboxylic acid (3, R = H, Me)

and the related 6-carboxylic acids (4) analogues of
aspirin, show analgesic, anti-inflammatory, and blood
platelet aggregation inhibition activity.   5-Acetoxy-
benzofuran-4-carboxylic acid (3, R = H) is obtained by
oxidising 4-acetyl-5-methoxybenzofuran with sodium
hypobromite, then demethylating the resulting acid and
forming the acetyl derivative of the phenolic product
(Royer *et. al.*, Eur. J. med. Chem. -Chim. Ther., 1979, $\underline{14}$,
223).

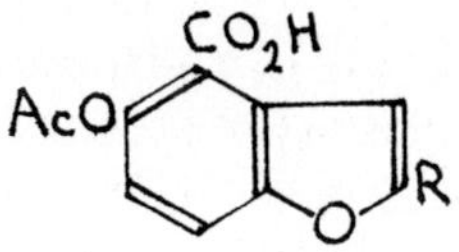

(3)

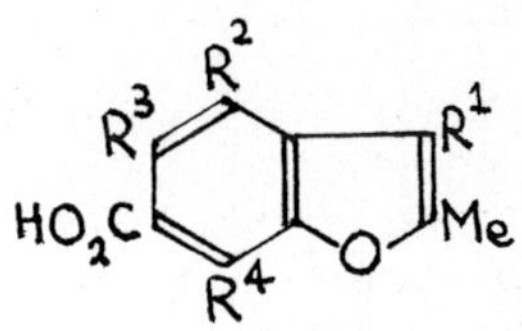

(4)

R = H, Me

R^1 = H, Me; R^2 = H, OAc ;
R^3, R^4 one is H the other OAc

Cyclization of 3-arylbenzofuran-2-acetic acid (R^1 =
R^2 = H, MeO; R^1 = H, R^2 = MeO) with phosphoric
anhydride in C_6H_6 gives the ester (5) (J.N. Chatterjea
et. al., J. Indian chem. Soc., 1976, $\underline{53}$, 295).

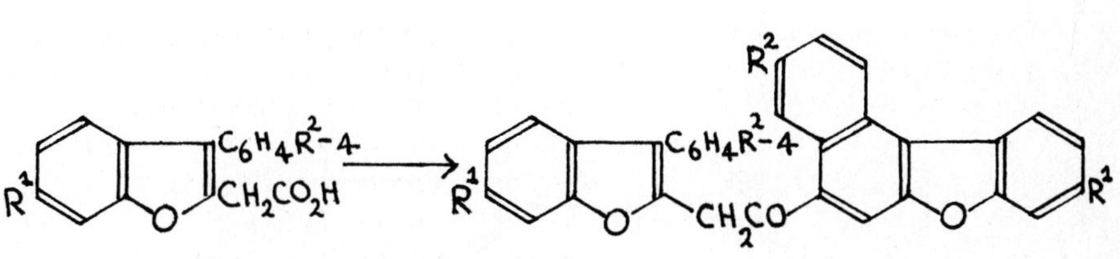

(5)

(b) Hydrobenzofurans

(i) 2,3-Dihydrobenzofurans (coumarans)

Benzofurans (1; R^1 = H, R^2 = Ph) and (2) on uv
irradiation in aliphatic amines are photoreduced to the
2,3-dihydro derivatives. Under these conditions benzofuran
(1; R^1 = R^2 = Ph) is photocyclized and benzofuran (1;
R^1 = R^2 = Me) in propylamine or triethylamine gives no
identifiable photoproducts. (1; R^1 = Ph, R^2 = H)
photodimerizes in propylamine (C. Parkanyi *et. al.*, J. org.
Chem., 1976, **41**, 151).

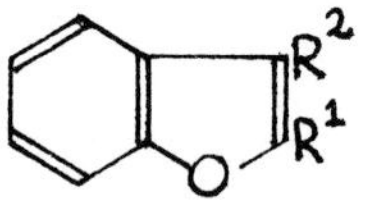

(1) 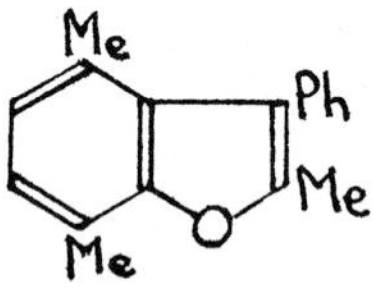   (2)

The electrochemical alkoxylation of benzofurans (3;
R^1 = H, R^2 = H, Me; R^1 = CO_2Et, R^2 = Me) in
methanol using sulphuric acid, potassium hydroxide, or
sodium perchlorate as a supporting electrolyte at -10°
yields a mixture of *cis*- and *trans*- 2,3-dimethoxy
derivatives. Benzofuran in ethanol affords 2,3-diethoxy-
-2,3-dihydrobenzofuran (J. Srogl *et. al.*, Synth., 1975,
717).

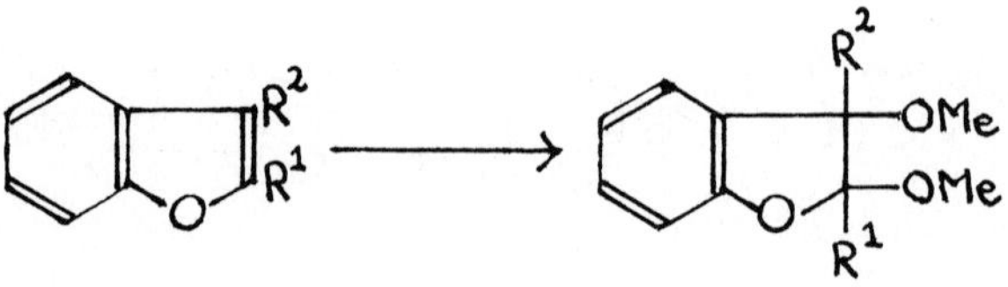

(3)

Treatment of benzofuran with thallium (III) nitrate in

methanol gives 2,3-dimethoxy-2,3-dihydrobenzofuran,
b.p.110°/7mm (A. McKillop *et. al.*, J. Amer. chem. Soc.,
1973, <u>95</u>, 3635).

2,3-Dihydrobenzofuran *(coumaran)* may be obtained by
the ionic hydrogenation of benzofuran with trifluoroacetic
acid-triethylsilane. Analogous hydrogenation of 2-methyl-,
and 3-methyl- benzofuran gives the respective 2,3-dihydro-
benzofurans (yields ∿88 and ∿65%) (E.A. Karakhanov *et. al.*,
Doklady Akad. Nauk SSSR, 1974, <u>214</u>, 584).

$$\text{benzofuran} \xrightarrow[\substack{20-60° \\ 3-10h}]{CF_3CO_2H - Et_3SiH} \text{dihydrobenzofuran}$$

6 – 55%

The passage of 2-allylphenol over alumina at 300°
gives 2-methyl-2,3-dihydrobenzofuran (82%), phenol (6%),
and 2-propylphenol (10%). The incorporation of zinc
chloride in the catalyst increases the yield of phenols and
decreases the yield of 2-methyl-2,3-dihydrobenzofuran (5%)
and also results in dehydrogenation of the latter
(Karakhanov, A.A. Freger, and E.A. Viktorova, Izv. Vyssh.
Ucheb. Zaved., Khim. Khim. Tekhnol., 1973, <u>16</u>, 586). The
decomposition of 2-ethoxybenzaldehyde tosylhydrazone at
258°/0.06mm affords 2-methyl-2,3-dihydrobenzofuran (76%),
2-methylphenyl vinyl ether (1.3%), and chroman (2%)
(W.D. Crow and H. McNab, Austral. J. Chem., 1979, <u>32</u>, 99).

$$\text{CH=NNHTos, OEt} \xrightarrow{\Delta} \text{CH:, OEt} \xrightarrow[0.06\,mm]{258°} \text{Me}$$

Preparation of some 2,3-dihydrobenzofurans have been
described (E.A. Titov and A.S. Grishchenko, Vopr. Khim.
Khim. Tekhnol., Resp. Mezhved, Temat. Nauchno-Tekh. Sb.,
1973, No.<u>29</u>, 36).

2,3-Dihydrobenzofuran can be dehydrogenated by nickel peroxide in benzene to give benzofuran (D.L. Evans *et. al.*, J. org. Chem., 1979, 44, 497). The dehydrogenation of dihydrobenzofurans using different catalysts under varying conditions and the analysis of the resulting products has been reported, for example, 2,3-dihydrobenzofuran and 2-methyl-2,3-dihydrobenzofuran over Pt-Al$_2$O$_3$ (Karakhanov *et. al.*, Vestn. Mosk. Univ., Khim., 1971, 12, 710); 2,3-dihydrobenzofuran, 2-alkyl-2,3-dihydrobenzofuran, and 3-methyl-2,3-dihydrobenzofuran treated with Pt, Pd, or Rh on carbon at 250-350° in a stream of nitrogen (Karakhanov, L.G. Saginova, and Viktorova, Geterogennyi Katal. Reakts. Poluch. Prevrashch. Geterotsikl. Soedin., 1971, 81); 2,2-dimethyl-2,3-dihydrobenzofuran with Pt or Pd at 250-300° (Viktorova, Karakhanov, and Saginova, *ibid.*, p.85); 2-ethyl-2,3-dihydrobenzofuran using ZnCl$_2$--Al$_2$O$_3$ gives 2-methylchroman *(idem, ibid.)*; 2,3--dihydrobenzofuran and 2-methyl-2,3-dihydrobenzofuran give the related octahydrobenzofurans and benzofurans, cyclo-hexane derivatives, phenols, and alkylbenzenes when treated with hydrogen over Pt on carbon at 200-350° (Karakhanov, Saginova, and Vikororova, Vestn. Mosk. Univ., Khim., 1972, 13, 86); under the same condition the octahydrobenzo-furans give similar products. 2,3-Dihydrobenzofurans may be dehydrogenated by triphenylmethyl perchlorate in acetic or formic acid, or by triphenylmethyl chloride in the presence of tin (IV) chloride, but chloranil requires drastic conditions, such as boiling xylene and affords poor yields of dehydrogenated products. With some dihydro-benzofurans isomerisation and ring expansion occurs, for instance, 2,2-dimethyl-2,3-dihydrobenzofuran and 2-ethyl--2,3-dihydrobenzofuran afford 2,3-dimethylbenzofuran, and 2-methylchroman respectively (Karakhanov, E.A. Dem'yanova, and Viktorova, Doklady Akad. Nauk SSSR, 1972, 204, 879). Dehydrogenation products are obtained on treating dihydrobenzofurans with Ph$_3^+$BF$_4^-$ in acetonitrile, nitro-methane, or dichloroethane. The reaction proceeds *via* a 1 electron transfer *(idem, ibid.,* 1977, 233, 369). The catalytic activity of activated carbon in dehydrogenation of 2,3-dihydrobenzofurans has been studied (Karakhanov, M.V. Vagabov, and Viktorova, Vestn., Mosk. Univ., Khim., 1972, 13, 493; Karakhanov *et. al.*, Isv. Vyssh. Ucheb. Zaved., Khim. Khim. Tekhnol., 1973, 16, 84). Pyrolysis of 3-methyl-2,3-dihydrobenzofuran at 305° in the presence of Al$_2$O$_3$ gives a mixture of 2-methyl-2,3-dihydrobenzofuran,

3-methylbenzofuran, and phenols. Increasing the acidity of
the Al_2O_3 by treatment with CCl_4 results in an
increase in phenolic products. Similar treatment of
2-methyl-2,3-dihydrobenzofuran and 2,3-dihydrobenzofuran
yields the related benzofurans and phenols (Freger
et. al., Vestn. Mosk. Univ., Khim., 1975, <u>16</u>, 353).
2,3-Dimethyl-2,3-dihydrobenzofuran in the presence of
aluminosilicate at 433° gives 2,3-dimethylbenzofuran
(17%) and by isomerisation 2-methylchroman (73%)
(Karakhanov, S.V. Lysenko, and Viktorova, *ibid.*, 1974, <u>15</u>,
500). Also 2-ethyl-2,3-dihydrobenzofuran isomerizes at
300-400° in the presence of activated carbon to yield
2-ethylbenzofuran and 2-methylchroman, and 2,2-dimethyl-
-2,3-dihydrobenzofuran affords 2,3-dimethyl-2,3-dihydro-
benzofuran and 2,3-dimethylbenzofuran (Karakhanov *et. al.*,
Khim. Geterotsikl. Soedin., 1975, 321).

The uv photoelectron spectra of 2,3-dihydrobenzofuran,
chroman, tetrahydrofuran and some other ethers have been
compared (J.M. Behan, F.M. Dean, and R.A.W. Johnstone,
Tetrahedron, 1976, <u>32</u>, 167); uv spectra (A.C.P. Alves,
J.M. Hollas, and B.R. Midmore, J. Mol. Spectrosc., 1979,
<u>77</u>, 124), and mass spectra of 2,3-dihydrobenzofuran have
been reported (W.J. Richter, J.G. Liehr, and
A.L. Burlingame, Org. mass. Spec., 1973, <u>7</u>, 479). The nmr
spectra of 2-methyl-, and 2-ethyl-2,3-dihydrobenzofuran,
and 2,3-dihydrobenzofuran have been used to indicate their
conformation (V.S. Petrosyan *et. al.*, Vestn Mosk. Univ.,
Khim., 1973, <u>14</u>, 717).

The reaction between dihydrobenzofuran derivatives
and Grignard reagents gives phenol in addition to products
previously described. The phenol results from dealkylation
of 2-alkylphenol, catalysed by the magnesium halide moiety
of the Grignard reagent (S. Cabiddu, R. Pirisi, and
A. Plumitallo, J. organometal. Chem., 1975, <u>88</u>, 129).

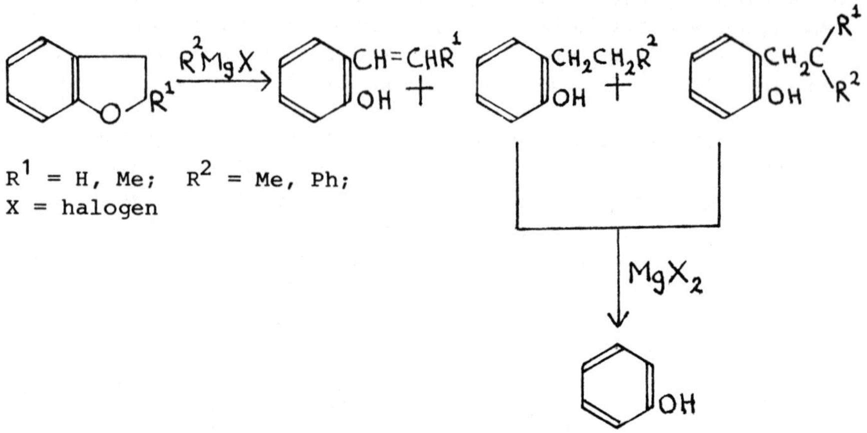

R^1 = H, Me; R^2 = Me, Ph;
X = halogen

Reaction of 2,2-dimethyl-2,3-dihydrobenzofuran with an alkylsodium RNa(R = Bu, C_5H_{11}, or C_6H_{13}) affords 2,2-dimethyl-2,3-dihydro-7-benzofurylsodium (A. Serban and P.K. Engel, Ger. Offen. 2,258,988/1973;   Austral. 465,512/1975). Treatment of the dihydrobenzofuran in hexane with lithium in paraffin wax gives the corresponding lithium derivative (R.F. Brown, Ger. Offen. 2,309,379/1973). Air oxidation of either metal derivative and acidification yields 2,3-dihydro-2,2-dimethyl-7-hydroxybenzofuran, which can be converted to the insecticidal carbofuran (4).

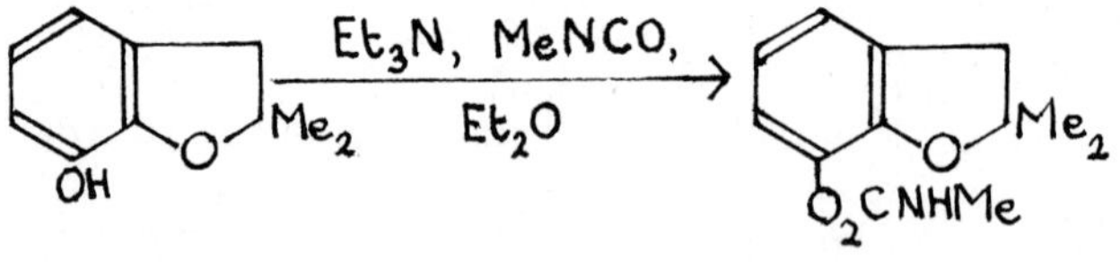

(4)

A number of derivatives of 2,3-dihydrobenzofuran-7-carbox-lic acid have been reported (W. Anderson *et. al.*, Brit. 1,314,325).   2,3-Dihydrobenzofuran-7-carboxylic acid and

5-methoxy-2-methyl-2,3-dihydrobenzofuran-7-carboxylic acid
have been prepared with a [14]CO_2H group by carbonation of
the corresponding Grignard reagent (L.F. Elsom *et. al.*, J.
Labelled Compd. Radiopharm., 1977, 13, 75).

   Treatment of 3,3-dimethyl-2-hydroxy-2,3-dihydro-
benzofuran with methyl dimethylsulphoxide (DMSOM) yields
4,4-dimethyl-3-hydroxychroman, which on pyrolysis of its
acetate affords 4,4-dimethylchrom-2-en (M.C. Sacquet,
B. Graffe, and P. Maitte, Tetrahedron Letters, 1972, 4453).

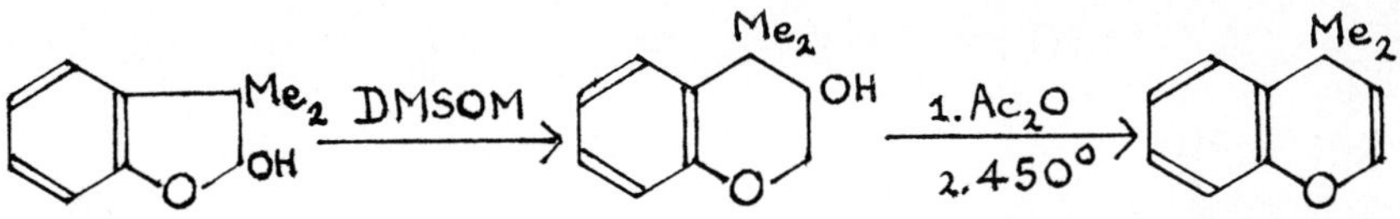

   2-Oxo-2,3-dihydrobenzofurans are readily obtained by
reacting glyoxal with phenols at high temperatures in the
presence of a relatively small amount of acid catalyst.
The 3-oxo-2,3-dihydrobenzofurans are not formed even
though, according to resonance theory they should be more
stable than the 2-oxo compounds (R.W. Layer, J. heterocyclic
Chem., 1975, 12, 1067).

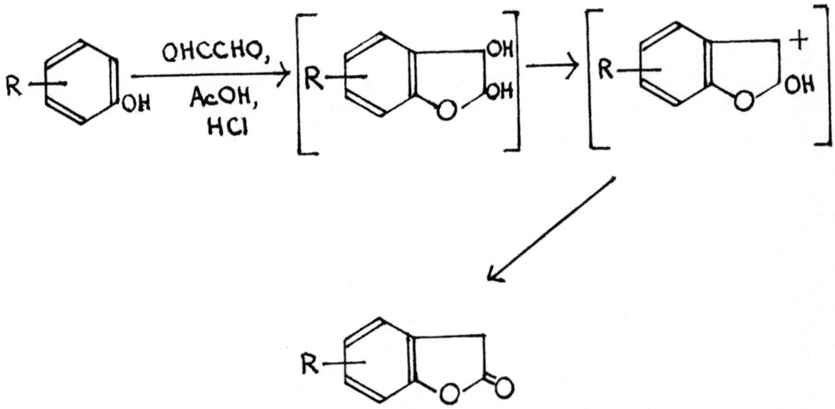

4,7-Di-*tert*-butyl-5-hydroxy-, m.p.209-211°; 5-methyl-
-7-*tert*-butyl-, m.p.179-180°; 5-chloro-4-methyl-,
m.p.132-133°; 5,7-di-*tert*-pentyl-, m.p.57-58°, 5,7-
-dimethyl-2-oxo-2,3-dihydrobenzofuran, m.p.125-127°.

The thermolysis of 2-oxo-2,3-dihydrobenzofuran at
750° and 10^{-3} - 10^{-1}mm gives carbon monoxide, fulvene
(6), benzene and a trimer of intermediate (5) (C. Wentrup
and P. Müller, Tetrahedron Letters, 1973, 2915).

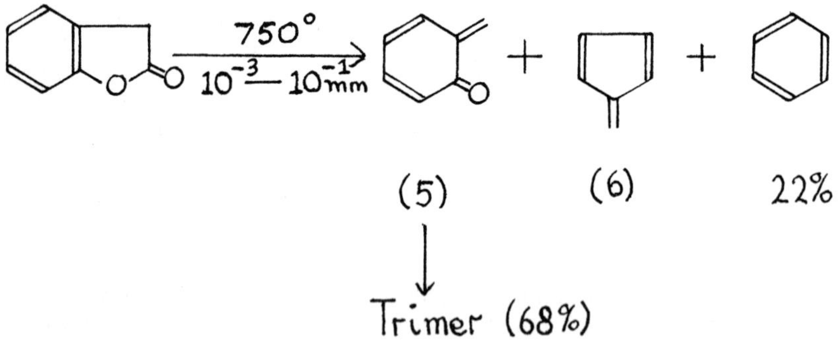

(5)   (6)   22%

Trimer (68%)

The photolysis of 2-oxo-2,3-dihydrobenzofuran in
methanol gives methyl 2-hydroxybenzyl ether
(B.A. M. Oude-Alink, A.W.K. Chan, and C.D. Gutsche, J. org.
Chem., 1973, 38, 1993).

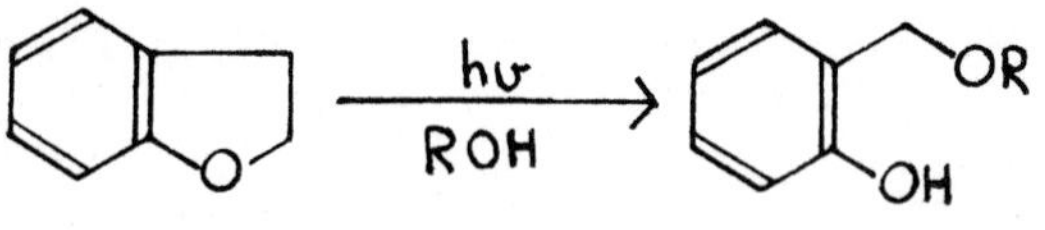

The photolysis of a number of 3,3-disubstituted 2-oxo-2,3-
-dihydrobenzofurans has been studied. 3,3-Dimethyl-,
b.p.43°/0.05mm; 3,3-diethyl-, b.p.68-70°/0.07mm;
3-phenyl-, m.p.117-118°; 3-methyl-3-phenyl-, b.p.130-
135°/0.04mm; and 3-ethyl-3-phenyl-2-oxo-2,3-dihydro-
benzofuran, m.p.69-70°. 3-Phenyl-2-oxo-2,3-dihydro-
benzofuran may be synthesised starting from 3-phenyl-
benzofuran-2-carboxylic acid (J.N. Chatterjea and
C. Bhata, J. Indian. chem. Soc., 1976, 53 293).

Methylation of 3-acyl-2-oxo-2,3-dihydrobenzofuran with diazomethane generally leads to a mixture of 3-acyl--2-methoxybenzofuran and 3-(α-methoxy)alkylidene-2-oxo--2,3-dihydrobenzofuran, hence providing the first evidence for the existence of 3-acyl-2-oxo-2,3-dihydrobenzofurans in the 3-acyl-2-hydrobenzofuran tautomeric form, as well as the widely accepted 3-(α-hydroxy)alkylidene-2-oxo-2,3--dihydrobenzofuran form.

The reaction of 3-acetyl-2-oxo-2,3-dihydrobenzofuran with methyl iodide and potassium carbonate yields a mixture of 3,3-dimethyl-2-oxo-2,3-dihydrobenzofuran and 3-methyl-2-oxo--2,3-dihydrobenzofuran. These products are also formed on methylating 2-oxo-2,3-dihydrobenzofuran under similar conditions (J.A. Elix and B.A. Ferguson, Austral. J. Chem., 1973, 26, 1079).

3-Oxo-2,3-dihydrobenzofuran may be synthesized from ethyl salicylate by first condensing it with 3.2 equivalents of the methylsulphinyl carbanion to obtain 2-hydroxy-2-[(methysulphinyl)acetyl]benzene (7), which cyclizes in boiling benzene in the presence of trifluoro-acetic acid to give 2-methylthio-3-oxo-2,3-dihydrobenzo-furan (8). Reductive cleavage of the thiomethyl group of (8), using deactivated W-2 Raney-nickel yields 3-oxo-2,3--dihydrobenzofuran (D.R. Amick, J. heterocyclic Chem., 12, 1051).

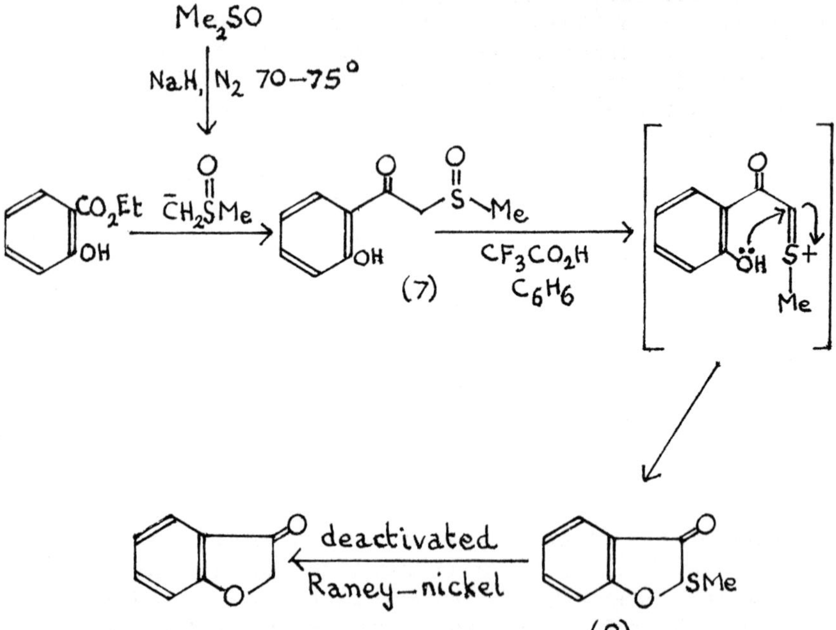

3-Oxo-2,3-dihydrobenzofurans undergo efficient Wittig reactions with resonance-stabilised phosphoranes at elevated temperatures, hence furnishing a convenient synthesis of benzofurans functionally substituted at the 3--position. The reaction of 3-oxo-2,3-dihydrobenzofurans with cyanomethylenephenylphosphorane affords 3-benzofuryl-acetonitriles, which on reduction with lithium tetrahydridc-aluminate yield the physiologically active 2-(3-benzofuryl)-ethylamines (J.H.T. Chan, Elix, and Feguson, Austral. J. Chem., 1975, 28, 1097). Some 2-aryl-3-oxo-2,3-dihydrobenzo-furans have been prepared (Chatterjea and Bhata, *loc. cit.*, p.86).

A number of 2,3-dihydrobenzofuryl chrysanthemates (9) have been prepared by reducing the appropriate 3-oxo-2,3--dihydrobenzofuran to the 3-hydroxy derivative followed by condensation with chryanthemoyl chloride. Related thiophenyl compounds (9; O = S) have been obtained in a similar manner (Y. Nakada *et. al.*, Agric. Biol. Chem., 1978, 42, 1767). The insecticidal activity of (9; R = 7-Me, 7-Cl, 4-Cl, 4,6-Me$_2$) on the American cockroach is much more potent than allethrin.

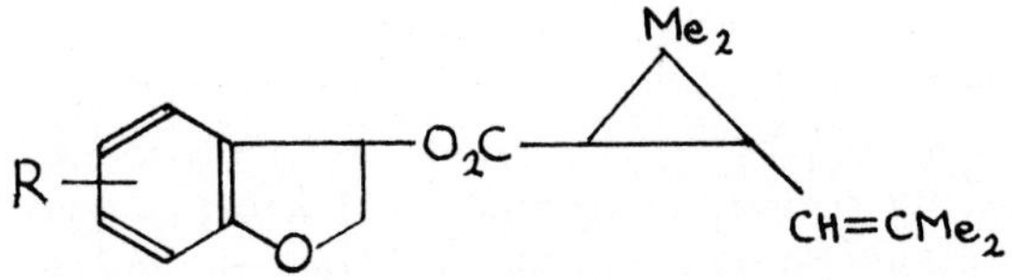

(9)

The 2-3-dioxo-2,3-dihydrobenzofuran ring-opens on treatment with aniline, phenylhydrazine, or semithio-carbazide in benzene to give 2-HOC$_6$H$_4$COCONHR (10; R = Ph, PhNH, NH$_2$CSNH). The reaction between the dioxodihydrobenzofuran and aniline or semicarbazide in aqueous alochol affords 2-HOC$_6$H$_4$C(=NPh)CO$_2$H and 2-HOC$_6$H$_4$C(=NNHCONH$_2$)CO$_2$H, respectively. Compounds (10; R = PhNH or H$_2$NCONH) on boiling in acetic acid yield the corresponding hydrazones (A.B. Tomchin, I.S. Ioffe, and E.A. Rusakov, Zh. Org. Khim., 1974, 10, 604).

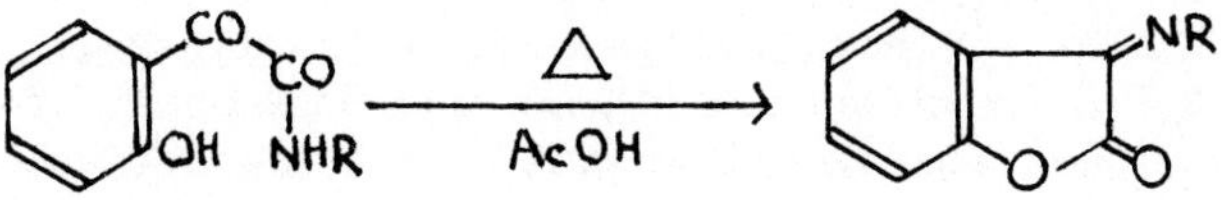

(10)

The pyrolysis of 2,3-dioxo-2,3-dihydrobenzofuran at 600° gives the ketoketene (11) also obtained by similar treatment of 2-hydroxybenzoic acid (R. Schulz and A. Schweig, Tetrahedron Letters, 1977, 59).

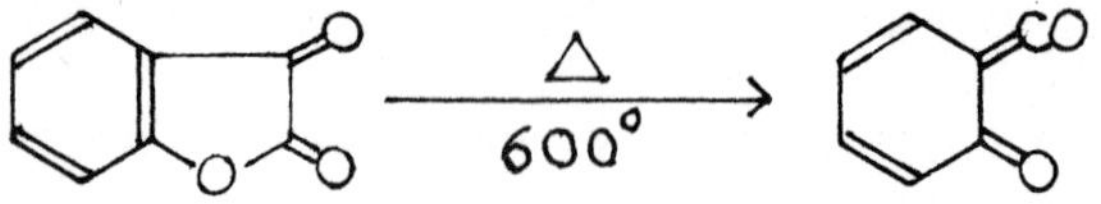

(11)

The reaction of 2,3-dioxo-2,3-dihydrobenzofuran with phosphorane (12) gives compound (13) (M.M. Sidky and L.S. Boulos, Phosphorus Sulphur, 1978, 4, 299).

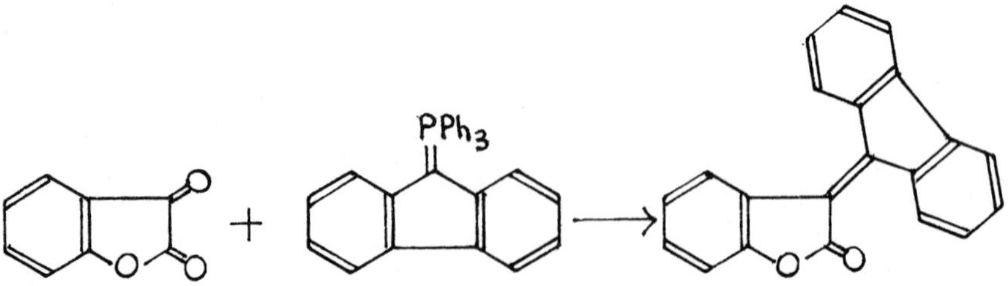

(12)   (13)

The photoelectron [He(I)] (V. Galasso, F.P. Colonna, and G. Distefane, J. Electron Spectrosc. Relat. Phenom., 1977, 10, 227), ^{1}H nmr (Galasso et. al., Org. mag. Reson., 1975, 7, 591), and ^{13}C nmr spectra of 2,3-dioxo--2,3-dihydrobenzofuran and related heterocycles have been reported (Galasso, G. Pellizer, and G.C. Pappalardo, ibid., 1977, 9, 401). The latter includes CNDO/S studies.

(ii) 4,5-Dihydrobenzofurans

The gas phase vacuum pyrolysis of 1-(2-furyl)-1,3--butadiene gives 4,5-dihydrobenzofuran (70%); also formed are 6,7-dihydrobenzofuran and benzofuran (B.I. Rosen and W.P. Weber, Tetrahedron Letter, 1977, 151).

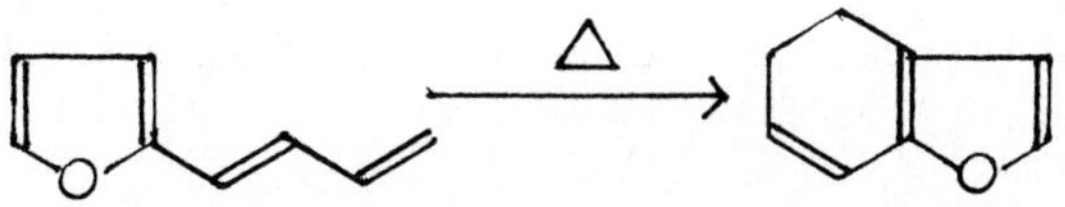

(iii) Tetrahydrobenzofurans

The cyclization of γ-methylthio-βγ-unsaturated ketones with titanium (IV) chloride gives furan derivatives in moderate yields. If the starting material is a derivative of cyclohexane the product is a 4,5,6,7-tetrahydrobenzofuran (S. Kano *et. al.*, Chem. Comm., 1979, 238).

LDA (lithium diisopropylamide)
R^1 = H, Me

R^2 = Me, Et

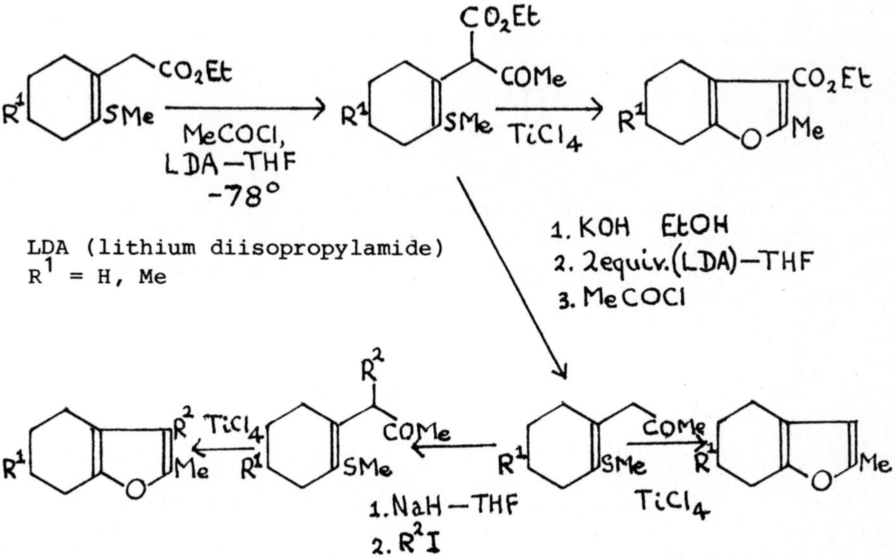

Menthofuran (3,6-dimethyl-4,5,6,7-tetrahydrobenzofuran) is prepared from pulegone oxide (Y. Watanabe and N. Iwata, Japan Pat. 79 73,766/1979).

The α-hydroxy-γ-keto diester (14) on treatment with

264

phosphorus pentoxide in methanesulphonic acid affords the
butenolide (15) (ethyl 2-oxo-2,4,5,6-tetrahydrobenzofuran-
-3-carboxylate) (A.G. Schultz and Y.K. Yee, J. org. Chem.,
1976, 41, 561).

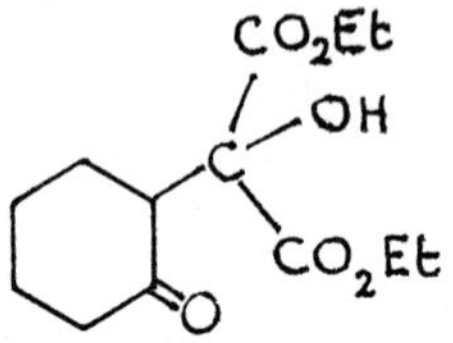

(14) 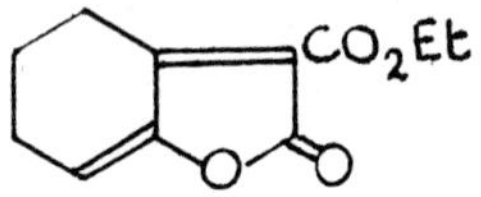   (15)

The ^{13}C nmr of some 4,5,6,7-tetrahydrobenzofurans
including furanosesquiterpenes have been reported (K. Tori
et. al., Tetrahedron Letters, 1975, 4583).

(iv) Hexahydrofurans

The thermal rearrangement of ketone (16; R^1 = Me,
R^2 = H) (a spiro[5.2]octanone) gives 2-allylcyclohexa-
none and a 1:1 mixture of 2- and 3-methyl-2,3,4,5,6,7-
-hexahydrobenzofuran.

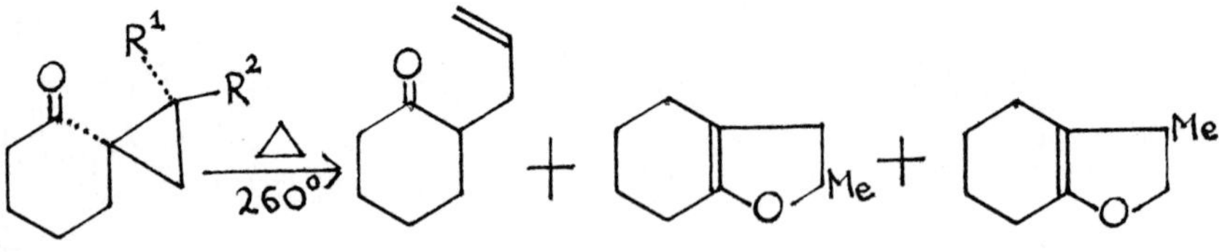

(16)   82%   10%

With the *anti*-isomer (16; R^1 = H, R^2 = Me) the amount
of the two hexahydrobenzofurans (54%) exceeds that of
2-allylcyclohexanone (30%). The thermal rearrangement of
(16; R^1 = R^2 = Me) at 200° yields besides other
products 2,2-dimethyl-2,3,4,5,6,7-hexahydrobenzofuran (12%)

(P.G. Khanazanie and E. Lee-Ruff, Canad. J. Chem., 1978, 56, 808).

(v) Octahydrobenzofuran

(2-Hydroxyethyl)hex-2-ene reacts with phenylseleneyl to produce an octahydrobenzofuran with a phenylseleno group *trans* and β to the ether oxygen and with *cis* fused rings (D.L.J. Clive, G. Chittattu, and C.K. Wong, *ibid.*, 1977, 55, 3894).

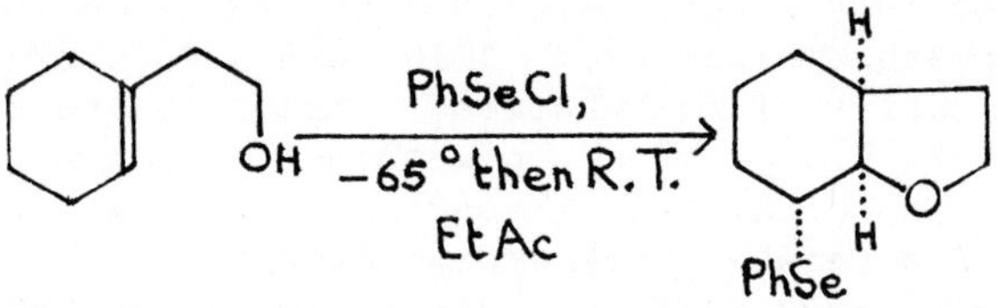

Treatment of the phenyl selenide with triphenyltin hydride in the absence of solvent gives octahydrobenzofuran. The phenyl selenide (17) on boiling with triphenyltin hydride in toluene affords 2-oxo-octahydrobenzofuran *(idem.*, Chem. Comm., 1978, 41).

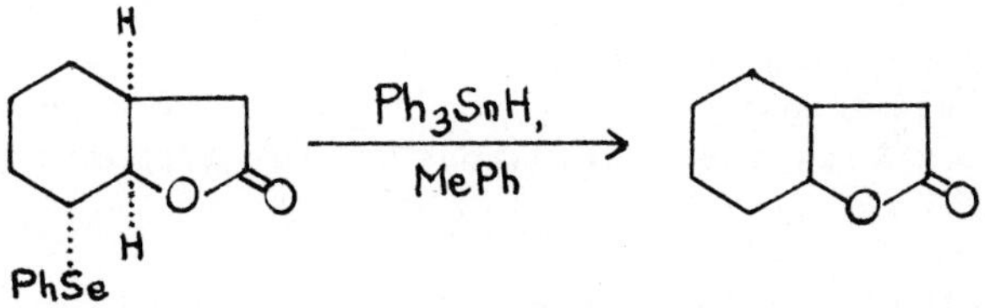

(17)

Lewis acid-catalyzed isomerisation of *cis*-2-methyl-
-octahydrobenzofuran at 200° yields the *trans*-isomer (11-23%) along with 2-methyl-2,3,3a,4,5,6-hexahydrobenzofuran (2-6%), 2-methyl-2,3-dihydrobenzofuran (1%) and cyclic

hydrocarbons (Karakhanov *et. al.*, Khim. Geterotsikl.
Soedin., 1973, 147). Treatment of octahydrobenzofuran with
phosphorus pentafluoride causes polymerisation (J.Kops,
E. Larsen, and H. Spanggaard, J. Polym. Sci., Polym. Symp.,
1976, 56, 91).

3. *Isobenzofuran (benzo[c]furan; 3,4-Benzofuran) and its Derivatives*

(a) *Isobenzofurans*

Indications that isobenzofuran is practically devoid
of aromatic character have been reported by M.J.S. Dewar
and his coworkers (Tetrahedron, 1970, 26, 4505), from the
results of a series of calculations, employing the semi-
empirical SCF-MO π approximation method. They predict, on
the basis of the calculated CC bond lengths, that it should
resemble the "polyene". Observation made by R.N. Warrener
(J. Amer. chem. Soc., 1971, 93, 2346) partly support this
prediction, but he also states that isobenzofuran now
appears more stable than the related carbocyclic systems
o-xylylene and isoindene. [1]H Nmr spectral data shows
that its ^{1}H chemical shifts are consistent with a
diamagnetic ring current and hence aromaticity. M.H. Palmer
and S.M.F. Kennedy (J. chem. Soc., Perkin II, 1976, 81)
report that non-empirical calculations of the electronic
ground and lowest triplet excited states for isobenzofuran,
benzo[c]thiophene, and isoindole, and N-methylisoindole
show that some resonance energy is present in these
systems, and that their instability can be attributed to a
combination of a low resonance energy and low-lying excited
state. Detailed analysis of the ^{1}H nmr spectra of these
compounds supports the conclusions of low aromatic
character.

The dihydropyridazine derivative (1) obtained from
3,6-di(2'-pyridyl)-*s*-tetrazine and 1,4-dihydro-1,4-
-endoxynaphthalene in dimethyl sulphoxide, on decomposition
is converted into 3,6-dipyridylpyridazine and isobenzofuran. Attemps to isolate the isobenzofuran by
conventional methods result in rapid polymerization, but
its formation may be confirmed by the addition of N-methyl-
maleimide giving a mixture of the *exo* and *endo* adducts, (2)
and (3). Isobenzofuran, m.p. *ca* 20° (tetracyanoethene
adduct, m.p.204°), may be collected at -80° on a cold

finger, following the controlled pyrolysis of (1) at 120°
and 0.1mm pressure (R.N. Warrener, J. Amer. chem. Soc.,
1971, 93, 2346).

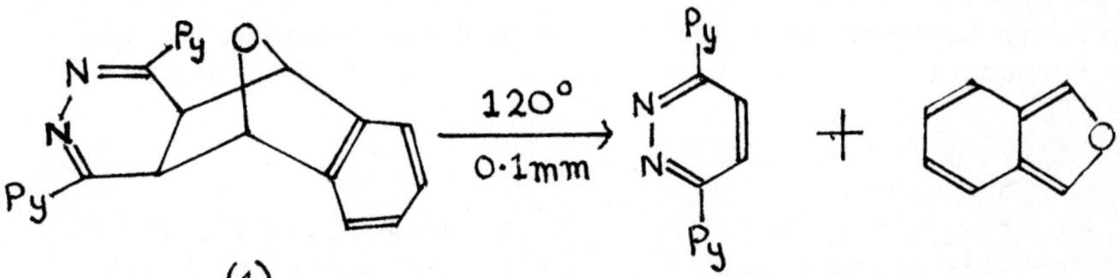

(1)

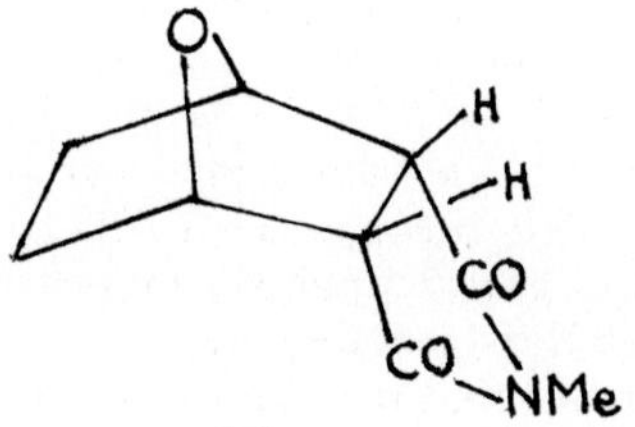

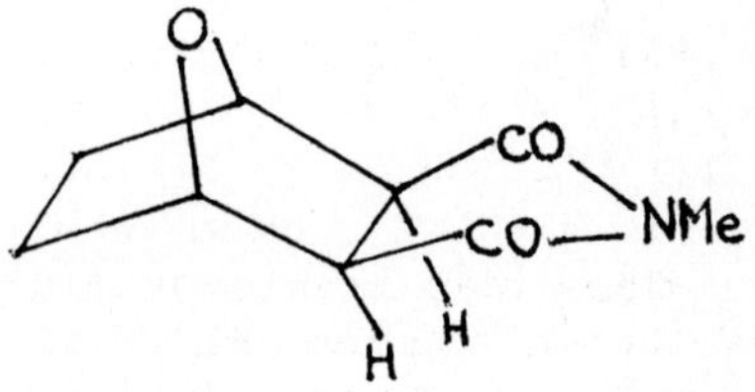

(2)   (3)

Isobenzofuran may be obtained in quantitative yield by
flash-vacuum thermolysis of 1,4-epoxy-1,2,3,4-tetra-
hydronaphthalene at 650° (U.E. Wiersum and W.J. Mijs,
Chem. Comm., 1972, 347) and also by the thermolysis of
lactone (4) (D. Wege, Tetrahedron Letters, 1971, 2337).

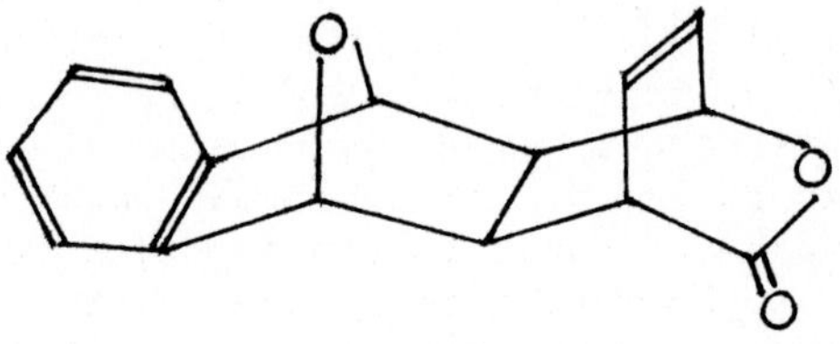

(4)

Isobenzofurans generated from 2-(α-bromoalkyl)benzophenones
have been intercepted (R. Faragher and T.L. Gilchrist, J.
chem. Soc., Perkin I, 1976, 336). Flash thermolysis of the
appropriate 1,2,3,4-tetrahydro-1,4-epoxynaphthalene (5)
affords 1-benzylisobenzofuran (R^1 = CH_2Ph, R^2 = H),
1-methylisobenzofuran (R^1 = Me, R^2 = H) and 1,3-
-dimethylisobenzofuran (R^1 = R^2 = Me) (E. Elsamma,
D.J. Sardella, and J. Barnstein, Tetrahedron Letters, 1976,
2507).

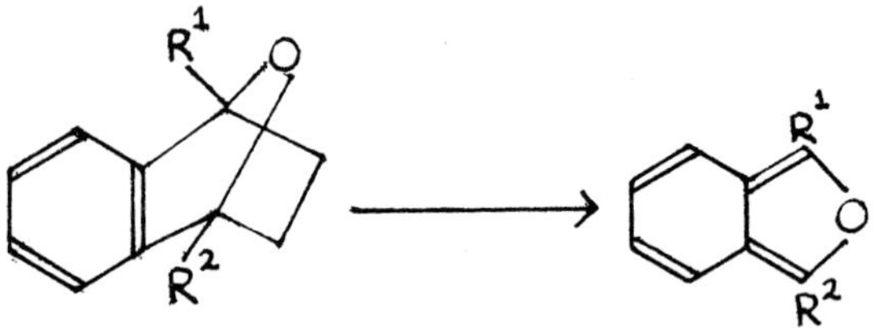

(5)

Sensitized photo-oxidation of 1,3-diphenylisobenzo-
furan at -50° gives the peroxide (6) m.p.112-114°
(decomp.), which on reduction yields the diketone (7) and
with methanol produces the methoxyhydroperoxide (8)
m.p.126-128°, also reducible to the diketone (7) (G. Rio
and M.-J. Scholl, Chem. Comm., 1975, 474).

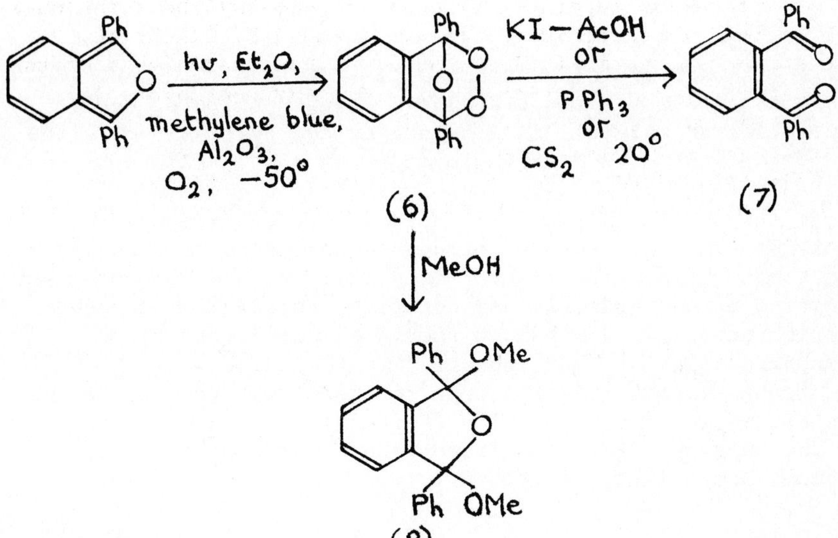

(6)

(7)

MeOH

(8)

Six isomers of the peroxide (6) have been obtained. The reactivity of isobenzofuran with singlet oxygen is found to be between that of furan and 1,3-diphenylisobenzofuran as predicted from their ionization potentials (R.H. Young and T.T. Feriozi, *ibid.*, 1972, 841).

1,3-Diphenylisobenzofuran on boiling in toluene with 3-methyl-2-phenyl-1-azirine yields on adduct, which on the basis of spectral and chemical evidence is assigned structure (9) (V. Nair, J. org. Chem., 1972, **37**, 2508).

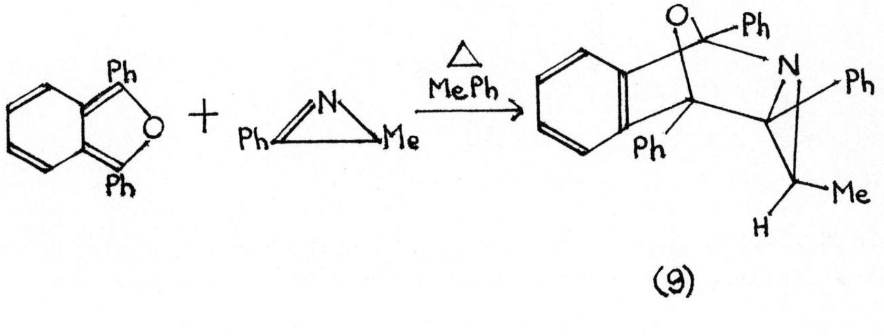

(9)

4. *Other Bi-, Tri, and Tetra-Cyclic Systems with a Furan Ring*

3-Oxabicyclo[3.2.0]hepta-1,4,6-triene (3,6-dehydro-oxepin a furocyclobutadiene) (1) an unsubstituted cyclic 8π-electron molecule is obtained by the pyrolysis of either *cis-* or *trans*-1,2-diethynyloxirane at 400°. Although stable for several days in solution it is extremely sensitive. It polymerizes instaneously on exposure to oxygen, does not survive either vapour phase or thin-layer chromatography, and reacts immediately with cyclopentadiene giving an air-sensitive adduct with tentatively proposed structure (2). Whether it should be considered to be "nonaromatic" or "antiaromatic" on the basis of ring current effects is still open to question, because its nmr spectrum is ambiguous with regard to the presence of a ring current (K.P.C. Vollhardt and R.G. Bergman, J. Amer. chem. Soc., 1972, 94, 8950).

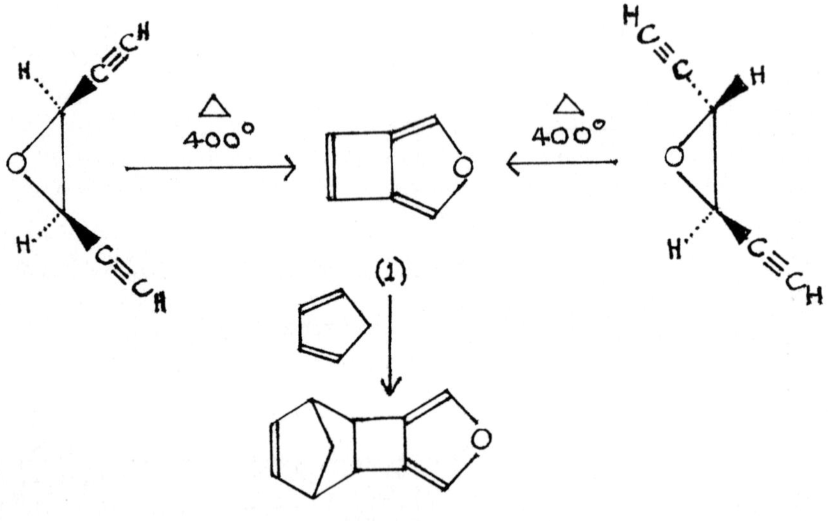

(1)

(2)

Attempts to prepare 6,7-diphenyl-3-oxabicyclo[3.2.0]hepta--1,4,6-triene and 3,4-phenylenefuran have failed.

3-Oxabicyclo[3.2.0]hepta-1,4-diene (3) is prepared by partial hydrogenation of 3-oxabicyclo[3.2.0]hepta-1,4,6--triene (1), (Bergman and Vollhardt, Chem. Comm., 1973, 214).

(1)   (3)

It is the first oxygen analogue of the hetero[5]annulene series and can be used as a suitable model with which to compare the properties of furocyclobutadiene (1) and related systems. It is more reactive than furan, but appears to be much less sensitive, for example, to oxygen and g.l.c., than its potentially anti-aromatic precursor (1), and its physical properties also differ dramatically from those of (1). The difference in spectral properties between (1) and (3) are ascribed to the fact that whereas the latter is a perturbed furan, the former is a truly antiaromatic planar 8π- system.

Bis(3-*tert*-butyl-2-propargyl)ether on treatment with potassium *tert*-butoxide in tetrahydrofuran at 50° rearranges to give 6,7-di-*tert*-butyl-3-oxabicyclo[3.2.0]--hepta-1,4-diene.

t-Bu Bu-t   t-Bu   Bu-t
 C   C
 ‖   ‖   KOBu-t,
 C   C   ———————→
                  THF
                  50°

Bis(3-phenyl-2-propargyl)ether and dipropargyl ether under similar basic conditions rearrange to give products (4) and (5) respectively (P.J. Garratt and S.B. Neoh, J. Amer. chem. Soc., 1975, <u>97</u>, 3255).

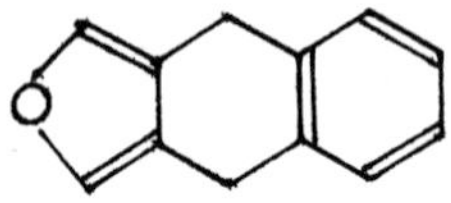

(4)

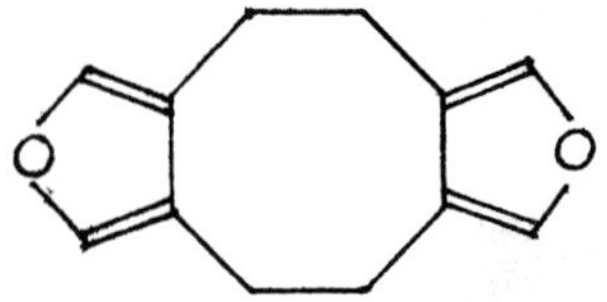

(5)

A mixture of *endo*- and *exo*-thiocarbonates (1) is readily accessible from furan and vinylene carbonate and its degradation to 7-oxanorbornadiene (2) on heating in excess triethylphosphite has been accomplished. 7-Oxanorbornadiene (2) on irradiation undergoes rapid and quantitative [$2\pi + 2\pi$] cycloaddition to form 3-oxaquadricylane (3), which on heating undergoes thermal stabilisation to give oxepin-benzene oxide (H. Prinzbach and H. Babsch, Angew. Chem. internat. Edn., 1975, <u>14</u>, 753).

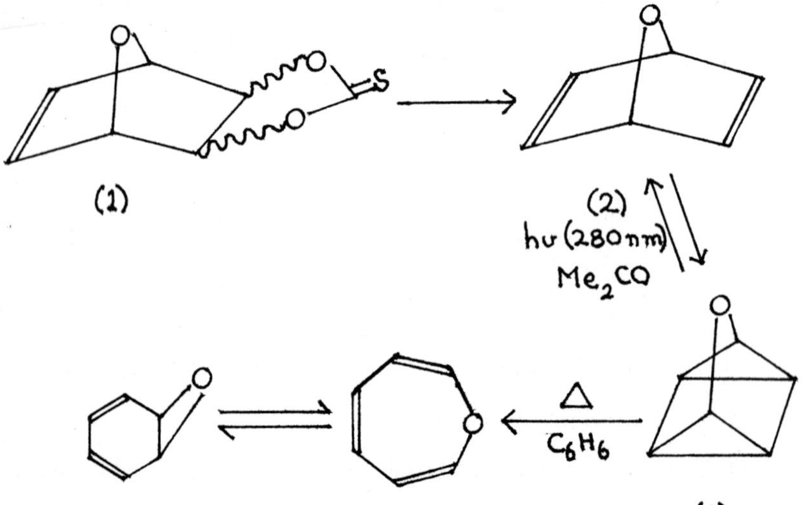

(1)

(2)
hυ (280 nm)
Me$_2$CO

$\triangle$
C$_6$H$_6$

(3)

Tetramethyl tricyclo[6.2.0.0^{3,6}]decane-2,7-dione-
-4,5,9,10-tetracarboxylate has been converted to *exo, exo*-
-11-oxatetracyclo[4.4.0.^{2,5}0^{7,10}]undeca-3,8-diene (4),
which contains a tetrahydrofuran ring (N.E.E Rowland *et.*
al., Tetrahedron Letters, 1970, 4769).

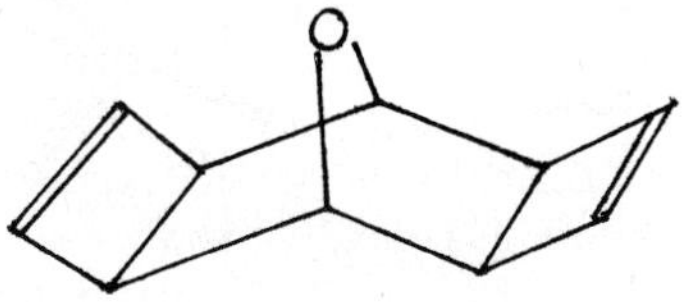

(4)

The irradiation of 9-oxatricyclo[4.2.1.0^{2,5}]nona-
dienes (5) in the presence of the photo-sensitizer benzophenone yields oxahomocubanes (6) as one of the products
(W. Eberbach and M. Perroud-Argüelles, Ber., 1972, 105,
3078).

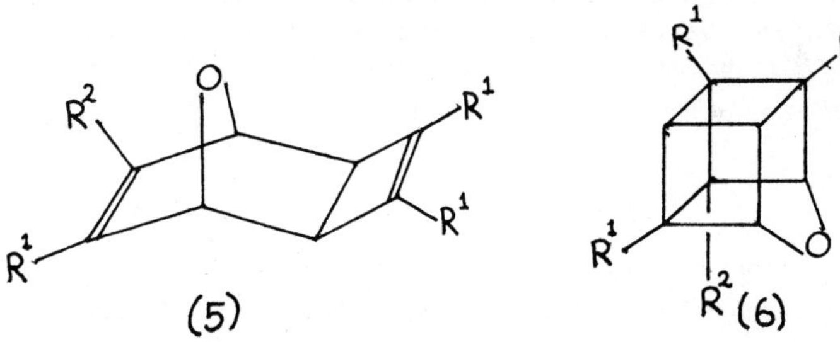

(5)　　　　　　　　　　　R^2(6)

R^1 = CO$_2$Me;　R^2 = CO$_2$Me, H, or D

5. *Dibenzofuran (Diphenylene Oxide) and its Derivatives*

(i) *General synthetic methods*

Photolysis of diaryl oxides in cyclohexane containing

iodine yields dibenzofurans (K.P. Zeller and H. Petersen, Synth., 1975, 532).

The dehydrogenation of 2-phenylphenol over Pt-C and Pd-C at 400-500° affords dibenzofuran (H. Yasui and H. Suzumura, Aromatikkusu, 1978, 30, 293). 2,2′-Dimethoxybiphenyls do not react with pyridinium chloride but with pyridinium bromide they are cyclized to dibenzofurans (J.P. Bachelet, P. Demerseman, and R. Royer, J. heterocyclic Chem., 1977, 14, 1409).

The benzologation of 2-(allyl or substituted allyl)-benzofurans by means of the Bradsher reaction gives rise to dibenzofuran derivatives, for example, the action of dichloromethyl alkyl ethers and tin (IV) chloride on 2-(allyl or substituted allyl)benzofurans gives the corresponding unsubstituted or 3-substituted derivatives of dibenzofuran. Of particular interest is the use of ethyl dichloro(ethoxy)acetate, which gives derivatives containing a 1-ethoxycarbonyl substituent (J. Ashby, M. Ayad, and O. Meth-Cohn, J. chem. Soc., Perkin I, 1974, 1744).

(ii) Properties and reactions

2-Chloro-, 1,2- and 2,3-dichloro-, and 2,3,8-tri-chlorodibenzofuran have been obtained by the photolysis of the appropriate chlorinated diphenyl oxides in acetone

(G.G. Choudry *et. al.*, Chemosphere, 1977, 6, 327),
dibenzofurans chlorinated in the 2,3,8-, 2,4,6-, 2,3,7,8-,
2,3,6,8-, 1,2,7,8-, 1,2,4,7,8- and 1,3,4,7,8-positions
(A.P. Gray *et. al.*, Trace Subst. Environ. Health, 1975, 9,
255) and these derivatives plus dibenzofurans chlorinated
in the 2,3,7-, 2,3,9- and 2,4,6,8- positions have been
prepared (Gray, V.M. Dipinto, and I.J. Soloman, J. org.
Chem., 1976, 41, 2428). Bromination of dibenzofuran with
bromine in boiling dichloroethane in the presence of ferric
chloride give heptabromodibenzofuran (H. Richtzenhain and
K. Schrage. Ger. Offen. 2,534,381/1977).

2-Acylbenzofurans may be obtained by Friedel-Crafts
acylation of 1,2,3,4-tetrahydrodibenzofurans, followed by
dehydrogenation of the acyl derivatives. The 1- and
4-acyldibenzofurans are prepared by the Friedel-Crafts
reaction, by the reaction of 1- and 4-(chlorocarbonyl)di-
benzofuran, respectively with the appropriate compounds
(T. Keumi *et. al.*, Fukui Daigaku Kogakuba Seni Kogyo Kenkyu
Shisetsu Hokokum 1976, 14, 71). Allylation of dibenzofuran
with allyl alcohol in the presence of zinc chloride at
140° gives 2-allyldibenzofuran (O.G. Akperov and
M.N. Agopyan, Uch. Zap., Azerb. Univ., Ser. Khim. Nauk,
1972, 64); synthesis of 2- and 2-styryl (T. Garmatter and
A.E. Siegrist, Helv., 1974, 57, 945) and 3-(4-substituted
benzoyl) derivatives of dibenzofuran have been described
(A.G. Khaitbaeva, V.G. Sidorova, and Kh. Yu. Yuldashev,
Khim. Geterotsikl. Soedin., 1977, 895).

Chloromethylation of dibenzofuran by paraformaldehyde-
-hydrogen chloride in acetic acid containing orthophosphoric acid gives 2,8-bis(chloromethyl)dibenzofuran (97%).
On reduction with zinc, followed by chromic oxide oxidation
it yields dibenzofuran-2,8-dicarboxylic acid, and on
hydrolysis it gives 2,8-bis(hydroxymethyl)dibenzofuran
(M.S. Ogii *et. al.*, Vop. Khim. Khim. Tekhnol., 1972, 87).
4-Dibenzofurylsulphinylacetic acid has been synthesized
starting from 4-dibenzofuryllithium and sulphur dioxide
(M. Janczewski and H. Maziarczyka, Rocz. Chem., 1977, 51,
891), and that of ethyl 2-(4-dibenzofuryloxy)-2-methyl-
propionate described (G. Bondesson *et. al.*, J. Med. Chem.,
1974, 17, 108). Heating dibenzofuran with trichlorosilane
in a sealed ampoule at 250-450° and initiating the
reaction by a beam of accelerated electrons gives 10,10-
-dichloro-10-sila-9-oxa-10,10-dihydrophenanthrene

(B.I. Vainshtein *et. al.*, Khim, Geterotsikl. Soedin., 1978, 1277).

^{1}H-nmr spectra of acetyl- and benzoyl-dibenzofuran (A.A. Pashchenko, Khaitbaeva and Yuldashev, Doklady. Akad. Nauk Uzb. SSR, 1978, 40); ^{13}C nmr spectra of dibenzo-furan and 2-bromo and 2,8-dibromo derivatives (T.N. Huckerby, J. mol. Struct., 1979, 54, 95), and dibenzo-furan (S. Florea, W. Kimpenhaus, and V. Farcasan, Org. mag. Res., 1977, 9, 133); ir spectra of chlorinated dibenzo-furans (J.Y.T. Chen, J. Ass. Off. Anal. Chem., 1973, 56, 962); uv spectra of dibenzofuran and derivatives (J. Gripenberg, Tetrahedron Letters, 1974, 619; C.-L. Chen and W.J. Conners, J. org. Chem., 1974, 39, 3877); ms of octachlorodibenzofuran and dibenzofuran [A. Curley *et. al.*, Mass Spectrom. NMR Spectrosc. Pestic. Chem. Proc. Symp., 1973 (Pub. 1974), 71].

Hydnuferrugin (1), $C_{18}H_{10}O_9$, a brown-violet pigment has been isolated from sporophores of *Hydnellum ferrugineum* (Fr.) Karsten and *H. zonatum* (Batsch) Karsten (Gripenberg, *loc. cit.*).

(1)

The Pummerer's ketone type trimeric ketone (2) and tetrameric ketone (3), which are dibenzofuran derivatives have been obtained, along with other products, by the dehydrogenation of *p*-cresol with alkaline potassium ferri-cyanide (Chen. and Conners, J. org. Chem., 1974, 39, 3877).

(2)   (3)

Chapter 3

COMPOUNDS WITH FIVE-MEMBERED RING HAVING ONE HETERO ATOM
FROM GROUP VI;   SULPHUR AND ITS ANALOGUES

R. LIVINGSTONE

1. *Monocyclic Thiophenes and Hydrothiophenes*

The results of many investigationa on the use of
various catalysts for the synthesis of thiophene and methyl-
thiophenes from suitable saturated and unsaturated hydrocarbons and hydrogen sulphide have been published, as have
studies on the catalytic preparation of thiophene from
butyl mercaptan and dibutyl sulphide. Other catalytic
syntheses involving really available starting materials
have been described. Reviews and reports have appeared
concerning trends in thiophene chemistry (L.I. Belen'kii
et. al., Nauka, Moscow, USSR, 1976), thiophenes and their
selenium and tellurium analogues (S. Gronowitz, Org. Compd.
Sulphur, Selenium, Tellurium, 1977, 4, 244;   1979, 5, 247),
and use of thiophene derivatives in the production of
chemicals (L.S. Fuller, J.W. Pratt, and F.S. Yates, Pharm.
Ind., 1979, 41, 979), especially in the area of medicinal
chemistry (*idem,* Manufacturing Chemist and Aerosol News,
1978, 49, May and June).

Two derivatives of thiophene of particular interest
are thiophene-2-acetic acid and thiophene-3-malonic acid
which have found use in the production of semisynthetic
β-lactam antibiotics cephalothin (1) and tricarcillin (2),
respectively.

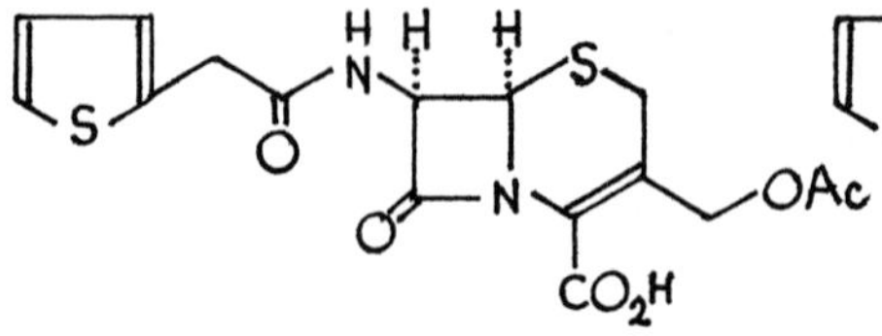

(1) 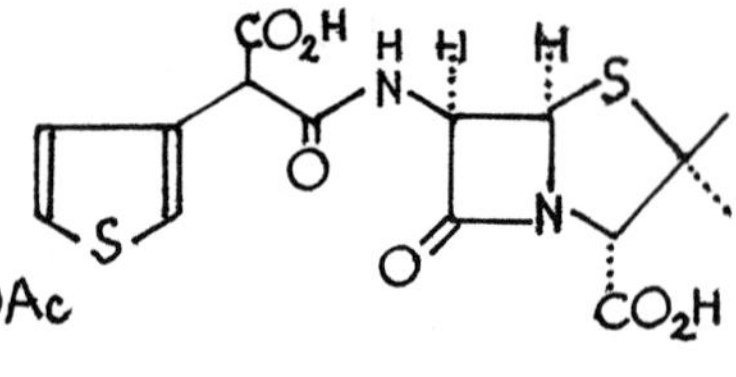   (2)

Contrary to the statement found in most textbooks that thiophene does not undergo Diels-Alder reactions, it has been found that a variety of thiophenes react smoothly with dicyanoacetylene in a Diels-Alder manner to produce phthalonitriles. Other examples have also been reported.

(a) Thiophene and its substitution products

(i) Synthetic methods

(1) The catalytic properties of rhenium sulphide facilitates the heterocyclization of pentane with hydrogen sulphide (ratio 1:2) at 550-560° to give thiophene and carbon disulphide, and the reduction of ketones, such as, ethyl 2-(2-thienyl)ethan-2-one-1-oate (1) to ethyl 2-thienylacetate (M.A. Ryashentseva Kh. M. Minachev, and E.P. Belanova, Tezisy Doklady Nauchn. Sess. Khim. Tekhnol Org. Soedin. Sery Sernistykh Neftei, 14th 1975, 221).

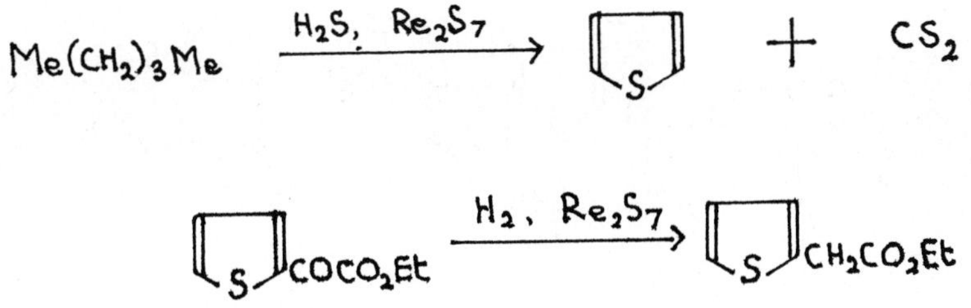

(1)

(2) A synthesis of thiophenes uses 1,3-dithiolylium-
-4-olates, which act as cyclic thiocarbonyl ylides (2) and
add dimethyl acetylenedicarboxylate in the 2,5-position.
The non-isolable adduct (3) decomposes quickly with the
elimination of carbonyl sulphide to give the thiophene-3,4-
-dicarboxylate (4) (H. Gotthardt, M.C. Weisshuhn, and
B. Christie, Ber., 1976, 109, 740, 753).

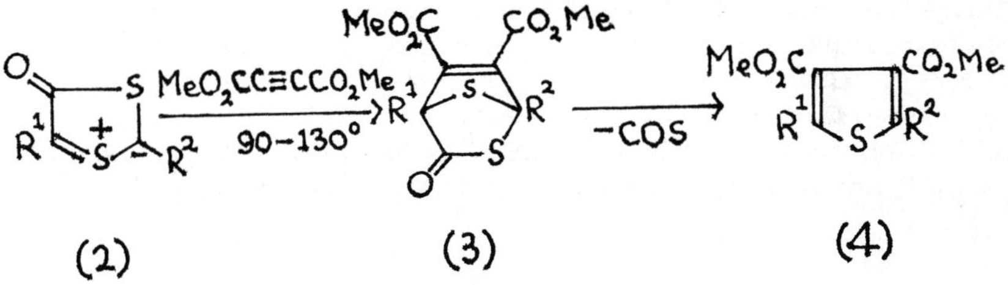

(2)   (3)   (4)

Analogous cycloaddition reactions of ylide (2) ($R^1 = R^2$
= Ph) with acetylene derivatives furnish thiophenes (5)
($R^3 = CO_2Me$, $R^4 = H$; $R^3 = Ph$, $R^4 = CO_2Me$, Ac, Bz, H).

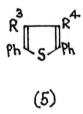

(5)

(ii) General properties and reactions

(1) The copper (II) chloride catalyzed reaction
between thiophene or certain derivative (2-methyl, 2-chloro,
2-bromo, 2-nitro, 2,5-dichloro, and 2,5-dibromo) and
diisopropylperoxydicarbonate in acetonitrile gives thienyl
isopropyl carbonates (1) (41-74%) in most cases, along with
by-products. In the absence of catalyst, different product
distributions are generally observed. Dealkylation with
aluminium chloride in *m*-xylene and decarboxylation of the
ester furnished hydroxythiophenes and hence, 2-oxo-2,5-
-dihydrothiophenes (A.P. Manzara and P. Kovacic, J. org.
Chem., 1974, <u>39</u>, 504).

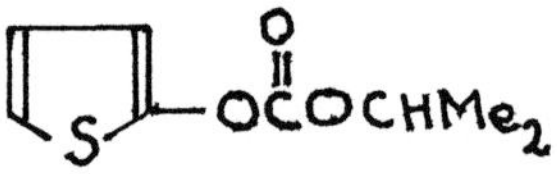

$$
\overset{O}{\underset{\|}{OCOCHMe_2}}
$$

(1)

(2) The electrochemical oxidation of 2,5-dimethyl-
thiophene in methanol results in three types of reactions,
depending on the electrolyte used. With ammonium bromide
the product is 3-bromo-2,5-dimethylthiophene exclusively;
with ammonium nitrate and sodium acetate, methoxide, and
perchlorate, the formation of 2-methoxymethyl-5-methyl-
thiophene is observed; and with sodium cyanide a mixture
of *cis*- and *trans*-2-cyano-5-methoxy-2,5-dimethyldihydro-
thiophenes, 3-cyano-2,5-dimethylthiophene, and 2-methoxy-
methyl-5-methylthiophene is obtained (K. Yoshida, T. Saeki,
and T. Fueno, *ibid.*, 1971, <u>36</u>, 3673).

(3) Diels-Alder addition of thiophene to maleic
anhydride in methylene chloride, at 100° and 15k bar
pressure gives the adduct (2) (37-47%); the reaction is
highly accelerated above 15k bar. The exo-configuration is
suggested for (2) and methylation affords derivative (3) (R
= H), which on esterification gives the diester (3) (R = Me)
(H. Kotsuki *et. al.*, *ibid.*, 1978, <u>43</u>, 1471; Bull. chem.
Soc. Japan, 1979, <u>52</u>, 544).

(2)   (3)

It has been reported that a variety of thiophenes react
smoothly with dicyanoacetylene in a Diels-Alder manner to
product phthalonitriles in reasonable yields (R. Helder and
H. Wynberg, Tetrahedron Letters, 1972, 605).

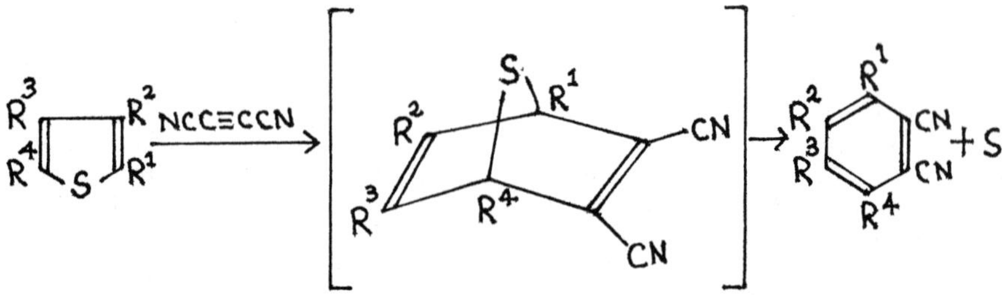

(4.)

Substituents = H, Me, t-Bu, Ph

Substituted benzenes are formed from thiophene or 2,5-
-dimethylthiophene and other disubstituted acetylenes.
Triplet sensitized or nonsensitized photoaddition with
dimethyl acetylenedicarboxylate gives the adduct (4)
($R^1 = R^4$ = H, Me; CN = CO_2Me; $R^2 = R^3$ = H)
(H.J. Kuhn and K. Gollnick, *ibid.*, p.1909).

Thiophene 1-oxides generated *in situ* by oxidation of
the thiophene with 3-chloroperbenzoic acid in the presence
of benzoquinones, give the napthoquinones (5) and the
Diels-Alder adducts (6). With 1,4-naphthoquinone only the
corresponding anthraquinone (7) is formed (K. Torssell,
Acta Chem. Scand., 1976, B30, 353).

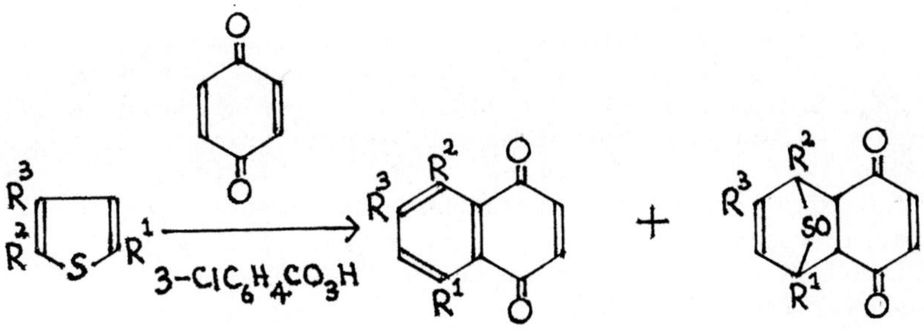

$$R^1 = R^2 = R^3 = H$$
$$R^1 = R^3 = H. \quad R^2 = Me$$
$$R^1 = R^2 = H, \quad R^3 = Me$$

(5) (6)

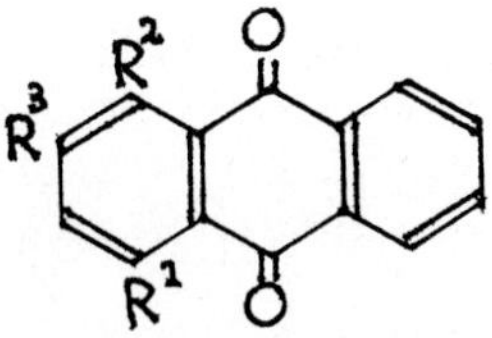

(7)

Reaction of α,α'-dibromo ketones, for example, 2,4-dibromo--2,4-dimethylpentan-3-one, and diiron nonacarbonyl with thiophene results only in the formation of α-substituted products and no cyclic adducts (*cf.* furan Ch. 2, p. :) (R. Noyori *et. al.*, Tetrahedron Letters, 1973, 1741).

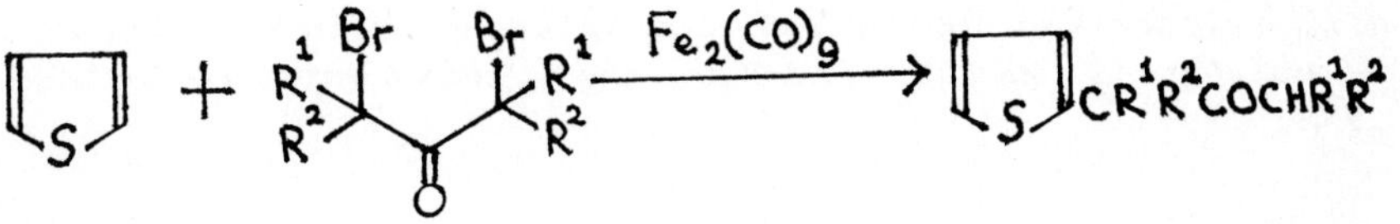

(4) The irradiation of cyclohexenone in thiophene besides the dimer of cyclohexenone also gives a small amount (8%) of an adduct (8) (R = H). Spectral data shows it to be the result of [2 + 2] addition of cyclohexenone to thiophene. Similarly, 2,5-dimethylthiophene affords a mixture of [2 + 2] adducts (43%) containing (8) (R=Me). The orientation of the adduct has not been determined, but is inferred by analogy with that for furan, which gives [2 + 2] and [4 + 2] addition products (T.S. Cantrell, J. org. Chem., 1974, 39, 3063).

R = H, Me $\qquad$ (8)

The photochemical reaction between thiophenes and 2,3--dichloro-1,4-naphthoquinone gives 2-chloro-3-(2-thienyl)--1,4-naphthoquinones, promising precursors of naturally occurring quinones (K. Maruyama and T. Otsukia, Chem. Letters, 1977, 851). Irradiation of thiophene and its methyl and phenyl derivatives with propylamine affords pyrrole derivatives. Related furans behave in a similar manner (A. Couture and A. Delevallee, Tetrahedron, 1975, 31, 785). Irradiation of 2-methylthiophene with dihalogeno-maleinimides (9) in acetone yields the 1:1 adduct (10). Corresponding adducts are obtained from furan and 2-methyl-furan. The 2:1 adduct (11) is produced by the irradiation of thiophene and N-methyl-dibromomaleinimide in the absence of solvent (H. Wamhoff and H.J. Hupe, Tetrahedron Letters, 1978, 125).

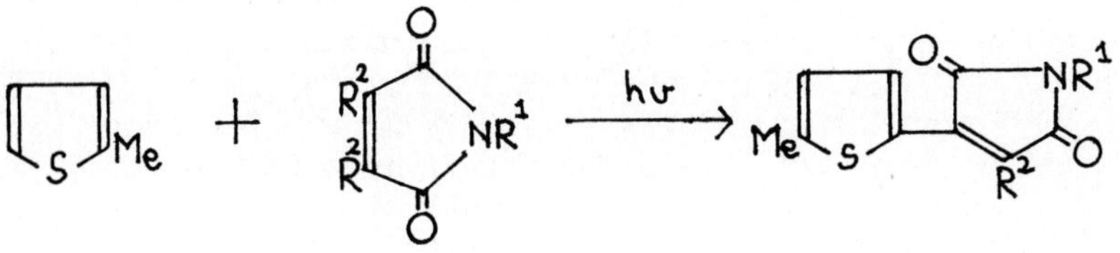

(9)   (10)

$$R^1 = H, \quad R^2 = Cl, I$$
$$R^1 = Me, \quad R^2 = Br$$

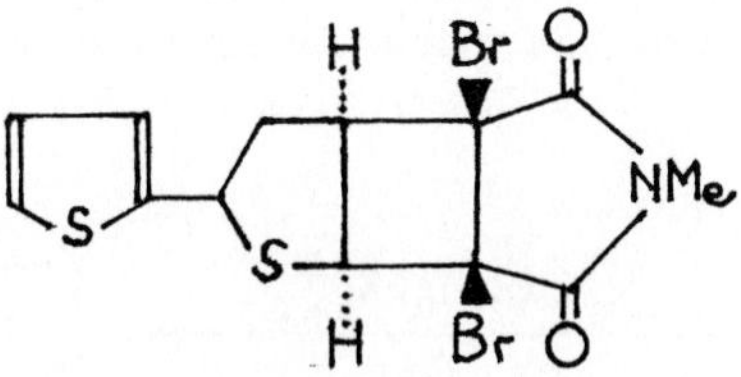

(11)

(5) Thiophene and tetraphenylthiophene unlike the corresponding furans and pyrroles are unreactive towards singlet oxygen. However, 2,5-dimethylthiophene reacts with singlet oxygen, when it is irradiated in methanol during oxygenation in the presence of methylene blue. Methanol is not incorporated in the product and the sulphine (12) m.p.44°, can be isolated (H.H. Wasserman and W. Strehlow, *ibid.*, 1970, 795).

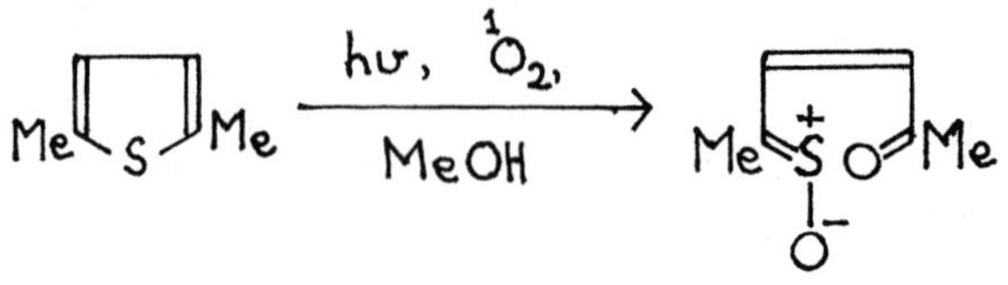

(12)

The rhodium (II) acetate-catalyzed addition of dimethyl diazomalonate to thiophene and its derivatives results in the formation of stable crystalline thiophenium ylides, for example, thiophenium bismethoxycarbonylmethylide (13), whose structure has been determined by X-ray analysis (R.J. Gillespie *et. al.*, Chem. Comm. 1978, 83).

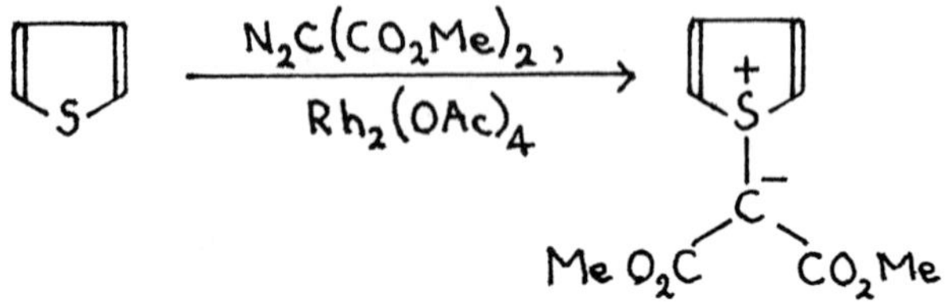

(13)

In the presence of copper or rhodium catalyst 2,5-dichloro-thiophenium bismethoxycarbonylmethylide undergoes fragmentation to produce 2,5-dichlorothiophene and bismethoxycarbonylcarbene or the corresponding metal stabilised carbenoid (Gillespie and A.E.A. Porter, *ibid.*, 1979, 50).

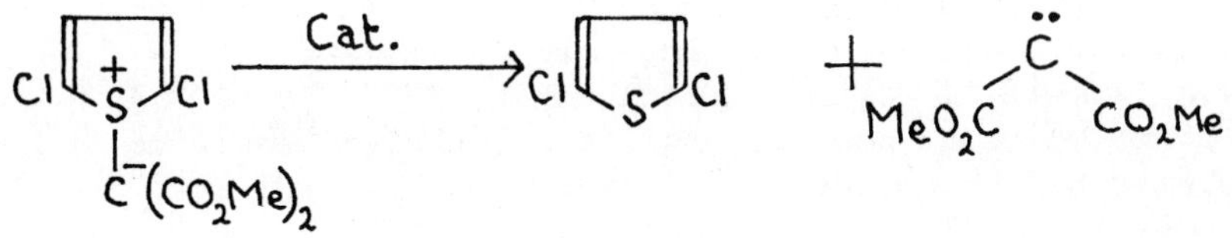

Although diazomalonic ester with thiophene forms a stable ylide other diazoalkanes may give two types of product. Simple diazoketones such as diazoacetophenone afford 2-substituted thiophenes (14) and diazoacetic esters react to give 2-thiabicyclo[3.1.0]hex-3-ene derivatives. Diazoacetoacetic esters yield a mixture of both types of product (15) and (16), with the former predominating (*idem*, J. chem. Soc., Perkin 1, 1979, 2624; Gillespie, Porter and W.E. Willmott, Brit. UK Pat. Appl. 2,013,652/1979).

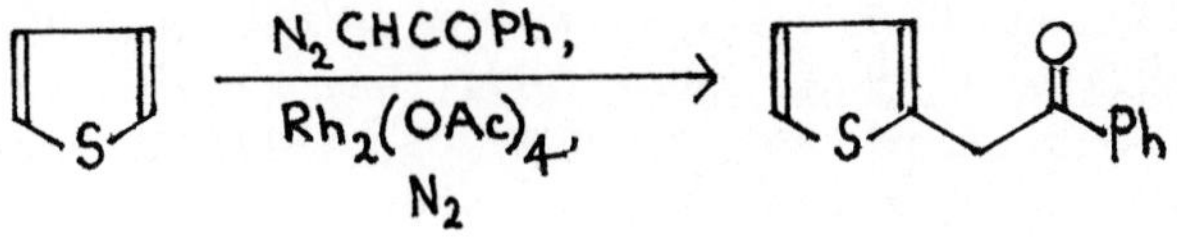

(14)

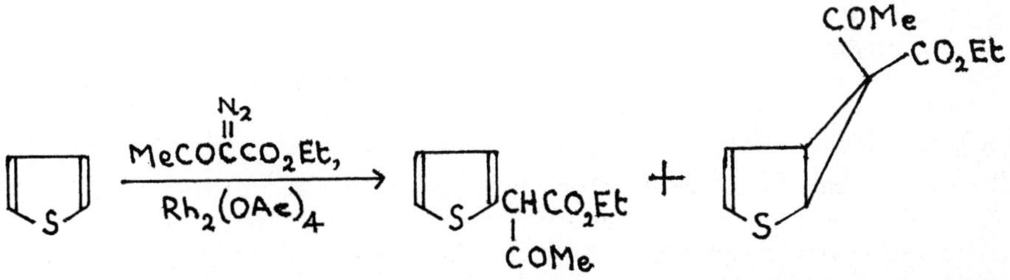

(15)   (16)

The ir spectra and intensities of ring stretching bands for 2-substituted thiophenes and furans (J.M. Angelelli *et. al.*, Tetrahedron, 1972, **28**, 2037); ^{13}C nmr spectra, chemical shifts and charge densities of substituted thiophenes (Y. Osamura, O. Sayanagi, and K. Nishimoto, Bull, chem. Soc. Japan, 1976, **49**, 845), discussion concerning the unusually high sensitivity of the C-5 position in 2-substituted thiophenes in relation to calculated charges (D.A. Forsyth and G.A. Olah, J. Amer. chem. Soc., 1979, **101**, 5309); uv and ^{1}H nmr spectra of mono-, di-, and tri-substituted acyl- and alkyl-thiophenes have been reported (J.G. Pomonis, C.L. Fatland, and F.R. Taylor, J. Chem. Eng. Data, 1976, **21**, 233).

(iii) Thiophene

Additional information to that already described on the preparation of thiophene; from acetylene and hydrogen sulphide [D.A. Sibarov *et. al.*, Zh. prikl. Khim. (Leningrad), 1970, **43**, 1767; M.A. Ryashentseva *et. al.*, U.S.S.R. 257,460/1969; V.F. Timofeev, Sibarov, and V.A. Proskuryakov, Chem. Abs., 1977, **86**, 29552t, 43480z; Timofeev and Sibarov, *ibid.*, 1976, **85**, 192473f, 192474g]; from butane and hydrogen sulphide over a rhenium catalyst (Ryashentseva *et. al.*, Neftekhimiya, 1973, **13**, 840); from butane-butene and hydrogen sulphide over chromic oxide - containing catalyst at 515-580° *(idem, ibid.*,1975, **15**, 458), best promoters are Cd_2O_3, SmO_3, and Ga_2O_3 (V.B. Abramovich *et. al.*, *ibid.*, 1977, **17**, 309); and thiophenes or their mixtures from $C \geqslant 4$ alkanes-alkenes, alcohols, or ketones and carbon disulphide over an appropriate catalyst (N.R. Clark and W.E. Webster, Ger. Offen. 2,225,443/1973). Other methods of obtaining thiophenes are; from C_4 hydrocarbons and sulphur dioxide over an alumina catalyst (Ryashentseva, Yu. A. Afanas'eva and Kh. M. Minachev, Izv. Akad. Nauk SSSR, Ser. Khim., 1970, 1067; catalysts, *idem*, Geterogennyi Katal. Reakts. Poluch. Prevrashch. Geterosikl. Soedin., 1971, 221); by dealkylating alkylthiophenes in the presence of a U-type zeolite catalyst in Ca or H form (V.I. Chernov and A.V. Mashkina, U.S.S.R. 259,906/1969); by reaction of vinyl chloride with hydrogen sulphide (M.G. Voronkov, E.N. Deryagina, and M.A. Kuznetsova, Khim. Geterotsikl. Soedin., 1976, 997); by the pyrolysis of divinyl sulphide at 450-560° (Voronkov *et. al.*, *ibid.*, 1975, 1579); by cyclization of butadiyne with Na_2S in dimethyl sulphoxide containing potassium hydroxide at 55°, yield 93.5% (idem, Ger. Offen. 2,818,580/1979). The catalytic synthesis of thiophene from thioethers or olefins proceeds by the same mechanism, involving the interaction of the hydrocarbon fragment of the starting material with sulphur present on the surface of the catalyst (A.V. Mashkina, T.S. Sukhareva, and G.L. Veitsman, Kinet. Katal., 1972, 13, 249). Vapour-phase catalytic dehydrocyclization of butyl mercaptan at 444° over Cr_2O_3-Al_2O_3-CuO-K_2O gives thiophene plus small amounts of substituted thiophenes (S. Trippler, T.A. Danilova, and A.F. Plate, Vestn. Mosk. Univ., Khim., 1971, 12, 625). Thiophene is also formed from butyl mercaptan and dibutyl sulphide in the presence of metal sulphides of Groups 3-6 (Sukhareva, L.V. Shepel, and Mashkina, Kinet. Katal., 1978, 19, 654). Thiophene may be obtained by the dehydrogenation of tetrahydrothiophene with sulphur dioxide in the presence of activated carbon or a catalyst of chromium trioxide-alumina at $500-600^\circ$ (J.H. Blanc, J. Tellier, and C. Thibault, Ger. Offen. 2,062,587/1971).

2-Methylthietane on ionic dehydrogenation in the presence of Ph_3C^+ BF_4^-, chloranil, or triphenylmethyl chloride and stannic chloride yields thiophene (E.A. Viktorova, A.A. Freger, and L.M. Petrova, Chem. Abs., 1976, 85, 159801p).

The physical and chemical properties of thiophene have been described (O. Meth-Cohn, Kirk-Othmer Encycl. Chem. Technol., 2nd Edn., 1969, 20, 219). Molecular reorientation processes in crystalline thiophene and furan have been studied using pulse nmr techniques, over a range of temperatures. Various inconsistencies with thiophene suggest that the published crystal determination should be repeated and that further work on the pulse nmr at lower temperatures is necessary (W.E. Sanford and R.K. Boyd,

Canad. J. Chem., 1976, 54, 2773). MINDO/3 calculations of
molecular vibration frequencies for thiophene (M.J.S. Dewar
and G.P. Ford, J. Amer. chem. Soc., 1977, 99, 1685).

Oxidation of thiophene with hydrogen peroxide in the
presence of hydrochloric acid proceeds with cleavage of the
thiophene ring and splitting off of sulphur as sulphur
dioxide. The oxidation products are organic peroxides,
carbonyl compounds, and dibasic acids [V.P. Gvozdetskaya,
Zh. Prikl. Khim. (Leningrad), 1977, 50, 900].

(iv) Metal and metalloid derivatives of thiophenes

n-Butylcesium and n-butylpotassium metallate
thiophene, 2-, and 3-methylthiophene at the α-position,
thus carbonation of metallated 2-methylthiophene yields 5-
methylthiophene-2-carboxylic acid (P. Benoit and
N. Colligan, Bull. Soc. chim. Fr., 1975, 1302). 2.5-
-Dilithio derivatives are otained on boiling thiophene or
3-methylthiophene with n-butyllithium and NNN'N'-tetra-
methylenediamine in hexane (D.J. Chadwick and C. Willbe, J.
chem. Soc., Perkin I, 1977, 887). Metallation of thiophene
with n-butyllithium shows that; 2-substituted thiophenes
are lithiated at the 5-position, i.e., the sulphur atom
controls the site of metallation; if the 5-position is
blocked, lithiation occurs at the 3-position; 3-substituted
thiophenes are lithiated at the 2-position, a site common
to the directing properties of both the substituents and
sulphur moiety. Each of these metallations is surprisingly
regiospecific (D.W. Slocum and P.L. Gierer, J. org. Chem.,
1976, 41, 3668).

The reaction of thiophene, 2-methyl-, and 2-tert-
-butyl-thiophene with n-butyllithium in hexamethylphosphoric
acid triamide (HMPA) causes deprotonation and ring-opening
of the thiophenes to give 1-substituted 1-methylthiopent-1-
-en-3-yne (1).

R = H, Me, t-Bu

(1)

The reaction between 2,5-bis(methylthio)thiophene (2) and
n-butyllithium affords different products namely, 1,1,4-
-tris(methylthio)but-1-en-3-yne (3), 2-methyl-5-(methylthio)thiophene (4), and 2-(ethylthio)-5-(methylthio)thio-
phene (5), depending on the reagents used and the reaction
conditions (R. Grafing and L. Brandsma, Rec. Trav. chim.,
1976, 95, 264).

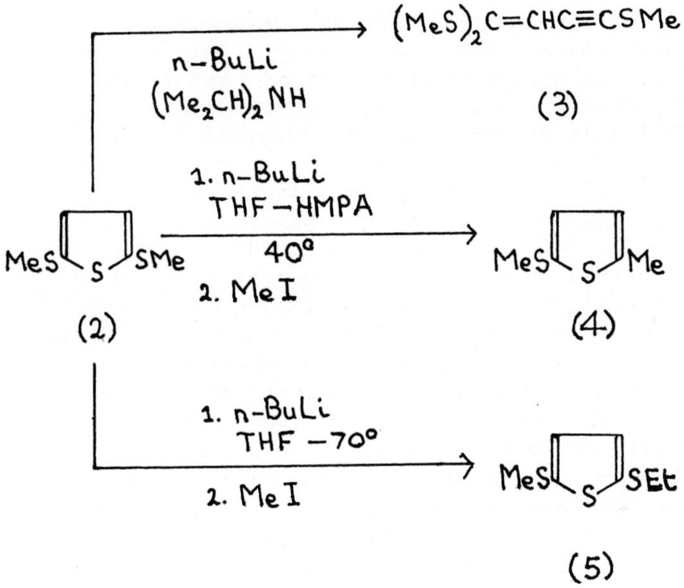

(v) Alkyl- and aryl- thiophenes

   Synthesis.   (1) *S*-Alkylthiophenium salts (1-alkyl-
thiophenium salts). 2,3,4,5-Tetramethylthiophene with the
methyl ester of fluorosulphonic acid and sodium hexafluorophosphate gives 1,2,3,4,5-pentamethylthiophenium
hexafluorophosphate (95%).

$$\text{Me}\underset{\text{Me}}{\overset{\text{Me}}{\bigsqcup_{S}}}\text{Me} \quad \xrightarrow[\text{NaPF}_6]{\text{MeOSO}_2\text{F},} \quad \text{Me}\underset{\text{Me}}{\overset{\text{Me}}{\bigsqcup_{\overset{+}{S}}}}\text{Me} \quad \underset{\text{Me}}{} \quad \text{PF}_6^-$$

The S-ethyl analogue is prepared in a similar manner. Thiophene and 2,5-dimethylthiophene on treatment with methyl iodide and silver tetrafluoroborate, followed by sodium hexafluorophosphate afford 1-methyl- and 1,2,5--trimethyl-thiophenium hexafluorophosphate, respectively. The pentamethyl salt reacts with butyllithium to give the ylide (1). The ^{13}C nmr spectra of these salts have been discussed (R.F. Heldweg and H. Hogeveen, Tetrahedron Letters, 1974, 75).

$$\text{Me}\underset{\text{Me}}{\overset{\text{Me}}{\bigsqcup_{\overset{+}{S}}}}\text{Me}$$
$$\overset{|}{\underset{\text{CH}_2}{|^-}}$$

(1)

(2) Thiacyclohexane isomerises over alumina, zinc chloride, or Al-Si to give 2-methylthiophene. Similarly 2-methylthiacyclobutane gives thiophene (A.K. Yus'kovich, T.A. Danilova, and E.A. Viktorova, Chem. Abs., 1976, 85, 159826a). Butadiene, piperylene, and isoprene react with hydrogen sulphide at 400-500° over a alumina — chromic oxide catalyst with potassium oxide (0.4%) as a promoter to give thiophene, 2- and 3-methyl-thiophene, respectively (V.N. Kulakov et. al., ibid., 1973, 79, 126197u). Catalytic synthesis of methylthiophene and thiophene from C_5 hydrocarbons and hydrogen sulphide (M.A. Ryashentseva, E.P. Belanova, and Kh. M. Minachev, Isv. Akad. Nauk SSSR,

Ser., Khim., 1978, 2756). Zinc chloride catalyzed
alkylation of thiophene with isobutene at 300° gives
2-*tert*-butylthiophene (B.L. Lebedev, O.A. Korytina, and
L.I. Petrovskaya, Khim. Geterotsikl. Soedin., 1973, 502).
Butadiene and isoprene on heating with sulphur and steam in
the gas phase affords thiophene and 3-methylthiophene,
respectively (T.Nishi *et. al.*, Japan Pat. 79
76,574/1979). Vapour-phase dehydrocyclisation of dipropyl
or diisopropyl sulphide over Al_2O_3-Cr_2O_3-CuO-K_2O
at 300-400° yields mainly 2,4- and 2,5-dimethylthiophene
and 2-ethylthiophene (S. Trippler, Danilova, and A.F. Plate,
Vestn. Mosk. Univ., Khim., 1971, 12, 625). Propionaldehyde
and hydrogen sulphide over alumina containing potassium
oxide give mainly 2,4-dimethylthiophene. Similarly
α,β-unsaturated carbonyl compounds afford thiophenes, for
example, crotonaldehyde gives thiophene [J. Barrault *et.*
al., J. chem. Res., (S), 1978, 207].

Alkylation of thiophene with isopropyl chloride in
the presence of aluminium chloride at -70° affords a
mixture of the 2- and 3-isopropylthiophenium salts, which
on treatment with water gives 2- and 3-isopropylthiophenes
(3:2). Similarly thiophene and *tert*-butyl chloride yield
2- and 3-*tert*-butylthiophene, but with ethyl and methyl
bromide higher temperatures, -20° and 3.6°,
respectively, are required and lower yields of products are
obtained (L. Belen'kii, A.P. Yakubov, and I.A. Bessonova,
Zh. Org. Khim., 1977, 13, 364).

Alkylation of thiophene by hexafluoroacetone
N-methanesulphonylimine yields N-methylsulphonyl(2-thienyl)-
ditrifluoromethylamine (G.F. Il'in, A.F. Kolomiets, and
G.A. Sokol'skii, *ibid.*, 1979, 15, 2220).

The *tert*-butylation of 2,5-octamethylenethiophene
with an excess of *tert*-butyl chloride — stannic chloride

in carbon disulphide results in the occurrence of a deep
seated rearrangement. *tert*-Butyl groups are introduced
into the 2- and 5-positions of the thiophene nucleus with a
concomitant migration of the octamethylene bridge to the 3-
and 4-positions. In addition to this product (2) two mono-
-*tert*-butylthiophenes (3) and (4) are also formed (R. Helder
and H. Wynbert, Tetrahedron, 1975, 31, 2551).

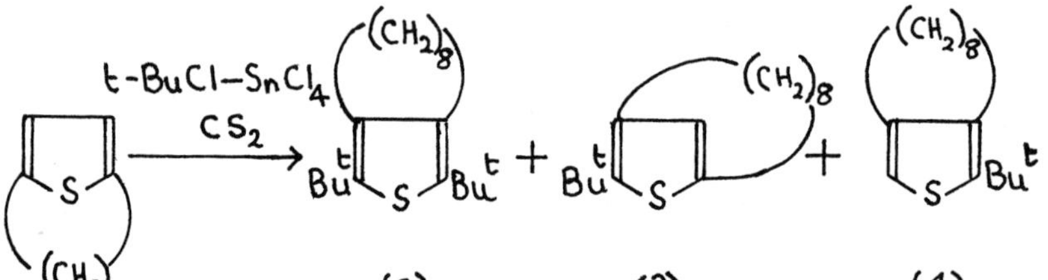

(2)   (3)   (4)

The reaction between thiophene and ethene in the
presence of a palladium salt at 25-100° gives
2-vinylthiophene, similarly furan affords vinylfuran
(I.V. Kozhevnikov, React. Kinet. Catal. Letters, 1976, 5,
439). The alkenylation of thiophene and furan with olefins
and palladium acetate — copper (II) acetate proceeds
regioselectively at the 2-position and when the
substituents on the olefin are bulky the products have
trans stereochemistry (O. Maruyama *et. al.*, Chem. Letters,
1979, 1229). Thiophene with styrene and palladium acetate
gives a mixture of the *trans*-styrylthiophenes (5) and (6)
(R. Asano *et. al.*, Bull. chem. Soc. Japan, 1973, 46, 663).

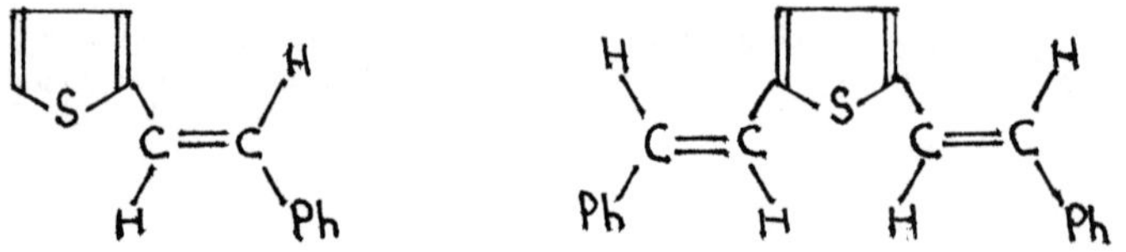

(5)   (6)

The reaction of 2-thienyllithium with 1,1-dichloro-
-2,2,-difluoroethene yields 2,2-dichloro-1-fluoro-1-(2-
-thienyl)ethene, which on treatment with n-butyllithium at
-40 to -60° affords the lithium compound (7), hydrolysed
by ice and hydrochloric acid to 2-ethynylthiophene b.p.55-
56°/30mm, (K. Okuhara, J. org. Chem., 1976, 41, 1487).

(7)

The phenylation of thiophene with several phenylating
agents gives in each case a mixture of 2- and 3-phenyl-
thiophene (90-95:10-5) (C.M. Camaggi *et. al.*, J. chem.
Soc., B, 1970, 1683). It has been shown that the
phenylation of thiophene and furan using aniline and amyl
nitrite follows an identical path and that, furan posesses
a susceptibility towards polarized radicals greater than
that observed for thiophene (L. Benati *et. al.*, J.
heterocyclic Chem., 1972, 9, 919). Under conditions in
which 2,4-dinitrobenenediazonium ions couple with anisole
to yield an azo-dye, the same ions interact with thiophene
with evolution of nitrogen to yield 2-(2,4-dinitrophenyl)-
thiophene. Similarly 2- and 3-methylthiophene give the 5-
and 2-substituted derivatives (M. Bartle *et. al.*, J. chem.
Soc., Perkin I, 1976, 1636), whereas 2-phenyl-, 2-*tert*-
-butyl-, 2,4-dimethyl- thiophene and 2- and 3-methylbenzo-
thiophene give the expected dinitrophenylazothiophene.
2,5-Dimethylthiophene besides 3-(2,4-dinitrophenylazo)-
thiophene yields an almost equal proportion of 5-methyl-
thiophene -2-carboxaldehyde 2,4-dinitrophenylhydrazone.

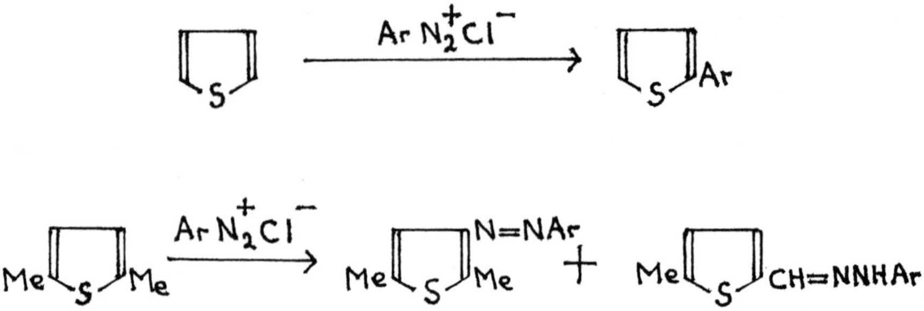

$$Ar = 2,4\text{-DiNO}_2C_6H_3$$

2,3,5-Trimethyl- and tetramethyl-thiophene on similar treatment yield exclusively the 2,4-dinitrophenylhydrazone by coupling at the α-methyl group. 2,3-Dimethylbenzo- thiophene does not react (S.T. Gore, R.K. Mackie, and J,M, Tedder, *ibid.*, p.1639). In the phenylation of thiophene using phenylazotriphenylmethane the percentage of 2-phenylthiophene relative to the 3-isomer is considerably lower than those obtained with other phenylating agents and is highly dependent on the experimental conditions and working up procedures. This is attributed to the trapping of the σ-complex leading to the 2-phenylthiophene by the triphenylmethyl radical to give two isomeric 2-phenyl-5- -triphenylmethyl-2,5-dihydrothiophenes (Camaggi *et. al.*, J. chem. Soc., B, 1969, 1251).

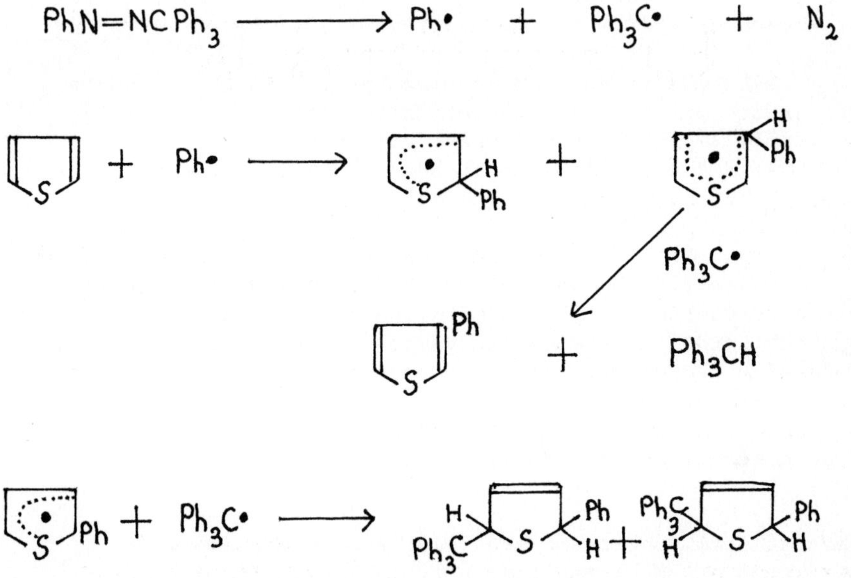

$$PhN=NCPh_3 \longrightarrow Ph\cdot \; + \; Ph_3C\cdot \; + \; N_2$$

Arylation of thiophene or furan using diazoaminobenzene
with a Lewis acid or isopentyl nitrite has been described
(L. Fisera, J. Kovac, and B. Hasova, Chem. Zvesti, 1976,
30, 480). Direct heteroarylation of thiophene occurs on
heating it with pyridine, quinoline or acridine in the
presence of activated sodium powder (A. Sheinkman *et. al.*,
Zh. org. Khim., 1973, 9, 2550). Heteroaryl radicals formed
by aprotic diazotization of the corresponding heterocyclic
amine in the presence of amyl or isoamyl nitrite,
substitute homolytically on thiophene to give 2-hetero-
arylthiophenes as the main reaction product. Thiophene
behaves somewhat unusually in these reactions, because its
reactivity towards heteroaromatic radicals (which are
slightly electrophilic) is lower than the reactivity
observed with nucleophilic radicals, such as cyclohexyl
and benzyl, and to a lesser extent, phenyl radicals
(G. Vernin and J. Metzger, J. org. Chem., 1975, 40, 3183).
Nitrobenzene with thiophene at 600° yields 2- and

3-phenylthiophene, *ca.* 3:1 and bithienyls (E.K. Fields and
S. Meyerson, *ibid.*, 1970, 35, 67).

2-Butylthiophene on heating at 550° in the presence
of chromic oxide — alumina gives benzothiophene (22-26%).
Similarly 2-pentyl- and 2-hexyl-thiophene yield 4-methyl-
(38-39%) and 4-ethyl-benzothiophene (30%), respectively,
and 3-butylthiophene gives benzothiophene (27%) (Danilova
and S.N. Petrov, Neftekhmiya, 1974, 14, 130).

2-Benzylthiophene is obtained by a modified Huang-
Minlon reduction of 2-benzoylthiophene using semicarbazide
hydrochloride in triethylene glycol in the presence of
potassium hydroxide. Direct benzylation of thiophene with
benzyl alcohol in the presence of zinc chloride gives a
mixture of 2- and 3-benzylthiophene (E. Maccarone, Boll.
Sedute Accad. Gioenia Sci., Natur. Catania, 1972, 11,
101).

The method of obtaining the thiophene 1,1-dioxides (8) by
oxidation of the corresponding thiophenes with 3-chloro-
perbenzoic acid has been improved. With the same reagent,
2-(1-methylnonyl)thiophene gives the dihydrobenzothiophene
(9) and 2-methylthiophene gives compound (10)
(W.J.M. Van Tilborg, Synth. Comm., 1976, 6 583).

$R^1 = R^3 = Me$, t-Bu, Ph, $R^2 = H$;
$R^2 = R^3 = $ t-Bu, $R^1 = H$

(vi) Halogenothiophenes

Chlorination of thiophene with sulphuryl chloride in the presence of iron gives a mixture of chlorinated bithienyls (T. Sone and Y. Abe, Bull. chem. Soc. Japan, 1973, **46**, 3603). 3-Chlorothiophene which is difficult to obtain may be prepared in good yield by a simple method, also applicable to the preparation of 2-, di-, tri-, and tetra- chlorothiophenes, by heating the corresponding bromothiophene with copper (I) chloride in dimethylformamide (S. Conde *et. al.,* Synth., 1976, 412). Tetrachlorothiophene 1,1-dioxide a reactive cheletropic Diels-Alder reagent has been used to annellate, with loss of sulphur dioxide, a large variety of olefinic compounds (M.S. Raasch, J. org. Chem., 1980, **45**, 856).

Electrochemical bromination of thiophene in methanol containing ammonium bromide gives a mixture of 2-bromo- and 2,5-dibromo-thiophene. No methylated derivatives are formed. Similarly 2-acetylthiophene yields 2-acetyl-5--bromothiophene (N. Nemec, J. Srogl, and M. Janda, Coll. Czech. chem. Comm., 1972, **37**, 3122). 3-Bromomethyl-thiophene may be obtained in consistent yields (70-75%) by adding 3-methylthiophene to *N*-bromosuccinimide in boiling carbon tetrachloride, containing a trace of azobis(iso-butyronitrile) as initiator (S. Gronowitz and T. Frejd, Synth. Comm., 1976, **6** 475).

2-Iodothiophene may be prepared by heating thiophene at 70-75° with iodine and periodic acid in 80% acetic acid containing a catalytic amount of sulphuric acid (H. Suzuki and Y. Tamura, Nippon Kagaku Zasshi, 1971, **92**, 1021) and by treating thiophene with iodine and silver trifluoromethanesulphonate in chloroform. The latter method is also applicable to bromination and is preferable to using silver perchlorate or silver trifluoroacetate and iodine, since the former may form explosive complexes and the latter iodotrifluoromethane [Y. Kobayashi, I. Kumadaki, and T. Yoshida, J. chem. Res. S (Synopses), 1977, 215]. The yield of 2-iodothiophene by direct iodination in dilute nitric acid is improved to 85-90% by adding the iodine to thiophene instead of *vice versa* (R. Arias *et. al.,* Ing. Cienc. Quim., 1979, **3**, 162).

Photostimulated reaction of 2-chloro-, 2-bromo-, and

3-bromo-thiophene in liquid ammonia with acetone enolate
ions gives mono-and di-thienylacetones. Treatment of
3-bromothiophene in liquid ammonia under stimulation by
solvated electrons affords the same products, but in lower
yields, and the corresponding secondary alcohol. In the
presence of excess potassium amide only debromination to
thiophene occurs. Solvated electron-stimulated reactions
of 2-chloro- and bromo-thiophene yield mainly thiophene
with little thienylation (J.F. Bunnett and B.F. Gloor,
Heterocycl. 1976, 5, 377). The reaction between 2,5-
-dihalogenothiophenes and tetracyanoethene oxide leads to
2,5-bis(dicyanomethylene)-3-thiolene (S. Gronowitz and
B. Uppstrom, Acta. Chem. Scand., 1974, B28, 981). The
reaction between tetracyanoethene oxide and thiophene,
furan, selenophene, benzothiophene and benzofuran has been
investigated (*idem, ibid.*, 1975, B2, 441).

(vii) Nitro- and amino-thiophenes

2-Nitro-thiophene undergoes smooth photosubstitution
of the nitro group by nucleophiles, for example, CN⁻, CNO⁻
MeO⁻, H_2O (M.B. Groen and E. Havinga, Mol. Photochem.,
1974, 6, 9). With methyl radicals generated from
dimethylsulphoxide — hydrogen peroxide, the nitro group of
2-nitrothiophene exhibits strong ortho-directing properties
to give mainly 3-methyl-2-nitrophene (U. Rudqvist and
K. Torssell, Acta Chem. Scand., 1971, 25, 2183).

Ethyl 2-aminothiophene-3-carboxylates rearrange on
treatment with base to give 3-cyanothiolen-2-ones, thus
showing the difference between aminothiophenes and anilines
(K. Gewald, H. Jablokoff, and M. Hentschel, J. pr. Chem.,
1975, 317, 861).

3-Bromothiophene in liquid ammonia containing
potassium amide under nitrogen gives 3-aminothiophene in

high yield. Little reaction occurs under the normal atmosphere (Bunnett and Gloor, *loc. cit.*).

(viii) Methoxythiophenes

Introduction of the electron releasing methoxy group into the thiophene ring increases its reactivity so that it undergoes the Mannich reaction. Hence 2- and 3-methoxythiophene react with formaldehyde and amines to yield 2-(5- and 3-methoxyl)thenylamines, respectively (J.M. Barker, P.R. Huddleston, and M.L. Wood, Synth. Comm., 1975, 5, 59).

R = NMe$_2$, piperidino, morpholino

Similarly with 3,4-dimethoxythiophene substitution occurs in the 2-position.

(ix) Sulphonic acids, thiols and related compounds

Thiophene is monosulphonated when heated with bis(trimethylsilyl)sulphate (M.G. Voronkov, S.V. Korchagin, and V.K. Roman, Isv. Akad. Nauk SSSR, Ser. Khim., 1977, 2340). Thiophene-2-sulphonic acids are prepared by sulphonating the thiophene with the complex from sulphur trioxide and ether or tetrahydrofuran in dichloroethane (T.K. Khanina and B.V. Passet, U.S.S.R. 707,916/1980).

The methoxycarbonylsulphenylation of thiophene followed by hydrolysis affords a route to thiophene-2-thiol (W. Schroth *et. al.*, Z. Chem., 1977, 17, 411).

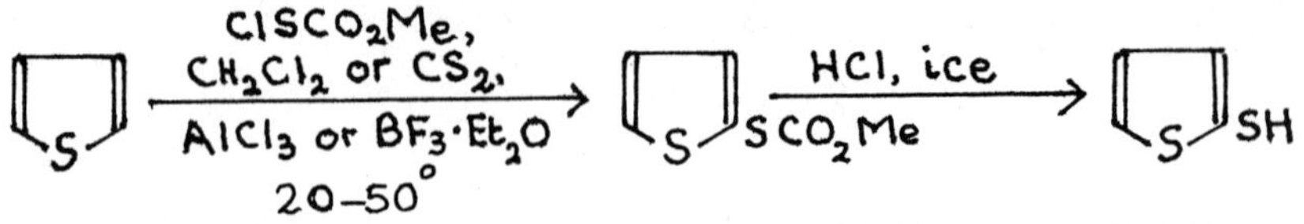

$$\text{thiophene} \xrightarrow[\substack{AlCl_3 \text{ or } BF_3 \cdot Et_2O \\ 20-50^\circ}]{\substack{ClSCO_2Me, \\ CH_2Cl_2 \text{ or } CS_2,}} \text{thiophene-}SCO_2Me \xrightarrow{HCl, \text{ice}} \text{thiophene-}SH$$

A number of 2-thienyl sulphides have been obtained by the reaction of thiophene-2-thiol with the appropriate compound containing a halogeno substituent (K.I. Sadykhov, S.M. Aliev, and M.M. Seidov, Tezisy Doklady Nauchn. Sess. Khim. Tekhnol. Org. Soedin. Sery Sernistykh Neftei, 13th, 1974, 115).

$$\text{thiophene-}SH \xrightarrow{RX} \text{thiophene-}SR$$

$$X = Cl, Br$$

$R = CH_2CH_2OH, CH_2CH_2Cl, CH_2CH_2Br, CH_2CO_2Me, CH_2CH\!=\!CH_2,$
$CH\!\equiv\!CHCO_2Me, CH_2CH(OH)CH_2NEt_2, CH_2CH(OH)CH_2NBu_2$

Treatment of thiophene with ethyl mercaptan in the presence of Fenton's reagent at $0-5^\circ$ gives 2,5-di(ethylthio)thiophene (Ya. L. Gol'dfarb, G.P. Pokhil, and L.I. Belen'kii, Chem. Abs., 1973, 79, 126188s).

$$\text{(structure)} \xrightarrow[\text{Reagent}]{\text{RSH,}} RS\text{(structure)}SR$$

R = Et, Bu

$$R^1\text{(structure)} \xrightarrow[\text{Reagent}]{R^2SH,} R^1\text{(structure)}SR^2$$

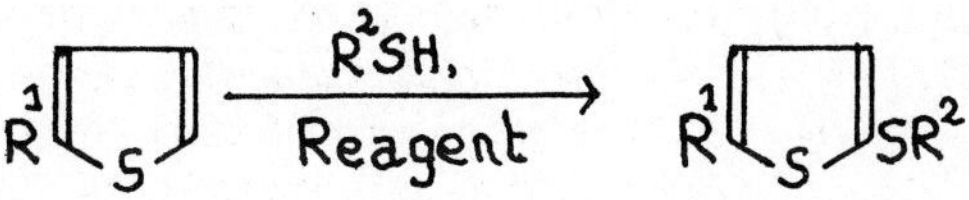

R^1 = Me, Bu R^2 = Et, Bu, PhCH$_2$

Reagent: t-BuOH, H$_2$O, 27% H$_2$O$_2$, FeSO$_4$, H$_2$SO$_4$

(x) Thiophene alcohols and related compunds

2,5-Di(aryloxymethyl)thiophenes useful as insecticides have been prepared by the cyclization of the approprite diacetylene derivative using hydrogen sulphide (M. Askaraliev *et. al.*, Chem. Abs., 1977, **86**, 43477d).

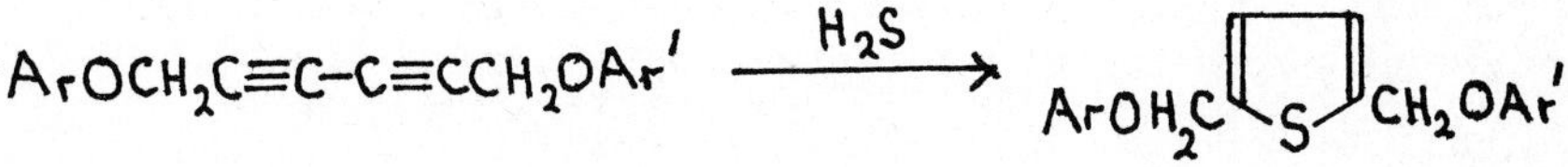

$$ArOCH_2C\equiv C-C\equiv CCH_2OAr' \xrightarrow{H_2S} ArOH_2C\text{(structure)}CH_2OAr'$$

(xi) Aldehydes and ketones

The reaction between thiophene and chloromethylene-malononitrile in the presence of aluminium chloride affords 2-thienylmethylenemalononitrile, which on treatment with aqueous potassium hydroxide gives thiophene-2-carboxalde-hyde (W. Ertel and K. Friedrich, Ber., 1977, 110, 86).

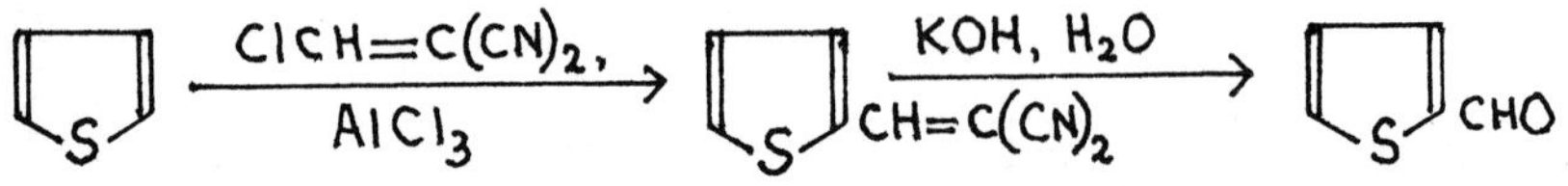

$$\underset{S}{\square} \xrightarrow[\text{AlCl}_3]{\text{ClCH=C(CN)}_2,} \underset{S}{\square}_{\text{CH=C(CN)}_2} \xrightarrow{\text{KOH, H}_2\text{O}} \underset{S}{\square}_{\text{CHO}}$$

Thiophene-2-carboxaldehyde on treatment with acetic anhydride in the presence of methanesulphonic aicd, phosphoric acid, or sulphuric acid gives the diacetate (F. Freeman and E.M. Karchefski, J. chem. Eng. Data, 1977, 22, 355).

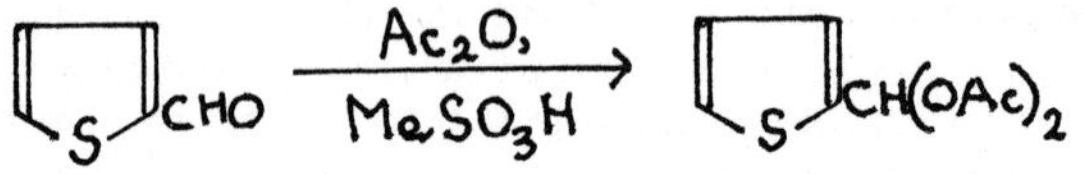

$$\underset{S}{\square}_{\text{CHO}} \xrightarrow[\text{Me SO}_3\text{H}]{\text{Ac}_2\text{O},} \underset{S}{\square}_{\text{CH(OAc)}_2}$$

The cycloaddition reaction between thiophene-2--carboxaldehyde and dichloroketene results only in the isolation of the decarboxylated product 1,1-dichloro-2--(2-thienyl)ethene from the reaction mixture (H.O. Krabbenhoft, J. org. Chem., 1978, 43, 1305).

The reduction of thiophene-2-carboxaldehyde with thiourea dioxide in alkaline ethanol gives 2-hydroxymethylthiophene [S.-L. Huang and T.-Y. Chen, J. Chin. chem. Soc. (Taipei), 1974, 21, 235], and its ammoxidation over a Mo-Bi-Sb catalyst at 400° yields thiophene-2-carbonitrile (P. Singh, Y. Miwa, and J. Okada, Chem. pharm. Bull., 1978, 26, 2838). Ir spectral data of the 5-deuterio-analogues shows that the multiple carbonyl absorptions of thiophene- and furan-2-carboxaldehydes are caused by Fermi resonance and/or rotational isomerism (D.J. Chadwick *et. al.*, Chem. Comm., 1972, 742); ^{13}C nmr spectra of thiophene-2- -carboxaldehyde and 2-acetylthiophene and related furans have been reported (T.N. Huckerby, J. Mol. Struct., 1976, 31, 161).

Bromination of the nucleus of alkylthiophenes may be suppressed and sometimes eliminated by the action of light and a radical catalyst, for instance, azobisisobutyronitrile (AZDN). Thus side-chain bromination occurs, and depending upon the alkylthiophene one, two, or three bromines may be introduced before ring bromination becomes a competing reaction. This method initiates a useful route to thiophenecarboxaldehydes, especially the 3-aldehyde and the 3-chloro-2-aldehyde, and to 3-vinyl- and 3-acetyl- -thiophenes (J.A. Clarke and O. Meth-Cohn, Tetrahedron Letters, 1975, 4705).

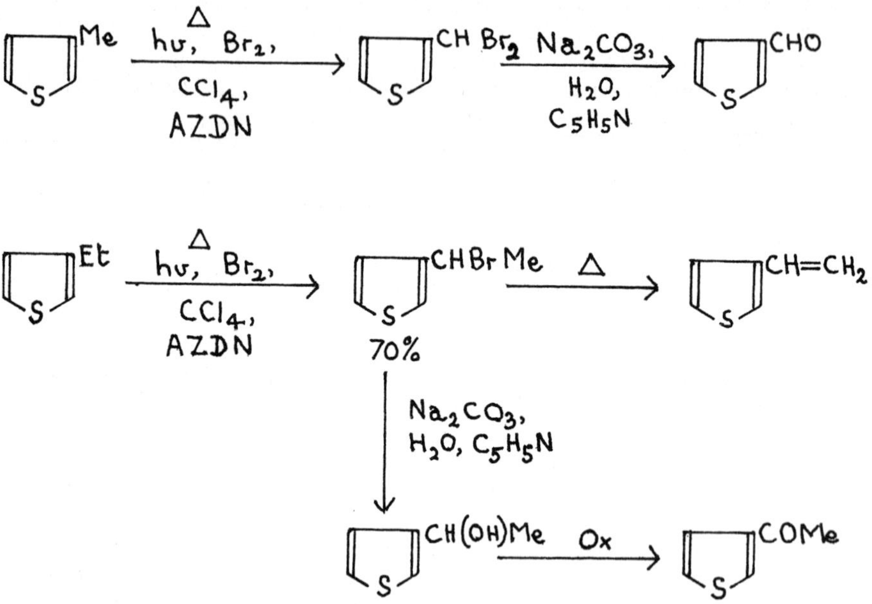

Pyrolysis of the sodium salt of thiophene-3-carboxaldehyde tosylhydrazone produces *cis*- and *trans*- 1,2-di(3-thienyl)-ethene (Ch. 2, p. 32) (R.V. Hoffman, G.G. Orphanides, and H. Shechter, J. Amer. chem. Soc., 1978, 100, 7927). Ir spectrum of thiophene-3-carboxaldehyde (C.G. Andrieu *et. al.*, Compt. rend., 1972, 275C, 559).

High yields of 2-acetyl- and 2-acetyl-3-methyl-thiophene may be obtained by acetylating the appropriate thiophene using acetyl toluene-4-sulphonate in acetonitrile or benzene (S.I. Pennanen, Heterocycl. 1976, 4, 1021).

2-Chloroacetylthiophene, m.p.47°, is obtained from thiophene and chloroacetyl chloride in carbon disulphide containing aluminium chloride; 2-bromoacetylthiophene, b.p.95°/1.5mm, is prepared from 2-acetylthiophene and bromine in carbon tetrachloride; and 2-iodoacetylthiophene, b.p.115°/1mm, is produced from the related chloro compound and sodium iodide in acetone. Trialkylphosphites

react with 2(α-halogenoketones) of thiophene and furan to
give either enol phosphates (Perkow reaction) or
β-ketophosphonates (Michaelis-Arbuzov reaction) or a
mixture of both isomers (A. Arcoria *et. al.*, J. heterocyclic
Chem., 1975, 12, 215). Some of the by-products formed
during the acetylation of a variety of substituted
thiophenes have been isolated and their mode of formation
rationalized. The bromination of a number of thienyl
ketones followed by their reduction and transformation
to thienylethanolamines has been reported (J.G. Bagli
and E. Ferdinandi, Canad. J. Chem., 1975, 53, 2598).

Fluorinating acylation of thiophene with fluorinated
immonium salts, obtained by action of boron trifluoride on
α-fluorinated amines, is regioselective giving the
2-acylated derivative (C. Wakselman and M. Tordeux, Chem.
Comm., 1976, 956).

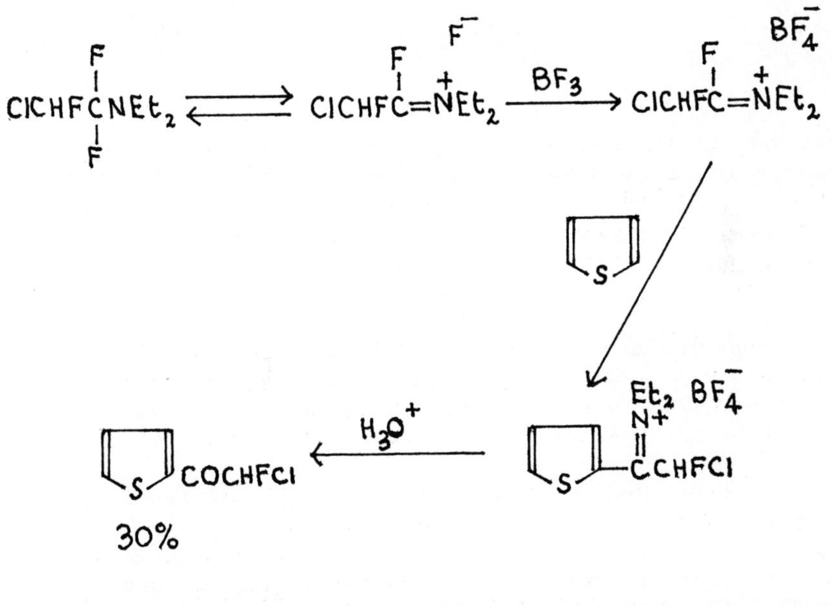

30%

5-Carbonyl derivatives (1) and (2) of 2-arylthiophenes are
obtained by Gomberg arylation of thiophene with
$4-RC_6H_4N_2{}^+Cl^-$ (R = H, Me, MeO, Cl, Br), followed by
formylation and acetylation, and condensation with

acetophenone and benzaldehyde, respectively. Condensation of the formyl derivatives with acetone yields the carbonyl derivatives (3) (V.K. Polyakov, Z.P. Zaplyuisvenka, and S.V. Tsukerman, Tezisy Doklady - Simp. Khim. Tekhnol. Geterotsikl. Soedin. Goryuch. Iskop., 2nd, 1973, 117).

$$4\text{-}RC_6H_4\text{-}S\text{-}CH=CHCOPh \qquad 4\text{-}RC_6H_4\text{-}S\text{-}COCH=CHPh$$

(1)   (2)

$$4\text{-}RC_6H_4\text{-}S\text{-}CH=CHCOCH=CH\text{-}S\text{-}C_6H_4R\text{-}4$$

(3)

Treatment of 3-methylthiophene with lauric acid in the presence of trifluoroacetic anhydride and phosphoric acid gives a mixture of the 3- and 4-methyl derivatives (4) and (5) of the 2-thienyl ketone (C. Galli, Synth., 1979, 303).

$$\text{S}\text{-}CO(CH_2)_{10}Me \text{ with } Me \qquad Me\text{ with }\text{S}\text{-}CO(CH_2)_{10}Me$$

(4)   (5)

2-Benzoylthiophene may be prepared by treating thiophene with 2-benzoyloxypyridine-trifluoroacetic acid (T. Keumi, R. Taniguchi and H. Kitajima, *ibid.*, 1980, 139), or benzoic acid in the presence of 2-(trifluoromethanesulphonyloxy)-pyridine in trifluoroacetic acid (Keumi *et. al.*, Chem.

Letters, 1977, 1099). 2-Benzoylthiophenes, for example, 2-benzoyl-3,5-diphenylthiophene (7, R = Ph), are prepared by condensing the appropriate thioacylacetophenone and α-bromo carbonyl compound, and dehydrating the resulting dihydrothiophenes (6) (M. Takaku, Y. Hayasi, and H. Nozaki, Bull. chem. Soc. Japan, 1970, **43**, 1917).

(6)   (7)

R = Me, Ph

Rates of reduction of 2-thienyl and 2-furyl ketones by sodium tetrahydridoborate in propan-2-ol have been measured (M. Fiorenza, A. Ricci, and G. Sbrana, J. chem. Soc., Perkin II, 1978, 1232). The 220M Hz nmr spectral studies of thiophene and furan carbonyl compounds show that the *cis* form of thiophene derivatives is favoured. Semi-empirical calculations (CNDO/2) have been reported and the influence of sulphur 3d-orbitals discussed (S. Nagata *et. al.*, Tetrahedron, 1973, **29**, 2545). Ir spectra and conformational analysis of 2-acylthiophenes and related furan compounds (Andrieu, C. Chatain-Cathaud, and M.C. Fournie-Zaluski, J. mol. Struct., 1974, **22**, 433).

(xii) *Thiophenecarboxylic acids*

The reaction of thiobenzoylacetophenone with ethyl bromoacetate in benzene in the presence of triethylamine yields compound (1) which on cyclization with sodium hydride, followed by dehydration of the product gives ethyl 3,5-diphenylthiophene-2-carboxylate (2) (Takaku, Hayasi, and Nozaki, *loc. cit.*).

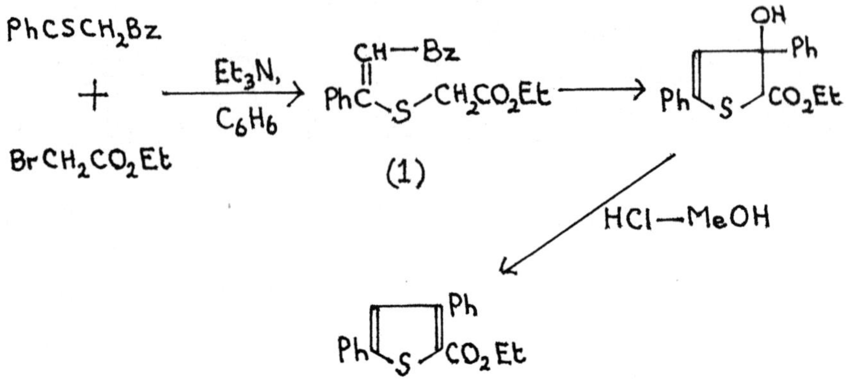

Ethyl 5-methyl-3-phenylthiophene-2-carboxylate is obtained by the above method from thioacetylacetophenone and ethyl bromoacetate.

1:1 Adducts of β-imino carbonyl derivatives and alkyl, aryl, or acyl isothiocynates react with phenacyl bromide in boiling propan-2-ol to produce tetra-substituted thiophenes, for example a tri-substituted thiophene-3--carboxylate (S. Rajappa and B.G. Advani, Tetrahedron Letters, 1969, 5057).

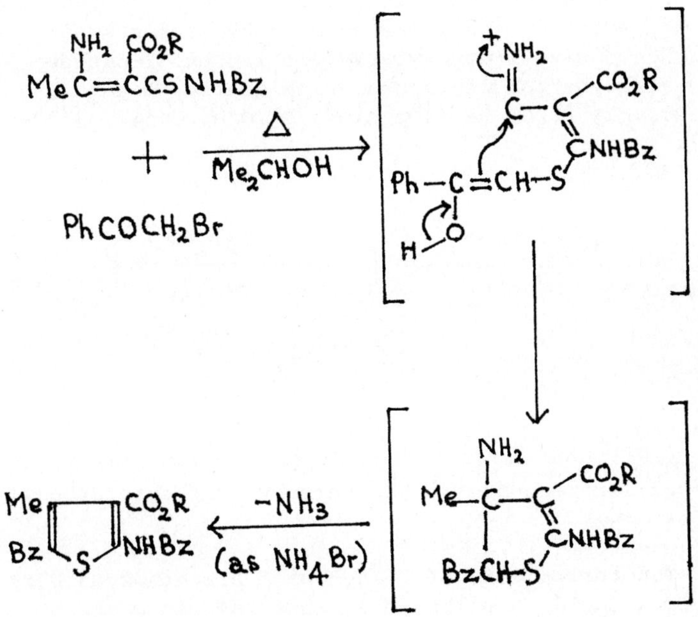

The reaction between appropriate thiophenes and tetracyanoethene oxide gives 2-thenoyl cyanides (3) (R^1 = H, R^2 = MeO, MeS; R^1 = MeO, R^2 = H) (S. Gronowitz and B. Uppstrom, Acta Chem. Scand., 1974, 28B, 339).

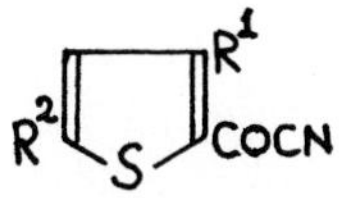

(3)

Thiophene reacts with ethoxycarbonyl isocynates and isothiocynates in the presence of anhydrous stannic chloride to give *N*-ethoxycarbonylthiophene-2-carboxamide and *N*-ethoxycarbonylthiophene-2-thiocarboxamide, respectively (E.P. Papadopoulos, J. org. Chem., 1974, **39**, 2540).

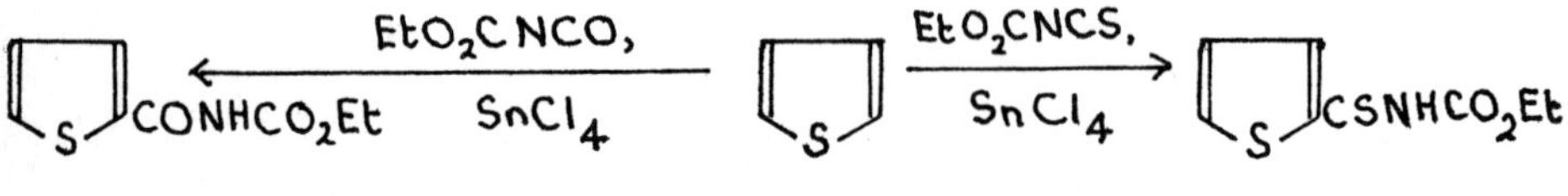

Both derivatives exhibit considerable reactivity towards nucleophilic reagents at the carbonyl and thiocarbonyl groups.

The thermolysis of thiophenium bismethoxycarbonyl-methylide (p.10) results in a ready rearrangement to dimethyl thiophene-2-malonate (Gillespie, Porter, and Willmott, Chem. Comm., 1978, 85; Brit. UK Pat. Appl. 2,008,570/1979).

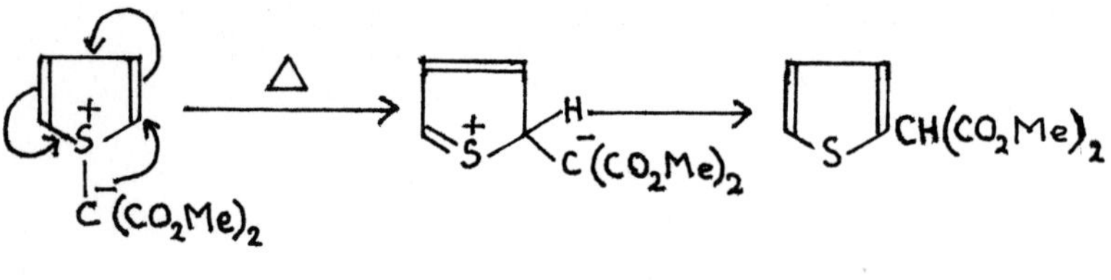

Dimethyl thiophene-3,4-dicarboxylates have been synthesized (p.3) and thiophene-2,3-dicarboxylic anhydrides converted into benzothiophene derivatives (p.45). Preparation of diethyl thiophene-2,5-diacetate (p.40).

(b) Thiolenes, dihydrothiophenes

A highly substituted 2-thiolene derivative is formed, by initially reacting benzoylthioacetophenone with ethyl bromophenylacetate and treating the resulting S-alkylated enolate with sodium ethoxide (Takaku, Hayasi, and Nozaki, *loc. cit.*).

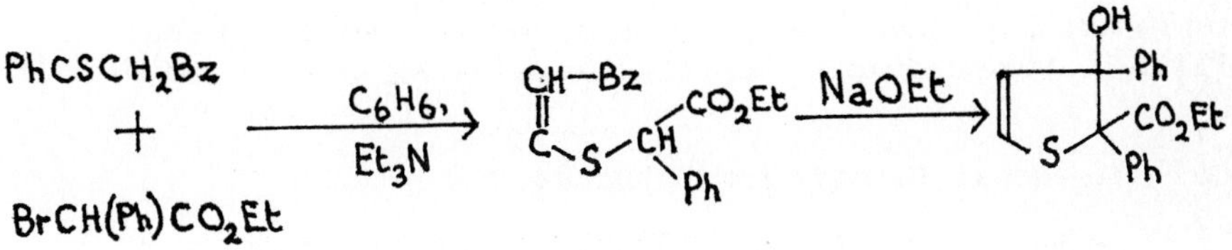

An excellent method of preparing 3-thiolenes with a number of substituents, involves the reaction of α-mercaptoketones with vinylphosphonium salts (J.M. McIntosh, H.B. Goodbrand and G.M. Masse, J. org. Chem., 1974, **39**, 202).

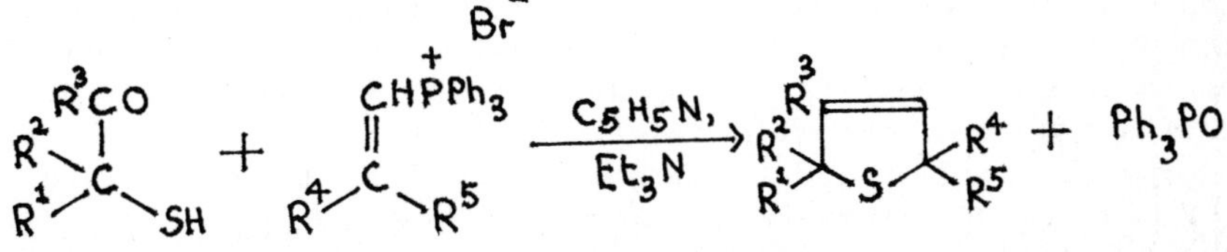

Starting with α-mercaptoaldehydes furnishes a method of preparing 3,4-unsubstituted 3-thiolenes (McIntosh and R.S. Steevensz, Canad. J. Chem., 1974, **52**, 1934).

Gas phase oxidation of thiophene over a 10% MoO_3-TiO_2 catalyst proceeds with 75% selectivity for 2,5-dioxo-3--thiolene (thiomaleic anhydride) and maleic anhydride (M. Blanchard and J. Goichon, Bull. Soc. chem. Fr., 1975, 289). Benzothiophene on similar treatment affords benzoic acid in almost quantitative yield.

 trans-2,5-Dimethyl-3-thiolene 1,1-dioxide, m.p.63-
63.5°; *cis*-isomer, m.p.43-44°. On thermolysis these
compounds give a hexadiene and sulphur dioxide. The *cis*-
isomer gives exclusively *trans*, *trans*-hexa-2,4-diene
and the *trans* gives only *cis*, *trans*-hexa-2,4-diene. In each
case, stereospecificity exceeds 99.9% and it therefore,
follows that the eliminations have proceeded entirely
suprafacially (W.L. Mock, J. Amer. chem. 1975, <u>97</u>, 3666,
3673). 3-Cyano-2-oxo-3-thiolenes see p. 24.

(c) Thiolanes, tetrahydrothiophenes

 A method of preparing thiolanes is by the ionic
hydrogenation of the related thiophene, using a mixture of
triethylsilane and trifluoroacetic acid. The 2- and 3-
-alkyl- and aryl- and 2,5-dimethyl-thiolanes, thiolanes
functionally substituted in the side chain and 2,3-
-dihydrobenzothiophenes have been obtained by this method
(D.N. Kursanov *et. al.*, Tetrahedron, 1975, <u>31</u>, 311). Ionic
hydrogenation of 3-phenyl-, 2,3-, 2,4-, and 3,4-diphenyl-
-thiophenes give the corresponding thiolanes in high yield
(A.N. Parnes, G.I. Bolestova, and Kursanov, Chem. Abs.,
1976, <u>85</u>, 159800n). It has been found that the rate of
ionic hydrogenation of thiophenes, for example, 2-ethyl-
thiophene, 2-acetylthiophene, and δ-(2-thienyl)valeric acid
is increased significantly by addition of a large excess of
trifluoroacetic acid or a small amount of toluene-*p*-
-sulphonic acid or its lithium salt, or lithium perchlorate
(Parnes *et. al.*, Izv. Akad. Nauk SSSR, Ser. Khim., 1977,
2526). Rhenium sulphide, Re_2S_7 is more active in the
hydrogenation of thiophene than the chloride, Re_3Cl_9
(M.A. Ryashentseva, Kh. M. Minachev, and E.P. Belanova,
ibid., 1976, 1183).

 2-Methyl-7-thiabicyclo[3.2.1]octane (2, R = Me),
containing a thiolane ring is obtained in a 3:1 mixture,
with 1-methyl-2-thiabicyclo[2.2.2]octane (3, R = Me), on
irradiating the hexene derivative (1, R = Me) in boiling
pentane. With compound (1, R = H) only 2-thiabicyclo[2.2.2]-
octane (3, R = H) is formed (J.M. Surzur *et. al.*,
Tetrahedron Letters, 1971, 2035).

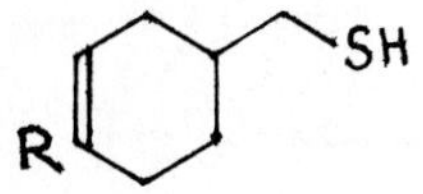

(1) 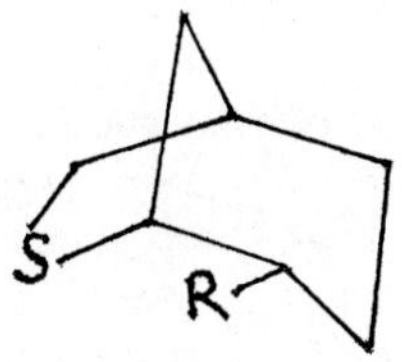 (2) 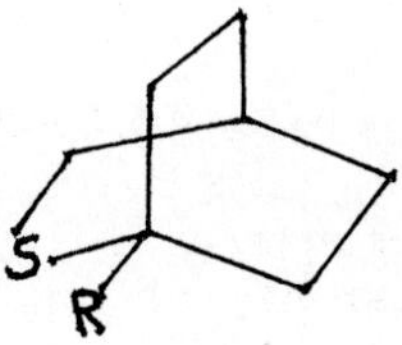 (3)

Highly selective β-chlorination occurs on treating thiolane 1,1-dioxide with sulphuryl chloride in the absence of light at 60° to give 3-chlorothiolane 1,1-dioxide, m.p. 54-54.5°(I.Tabushi, Y.Tamaru, and Z.Yoshida, *ibid.*, p.3893).

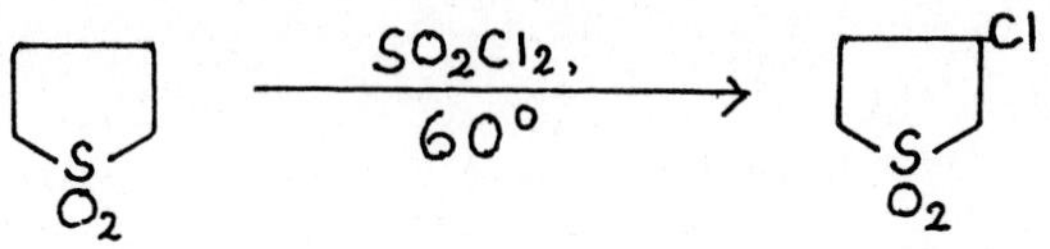

Unsuccessful attempts have been made to dehydrohalogenate bis-*endo*-1,5-dichloro-7-thiabicyclo[2.2.1]heptane (4) with a variety of rather non-nucleophilic bases to obtain 7-thianorbornadiene (5), for example, with 1,5--diazobicyclo[5.4.0]undec-2-ene no 7-thianorbornadiene is observed, but benzene is formed presumably from this intermediate (T.J. Barton, M.D. Martz, and R.G. Zika, J. org. Chem., 1972, 37, 552).

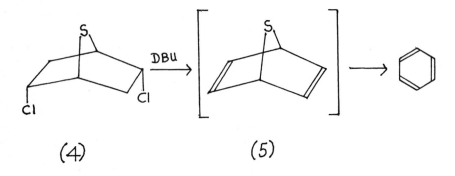

(4)   (5)

Mutual isomerisation occurs between 2-methylthietane and thiolane, for instance, 2-methylthietane rearranges in acetonitrile containing Ph_3C^+ BF_4^- or nitrobenzene containing chloroanil to give thiophene and thiolane, whereas thiolane with the same reagents in acetonitrile affords 2-methylthietane, thiophene, and thiolene, and in nitrobenzene containing $Ph_3CCl-SnCl_4$ yields only 2-methylthietane and thiophene (L.M. Petrova, A.A. Fregar, and E.A. Viktorova, Vestn. Mosk. Univ., Khim., 1974, 15, 498).

The reaction of 2,5-dioxothiolane with the stabilised phosphorane, ethoxycarbonylmethylene triphenylphosphorane gives diethyl thiophene-2,5-diacetate and with cyanomethylenetriphenylphosphorane it gives the stereoisomers (6) and (7), existing mainly as the non-aromatic tautomers, i.e. as 2,5-bis(cyanomethylene)thiolanes (W. Flitsch, J. Schwieger, and U. Strunk, Ann., 1975, 1967).

$$EtO_2CCH_2 S CH_2CO_2Et \xleftarrow{\overset{\triangle}{Ph_3P=CHCO_2Et}} \underset{O S O}{\quad} \xrightarrow{Ph_3P=CHCN} \quad (6) \; + \; (7)$$

(d) Compounds having two or more unfused thiophene rings

Coupled heterocycles (bithienyls, bifuryls, furyl-thiophenes) are obtained in high yield *via* the reaction of thiophene and furan with palladium salts in solution at 50-100° (I.V. Kozhevnikov, React. Kinet. Catal. Letters, 1976, 4, 451). The oxidative coupling of thiophene in the presence of $PdCl_2$, $PdCl_2$-$CuCl_2$, $Pd(OAc)_2$-$Cu(OAc)_2$ and $CuSO_4$-$Fe_2(SO_4)_3$ on Al_2O_3, C, or SiO_2 catalysts yields 2,2'-bithienyl and 2,3'-bithienyl. It is found that the activity and selectivity of the reaction depends on the anion and decreases in the order $Cl^- > SO_4^{2-} > AcO^-$, and that the most active catalysts are mixed salts with different anions (A.M. Osipov, L.P. Metlova, and T.I. Emel'yanova, Ukr. Khim. Zh., 1978, 44, 660). Catalysts containing $PdCl_2$-$FeCl_3$-$CuCl_2$-MnO_2 have been recommended for the synthesis of 2,2'- and 2,3'-bithienyl, and it has been shown that their ratio in the reaction product, varies with the composition of the catalyst and the reaction conditions (E.S. Rudakov and R.I. Rudakov, Dopov. Akad. Nauk Ukr. RSP, Ser. B: Geol., Khim. Biol. Nauki, 1977, 815). The ratio of 2,2'- to 2,3'-bithienyl from the reaction in aqueous H_2SO_4, $HClO_4$, or HCl at 51-120° increases with increasing temperature and acid and Fe^{3+} concentration. The combined yield increases with temperature and with decreasing Pd^{2+}, Fe^{3+}, and Cl^- concentration (Rudakov and V.N. Ignatenko, *ibid.*, p.1003). The oxidative coupling of thiophene to give 2,2'- and 2,3'-bithienyl has also been carried out using $Pd(OAc)_2 - H_5PMo_{10}V_2O_{40}$ (HPA-2) in dimethylformamide (V.E. Tarabanko, Kozhevnikov, and K.I. Mateev, React. Kinet. Catal. Letters, 1978, **8**, 77).

Dehydrodimerization of thiophene occurs when it is pyrolysed at 780-800° in a continuous system, in an atmosphere of hydrogen sulphide. The resulting products, in high yield, are 2,2'-, 3,3'-, and 2,3'-bithienyl (M.G. Voronkov *et. al.*, Khim. Geterotsikl. Soedin., 1976, 1186). Irradiation of 2,2'-bithienyl leads to 2,3'--bithienyl and a small amount of benzothiophene. Similarly 2,3'-bithienyl gives 3,3'-bithienyl and benzothiophene as major products (R.M. Kellog *et. al.*, J. org. Chem., 1970, **35**, 2737).

On treatment with sulphuryl chloride in the presence of iron powder, thiophene undergoes chlorinative coupling to give chlorinated 2,2'-bithienyls as the main products. On separation the reaction product affords 5,5'-dichloro-, 3,5,5'-trichloro-, and 3,3',5,5'-tetrachloro-2,2'-bithienyl. The reaction is also catalyzed by Friedel-Crafts catalysts, the efficiency of the catalysts being $AlCl_3 > FeCl_3 > SnCl_4 > ZnCl_2$. The above reaction without a catalyst gives chlorothiophenes (T. Sone, K. Sakai, and K. Kuroda, Bull. chem. Soc., Japan, 1970, **43**, 1411).

Chlorination of 2,2'- and 2,3'-bithienyls with sulphuryl chloride gives 5-chloro-, 5,5'-dichloro-, 3,5,5'-trichloro-2,3'-bithienyl, 2'-chloro-, 2',5-dichloro-, and 2',3,5-trichloro-2,3'-bithienyl, respectively (Sone and Y. Abe, *ibid.*, 1973, **46**, 3603).

Nitration of 2,3'-bithienyl with fuming nitric acid in acetic anhydride at 0° gives a mixture of 3-nitro-, m.p.77-78°, 2'-nitro-, m.p.50°, and 5-nitro-2,3'-bithienyl, m.p.92-93° with relative yields of 38.7, 34.8, and 26.5%. When the nitration is carried out with fuming nitric acid in acetic acid at 20° the same nitro derivatives are obtained, but with different relative yields of 20.4, 36.5, and 43.1%, respectively. On further nitration with nitric acid in acetic anhydride at room temperature, the 3-nitro derivative affords 2',3-dinitro-, m.p.194-195°, and 3,5'--dinitro-2,3-bithienyl, m.p.132-133°; the 2'- nitro derivative affords 2',3-dinitro-, 2',5-dinitro-, m.p.163-164°, and 2',4-dinitro-2,3'-bithienyl, m.p.135-136°; and 5-nitro derivative affords 2'5-dinitro-, and 5,5'--dinitro-2,3'-bithienyl, m.p.238-239° (C. Dell'Erba, G. Guanti, and G. Garbarino, J. heterocyclic Chem., 1974, **11**, 1017).

2. Benzo[b]thiophenes (Thianaphthenes)* and Related Compounds

(a) Benzo[b]thiophenes

(i) Synthetic methods

(1) **Unsaturated ketones** (e.g.1) prepared by the reaction of substituted benzaldehydes with acetone in the presence of sodium hydroxide, on treatment with bromine or sulphuryl chloride give the dihalogeno compounds, which on boiling in pyridine cyclize to yield the benzothiophenes (2) (A. Ruwet and M. Renson, Bull. Soc. chim. Belg., 1970, **79**, 593).

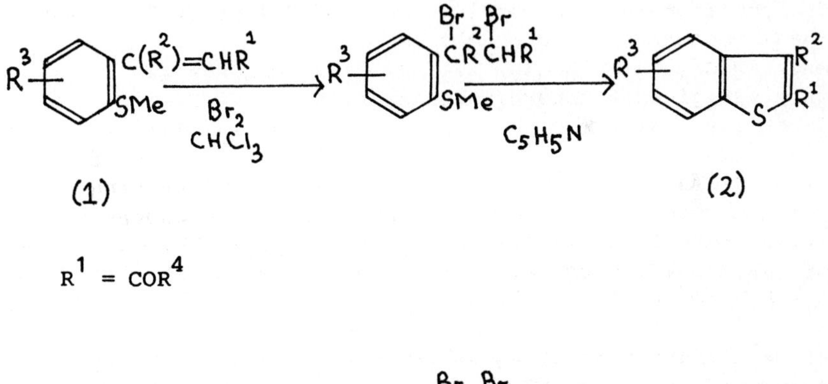

(1)   (2)

$$R^1 = COR^4$$

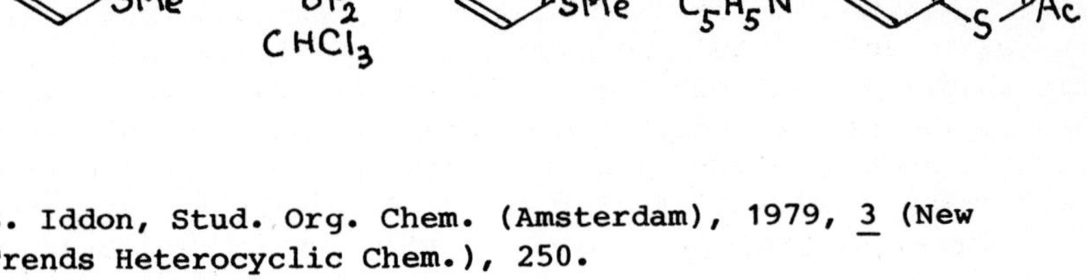

* B. Iddon, Stud. Org. Chem. (Amsterdam), 1979, **3** (New Trends Heterocyclic Chem.), 250.

Benzo[b]selenophenes are similarly prepared from the appropriate aryl methyl selenoethers.

(2) A new approach to the synthesis of 2-substituted benzothiophenes involves the ozonolysis of benzothiophene when one of the products is the aldehyde (3), which on interaction with compounds containing an 'active' methyl or methylene group yields 2-substituted benzothiophenes (K.J. Brown and O. Meth-Cohn, Tetrahedron Letters, 1974, 4069).

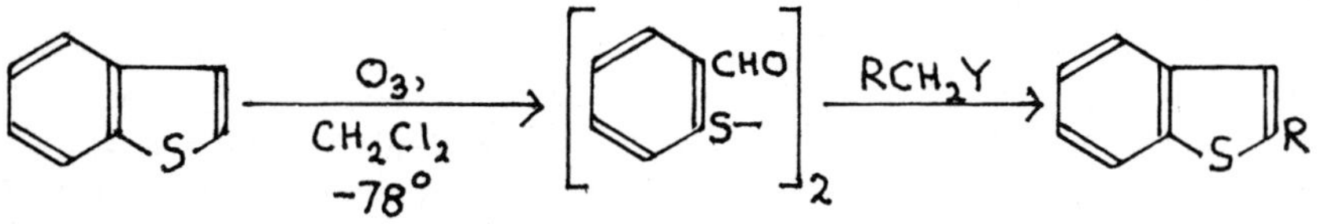

(3)

$$Y = H, \ CO_2H$$
$$R = CO_2H, \ CN, \ NO_2, \ Ph$$

The aldehyde (3) is also otained using diazotized anthranilic acid as starting material.

(3) An alternative approach to the synthesis of benzothiophene, i.e., annellation of the benzene ring on preformed thiophenes has been described and a number of benzothiophenes prepared (H.J.J. Loozen and E.F. Godefroi, J. org. Chem., 1973, 38, 1056).

(4) Phenyl prop-2-ynyl sulphoxide (4) on heating in dioxan at 100° gives the hemithioacetal (5), which on warming with a suitable protic solvent and a catalytic amount of toluene-p-sulphonic acid affords the benzothiophene (6) (Y. Makisumi and S. Takada, Chem. Comm., 1974, 848).

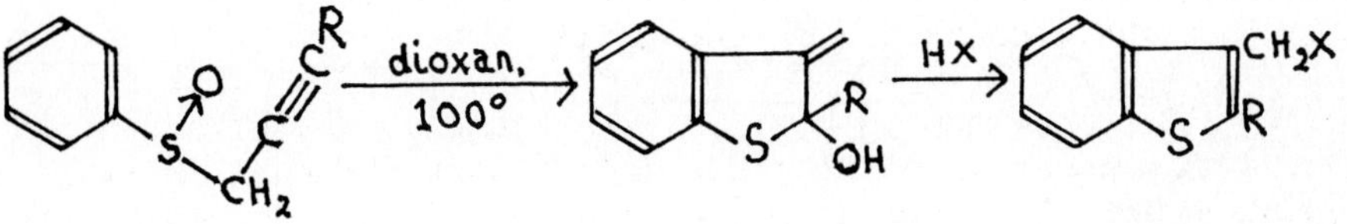

$$(4) \qquad\qquad (5) \qquad\qquad (6)$$

(5) Flash vacuum thermolysis of thiophene-2,3-
-dicarboxylic anhydride in the presence of several diene
traps leads to benzothiophenes substituted in the benzene
ring, by aromatization of an initially formed diene-
-thiophyne Diels-Alder adduct (M.G. Reinecke and
J.G. Newson, J. Amer. chem. Soc., 1976, **98**, 3021).

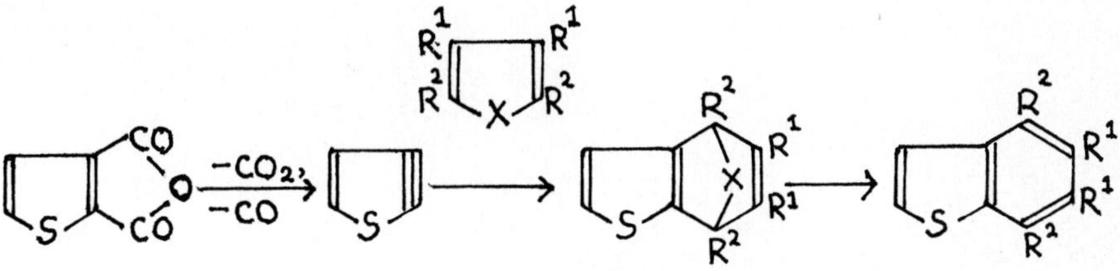

X	R^1	R^2
S	H	H
-H,H-	Me	H
$-CH_2-CH_2$	H	H
O	H	H, OH
$-CH_2-$	H	H, Me

(6) The thermal extrusion of dinitrogen from the
cinnoline derivative (7) gives benzo[3.4]cyclobuta[1,2-b]-
thiophene (8) (J.W. Barton and D.J. Lapham, Tetrahedron
Letters, 1979, 3571).

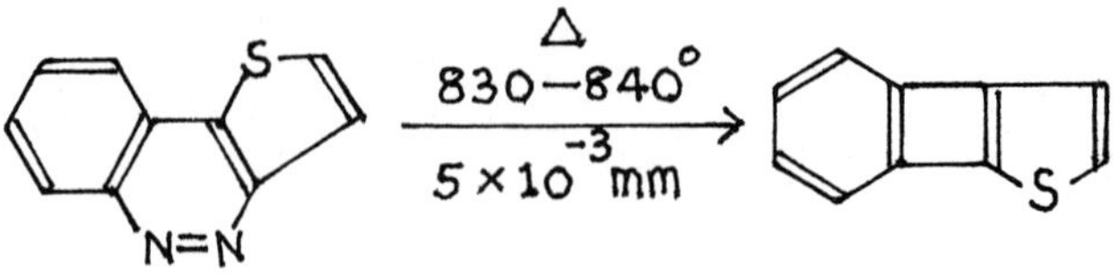

(7)   (8)

(ii) General properties and reactions

(1) The photoaddition of diphenylacetylene to benzothiophene results in the formation of the 1:1 photo--adducts (1) and (2). Irradiation of (1) produces adduct (2) rapidly and quantitatively (W.H.F. Sasse, P.J. Collin, and D.B. Roberts; Tetrahedron Letters, 1969, 4791).

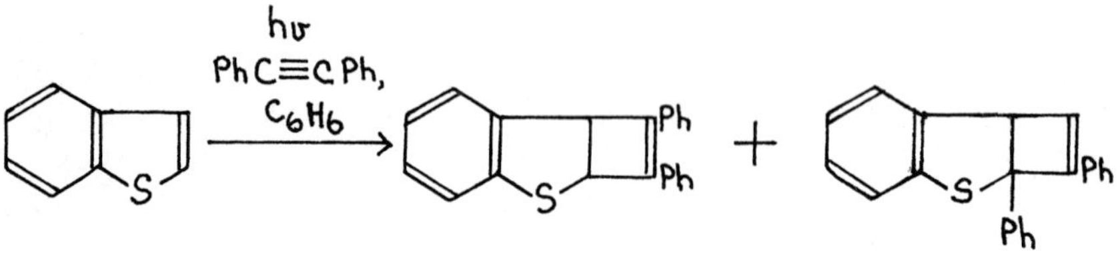

(1)   (2)

An adduct (2, Ph = CO_2Me) from benzothiophene and dimethyl acetylenedicarboxylate has been reported (D.C. Neckers, J.H. Dopper, and H. Wynberg, *ibid.*, p.2913).

(2) 1-Oxides are obtained on oxidising benzothiophenes (3) with *tert*-butyl hypochlorite in methanol at $-70°$ (P. Geneste *et. al.*, *ibid.*, 1975, 2345).

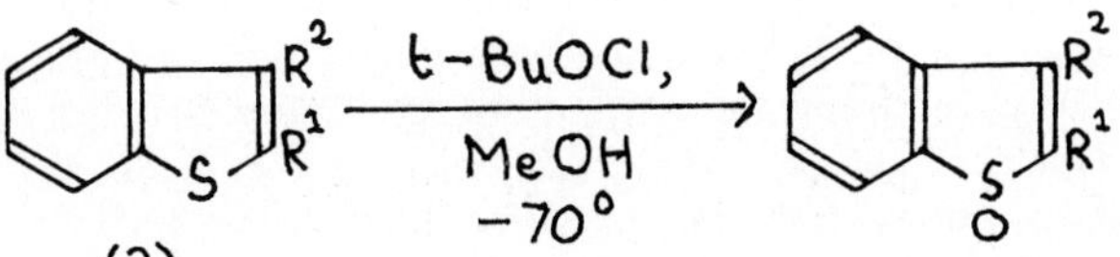

(3)

$$R^1 = H, \quad R^2 = Br, \; Me$$
$$R^1 = Br, \quad R^2 = H$$

With benzothiophene under the same conditions the *tert*-butyl hypochlorite acts as an oxidising and chlorinating reagent to give 3-chlorobenzothiophene, 2-chlorobenzothiophene 1-oxide and the isomeric 2,3-dichloro-2,3-dihydrobenzothiophene 1-oxides (4,5 and 6) (Geneste and J.L. Olive, J. heterocyclic Chem., 1977, 14, 449).

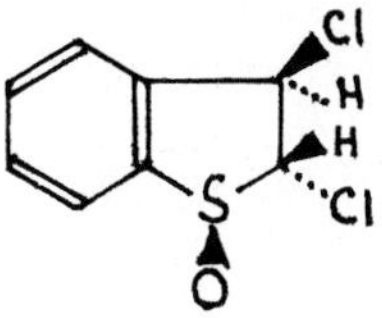

(4) 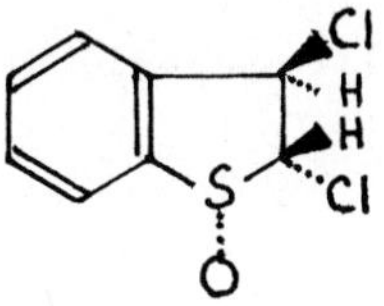   (5) 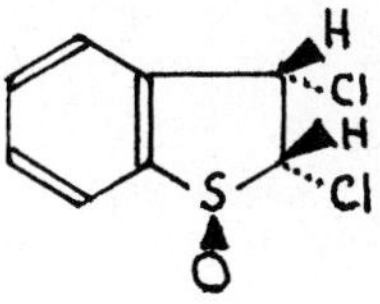   (6)

Oxidation of benzothiophene over a 10%MoO$_3$-TiO$_2$ catalyst yields benzyl alcohol (M. Blanchard and J. Goichon, Bull. Soc. chim. Fr., 1975, 289). Benzothiophene and 3-methylbenzothiophene have been oxidised to the corresponding 1,1-dioxides by a triple excess of hydrogen peroxide in acetic acid at 40-80° and a spectrophotometric study made of the reaction (N.P. Anashkina, S.M. Maksimov, and V.M. Dzyuba, Khim. Seraorg. Soedin., Soderzh. Neftyakh Nefteprod., 1972, 9, 160).

The hydrosulphurization of benzothiophene and 2,3-
-dihydrobenzothiophene under low-pressure conditions, over
cobalt molybdate on alumina and over an unsupported MoS_2
catalyst promoted with Co gives styrene as one of the
products. This supports cleavage of the hetero-ring in
benzothiophene as the initial step in the reaction
sequence. Ethylbenzene is the predominant product from
both benzothiophene and dihydrobenzothiophene. With rising
temperatures increasingly significant amounts of dihydrobenzothiophene are dehydrogenated to benzothiophene (E.
Furimsky and C.H. Amberg, Canad. J. Chem., 1976, 54, 1507),
also with increased hydrogen sulphide concentration levels,
a back reaction occurs between styrene and hydrogen
sulphide to give benzothiophene (F.P. Daly, J. Catal.,
1978, 51, 221).

(iii) Benzo[b]thiophene

Benzothiophene may be prepared by passing a mixture
of styrene or ethylbenzene and hydrogen sulphide over a
CrO_3-Al_2O_3 catalyst at 690° (V.N. Kulakov,
M.F. Pankratova, and Yu. M. Pinegina, Khim. Seraorg.
Soedin. Soderzh. Neftyakh Nefteprod., 1972, 9, 25). The
dehydrogenation of 2,3-dihydrobenzothiophene and
substituted 2,3-dihydrobenzothiophene, and the
dehydroisomerisation of thiochroman give benzothiophene and
substituted benzothiophenes, respectively (A.N. Korepanov,
T.A. Danilova, and E.A. Viktorova, Neftekhimiya, 1976, 16,
909).

Benzothiophene is decomposed on sodium in hexane and
benzene at >220° and naphthalene >180° to give ethylbenzene, styrene, and phenylacetylene (Y. Kurauchi *et. al.*,
Aromatikkusu, 1978, 30, 384).

When an ethereal solution of 3-benzothienyllithium
prepared by treating 3-bromobenzothiophene with n-butyllithium at -70° is kept for 1h at room temperature, prior
to treatment of the product with dimethyl sulphate, it
gives a mixture of benzothiophene, 3-bromo-2-methylbenzo-
thiophene, *o*-(methylthio)phenylacetylene, and methyl-[*o*-
-(methylthio)phenyl]acetylene, thus indicating the
occurrence of ring-cleavage reactions (R.P. Dickinson and
B. Iddon, J. chem. Soc., C, 1971, 3447).

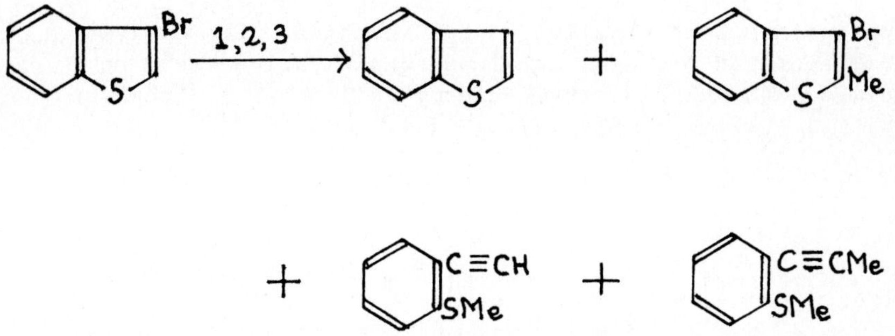

Reagents: (1) BuLi -70°/Et$_2$O, (2) 1h. R.T., (3) Me$_2$SO$_4$

(iv) Alkyl- and aryl-benzothiophene

(1) 2-(4-Iodobutyl)3-methylbenzothiophene (1)
undergoes an intramolecular cyclization *via* ring closure at
thiophenic sulphur (cyclo-S-alkylation) to give 6,7,8,9-
-tetrahydro-10-methylthiopyrano[1,2-a]benzothiophenium ion
(2), when it is treated with either anhydrous silver
hexafluorophosphate in methylene chloride at 0°, or
anhydrous silver perchlorate in toluene at 0°, followed by
anion exchange with NH$_4$PF$_6$ in water; hexafluorophosphate of (2), m.p.130-131° (J.A. Cotruvo and
I. Degani, Chem. Comm., 1971, 436).

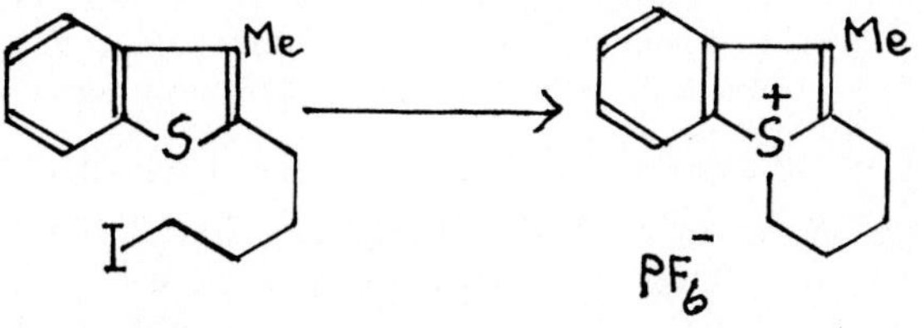

(1)

(2) 2-Methylbenzothiophenes may be pepared from the appropriate 2-chloroallyl phenyl sulphide. The Claisen rearrangement proceeds in almost quantitative yield on boiling the sulphide in *NN*-diethylaniline (W.K. Anderson and E.J. LaVoie, *ibid.*, 1974, 174).

R = H, 4-OMe, 4-Cl

Better yields are obtained by using orthophosphoric acid and phosphorus pentoxide when alkylating benzothiophene with primary alcohols (I.M. Nasyrov and R. Usmanov, Tezisy Doklady Nauchn. Sess. Khim. Tekhnol. Org. Soedin. Sery Sernistykh Neftei, 13th, 1974, 170). Contrary to previous reports the alkylation of benzothiophene with 2-methylpropene in the presence of polyphosphoric acid does not give entirely 3-*tert*-butyl-benzothiophene; a mixture is obtained containing 2-(22%) and 3-(71%)-*tert*-butylbenzothiophene and an unidentified component (7%). Alkylation with *tert*-butanol and concentrated sulphuric acid yields mainly 2-(6%) and 3-(89%)-*tert*-butylbenzothiophene (J. Cooper and R.M. Scrowston, J. chem. Soc., Perkin I, 1972, 414). A number of alkylated benzothiophenes have been prepared (Nasyrov, I.U. Numanov, and Usmanov, Doklady Akad. Nauk Tadzh. SSR, 1974, 17, 35).

The allylation of benzothiophene with allyl iodide in the presence of silver trichloracetate gives 3-allylbenzo-thiophene in 5 minutes. A longer reaction time leads to diallylbenzothiophene, with the second allyl group being attached to the benzene ring (A.V. Anisimov *et. al.*, Khim. Geterotsikl. Soedin., 1977, 1625). Alkenylation of benzothiophene with 2-methylallyl bromide using the same catalyst at 20° yields 3-(2-methylallyl)benzothiophene, which on reduction affords 3-isobutylbenzothiophene (*idem,*

Zh. Org. Khim., 1979, 15, 172).

The Heck reaction of benzothiophene with RHgCl
(R = Ph, 4-MeC$_6$H$_4$, 4-MeOC$_6$H$_4$) and Li$_2$PdCl$_4$ in
ethanol at room temperature yields 2- and 3-arylbenzo-
thiophenes with a large amount of biaryls. Benzothiophene
on boiling with styrene and Pd(OAc)$_2$ in acetic acid
affords 2-(9%) and 3-(49%)-styrylbenzothiophene (T. Izumi,
T. Takeda, and A. Kasahara, Yanagata Daigaku Kiyo, Kogaku,
1975, 13, 107). The homolytic phenylation of
benzothiophene may be carried out, using the thermal
decomposition of N-nitrosoacetanilide as the souce of
phenyl radicals. All the available positions of benzothiophene show a comparable reactivity to substitution by
phenyl radicals, in contrast with benzofuran, in which
homolytic substitution takes place almost exclusively in
the heterocyclic ring. The 4-, 6-, and 7-phenylbenzothio-
phene have m.p.46-47°, 43-45°, and 37.5-38.5°,
respectively (P. Spagnolo, M. Tiecco, and A. Tundo, J.
chem. Soc., Perkin I, 1972, 556).

The photocyclization-oxidation reaction of 1-phenyl-
-4-(2-thienyl)buta-1,3-diene (1) in benzene containing
iodine gives 4-phenylbenzothiophene and irradiation of
1-phenyl-4-(3-thienyl)buta-1,3-diene (2) gives 7-phenyl-
benzothiophene (C.C. Leznoff, W. Lilie and C. Manning,
Canad. J. Chem., 1974, 52, 132).

(1)   (2)

Vinyl bromides with three aryl substituents and
containing a sulphur atom at the *ortho* position of one of
the β-aryl substituents may be converted to 2,3-
diarylthiophenes (see benzofurans Ch. 2, p. 62).

Both 2- and 3-methylbenzothiophene on heating at 350-400° in the presence of zinc chloride-alumina disproportionate to give 2,3-dimethylbenzothiophene and benzothiophene; isomerisation also occurs (A.N. Korepanov, T.A. Danilova, and E.A. Viktorova, Zh. Org. Khim., 1973, 9, 641; Khim. Geterotsikl. Soedin. 1977, 1079.

(v) Halogenobenzothiophenes

Passing chlorine through benzothiophene in water-carbon tetrachloride (1:1 vol.) at 0° results in oxidative chlorination of benzothiophene to give 2,3-dichlorbenzothiophene (40%). Similar treatment of 3-methylbenzothiophene in water at 5-10° yields 2-chloro-3-methylbenzothiophene (33%) (V.I. Dronov *et. al.*, Khim. Seraorg. Soedin., Soderzh. Neftyakh Nefteprod., 1972, 9, 158). Treatment of benzothiophene with a large excess of *tert*-butyl hypochlorite affords 2,3-dichlorobenzothiophene, 2,3-dichlorobenzothiophene 1-oxide, 2,2,3-trichloro-2,3-dihydrobenzothiophene 1-oxide, and 2,2,-dichloro-3-oxo-2,3-dihydrobenzothiophene 1-oxide (Geneste and Olive, *loc. cit.*) (p. 47).

Bromination of 2,3-dibromobenzothiophene in chloroform or acetic acid gives 2,3,6-tribromobenzothiophene m.p. 122-123°. 5-Methyl-2,3,6-tribromobenzothiophene, m.p. 119-121°; 2,3,5-tribromobenzothiophene, m.p. 154-156° (Cooper *et. al.*, J. chem, Soc., C, 1970, 1949).

2-Halogenobenzothiophenes react with metal amides in liquid ammonia to give the 3-halogenobenzothiophene, which are stable under the reaction conditions. No amines or polyhalogeno derivatives are detected and 2,3-dibromobeno-thiophene is converted into 3-bromobenzothiophene (M.G. Reinecke and T.A. Hollingworth, J. org. Chem., 1972, 37, 4257).

(vi) Nitrobenzothiophenes

Nitration of 2,3-dibromobenzothiophene in acetic acid-sulphuric acid with nitric acid yields a mixture of 2,3,6-tribromo-(25%), 2,3,4-tribromo-(3%), 2,3-dibromo-(32%), 6-nitro-benzothiophene, m.p. 154-156°, and 2,3-dibromo-4-nitrobenzothiophene(36%), m.p. 139-140°.

Analogous results are obtained for the nitration of 2,3-
-dibromo-5-methylbenzothiophene except that the presence of
the 5-methyl group increases the amounts of the
4-substituted products. 2,3-Dibromo-5-methyl-4-nitrobenzo-
thiophene, m.p. 185-186°; 2,3-dibromo-5-methyl-6-nitro-
benzothiophene, m.p. 137-138°; 4-nitro-2,3,6-tribromo-
benzothiophene, m.p. 201-202°. Also reported ^{1}H nmr
spectra (Cooper *et. al., loc. cit.*). Nitration of 3-*tert*-
-butylbenzothiophene-2-carboxylic acid in acetic acid-
sulphuric acid at 60° or in acetic anhydride - acetic acid
at 0° gives differing proportions of the following acidic
products in each case: benzothiophene-2-carboxylic acid,3-,
4-, 6-, and 7-nitrobenzothiophene-2-carboxylic acid, and
3-*tert*-butyl-4-, 3-*tert*-butyl-6-, and 3-*tert*-butyl-7-nitro-
benzothiophene-2-carboxylic acid (J. Cooper and
R.M. Scrowston, J. chem. Soc., Perkin I, 1972, 414).

(vii) Aminobenzothiophenes

The reaction between 2-oxo-2,3-dihydrobenzothiophene
and hexamethylphosphoric acid triamide at 160-240° gives
2-dimethylaminobenzothiophene. Other 2-aminobenzothiophenes
are obtained by boiling 2-oxo-2,3-dihydrobenzothiophene and
hexamethylphosphoric acid triamide in the presence of excess of
the corresponding amine (N.O. Vesterager, E.B. Pedersen,
and S.-O. Lawesson, Tetrahedron, 1973, 29, 321).

(viii) Alcohols, aldehydes, ketones

Benzothiophene-2- and-3-carboxaldehydes have been
converted into the respective 2- and 3-methyl and 2- and
3-hydroxymethyl-benzothiophenes and their related 1,1-
-dioxides (T.V. Shchedrinskaya *et. al.*, Khim. Geterotsikl.
Soedin., 1973, 1026); benzothiophene-2- and -3-carbox-
aldehyde thiosemicarbazones (R.P. Dickinson, B. Iddon, and
R.G. Sommerville, Int. J. Sulphur Chem., 1973, 3, 233).

3-Chloromethylbenzothiophene obtained by treating
benzothiophene with formaldehyde and hydrogen chloride,
reacts with nucleophiles, thus replacing the Cl, for
example, by CN, OEt (R. Neidlein and E.P. Mrugowski, Arch.
Pharm., 1975, 308, 513).

Acylation of benzothiophene with alkanoic acids
(RCO_2H, R = Me, Et, Pr, Bu, $PhCH_2$) in the presence of

phosphoric acid-phosphorus pentoxide gives the 3-acylated
derivative. 2-Methyl- and 3-ethyl-benzothiophene with
acetic acid under the same conditions yield 3-acetyl-2-
-methyl- and 2-acetyl-3-ethyl-benzothiophene, respectively
(R. Usmanov, I.U. Numanov, and I.M. Nasyron, Doklady Akad.
Nauk Tadzh. SSR, 1975, 18, 11). Acylation of
benzothiophene with substituted benzoyl chlorides (4-Me,
OMe, H) in the presence of ferric chloride gives a mixture
of 2- and 3-benzoylated products (72-75%), with the main
product being the 3-isomer (Kh. Yu. Yuldashev, Khim.
Geterotsikl. Soedin., 1978, 1039). Phenyl 2-thienyl ketone
reacts with BrMgC≡CMgBr in ether to give 1,4-bis(2-
-thienyl)-1,4-diphenylbut-2-yne-1,4-diol (J. Krupowicz,
K. Sapiecha, and R. Gaszczyk, Rocz. Chem., 1974, 48, 2067).
Reaction of benzothiophene with butyllithium and dimethylacetamide affords 2-acetylbenzothiophene (Shchedrinskaya
et. al., loc. cit.).

(ix) Carboxylic acids

Benzothiophene-2-carboxylic acid preparation see (2)
(p. 44) and for preparation by carbonation of 2-benzothienyl-
lithium see Shchedrinkskaya *et. al. (loc. cit.).* 5-Chloro-
and 5-fluoro-benzothiophene-2-carboxylic acid (S. Gronowitz
et. al., Acta Pharm. Suec., 1978, 15, 368).

β-Phenylpropanoic acid on treatment with thionyl
chloride yields 3-chlorobenzothiophene-2-carbonyl chloride.
The mechanism of the reaction has been discussed
(A.J. Krubsack and T. Higa, Tetrahedron Letters, 1973, 125,
4515).

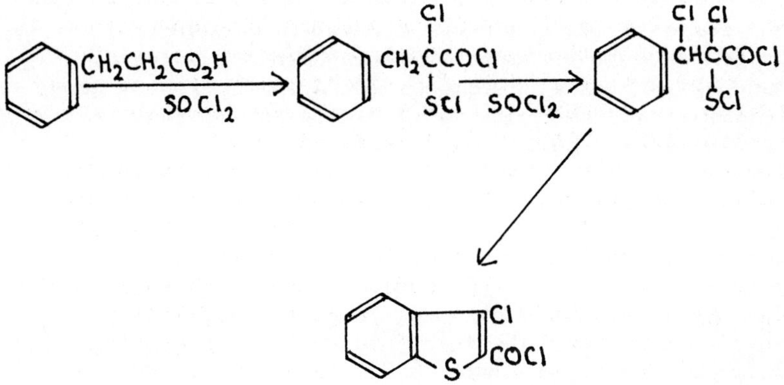

Benzothiophene-3-acetic acid ethyl ester is obtained in low yield along with 2,3-cyclopropanation products, by the thermolysis of ethyl diazoacetate in benzothiophene (E. Wenkert *et. al.*, J. org. Chem., 1977, **42**, 3945).

(b) Reduced benzothiophenes

(i) Dihydrobenzothiophenes, benzothiolenes

(1) 2-Ethylthiobenzaldehyde hydrazone tosylate on thermal decomposition under reduced pressure (0.06mm) gives 2-methyl-2,3-dihydrobenzothiophene (37%) and thiochroman (7%) *via* carbene insertion (W.D. Crow and H. McNab, Austral. J. Chem., 1979, **32**, 99).

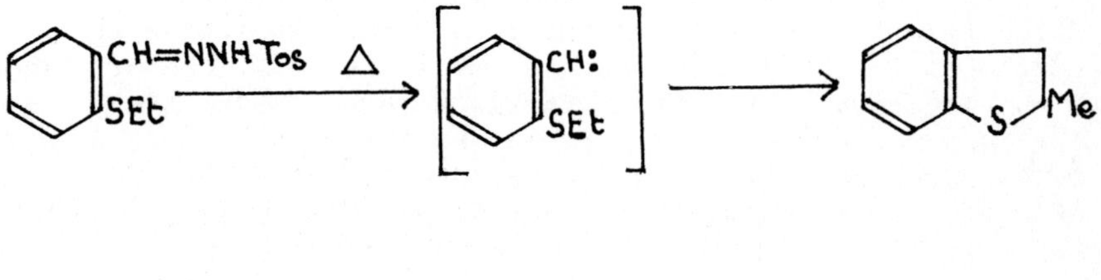

4-Aryloxy-1-arylsulphinylbut-2-yne (1) on boiling in carbon tetrachloride undergoes a novel mild thermal rearrangement (*ortho*-Claisen rearrangement) to give compound (2), which with aqueous potassium hydroxide gives the 2,3-dihydrobenzothiophene (3) (K.C. Majumdar and B.S. Thyagarajan, Chem. Comm., 1972, 83).

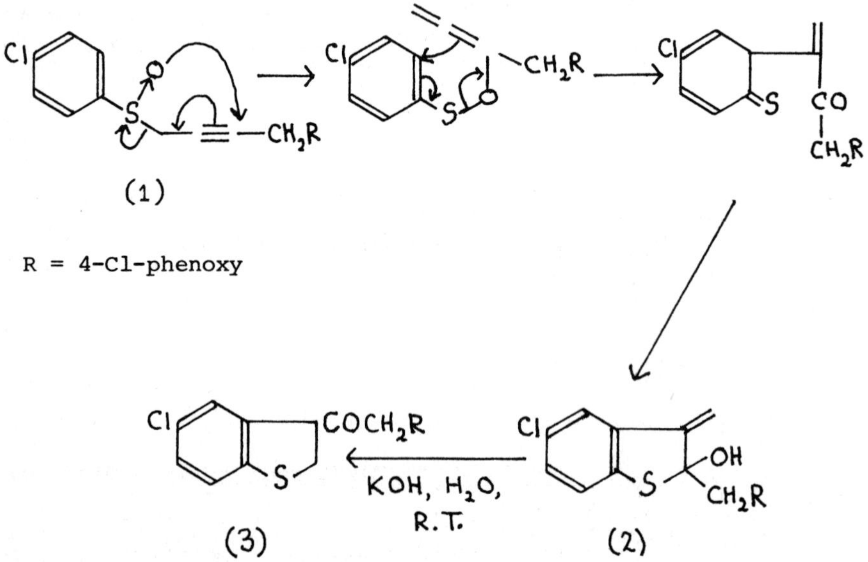

(1)

R = 4-Cl-phenoxy

(3)   KOH, H$_2$O, R.T.   (2)

2,3-Dihydrobenzothiophenes on treatment with triphenylmethyl tetrafluoroborate and chloranil give the corresponding benzothiophenes (L.M. Kedik, A.A. Freger, and E.A. Viktorova, Khim. Geterotsikl. Soedin., 1976, 328). The dehydrogenation of 2,3-dihydrobenzothiophene over a Cr$_2$O$_3$-BeO-Al$_2$O$_3$-SiO$_2$ catalyst at 400-500° gives benzo-thiophene (S.I. Dolganskaya, Zh. Vses. Khim. O-va., 1974, 19, 474, 475). 2-Methyl-2,3-dihydrobenzothiophene on heating at 350-400° in the presence of zinc chloride-alumina gives 2- and 3-methyl-benzothiophene, benzo-thiophene and 2,3-dimethylbenzothiophene (Korepanov, Danilova, and Viktorova, *loc. cit.*). Acylation of 2,3-dihydrobenzothiophene or its 2-methyl derivative with

acetic acid in presence of phosphoric acid-phosphorus
pentoxide, acetyl chloride, or acetic anhydride gives the
respective 5-acetyl derivative, whereas 2,5-dimethyl-2,3-
-dihydrobenzothiophene affords 7-acetyl-2,5-dimethyl-2,3-
-dihydrobenzothiophene (R.Usmanov, I.U.Numanov, and
I.M. Nasyron, Doklady Akad. Nauk Tadzh. SSR, 1975, 18,
11).

4,5-Benzo-3-thiatricyclo$[4.1.0.0^{2,7}]$heptene (4)
undergoes isomerisation exclusively to 3,4-benzo-2-
-thiabicyclo[3.2.0]hepta-3,6-diene (5) by a thermal,
photochemical, or silver (I)-promoted reaction (I. Murata,
T. Tatsuoka, and Y. Sugihara, Tetrahedron Letters, 1974,
199).

(4)   (5)

(2) 2-Oxo-2,3-dihydrobenzothiophene is prepared by
the dealkylation of 2-*tert*-butoxybenzothiophene on heating
with toluene-*p*-sulphonic acid at 150-160°. Alkylation of
sodium, thallium, and tetrabutylammonium salts of 2-oxo-
-2,3-dihydrobenzothiophene produces both *C*- and
O-alkylation product along with products due to ring-
opening (N.O. Vesterager, E.B. Pedersen, and S.-O. Lawesson,
Tetrahedron, 1973, 29, 321).

(ii) 4,5-Dihydrobenzothiophene

The gas phase vacuum pyrolysis of 1-(2-thienyl)buta-
-1,3-diene yields 4,5-dihydrobenzothiophene (72%) (I. Rosen
and W.P. Weber, Tetrahedron Letters, 1977, 151).

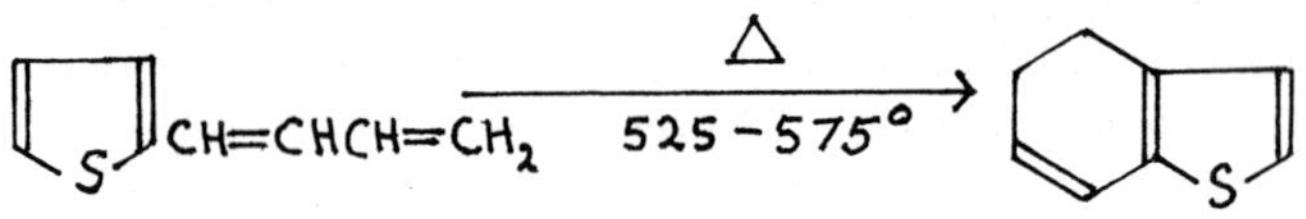

(iii) Tetrahydrobenzothiophenes

It has been found that when thiophenes are substituted with alkyl groups, they react with singlet oxygen. 2,3-Dimethyl-4,5,6,7-tetrahydrobenzothiophene on irradiation in methanol during oxygenation in the presence of methylene blue affords compound (2) *via* the intermediate sulphine (1) (H.H. Wasserman and W. Strehlow, *ibid.*, 1970, 795).

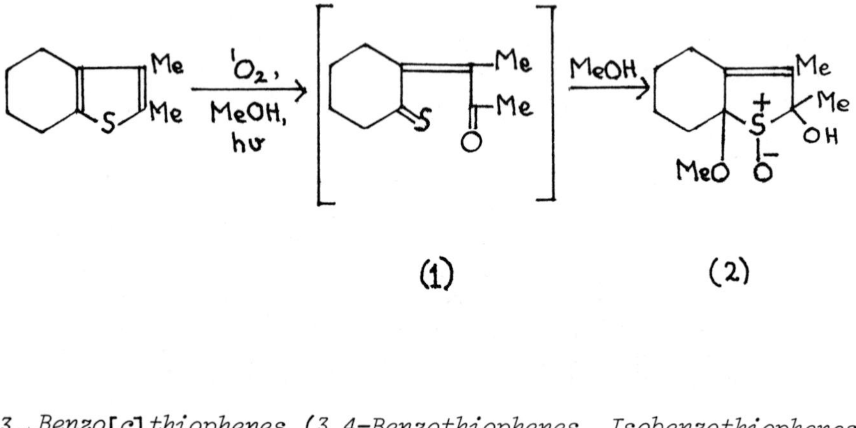

(1)   (2)

3. Benzo[c]thiophenes (3,4-Benzothiophenes, Isobenzothiophenes) and Related Compounds

(a) Benzo[c]thiophenes

Benzo[c]thiophene is obtained when 1,3-dihydrobenzo-[c]thiophene 2-oxide is heated with alumina to 120-130°. Naphtho[1,2-c]thiophene, m.p.110-112° is prepared in a similar manner (M.P. Cava *et. al.*, J. org. Chem., 1971, 36, 3932). The former 2-oxide on heating to 100° with 30%

aqueous sodium hydroxide undergoes sulphoxide dehydration but gives benzo[c]thiophene only in low yield. Similarly the related oxide affords naphtho[1,2-c]thiophene (38%) (C.J. Horner *et. al.*, Tetrahedron Letters, 1976, 2581).

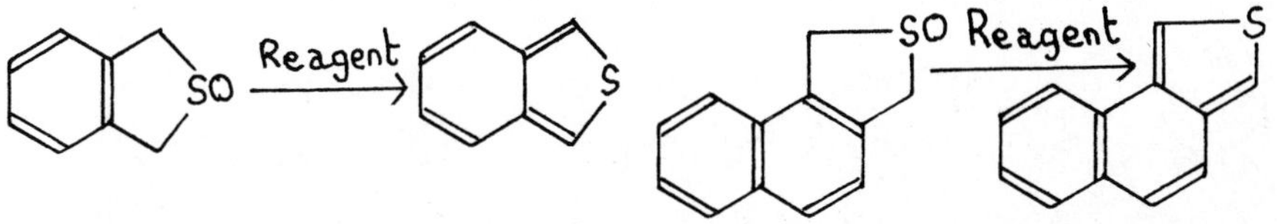

Reagent: $Al_2O_3/120-130^\circ$ or $NaOH/H_2O/100^\circ$

Both compounds form exo and endo adducts with *N*-phenyl-maleimide (Cava *et. al., loc. cit.*).

The pyrolysis of isothiochromanone enamines (1) results in a novel rearrangement which leads to the synthesis of benzo[c]thiophenes (F.H.M. Deckers, W.N. Speckamp, and H.O. Huisman, Chem. Comm., 1970, 1521).

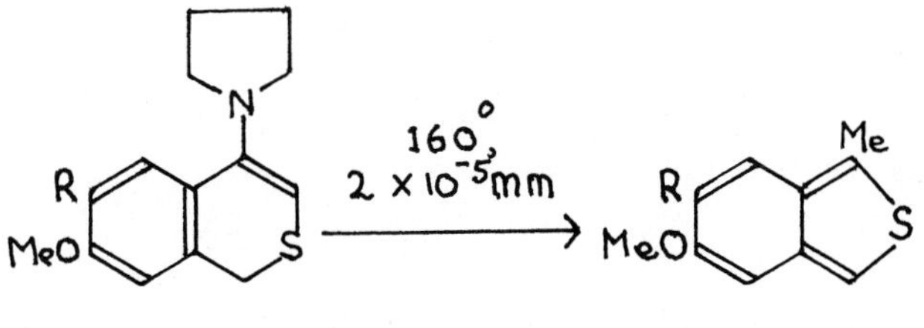

(1)

R = H, OMe

Racemic dimethyl 2,3-benzodithiane-1,4-dicarboxylate in methanolic sodium methoxide gives dimethyl benzo[c]-

-thiophene-1,3-dicarboxylate m.p.160°, which on alkaline hydrolysis affords benzo[c]thiophene-1,3-dicarboxylic acid, m.p.280-285° (decomp.) (G. Cignarella and G. Cordella, Tetrahedron Letters, 1973, 1871).

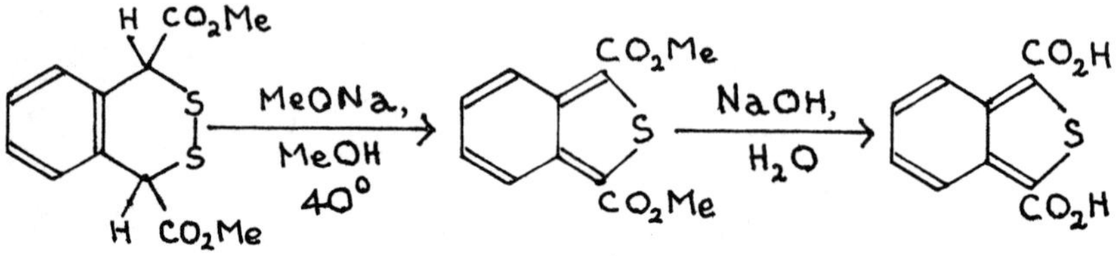

(b) Compounds related to benzo[c]thiophene

The cis- and trans- sulphoxides (2) and (3) on being subjected to base dehydration in benzene produce 1,3,4,6--tetraphenylthieno[3,4-c]thiophene (4), m.p.245-247°, a bicyclic heterocycle containing 10π electrons and a "tetravalent sulphur" atom (Horner et. al., loc. cit.).

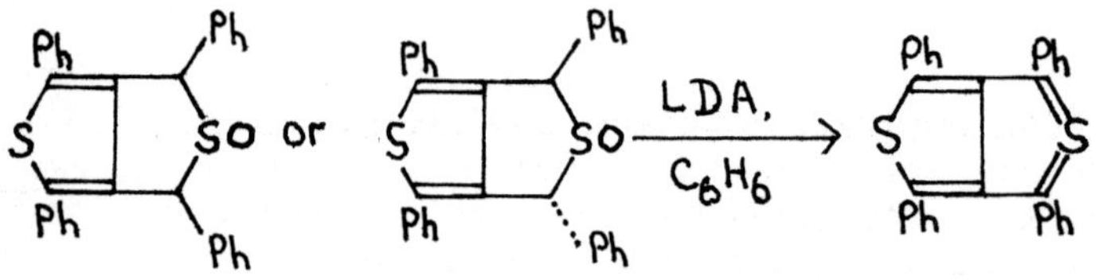

LDA = lithium diisopropylamide

Compound (4) is also obtained by treating 3,4-dibenzoyl--2,5-diphenylthiophene with phosphorus pentasulphide in boiling pyridine (K.T. Potts and D. McKeough, J. Amer. chem. Soc., 1973, 95, 2750).

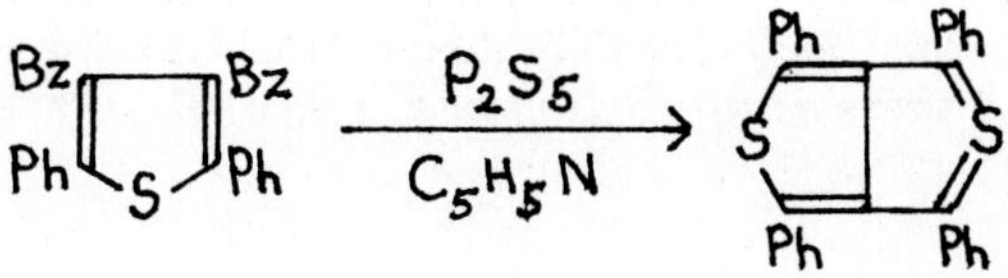

(4)

1,3,4,5,7,8-Hexaphenylthieno[3,4-f]benzo[c]thiophene (6),
m.p.348-350°, an example of a stable 14π-electron system
containing a "tetravalent sulphur" atom in a five-membered
ring is obtained from compound (5).

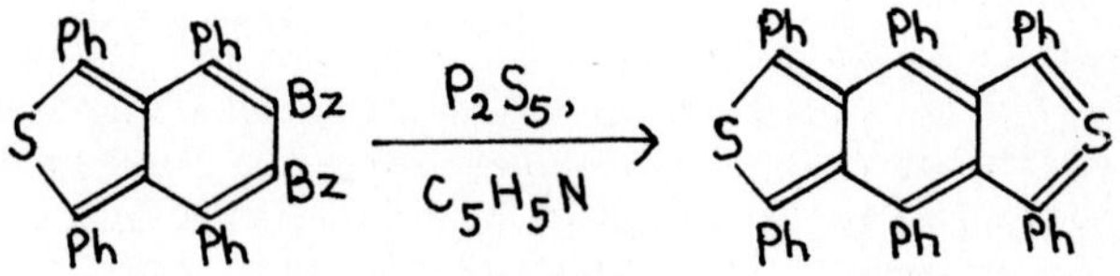

(5)   (6)

The Wittig reaction between 1,2-benzocyclobutadiene-
quinone and the bis-ylide, bis(triphenylphosphoranylidene-
methyl) sulphide (7) gives benzo[3,4]cyclobuta[1,2-c]-
thiophene (2-thianorbiphenylene) (8), m.p.98-98.5°, the
first know analogue of biphenylene. Oxidation of (8) with
6% hydrogen peroxide in acetic acid affords the 2-oxide
(sulphoxide), m.p.143-145° and then the 2,2-dioxide
(sulphone), m.p.213-215°, which on photo-irradiation
undergoes a 2π + 2π addition to give a dimer. A
comparison of the nmr proton chemical shifts of (8) and
its 2-oxide (sulphoxide) and 2,2-dioxide (sulphone)
suggests that the four-membered ring in (8) sustains a
paramagnetic ring current (P.J. Garratt and
K.P.C. Vollhardt, Chem. Comm., 1970, 109; J. Amer. chem.
Soc., 1972, 94, 7087).

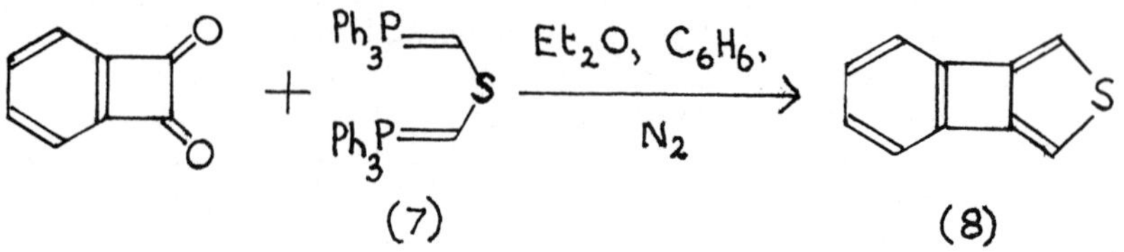

(7)   (8)

On bromination of (8), the thiophene ring adds four atoms
of bromine to give a mixture of tetrabromo derivatives,
which can be debrominated to give (8) or dehydrobrominated
to give 1,3-dibromobenzo[3,4]cyclobuta[1,2-c]thiophene.

The addition of 3,4-diphenylcyclobutadienequinone in
tetrahydrofuran to an ethereal solution of ylide (7) at -
78°, gives 6,7-diphenyl-3-thiabicyclo[3.2.0]heptatriene
(9), m.p.134-136° (decomp.), a thienocyclobutadiene
(*idem*, *ibid.*, p.1022).

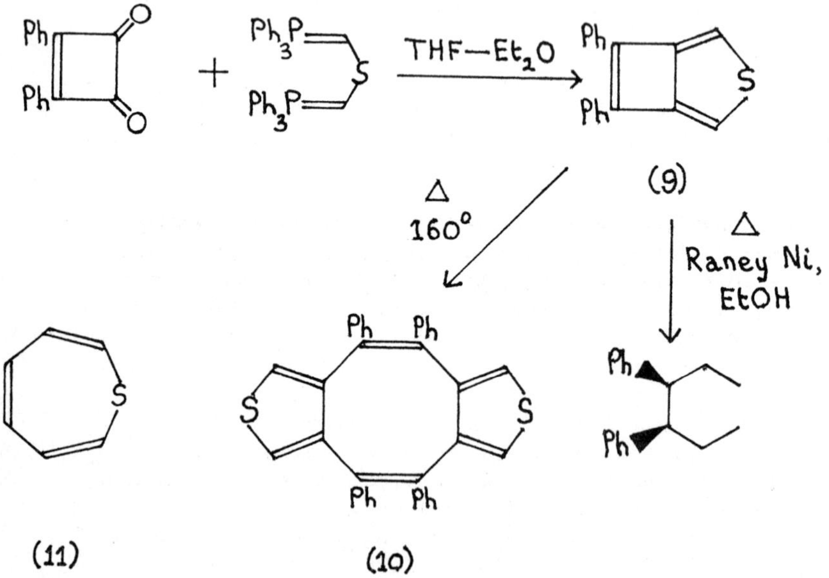

(9)

(11)   (10)

Treatment of compound (9) with Raney nickel in boiling
ethanol yields *meso*-3,4-diphenylhexane and heating (9)
under nitrogen at 160° gives the dimer (10). Compound
(9) may be considered to be derived from the 8π-electron
system thiepin (11) by formation of a 3,6-transannular
bond. The Wittig reaction between cyclobutane-1,2-dione
and ylide (7) in ether at -65° yields 3-thiabicyclo-
[3.2.0]hepta-1,4-diene (12) the parent system of (9); (12)
3,3-dioxide m.p.127.5-128.5° (Garratt and D.N. Nicolaides,
Chem. Comm., 1972, 1014). The *cis*- and *trans*-5,6-dibromo-
-2,4-dihalogeno-3-thiabicyclo[3.2.0]hepta-1,4-diene

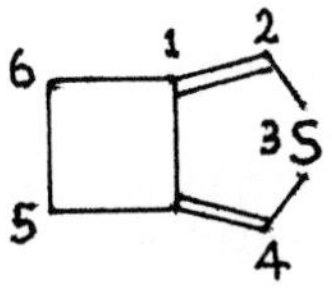

(12) 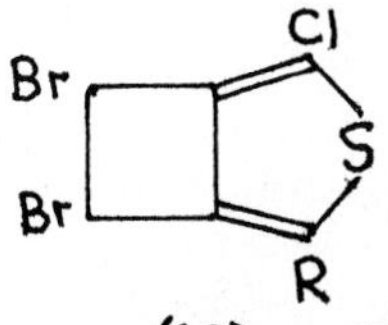   (13)   R = Cl, Br

(13) have been prepared by the action of sodium iodide on
3,4-bisdibromomethyl-2,5-dihalogenothiophenes in
dimethylformamide (S.W. Longworth and J.F.W. McOmie, *ibid.*,
p.623).

Photolysis of both the *cis*- and the *trans*-dihydro-
thienothiophene (14) and (15) yields primarily 2,4,6,7-
-tetraphenyl-3-thiabicyclo[3.2.0]hepta-1,4-diene (16)
(M.P. Cava, M.V. Lakshmikantham, and M. Behforouz, J. org.
Chem., 1974, 39, 206).

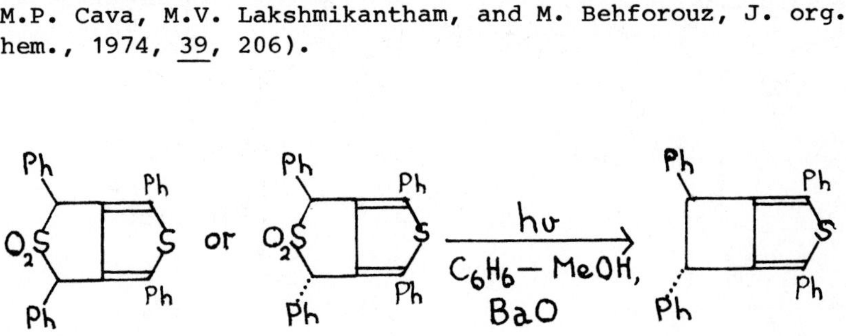

(14)   (15)   (16)

The acetylene derivative (17) with sodium sulphide in
acetone gives the thienocyclobutene derivative (18)
(H. Hauptmann, Tetrahedron Letters, 1974, 3589).

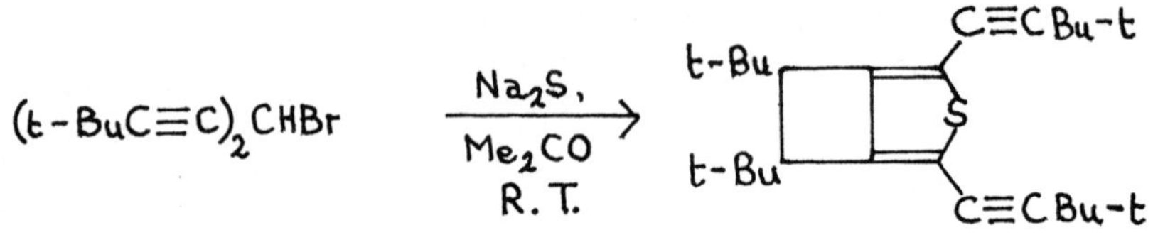

$(t-BuC\equiv C)_2 CHBr$ $\xrightarrow[\text{R. T.}]{\text{Na}_2\text{S,}\ \text{Me}_2\text{CO}}$

(17)   (18)

4. *Cycloalkanothiophenes*

4-(2-Phenyl)thiophene-3-carboxylic acid on treatment
with polyphosphoric acid at 100° gives 9,10-dihydrobenzo-
[4,5]cyclohepta[1,2-c]thiophene-4-one (1), but at 160°
the reaction yields the isomeric ketone, 4,5-dihydrobenzo-
[5,6]cyclohepta]1,2-b]thiophene-10-one (2), also obtained
by heating ketone (1) at 160° (D.E. Ames and O. Ribeiro,
J. chem. Soc., Perkin 1, 1975, 1390).

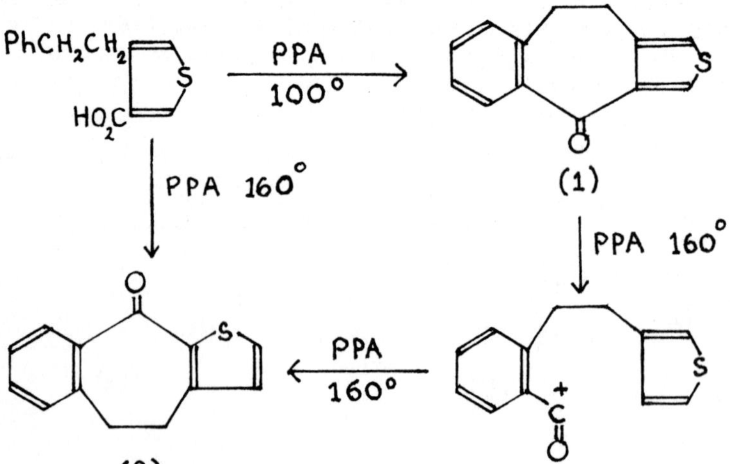

(1)

(2)

The following cycloalkanothiophenes have been synthesised;
3-isobutyl-3-methylthieno[6,7-b]cycloheptane (V.K. Chhatwal
and G.S. Saharia, Indian J. Chem., 176, 14B, 630);
cyclohepta[b]thiophene derivatives (3) (P.K. Sen, B. Kinda,
and T.K. Das, J. Indian chem. Soc., 1978, 55, 847);
cyclohepta[cd]benzo]b]thiophen-6-ones (4) (R. Neidlein and
N. Kolb, Arch. Pharm., 1979, 312, 338).

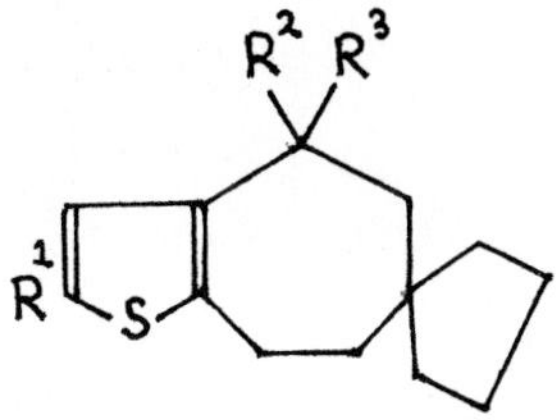

(3) 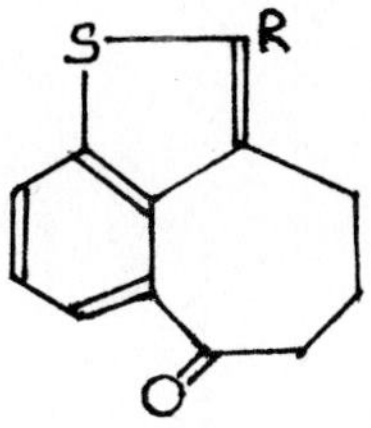   (4)

R^1 = H, Me, Et;   R = Et, Br
R^2 = R^3 = H; R^2R^3 = O

There are a number of interesting reactions
concerning some tricyclic compounds containing a
tetrahydrothiophene ring. One is a novel structural
rearrangement which occurs on the dehydrohalogenation of
the unsaturated cyclic α-halogeno sulphones (5). Treatment
of sulphone (5) with potassium *tert*-buoxide in dimethyl
sulphoxide gives the bridged tricyclic sulphone (6) and a
trace amount of "normal" Ramberg-Bäcklund product (7).
Gas-phase pyrolysis or photolysis of (6) affords the fused
cyclooctatetraene (7) (L.A. Paquette, R.E. Wingard Jr., and
R.H. Meisinger, J. Amer. chem. Soc., 1971, 93, 1047).

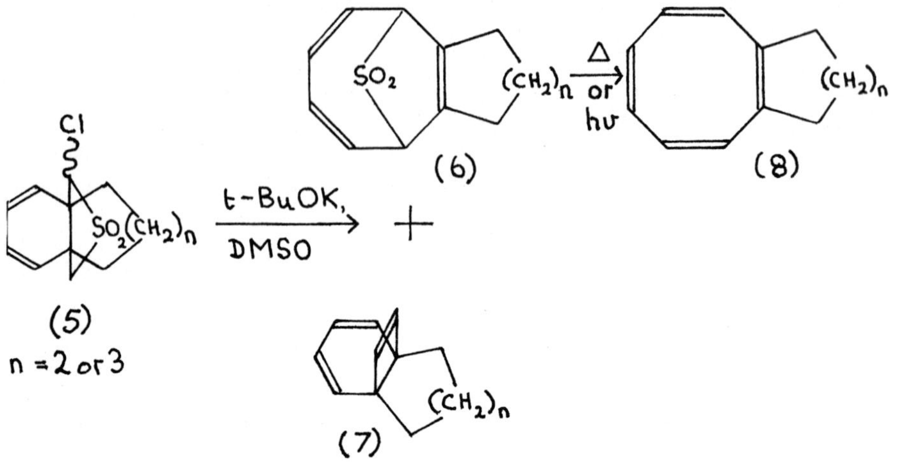

(6)

(8)

(5)

n = 2 or 3

t-BuOK,
DMSO

+

(7)

The reaction of *cis*-1,6-bis(methanesulphonyl-oxymethyl)bicyclo[4.2.0]oct-3-ene (9) with sodium sulphide in anhydrous hexamethylphosphoramide gives 8-thia[4.3.2]-propell-3-ene (10), which on chlorination with *N*-chloro-succinimide followed by oxidation with 3-chloroperbenzoic acid affords a mixture of α-chlorosulphones (11) and (12). Treatment of (11) and (12) with n-butyllithium in ether gives [4.2.2]propella-3,7-diene (13) in low yield (Paquette and R.W. Houser, *ibid.*, p.4522).

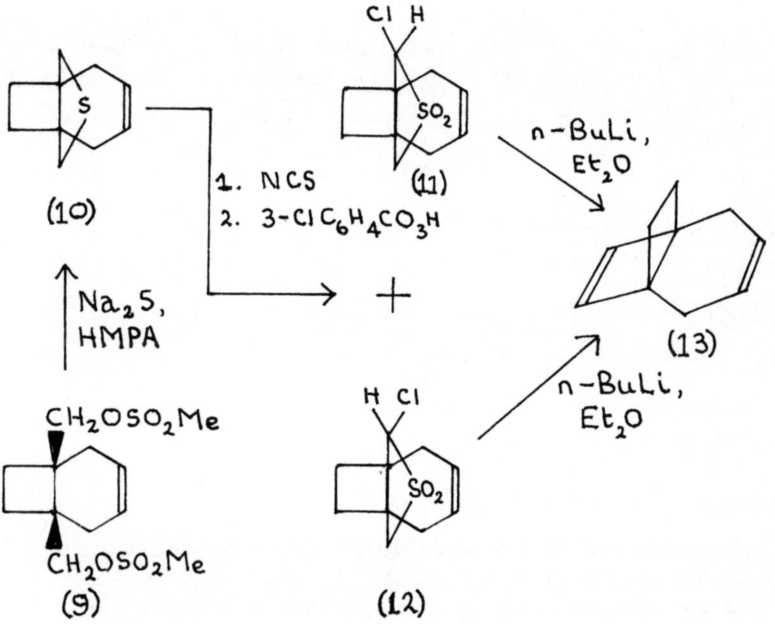

A number of multilayered cyclophanes containing heteroaromatic nuclei, thiophene and furan, have been prepared and their nmr spectral data discussed. These include [2.2](1,4)naphthaleno(2,5)thiophenophane, *anti*-form (14) and *syn*-form (15); [2.2](1,4)anthraceno(2,5)-thiophenophane, *anti*-form (16); [2.2](9,10)anthraceno(2,5)-thiophenophane (17); and related furan compounds (Y. Sakata and S. Misumi, Tetrahedron Letters, 1974, 799).

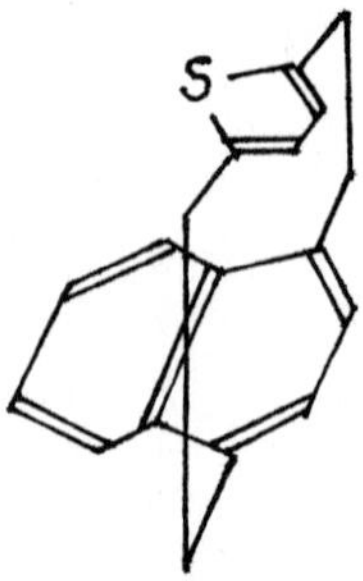

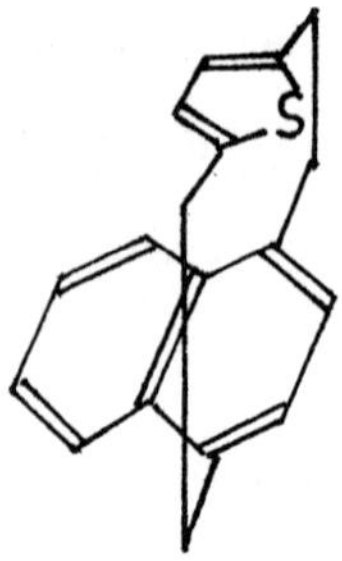

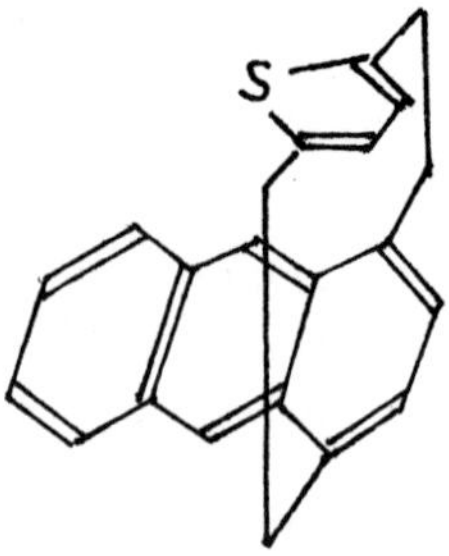

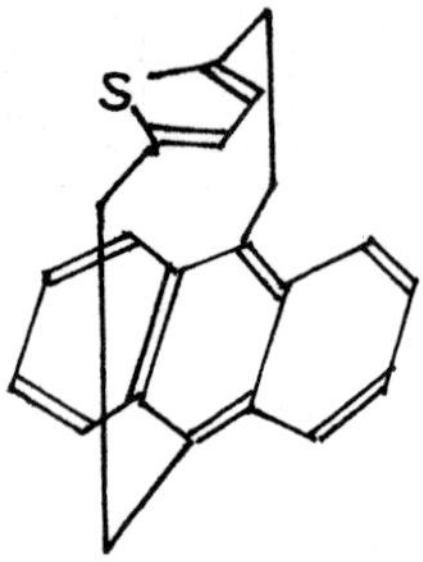

(14)   (15)   (16)   (17)

Other analogues have been reported (S. Mizogami *et. al.*, Chem. Letters, 1974, 515).

5. *Systems of Two Fused Thiophene Rings*

Preparations of derivatives of thieno[2,3-b]thiophene--2-carboxylic acid (1), benzothieno[3,2-b]thiophene-2--carboxylic acid (2), and benzothieno[3,2-b]thiophene-2--carboxylic acid (3) have been described (S. Gronowitz *et. al.*, Acta Pharm. Suec., 1978, **15**, 368).

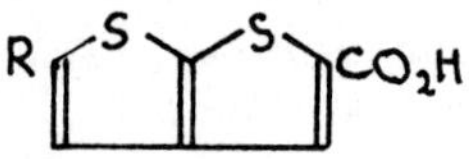

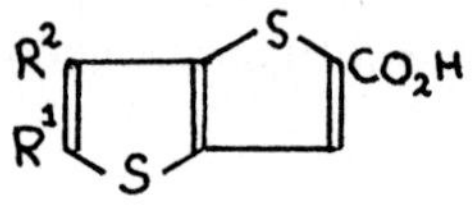

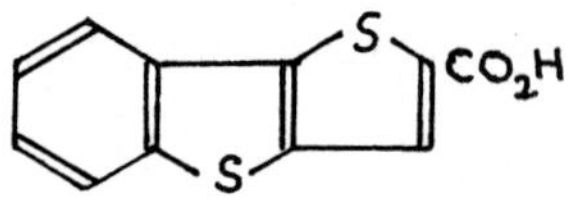

(1)   (2)   (3)

R = H, Cl

R^1 = Cl, F, Me, R^2 = H
R^1 = H, R^2 = Cl

6. *Dibenzothiophene*

Photolysis of diaryl sulphides in cyclohexane containing iodine yields dibenzothiophenes (K.P. Zeller and H. Petersen, Synth., 1975, 532).

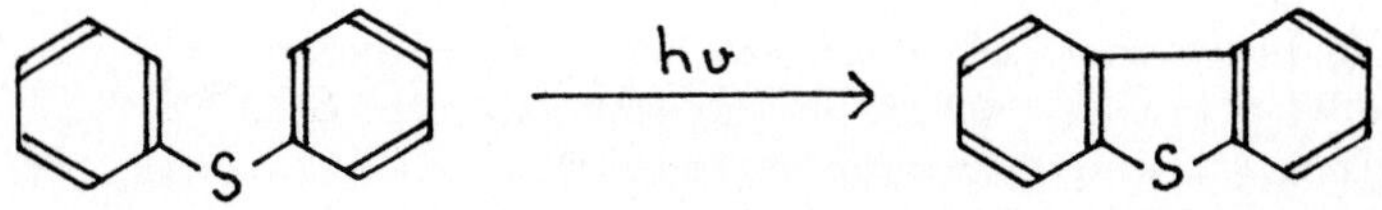

The thermolysis of biphenyl-2-thiol with methanol in a packed flow reactor at 450-550° affords small yields of dibenzothiophene and biphenyl. With sulphided alumina, products vary with reaction temperature from 82% methyl 2-biphenylyl sulphide 250° to only dibenzothiophene plus biphenyl at 550°. The use of other catalysts has been studied (L.H. Klemm and J.J. Karchesy, J. heterocyclic Chem., 1978, 15, 281).

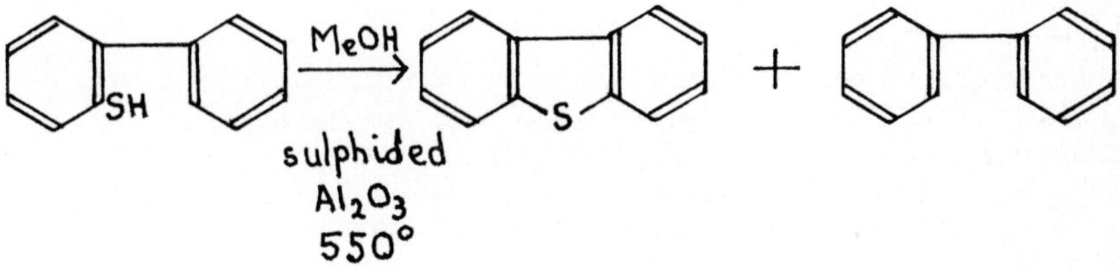

Dibenzothiophene 5-oxide is reduced by stannous chloride--hydrochloric acid to dibenzothiophene. The 5,5-dioxide does not react under these conditions (T.-L. Ho and C.M. Wong, Synth., 1973, 206).

Bond lengths and bond orders in π-electron heterocycles, including dibenzothiophenes have been discussed (C. Parkanyi and W.C. Herndon, Phosphorus Sulphur, 1978, 4, 1). ^{13}C nmr spectra of dibenzothiophene, dibenzofuran, and some bromo derivatives (T.N. Huckerby, J. Mol. Struct., 1979, 54, 95); and high-resolution ^{1}H nmr spectra of dibenzothiophene and benzothiophene have been reported (E.D. Bartle, D.W. Jones and R.S. Matthews, Tetrahedron, 1971, 27, 5177; P.M. Nair and V.N. Gogte, Indian J. Chem., 1974, 12, 589).

The hydrodesulphurization of dibenzothiophene in the presence of hydrogen and a sulphided cobalt oxide- -molybdenum trioxide/γ-alumina catalyst at 350-450° affords biphenyl and hydrogen sulphide (D.R. Kilanowski *et. al.*, J. Catal., 1978, 55, 129). Dibenzothiophene is oxidised with *tert*-butyl hydroperoxide in boiling toluene in the presence of molybdenum hexacarbonyl as catalyst, or in benzene-tetralin with air under pressure at 200° to give dibenzothiophene 5,5-dioxide, which on treatment with aqueous alkali at 300° is converted into sodium 2-phenylphenolate (R.B. LaCount and S. Friedman, J. org. Chem., 1977, 42, 2751).

Passing chlorine through a solution of dibenzothiophene in 90% methanol at 12-14° gives the 5,5-dioxide (V.I. Dronov *et. al.*, Khim. Seraorg. Soedin., Soderzh. Neftyakh Nefteprod., 1972, 9, 158).

Direct formylation of dibenzothiophene with dichloro-methyl methyl ether in methylene chloride containing aluminium chloride at 0° affords the 2- and 3-carbox-aldehydes and the 2,8-dicarboxaldehyde (J.N. Chatterjea and R.S. Gandhi, J. Indian chem. Soc., 1977, 54, 1151). Acetylation of dibenzothiophene using acetyl chloride in methylene chloride in the presence of aluminium chloride gives 2,8-diacetyldibenzothiophene, m.p.204-205.5°, 208-209°, 5,5-dioxide m.p.272-277°. If the solvent is tetrachloroethane then a mixture of 2,8- and 2,6-diacetyl-dibenzothiophene, m.p.205-206°, 5,5-dioxide m.p.303° (decomp.), is obtained (W.L. Albrecht, D.H. Gustafson, and S.W. Horgan, J. org. Chem., 1972, 37, 3355). [1]H nmr spectral data for 2-acetyl- and 2,8-diacetyl- and dibenzoyl- -dibenzothiophene and -dibenzofuran have been reported (A.A. Pashchenko, A.G. Khaitbaeva, Kh. Yu. Yuldashev, Doklady Akad. Nauk Uzb. SSR, 1978, 40).

Ethyl 2- and 3-dibenzothienyloxypropionates have been prepared by esterification of the corresponding hydroxydibenzothiophenes with 1,1-dimethyl-2,2,2-trichloro-ethanol with concomitant hydrolysis followed by esterification of the resulting acids (Gronowitz *et. al.*, Acta. Pharm. Suec., 1978, <u>15</u>, 337).

7. *Other Tri- and Poly-cyclic Systems Containing the Thiophene Ring*

Naphtho[2,1-b]thiophenes (3) may be obtained by heating the appropriate 2-naphthyl prop-2-ynyl sulphoxide (1) in dioxan at 100° to form the hemithioacetal (2) which is transformed into the required naphthothiophene by treatment with a suitable protic solvent containing a small amount of toluene-4-sulphonic acid. Reaction of sulphoxide (1, R = H or Me) with dimethylaniline at 80° yields either 1,2-dihydronaphtho[2,1-b]thiophene-1-carboxaldehyde (4, R = H) or 1-acetyl-1,2-dihydronaphtho[2,1-b]thiophene (4, R = Me), respectively (Y. Makisumi and S. Takada, Chem. Comm., 1974, 848).

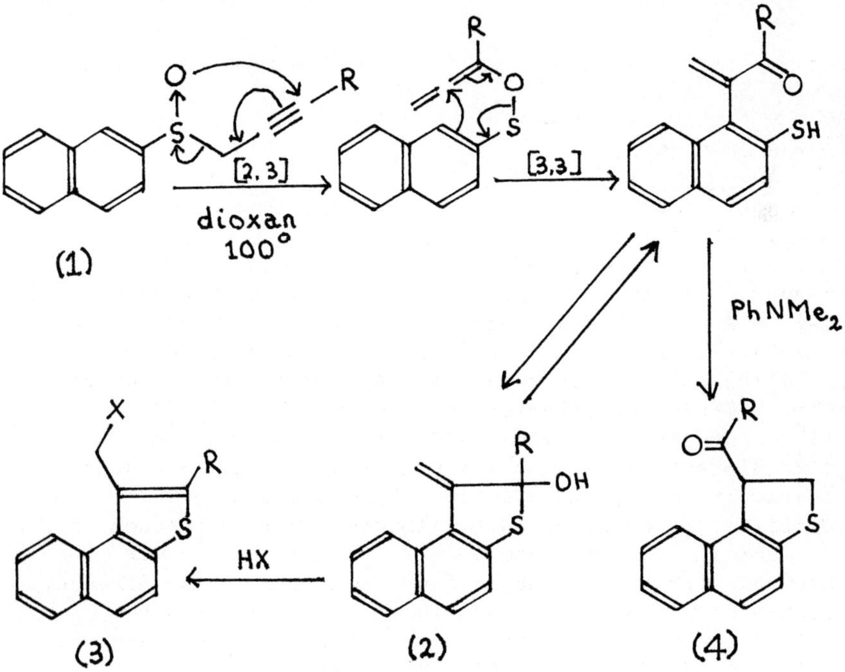

Allyl 2-naphthyl sulphoxides (5) on heating in dimethylaniline or dimethylformamide, undergo a thio-Claisen rearrangement to give quantitatively the 2,3-dihydro-naphtho[2,1-b]thiophene 3-oxides (6). Heating (5) at 110° in acetic acid-dimethylformamide gives 2-acetoxymethyl-2,3--dihydronaphtho[2,1-b]thiophene (7), 3,3-dioxide, m.p.129-130° (Makisumi, Takada, and Y. Matsukura, *ibid.*, p.850).

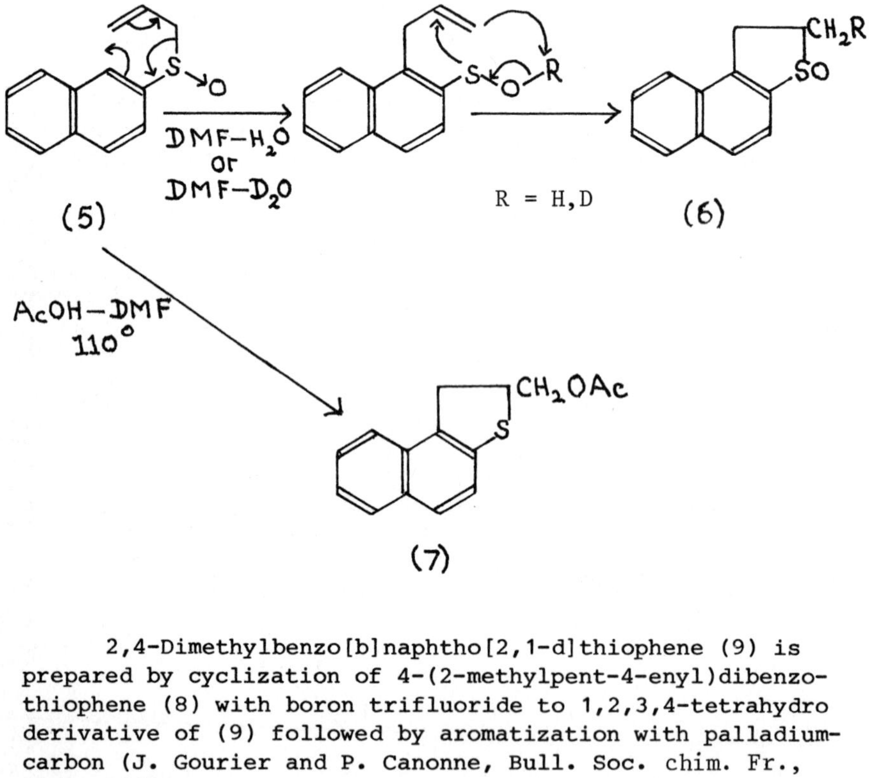

(5)

DMF–H₂O
or
DMF–D₂O

R = H,D

(6)

AcOH–DMF
110°

CH₂OAc

(7)

2,4-Dimethylbenzo[b]naphtho[2,1-d]thiophene (9) is prepared by cyclization of 4-(2-methylpent-4-enyl)dibenzo-thiophene (8) with boron trifluoride to 1,2,3,4-tetrahydro derivative of (9) followed by aromatization with palladium-carbon (J. Gourier and P. Canonne, Bull. Soc. chim. Fr., 1973, 3110).

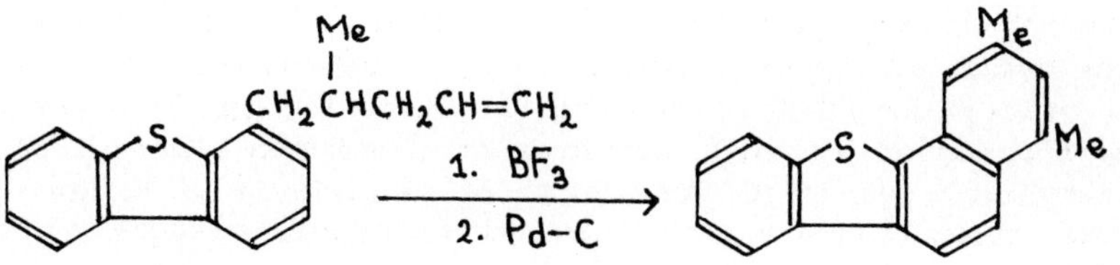

(8)   (9)

Also prepared by similar routes from appropriate starting materials are 7,9-dimethylbenzo[b]naphtho[2,3-d]thiophene and 1,3,6,7-tetramethylbenzo[b]phenanthro[9,10-d]thiophene.

Benzo[b]anthra[2,3-d]thiophene (10) and benzo[b]-anthra[2,1-d]thiophene (11) have been prepared from dibenzothiophene by initial acylation with phthalic anhydride followed by cyclization of the products, and reduction with tin-hydrochloric acid (D.D. Gverdtsiteli and V.P. Litvinov, Soobshch. Akad. Nauk Gruz. SSR, 1970, 58, 333).

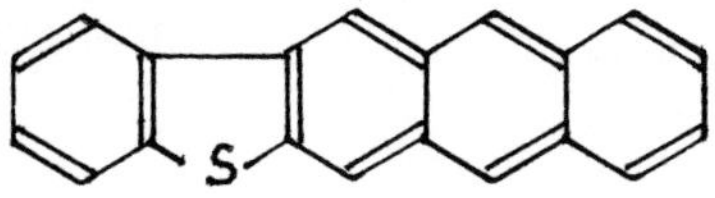

(10)

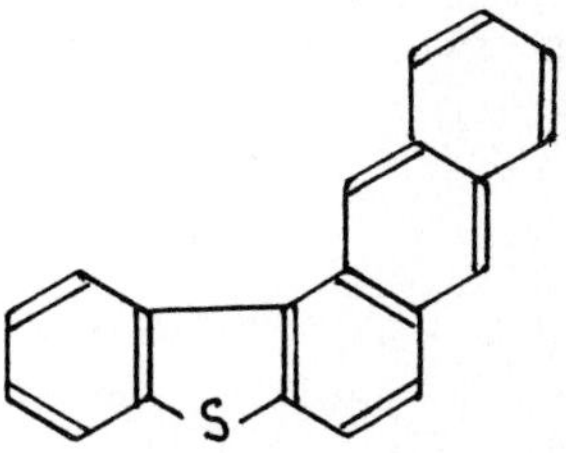

(11)

2-Phenylbenzo[b]thiophene-3-carboxaldehyde 4-tolysulphonylhydrazone (12) reacts with sodium methoxide in hot bis(2-methoxyethyl)ether to give 10H-benzo[b]indeno-[1,2-d]thiophene (13) (30%), together with NN'-bis(2-phenyl-benzo[b]thiophene-3-ylidene)hydrazine (<5%) and 2-phenyl-3--(4-tolysulphonylmethyl)benzo[b]thiophene (10.5%). Under similar conditions 3-phenylbenzo[b]thiophene-2-carboxaldehyde 4-tolylsulphonylhydrazone gives a complex mixture of products, which contains 2-hydroxymethyl-3-phenylbenzo[b]-thiophene (18%) (B. Iddon, H. Suschitzky, and D.S. Taylor, J. chem. Soc., Perkin I, 1974, 2505).

(12)   (13)

2,2'-Methylenebis(thiophene) and 2,2'-methylenebis-(benzo[b]thiophene) with dichloromethyl methyl ether --stannic chloride give by direct Bradsher reaction the corresponding linear, polycyclic, heteroaromatic compounds, benzo[1,2-b;5,4-b']dithiophene (14) and dibenzo[b][b']benzo-[1,2-d;5,4-d']dithiophene (15) (M. Ahmed *et. al.*, J. chem. Soc., Perkin I, 1973, 1099).

(14)   (15)

A new polycyclic system (16) containing thiophene has been synthesized and the name heterocirculene suggested; (16) is a [7]heterocirculene, the number between the

brackets indicating the number of (external) aromatic rings
and hence the size of the cavity (J.H. Dopper and
H. Wynberg, Tetrahedron Letters, 1972, 763).

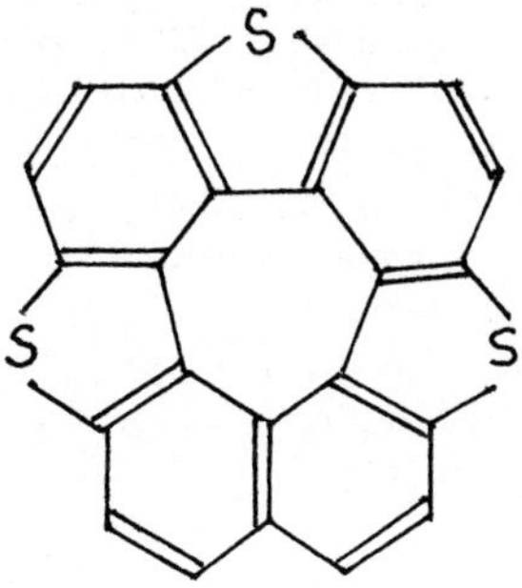

(16)

The electronic absorption and fluorescence spectra of
condensed thiophene-containing systems have been discussed
and compared with those of the polycyclic aromatic
analogues (A.N. Nikitina *et. al.*, Khim. Geterotsikl.
Soedin., 1972, 925).

The preparation and resolution of the hexahetero-
helicene benzo[d]naphtho[1,2-d']benzo[1,2-b:4,3-b']-
dithiophene has been described and thieno[3,2-e]benzo-
[1,2-b:4,3-b']bis[1]benzothiophene synthesized (Wynberg and
M.B. Groen, Chem. Comm., 1969, 964; J. Amer. chem. Soc.,
1968, 90, 5339). Synthesis of a related optically active
undecaheterohelicene, starting from the partially resolved
heptaheterohelicene (*idem, ibid.*, 1970, 92, 6664) and the
construction of helicenes with alternatively fused benzene
and thiophene rings involving, five, seven, nine and eleven
members have been described (P.G. Lehman and Wynberg,
Austral. J. Chem., 1974, 27, 315).

8. Compounds With Five-Membered Heterocyclic Ring Having One Selenium Atom

(a) Selenophene* and related compounds

Selenophene has been obtained from butadiene and selenium dioxide using zeolite - rare earth catalysts at 480-520° (E. Sh. Namedov *et. al.*, Zh. Org. Khim., 1979, 15, 1554; Doklady Akad. Nauk Az. SSR, 1978, 34, 41). It forms an adduct with tetracyanoethene oxide (S. Gronowitz and B. Uppstrom, Acta. Chem. Scand., 1975, B29, 441). A comparative study has been made of the aromatic character of furan, thiophene, selenophene, and tellurophene (F. Fringuelli, *et. al.*, J. chem. Soc., Perkin II, 1974, 332).

2- And 3-chloroselenophenes are prepared by lithiation of selenophene and its 3-bromo derivative, respectively, with ethyllithium or butyllithium and treating the metallated derivative with hexachloroethane (Gronowitz and A.B. Hornfeldt, Synth. Comm., 1973, 3, 213). 3-Bromo-2,5-dialkylselenophenes are lithiated in the 3-position by halogen-lithium exchange with ethyllithium; the ring opening of these compounds has been investigated (Gronowitz and T. Frejd, Acta Chem. Scand., 1976, B30, 313). Metallation of 2,5-dichloroselenophene by lithium di-isopropylamide gives the stable 2,5-dichloro-3-seleno-phenyllithium, but 2,5-dichloro-3-iodoselenophene on treatment with ethyllithium at -70° ring opens to give to compound (1) (*idem, ibid.*, p.439).

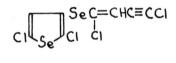

(1)

---

*Thiophenes and their selenium and tellurium analogues, S. Gronowitz, Org. Compd. Sulphur, Selenium, and Tellurium 1979, 5, 247.

For the reduction and formylation of 3-nitroseleno-
phene see G. Ah-Kow, C. Paulmier, and P. Pastour (Bull.
Soc. chim. Fr., 1976, 151). The kinetics of the reaction
of selenophene-2-sulphonyl chloride and 2-chloromethyl-
selenophene with aniline, and of selenophene-2-
-carboxaldehyde with aniline and with a phosphorus ylide
have been investigated and their reactivities compared with
related thiophene compounds (A. Arcoria, E. Maccarone, and
A. Mamo, J. chem. Soc., Perkin II, 1979, 1347).
Formylation or acetylation of 3-acetylaminoselenophene
affords 3-acetylaminoselenophene-2-carboxaldehyde or
2-acetyl-3-acetylaminoselenophene, respectively (Ah-Kow,
Paulmier, and Pastour, Compt. rend., 1974, 278C, 1513).

Ir (R. Cataliotti and G. Paliani, Canad. J. Chem.,
1976, 54, 2451; A. Poletti, Cataliotti, and Paliani, Chem.
Phys., 1974, 5, 291), ^{13}C - ^{77}Se nmr (G.A. Kalabin and
D.F. Kushnarev, Zh. Strukt. Khim., 1979, 20, 617), ^{77}Se
nmr (W.H. Pan and J.P. Fackler, Jr., J. Amer. chem. Soc.,
1978, 100, 5783; Gronowitz, I. Johnson, and Hornfeldt,
Chem. Scr., 1973, 3, 94), ^{13}C and 'H nmr *(idem, ibid.,*
1975, 7, 111; Fringuelli *et. al.,* Acta Chem. Scand.,
1974, 28B, 175), and 'H nmr of selenophene and derivatives
and some related heterocycles have been reported
(C.L. Khetrapal, A.C. Kunwar, and K.P. Sinha, Chem. Phys.
Letters, 1979, 67, 444; V. Galasso and A. Bigotto, Org.
mag. Res., 1974, 6, 475; Fringuelli, G.Marino, and
A. Taticchi, Gazz., 1973, 103, 1041). The magnetic
circular dichroism of selenophene and derivatives and
related heterocycles has been observed (B. Norden
Chem. Phys., 1978, 33, 355) and MO-LCAO calculations
carried out for selenophene and other five-membered
conjugated heterocycles (J. Fabian, Z. phys. Chem., 1979,
260, 81). Calculations relating to the electronic
structure of selenophene have been reported (R.H. Findlay,
J. chem. Soc., Faraday Trans. II, 1975, 1397).

*(b) Benzo[b]selenophenes (selenanaphthenes, 2,3-benzo-
selenophenes)**

Benzo[b]selenophene may be obtained in 65% yield from

---

*Recent aspects of the chemistry of benzo[b]selenophene and
 benzo[b]tellurophene, M. Renson, Chem. Scr., 1975, 8A, 29.

$2\text{-EtO}_2\text{CC}_6\text{H}_4\text{SeCH}_2\text{CO}_2\text{Et}$ by the Dieckmann reaction followed by sodium tetrahydridoborate reduction and acidification (T.Q. Minh, L. Christiaens, and M. Renson, Tetrahedron, 1972, 28, 5397). Benzene and acetylene when passed over molten selenium at 540° give benzo[b]selenophene (18%) (T. Lesiak, Rocz. Chem., 1971, 45, 1589). The rates of the base-catalyzed hydrogen — deuterium exchange at C-2 and the reverse have been reported for benzo[b]selenophene and related heterocycles (O. Attanasi, P. Battistoni, and G. Fava, Phosphorus Sulphur, 1979, 5, 305).

Bromination of benzo[b]selenophene with one mole of bromine gives 2- and 3-bromobenzo[b]selenophene and with two mole of bromine, 2,3-dibromobenzo[b]selenophene. Also reported are the preparations of 2-acetylbenzo[b]seleno-phene and benzoselenophene-2-carboxylic acid and its methyl ester and acid chloride (N.N. Magdesieva and V.A. Vdovin, Khim. Geterotsikl. Soedin., 1973, 15). Bromination of either 2- or 3-methylbenzo[b]selenophene yields 3-bromo-2--methyl- or 2-bromo-3-methyl-benzo[b]selenophene, respectively (Christiaens, R. Dufour, and Renson, Bull. Soc. chim. Belg., 1970, 79, 143). 5-Chloro- and 5-chloro--2-methyl-benzo[b]selenophene (G. Kirsch and P. Cagniant, Compt. rend., 1972, 275C, 1029). The reaction of 3-bromo-benzo[b]selenophene with butyllithium, at 0° results in ring-opening *via* the 2,3-dilithio derivative; the 3-lithio derivative is stable under these conditions (Christiaens and Resnon, Bull. Soc. chim. Fr., 1974, 2244). Acylation of benzo[b]selenophene with acetyl chloride or benzoyl chloride in the presence of aluminium chloride yields the 2-acylbenzo[b]selenophene with small amounts of the 3-acyl derivative. Increased reaction favours formation of the latter derivatives (Minh *et. al.*, Bull. Soc. chim. Fr., 1972, 3955).

The reaction of benzisoselenazolinone (1) with acetic anhydride — potassium acetate yields 3-acetoxy- and 3-hydroxy-benzo[b]selenophene besides other products (R. Weber and Renson, Bull. Soc. R. Sci. Liege, 1979, 48, 146).

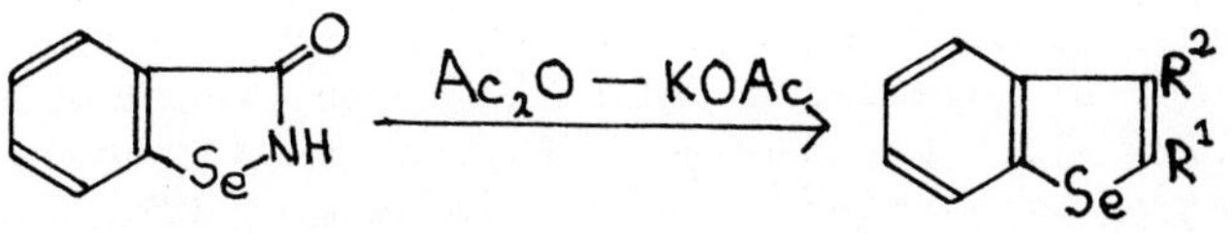

(1)

R^1 = H, R^2 = OAc, OH, NHAc;
R^1 = Ac, R^2 = H; R^1 = CONH$_2$, R^2 = H

2,3-Dihydro-3-oxobenzo[b]selenophene on oxidation with potassium ferricyanide in alkali yields a mixture of *cis*- and *trans*-selenoindigo (2 and 3) (D.L. Ross, J. Blane, and F. Matticoli, J. Amer. chem. Soc., 1970, 92, 5750).

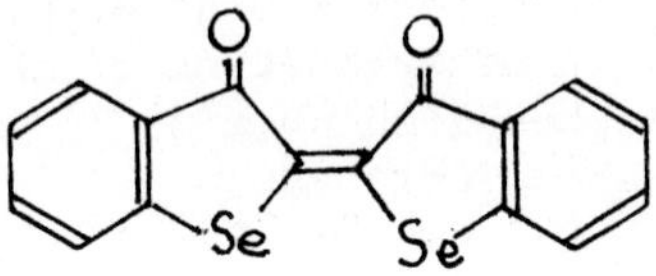

(2)

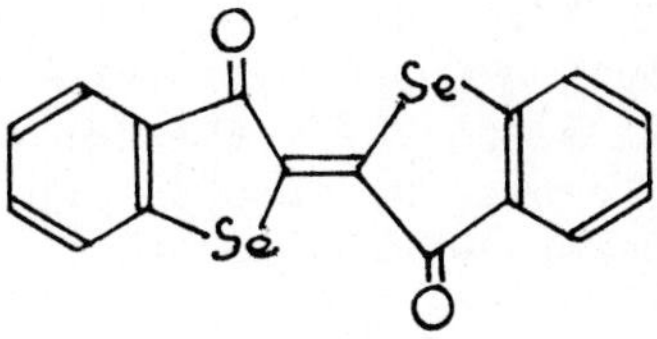

(3)

Photoelectron (J.F. Muller, Helv., 1975, 58, 2646), ^{1}H nmr (V. Galasso and A. Bigotto, Org. mag. Res., 1974, 6, 475; P. Faller and J. Weber, Bull. Soc. chim. Fr., 1972, 3193; G. Llabres *et. al.*, Tetrahedron Letters, 1972, 3177; N.N. Magdesieva, V.A. Vdovin, and N.M. Sergeev, Khim. Geterotsikl. Soedin., 1971, 7, 1382), and mass spectra (N.P. Buu-Hoï *et. al.*, J. chem. Soc., B, 1969, 971) of benzo[b]selenophene and some derivatives and related heterocycles have been reported.

(c) Dibenzoselenophene, diphenylene selenides

Dibenzoselenophene is obtained on boiling biphenylene and selenium together at *ca.* 275° for several hours

(J.M. Gaidis, J. org. Chem., 1970, 35, 2811).

$$\text{(structure)} \xrightarrow[275^{\circ}]{Se} \text{(structure Se)}$$

A synthesis of benzo[b]naphtho[1,2-d]selenophene from ethyl dibenzoselenophene-1-carboxylate (P. Cagniant, N. Bellinger, and D. Cagniant, Compt. rend., 1973, 277C, 779) and the mass spectrum of octafluorodibenzoselenophene have been reported (S.C. Cohen, Org. Mass Spectrom., 1972, 6, 373).

Irradiation of dibenzoselenophene 5-oxide with methyl phenyl sulphide in methanol results in deoxygenation of the 5-oxide and the formation of methyl phenyl sulphoxide (T. Tezuka, H. Suzuki, and H. Miyazaki, Tetrahedron Letters, 1978, 4885).

9. *Heterocyclic Five-Membered Ring Compounds Having One Tellurium Atom*

(a) *Tellurophenes**

The preparation of tellurophene from butadiyne and sodium telluride in methanol has been described in detail. 2-Methyltellurophene, b.p. $108-110^{\circ}/100$mm, tellurophene--2-carboxaldehyde, b.p. $90-92^{\circ}/2$mm, 2-hydroxymethyltellurophene, tellurophene-2-carboxylic acid, m.p. $110-111^{\circ}$, methyl ester, b.p. $118-120^{\circ}/13$mm, n_D^{24} 1.6358, 5-methyltellurophene-2-carboxylic acid, m.p. $149-150^{\circ}$, 2-acetyltellurophene, b.p. $134-136^{\circ}/15$mm, 1-(tellurophen-2-yl)-ethanol, b.p. $133-135^{\circ}/15$mm, 5-acetyltellurophene-2--carboxylic acid, m.p. $190-191^{\circ}$, methyl ester, m.p. $80-81^{\circ}$ (F. Fringuelli and A. Taticchi, J. chem. Soc., Perkin I,

*Thiophenes and their selenium and tellurium analogues, S. Gronowitz, Org. Compd. Sulphur, Selenium, Tellurium, 1977, 4, 244; 1979, 5, 247.

1972, 199). Spectral data has been reported and the effect
of substituents on the ring-proton nmr chemical shifts
examined. ^{1}H nmr relating to the determination of covalent
radii (C.L. Khetrapal, A.C. Kunwar, and K.P. Sinha, Chem.
Phys. Letters, 1979, 67, 444). and the mass spectrum of
tellurophene have been described (Fringuelli and Taticchi,
J. heterocyclic Chem., 1978, 15, 137). For magnetic
circular dichroism of, and MO–LCAO calculations for
tellurophene see p.77. Tellurophene has been
synthesised by an improved method from 1,4–bis(tri–
methylsilyl)butadiyne (W.Lohner and K.Praefcke,
Ber., 1978, 111, 3745). Comparison has been made between
conventional electrophilic substitution and side-chain
solvolyses of tellurophenes and related heterocycles
(Clementi *et. al.*, Gazz., 1977, 107, 339). Ir spectra of
tetrahydrotellurophene, tetrahydroselenophene, and tetrahydrothiophene as liquids and solids at 100K and their
Raman spectra (liquid) have been reported (M. Grazia
Giorgini, G. Paliani, and R. Cataliotti, Spectrochim.
Acta., Part A, 1977, 33A, 1083).

(b) Benzotellurophenes

Benzo[b]tellurophene may be obtained by the
decarboxylation with copper in quinoline of benzo[b]-
tellurophene-2-carboxylic acid, prepared by the cyclization
of 2-(carboxymethyltelluro)benzaldehyde (J.L. Piette and
M. Renson, Bull. Soc. chim. Belg., 1971, 80, 521).

Its mass spectrum has been reported (N.P. Buu-Hoï *et. al.*,
J. heterocyclic Chem., 1970, 7, 219).

The ^{125}Te nmr spectra have been determined for 2-
and 3-monosubstituted benzo[b]tellurophenes and compared
with those of some tellurophenes and with the ^{77}Se nmr
spectral data of related benzo[b]selenophenes

(T. Drakenberg *et. al.*, Chem. Scr., 1979, **13**, 152).

3-Chloro-2-phenylbenzo[b]tellurophene is obtained
from 2-phenyl-1,1,3-trichlorobenzo[b]tellurophene, prepared
by boiling 2-chloro-1,2-diphenylvinyltellurotrichloride in
1,2,4-trichlorobenzene (I.D. Sadekov and V.I. Minkin, Khim.
Geterotsikl. Soedin., 1971, **7**, 138).

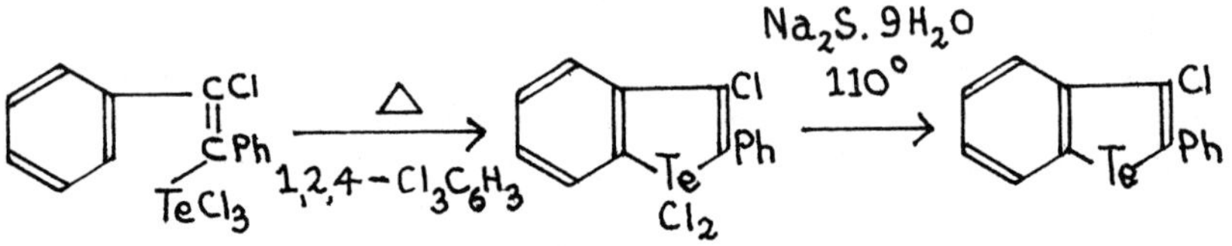

3-Halogenobenzo[b]tellurophenes may be obtained by a
one-step synthesis by the reaction of the appropriate
arylacetylene in the presence of tellurium dioxide with
lithium halide in acetic acid (J. Bergman and L. Engman,
Tetrahedron Letters, 1979, 1509).

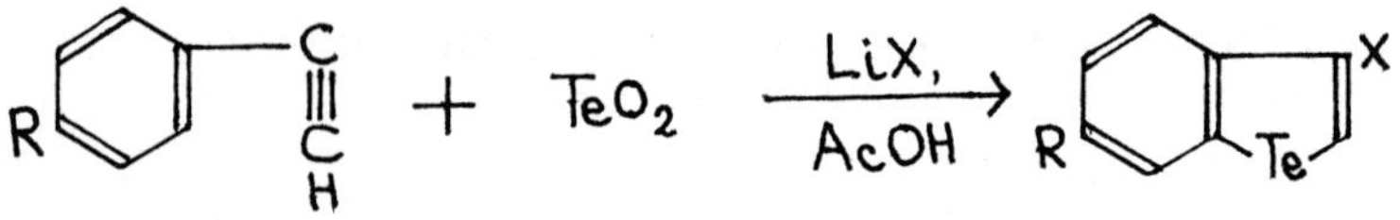

R = H, Me, Br   X = Cl, Br, I

The synthesis and spectral data of 1,1-diiodo-2,3-dihydro-
benzo[b]tellurophene have been reported (R.F. Ziolo and
W.H.H. Guenther, J. organometal. Chem.. 1978, **146**, 245).

Tellurium on treatment with sodium tetrahydridoborate
followed by phthalic acid dichloride affords benzo[c]-
tellurophene-1,3-dione. Benzo[c]selenophene-1,3-dione may
be obtained in a similar manner from selenium (Bergman and
Engman, Org. Prep. Proced. Int., 1978, **10**, 289).

Chapter 4

COMPOUNDS CONTAINING A FIVE-MEMBERED RING WITH ONE HETERO
ATOM FROM GROUP V;   NITROGEN

R. LIVINGSTONE

1. Pyrroles[1]

In recent years there has not been a great expansion
in the chemistry of pyrrole, but a few derivatives of
pyrrole have been synthesized, some for the first time
and others by alternative routes. Additional properties
and reactions of known derivatives have also been
described.

(a) General synthetic methods

[14]C-Labelling experiments have been used to
investigate the Knorr synthesis of pyrroles (J.W. Harbuck
and H. Rapoport, J. org. Chem., 1971, 36, 853) and a
modification of the synthesis to give dialkyl 2-methyl-4-
alkylpyrrole-3,5-dicarboxylates has been developed
(H. Plieninger and H. Husseini, Synth., 1970, 587). The
use of the Hantzch synthesis has also been extended
(M.W. Roomi and S.F. MacDonald, Canad. J. Chem., 1970,
48, 1689) and hydroxylamine has been used in the Paal-Knorr
synthesis instead of an amine, thus leading to 1-hydroxy-
pyrroles (R. Ramasseul and A. Rassat, Bull. Soc. chim.
Fr., 1970, 4330).

(1) The addition reaction between acetophenone oxime
and dimethyl acetylenedicarboxylate in boiling methanol
in the presence of sodium methoxide yields adduct (1),
a *O*-vinyloxime, which on heating to 170-180° rearranges
to give dimethyl 5-phenylpyrrole-2,3-dicarboxylate,
probably *via* the iminoketone (2) (T. Sheradsky, Tetrahedron
Letters, 1970, 25).

[1] A. Gossauer, 'Die Chemie der Pyrrole', Springer-
Verlag, Berlin 1974;   R.A. Jones and G.P. Bean, 'The
Chemistry of Pyrroles', Academic Press, London, 1977.

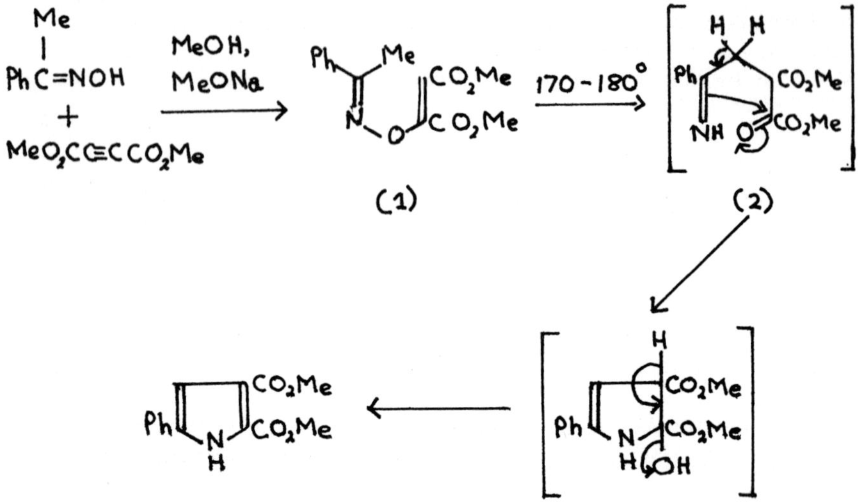

(1) (2)

(2) The photochemical cyclisation of phenyl
β-dialkylaminovinyl ketone yeilds pyrroles, for example,
phenyl β-dimethylaminovinyl ketone (3, $R^1 = R^2$ = Me) gives
1-methyl-2-phenylpyrrole (4, $R^3 = R^4$ = H). Similarly (3)
($R^1 = R^2$ = Et) and (3) ($R^1 = R^2$ = Pr^n) yield (4)
($R^3 = R^4$ = Me) and (4) ($R^3 = R^4$ = Et), respectively,
whereas (3) (R^1 = Me, R^2 = Et) gives a 3:1 mixture of (4)
(R^3 = H, R^4 = Me) and (4) (R^3 = Me, R^4 = H) (H. Aoyama
et. al., Chem. Comm., 1972, 775).

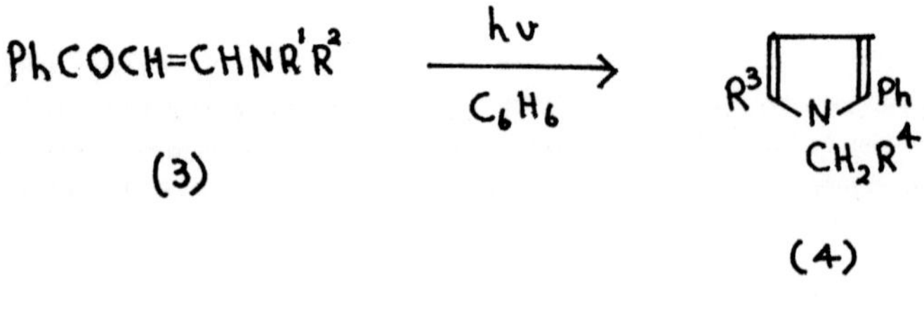

(3) (4)

(3) Heating 1-bromo-1-alkoxy-3-chloromethylallenes and primary amine in acetonitrile in a sealed tube at 120° gives a pyrrole (M.V. Mavrov, A.P. Rodionov, and V.F. Kucherov, Tetrahedron Letters, 1973, 759).

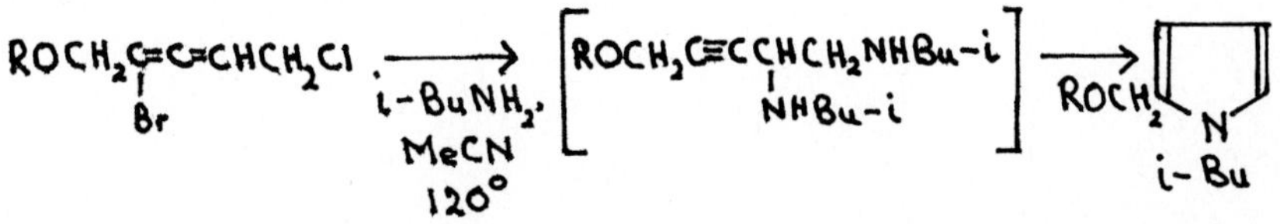

(4) Pyrrole derivatives may be obtained from the reaction of Schiff base anions with α-halogeno ketones, for example, lithium ethylidenecyclohexylamine reacts with bromomethyl phenyl ketone in ether at -78° to give 1-cyclohexyl-2-phenylpyrrole (G. Wittig, R. Röderer, and S. Fischer, *ibid.*, 3517).

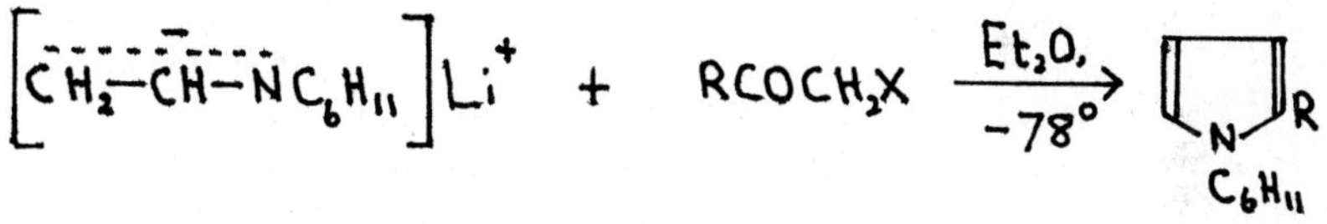

R = Ph, Me, X = Br;
R = Ph, X = Cl

The use of a cyclic α-halogeno ketone enables the preparation of compounds with a pyrrole ring fused to an alicyclic ring.

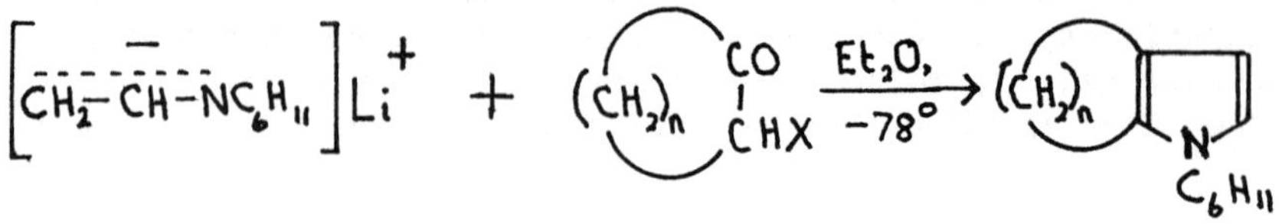

$$\left[\overline{CH_2-CH-NC_6H_{11}} \right] Li^+ \; + \; (CH_2)_n \begin{matrix} CO \\ | \\ CHX \end{matrix} \xrightarrow[-78°]{Et_2O_1} (CH_2)_n \underset{\underset{C_6H_{11}}{N}}{\bigcirc}$$

$$X = Cl, \; Br, \; n = 4, \; 5, \; 6$$

The action of lithium on α-bromoaldimines affords mainly
1,3,4-trisubstituted pyrroles and that of isopropylmagnesium chloride affords mainly 1,2,4-trisubstituted
pyrroles (P. Duhamel, L. Duhamel, and J.-Y. Valnot, *ibid.*,
p.1339).

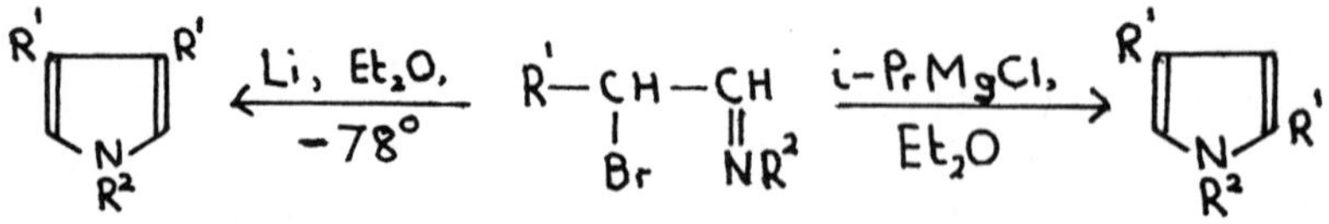

$$R^1 \underset{R^2}{\overset{R^1}{\boxed{N}}} \xleftarrow[-78°]{Li, \; Et_2O_1} R^1-\underset{Br}{\overset{}{CH}}-\underset{NR^2}{\overset{}{CH}} \xrightarrow[Et_2O]{i-P_rMgCl,} R^1 \underset{R^2}{\overset{R^1}{\boxed{N}}} R^1$$

(5) The electrolysis of enamine ketones and esters
(5) in methanolic sodium perchlorate gives pyrroles (6),
presumably by dimerisation of intermediately formed radical
cations (D. Koch and H. Schäffer, Angew. Chem. internat.
Edn., 1973, **12**, 245).

$$R^2CH=CHMeNHR^1$$

(5)

$$Me \underset{\underset{R^1}{N}}{\overset{R^2 \quad R^2}{\boxed{\quad\quad}}} Me$$

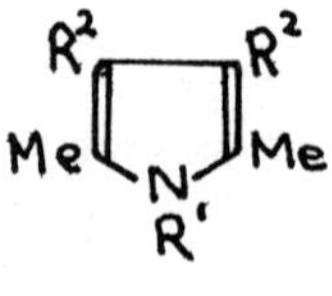

(6)

$R^1 = H, \; Me, \; PhCH_2, \; Ph$
$R^2 = CO_2Me, \; COMe$

(6) Thermolysis of 1,2-dialkyl-5,6-dimethoxycarbonyl-
-2,4-diphenyl-2a,6a-dihydroazetidino[3,2-b]pyridines in
benzene gives pyrroles (7) and the ketenimines (8)
(J.W. Lown and M.H. Akhtar, Tetrahedron Letters, 1973,
3727).

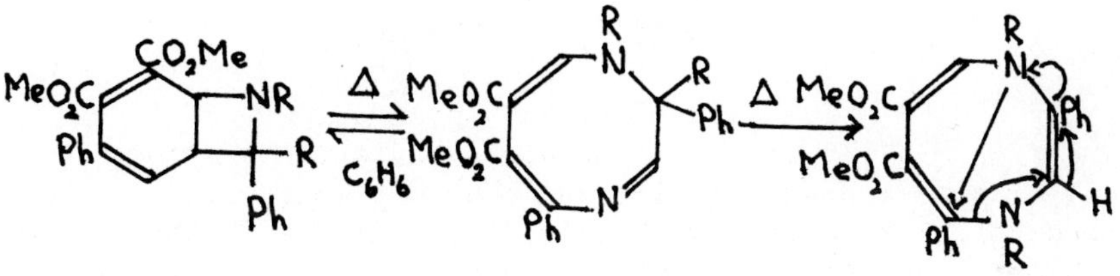

$R = C_6H_{11}, C_{12}H_{23}$

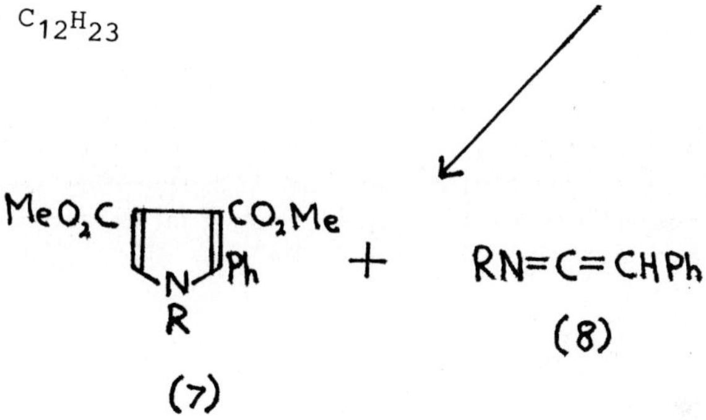

(7) Dialkylketazins (9) on heating in the presence
of zinc chloride rearrange to give pyrroles (Piloty
synthesis) (R. Baumes, R. Jacquier, and G. Tarrago, Bull.
Soc. chim. Fr., 1974, 1147).

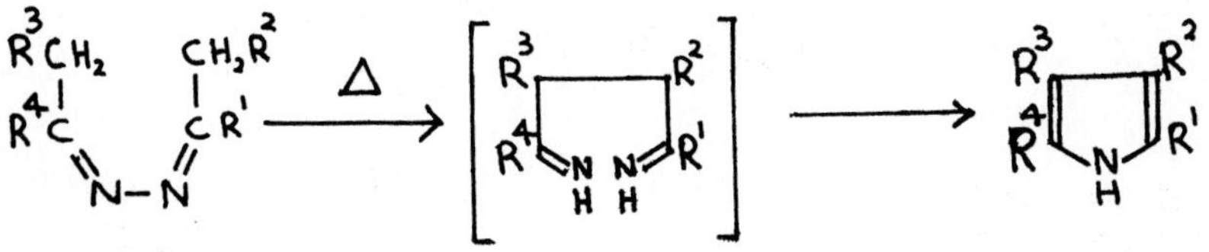

(8) Pyrroles may be prepared by treating dioxins (10) with amines in strong base (K. Kondo and M. Matsumoto, Chem. Letters, 1974, 701).

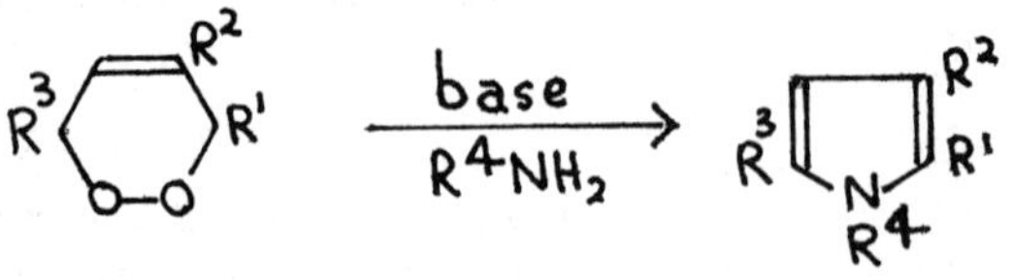

(10)

R^1, R^2, R^3 = H, Me, Ph; R^4 = H, Et, Ph

(9) Photolysis of 3,3-dimethyl-2-phenyl-1-azirine and triphenylvinylphosphonium bromide in acetonitrile — triethylamine produces 2,2-dimethyl-5-phenyl-2H-pyrrole (11). The triphenylvinylphosphonium bromide reacts as a dipolarophile with the ylide from the azirine (N. Gakis, H. Heimgartner, and H. Schmid, Helv., 1974, 57, 1403).

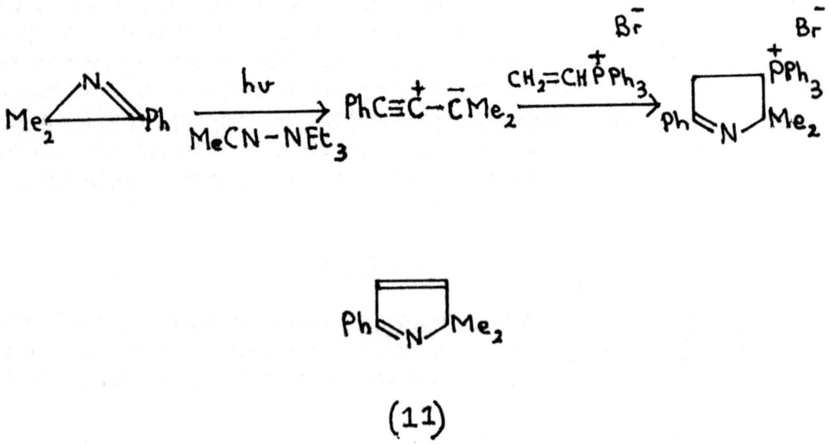

(11)

A review of synthetic methods for pyrroles and 2H- or 3H-
-pyrroles has been published (J.M. Patterson, Synth., 1976,
281).

(b) General properties and reactions

(1) 1-methylpyrrole on treatment with C-acetyl-N-
-phenylnitrilimine (N-phenyl-pyruvonitrilimide) (1)
prepared in situ, by the reaction between α-chloro-α-
-(N-phenylhydrazono)acetone and triethylamine in tetrahydrofuran, yields a complex reaction mixture from which
double addition products (2) and (3) (3:1) can be isolated.
These are formed by addition of the 1,3-dipole (1) to the
2,3-position of 1-methylpyrrole (M. Ruccia, N. Vivona,
and G. Cusmano, Tetrahedron Letters, 1972, 4073).

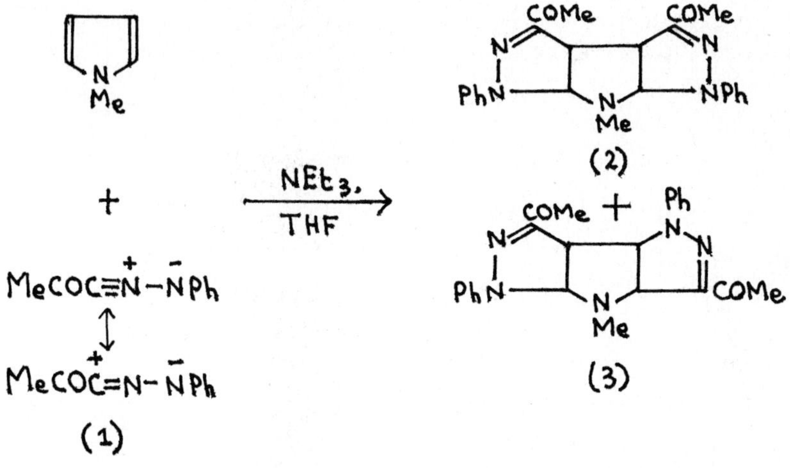

The addition reactions of 1,2-dimethylpyrrole, methyl
1-methylpyrrole-2-carboxylate (idem., J. heterocyclic
Chem., 1978, 15, 293), and pyrrole with C-acetyl-N-
phenylnitrilimine have been investigated and the products
identified (Ruccia et. al., ibid., p.1485).

(2) Pyrrole on treatment with hydrogen peroxide in
either an acidic or an alkaline medium affords a mixture
of condensed and partly oxidised pyrroles, containing 2,3,

or 4 pyrrole nuclei. However, in a neutral medium, pyrrole can be converted into 3-pyrrolin-2-one (4) (G.P. Gardini, Chem. Abs., 1970, 72, 132427r).

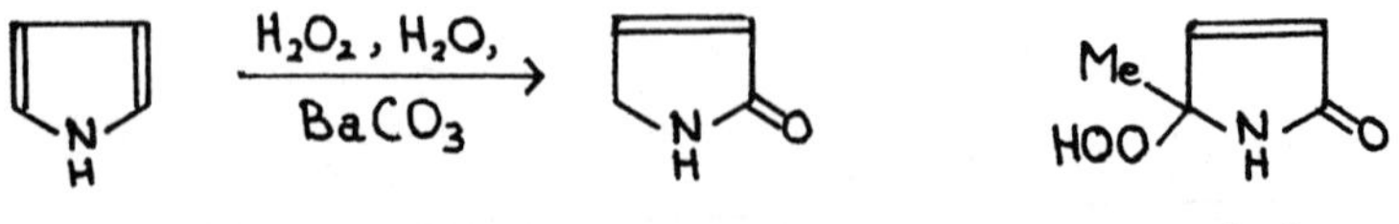

(4)   (5)

2-Methylpyrrole may undergo further oxidation to hydroperoxide (5) (Gardini and V. Bocchi, Gazz., 1972, 102, 91).

(3) The reaction of pyrrole with enamines of types (6) and (7) yield Michael type adducts (8) and (9) (O. Tsuge, M. Tashiro, and Y. Kiryu, Chem. Letters, 1974, 795).

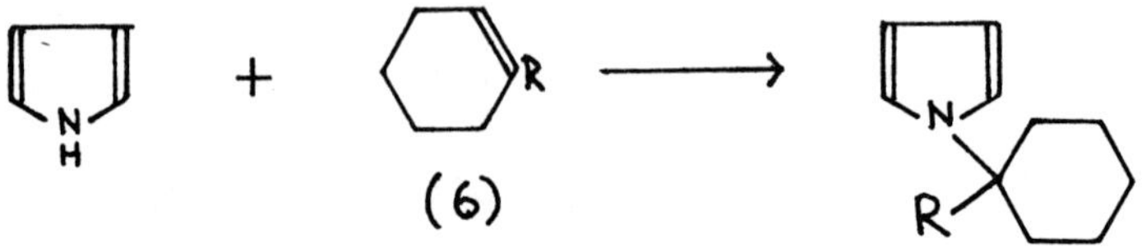

(6)   (8)

R = 1-pyrrolidinyl, piperidino, morpholino

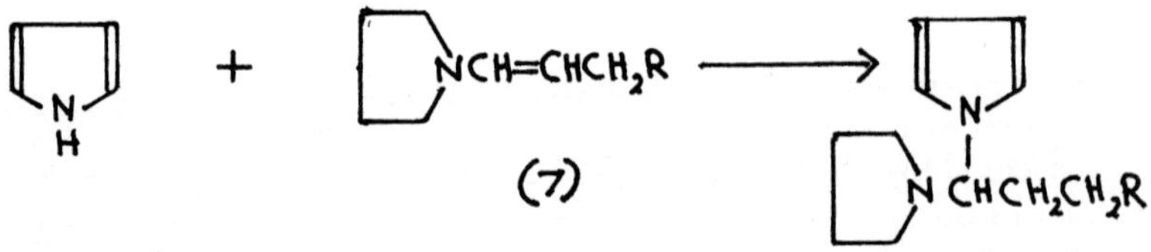

(7)   (9)

R = H, Me

(4) The reaction between pyrrole and ethoxycarbonyl isothiocyanate gives *N*-ethoxycarbonylpyrrole-2--thiocarboxamide (10), which is converted to 2-thiopyrrole--1,2-dicarboximide (11) on heating with quinoline (E.P. Papadopoulos, J. org. Chem., 1973, 38, 667).

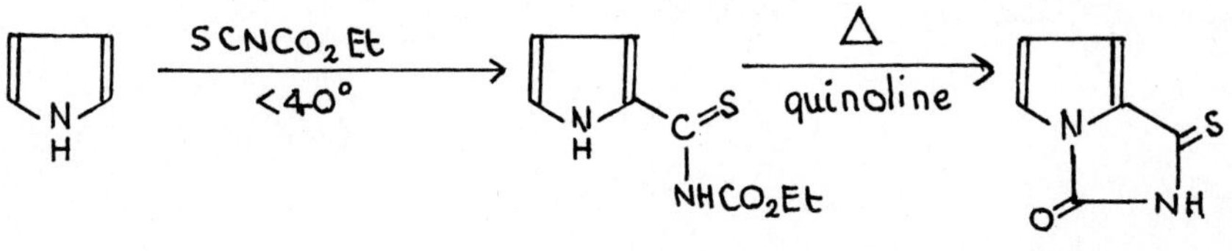

(5) A high-melting solid with a tetrazaquaterene structure (12) is formed by the condensation of equimolar quantities of pyrrole and cyclohexanone in the presence of hydrochloric acid. When pyrrole and cyclohexanone, in a mole ratio of 4:1, are treated with hydrochloric acid, compound (13) is formed (W.H. Brown, B.J. Hutchinson, and M.H. MacKinnon, Canad. J. Chem., 1971, 49, 4017).

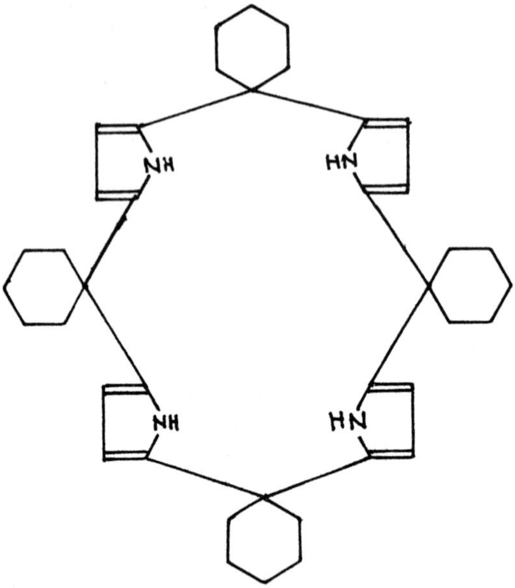

(12)

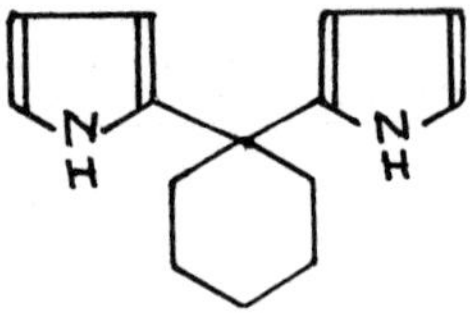

(13)

Furan also gives related condensation products when treated with cyclohexanone under acid conditions.

(6) The reaction of pyrroles with common dienophiles seems to follow two different pathways, that is [4+2] cycloaddition or a Michael-type addition at the α-position of pyrroles. Pyrroles with an aryl or electron-withdrawing substituent at position-1, give with dimethyl acetylene-dicarboxylate 1:1 adducts of type (14). With a 1-alkyl group the 1:1 adduct (14) reacts further with dimethyl acetylenedicarboxylate to give a 1:2 adduct (15). It is reported that pyrrole itself gives a Michael-type 1:1 adduct (16), though the structure has not been thoroughly established. However, pyrrole on boiling for several days with dimethyl acetylenedicarboxylate in ether gives 1:2 (15) (R = H), tetramethyl 3a,7a-dihydroindole-2,3,3a,4-

-tetracarboxylate (C.K. Lee and C.S. Hahn, J. org. Chem.,
1978, _43_, 3727).

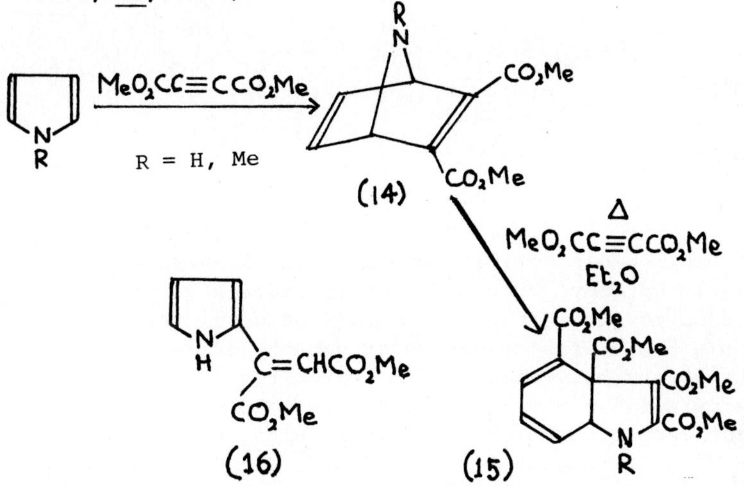

The difference in chemistry between adducts (15) (R = H
and R = Me) has been discussed. The aluminium chloride
catalysed addition of dimethyl acetylenedicarboxylate to
N-methoxycarbonylpyrrole can be achieved in high yield
(R.C. Bansal, A.W. McCulloch and A.G. McInnes, Canad. J.
Chem., 1969, _47_, 2342).

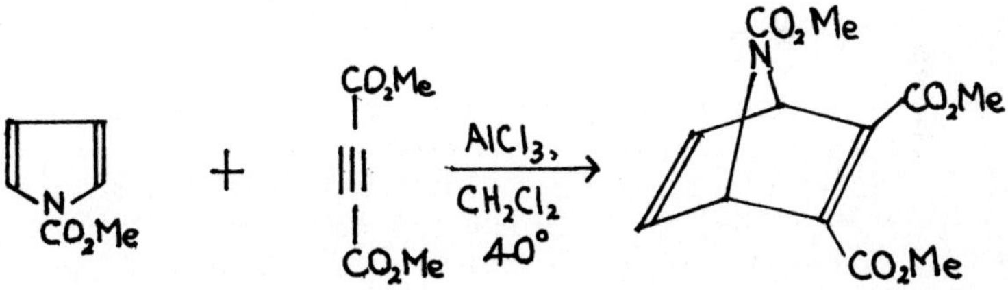

(7) Tetrafluorobenzyne adds smoothly to 1-trimethyl-
silylpyrrole in tetrahydrofuran to establish the 1,4-
dihydronaphthalen-1,4-imine system. The addition of water
to the anhydrous reaction mixture removes the
trimethylsilyl protecting group to give 1,4-dihydro-5,6,7,8,-
-tetrafluoronaphthalen-1,4-imine (17) (P.S. Anderson _et._
_al.,_ Tetrahedron Letters, 1974, 2553).

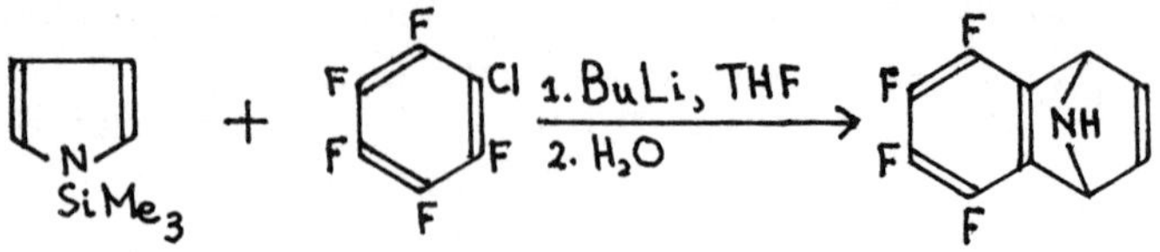

(17)

(8) The regioselective photoaddition of pyrroles and aliphatic carbonyl compounds occurs under a variety of conditions to give either 1:1 or 1:2 adducts, for example, the photolysis of equimolar amounts of 1-methylpyrrole and butyraldehyde in acetonitrile gives virtually pure (18), whereas extended radiation in the presence of excess carbonyl compound results in the formation of a 1:2 adduct (19) (G. Jones, II, H.M. Gilow, and J. Low, J. org. Chem., 1979, 44, 2949).

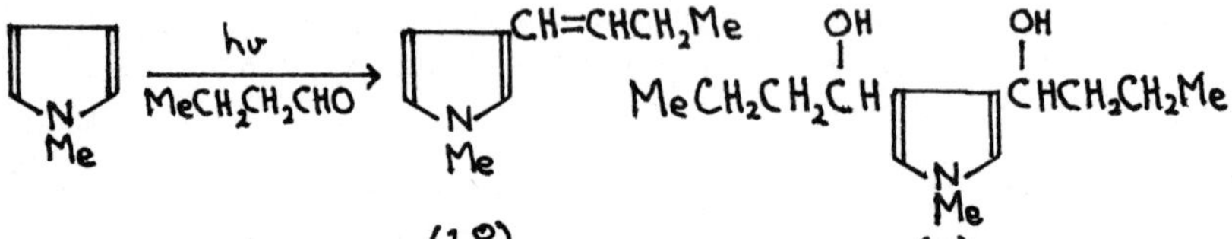

(18)   (19)

(9) 1-Alkylpyrroles (20, R = Me, Bu) reacts with diethyl azodicarboxylate in protic and aprotic solvents to afford Michael-type 1:2 adducts at both α-positions, but 1-arylpyrroles (20, R = Ph, 4-MeOC$_6$H$_4$, 4-NO$_2$C$_6$H$_4$, 2,6-diMeC$_6$H$_3$) do not react in aprotic solvents (Lee, S.J. Kim, and Hahn, *ibid.*, 1980, 45, 1692).

(20)

Ring expansion occurs, when chloroform reacts with
pyrrole and with some methylpyrroles in the vapour phase
at 550° to give high yields of chloromethylpyridines
(R.E. Busby *et. al.*, J. chem. Soc., Perkin I, 1979, 1578).
Addition reactions of pyrrole have been reviewed (Lee,
Hwahak Kwa Kongop Ui Chinbo, 1977, 17, 77). For the
reactions of some metal derivatives of pyrrole see
N.-C. Wang (Diss. Abs. Int. B, 1977, 38, 1222). The
physiochemical properties of pyrroles have been reviewed
(R.A. Jones, Adv, heterocyclic Chem., 1970, 11, 383).
^{1}H nmr spectra (R.J. Abraham and H.J. Bernstein, Canad.
J. Chem., 1969, 37, 1056; 1971, 39, 905; A. Gossauer,
'Die Chemie, der Pyrrole', Springer-Verlag, Berlin, 1974,
83, 93; G. Marino, Adv. heterocyclic Chem., 1971, 13.
235) and ^{13}C nmr spectra of pyrrole and substituted
pyrroles have been reported (Gossauer, *loc. cit.*, p.89;
E. Lippmaa *et. al.*, Org. mag. Res., 1972, 4, 153, 197;
Abraham *et. al.*, J. chem. Soc., Perkin II, 1974, 1004).
For the nmr and uv spectra of salts obtained by protonation
of alkylpyrroles in the α-position by hydrogen chloride
in ether see G.P. Bean, Chem. Comm., 1971, 421.

(c) *Pyrrole and its substitution products*

(i) *Pyrrole*

The catalytic dehydrogenation of pyrrolidine to give pyrrole can now be carried out successfully at 400-500° in the presence of hydrogen using a catalyst containing palladium on silica gel. Previously the dehydrogenation had not been performed satifactorily due to catalyst poisoning by pyrroline and its polymerization by-products (P. Guyer and D. Fritze, Fr. Pat. 1,579,449/1969). Pyrrole is obtained by the vapour phase reaction of diethylamine diluted with nitrogen, and iodine at <800° in the presence of calcium oxide (W.H. Bell and P.M. Quan, Brit. Pat. 1,183,958/1970); by heating 2-hydroxyethylamine and bis(2-hydroxyethyl)amine in the presence of Al_2O_3-ZnO- -Cr_2O_3 and by passing morpholine over a Cr_2O_3- -Al_2O_3 catalyst at 400-440° (K.M. Akhmerov *et. al.*, Chem. Abs., 1974, **81**, 49501a; 1975, **82**, 156001e).

The photo-oxidation of pyrrole in methanol gives mainly 5-methoxy-3-pyrrolin-2-one with a low yield of maleimide, but in aqueous solution, only 5-hydroxy-3- -pyrrolin-2-one is formed.

The photo-oxidation of 3,4-diethylpyrrole in methanol gives mainly diethylmaleimide and 3,4-diethyl-5-methoxy-3- -pyrrolin-2-one (G.B. Quistad and D.A. Lightner, Chem. Comm., 1971, 1099).

Pyrrole on treatment with equimolar proportions of n-butyllithium and *NNN'N'*-tetramethylethylenediamine in hexane, in attempts to prepare dilithio derivatives, yields 1-lithio-(80%), 2-lithio-(4%), 3-lithio-(4%)-, and 2,5-dilithio-pyrrole(2%) (D.J. Chadwick and C. Willbe, J. chem. Soc., Perkin I, 1977, 887).

(ii) N-Derivatives of pyrrole (1-substituted pyrroles)

1-Alkylpyrroles are prepared by the alkylation of pyrrole with alkyl halides in the presence of a crown ether, thus pyrrole and potassium hydroxide in boiling benzene containing 18-crown-6 ether, on treatment with benzyl bromide affords 1-benzylpyrrole (E. Santaniello, C. Farachi, and F. Ponti, Synth., 1979, 617).

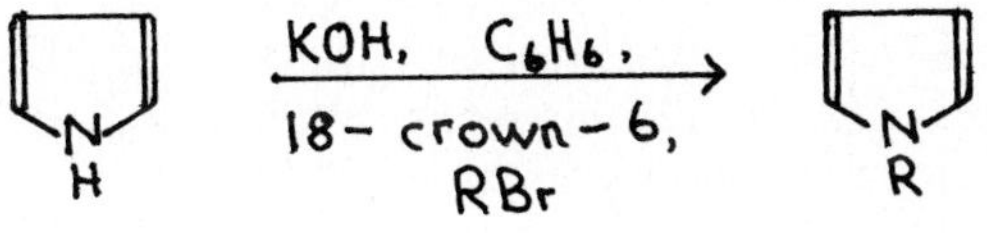

$$R = Et, Bu, PhCH_2$$

1-Alkylation of pyrrole and several 2-substituted pyrroles is readily achieved by phase transfer catalysis with primary alkyl halides, for example, a solution of pyrrole, butyl bromide, and tetrabutylammonium bromide in methylene chloride when treated with a 50% aqueous solution of sodium hydroxide gives 1-butylpyrrole (84%) (N.-C. Wang, K.-E. Teo, and H.J. Anderson, Canad. J. Chem., 1977, 55, 4112). Grignard derivatives of polyalkylated pyrroles alkylate at both the 2- and the 3- positions, although the former predominates. If there is a substituent at the 2- or the 3-position the pyrrolenines (1) or (2) may be formed (J.L. Wong and M.H. Ritchie, Chem. Comm., 1970, 142; Wong, Ritchie, and C.M. Gladstone, ibid., 1971, 1093).

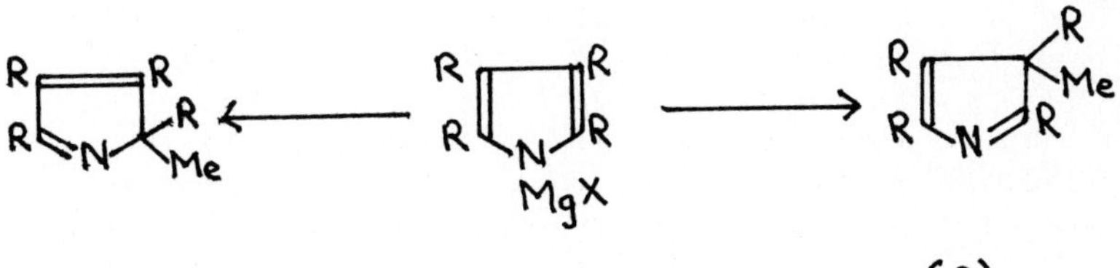

(1)

(2)

1-Alkoxycarbonylpyrroles with diazomethane or diazoacetic ester and copper (I) salts give 2,3-homopyrroles and bis--adducts (F.W. Fowler, Angew. Chem., 1971, 83, 147); pyrrole also affords the bis-homopyrrole derivatives.

Heating 1-methylpyrrole with n-butyllithium in hexane in the presence of *NNN'N'*-tetramethylethylenediamine leads mainly to 1-methyl-2,5-dilithiopyrrole, some 2,4--dilithio derivative is also formed. It has been shown that the relative proportions of the two derivatives depends upon quantities of reagents and reaction time, and that the quantities of both increase with time, but after a certain point the concentration of the 2,4-dilithio derivative increases at the expense of the 2,5-compound: the initial rate of production of the former is, however, much less than that of the latter (Chadwick and Willbe, J. chem. Soc., Perkin I, 1977, 877). Although 1-lithium salts and Grignard derivatives tend to undergo C-alkylation, it has been shown that in more polar solvents, for example, hexamethylphosphoramide, substitution occurs on the nitrogen, due to complexing of the metal by the solvent (N. Gjøs and S. Gronowitz, Acta Chem. Scand., 1971, 25, 2596). For photoreactions of 1-substituted pyrroles and 2-substituted 2H-pyrroles see R.L. Beine (Diss. Abs. Int. B, 1977, 37, 4458).

(iii) C-Alkyl- and C-aryl- pyrroles

2,3-Disubstituted pyrroles (2) may be prepared by condensing ketones or aldehydes containing an active methylene group, with glyoxal monohydrazones and reducing the product (1) with sodium dithionite. Product (1) on reduction with sodium tetrahydridoborate, followed by cyclization with acid gives the 2,3-disubstituted 1-aminopyrrole (3) (T. Severin and H. Poehlmann, Ber., 1977, 110, 491).

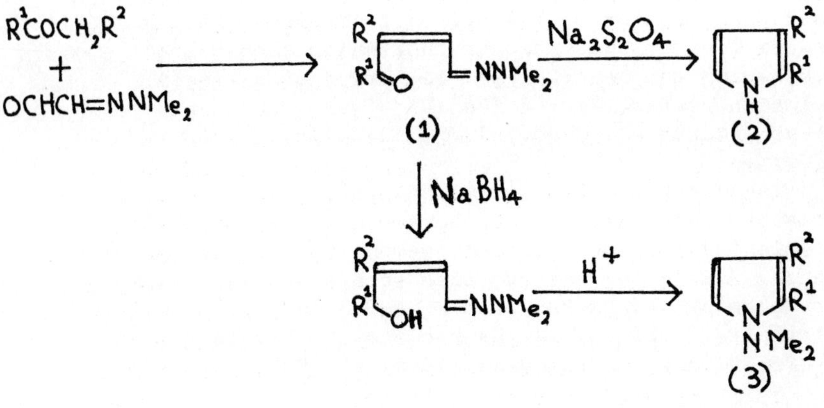

$$\begin{array}{c} \overset{1}{R}COCH_2\overset{2}{R} \\ + \\ OCHCH=NNMe_2 \end{array} \longrightarrow$$

(1)

$$\xrightarrow{Na_2S_2O_4}$$

(2)

$$\downarrow NaBH_4$$

$$\xrightarrow{H^+}$$

(3)

R^1 = 4-MeC$_6$H$_4$, 4-MeOC$_6$H$_4$, 3-pyridyl, R^2 = H;
R^1 = Et, R^2 = Me; R^1 = H, R^2 = Ph;
R^1 = Ph, R^2 = MeO; R^1R^2 = (CH$_2$)$_4$

The reaction between N-sulphinyl compounds (4) and 3,4-
-dimethylbuta-1,3-diene gives a cyclo adduct (5), which
on heating with sodium methoxide in methanol affords 3,4-
-dimethylpyrrole (K. Ichimura, S. Ichikawa, and K. Imamura,
Bull. chem. Soc. Japan, 1976, 49, 1157).

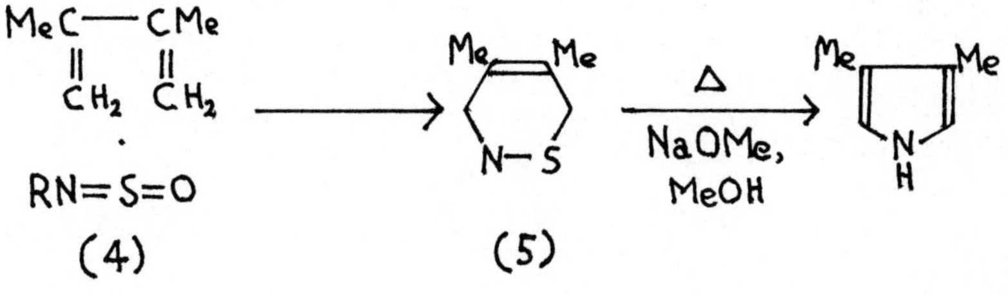

$$\begin{array}{c} MeC \!\!-\!\! CMe \\ \parallel \quad \parallel \\ CH_2 \quad CH_2 \\ \cdot \\ RN\!=\!S\!=\!O \end{array} \longrightarrow$$

(4)

$$\xrightarrow[\substack{NaOMe, \\ MeOH}]{\Delta}$$

(5)

R = CO$_2$Et, CO$_2$Ph,
PO(OEt)$_2$

Investigations to find the optimum conditions for C- and N-alkylation of the pyrrolyl ambient anion have been carried out and indicate that while almost total C-alkylation can be obtained, isolation of a single alkylation product is not feasible (Wang, Teo, and Anderson, Canad. J. Chem., 1977, 55, 4112).

The reaction of 2,5-dimethylpyrrole with chloroform and strong bases gives 2-dichloromethyl-2,5-dimethyl--pyrrolenine (6) and no rearrangement to a chloropyridine, however, with a very strong base such as butyllithium rearrangement occurs to yield 2- and 3-chlorodimethyl-pyridines (7) and (8) (A. Gambacorta, R. Nicoletti, and M.L. Forcellese, Tetrahedron, 1971, 27, 985).

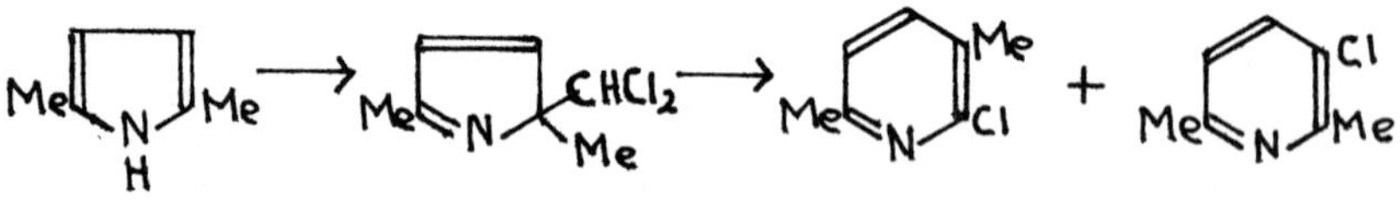

(6)   (7)   (8)

1-Alkoxycarbonyl and 1-alkyl-pyrroles react with benzynes to yield azanorbornadiene derivatives (9) (M.G. Barlow, R.N. Hazeldine and R. Hubbard, J. chem. Soc. C, 1971, 90).

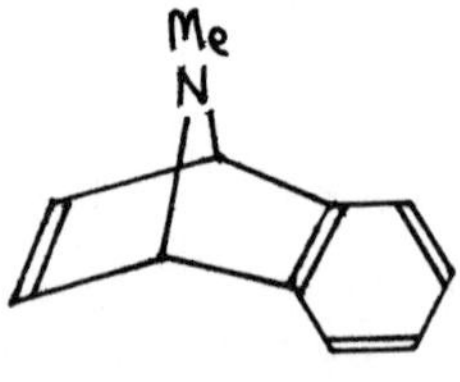

(9)

Pyrrole reacts with butadiene or isoprene in toluene in the presence of a palladium acetylacetonate-PPh$_3$-AlEt$_3$ catalyst at 100° to give the 2,5-disubstituted pyrrole

(10) (U.M. Dzhemilev, F.A. Selimov, and G.A. Tolstikov, Izvest. Akad. Nauk SSSR, Ser. Khim., 1979, 2652).

(10)

R = H, Me

(iv) Halogenopyrroles

Consideration has been given to the instability of 2-chloropyrrole and it has been shown that this can be mitigated by the preparation of N-alkyl derivatives and that due to electronic effects C-alkylation decreases the stability of 2-halogenopyrrole. 2-Chloro-1-methylpyrrole, b.p. 30-32°/10mm, is obtained along with 2,5-dichloro-1--methylpyrrole, b.p. 40-42°/8mm on reacting 1-methyl-pyrrole with sulphuryl chloride in ether. The following 1-alkyl-2-halogenopyrroles, 2-bromo-1-methyl-, 1-benzyl-2--chloro-, 1-benzyl-2-bromo-, 5-chloro-1,2-dimethyl-, and 5-chloro-1,2,3,4-tetramethyl-pyrrole, and the C-alkyl-2--halogenopyrroles, 5-chloro-2-methyl-, 5-chloro-2-*tert*--butyl-, 5-chloro-2,3,4-trimethyl-, and 5-bromo-2,3,4--trimethyl-pyrrole have been synthesised. Electrophilic substitution of 2-chloro- and 2-bromo-pyrrole under the conditions for formylation, *i.e.*, using phosphoryl chloride and dimethylformamide gives in the case of the former the *a*-substituted derivative, but the latter gives a product arising from the displacement of bromine, 5-chloropyrrole--2-carboxaldehyde, in addition to 5-bromopyrrole-2--carboxaldehyde. Diazo coupling of 2-chloro-1-methyl-pyrrole gives rise to exclusive *a*-substitution (G.A. Cordell, J. org. Chem., 1975, **40**, 3161).

2,3-Dichloropyrrole is obtained by boiling ethyl 2,3-dichloropyrrole-4-carboxylate, m.p. 141-144°, in

ethylene glycol containing sodium hydroxide and water.
The above ester is prepared by chlorinating ethyl pyrrole-
-2-carboxylic acid-3-carboxylate with chlorine in acetic
acid and boiling the resulting ethyl 4,5-dichloro-
-pyrrole-2-carboxylic acid-3-carboxylate, m.p. 193-195°,
with dimethylformamide (D.G. Durham, C.G. Hughes, and
A.H. Rees, Canad. J. Chem., 1972, 50, 3223).

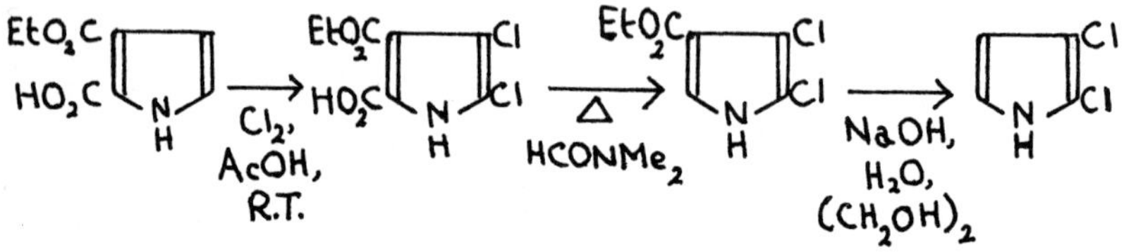

Deoxypyoluteorin (1) and pyoluteorin (2) have been
synthesized and a number of analogues made and their
microbiological activities compared. The monodeoxy series
is generally more active. Related dibromo and diiodo
derivatives have been prepared.

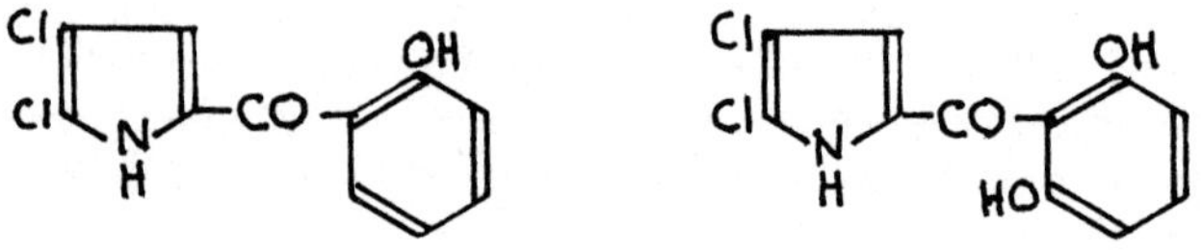

(1)   (2)

No ring-opening products are formed when 3-bromo-1-
-methyl-2,4-5-triphenylpyrrole is treated with butyllithium
in hexane and when water is added 1-methyl-2,4,5-triphenyl-
pyrrole is isolated in nearly quantitative yield
(T.L. Gilchrist and D.P.J. Pearson, J. chem. Soc., Perkin
I, 1976, 989).

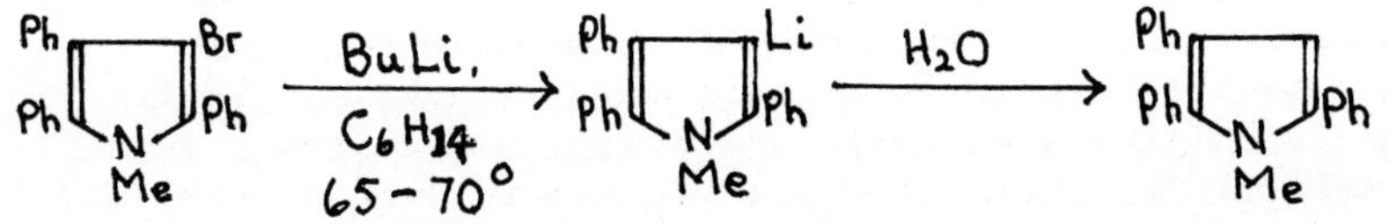

(v) Nitro-, nitroso-, diazo-, amino-, and cyano-pyrroles

The nitration of pyrrole has been investigated for a variety of reagents and over a wide temperature range. Nitration in acetic anhydride gives nitropyrroles in the ratio of 2-nitro:3-nitro of *ca.* 4:1 over a wide temperature range. It has been estimated that the reactivities of the 2- and 3- positions of the pyrrole ring are 130,000 and 30,000 respectively (with respect to benzene = 1) (A.R. Cooksey, K.J. Morgan, and D.P. Morrey, Tetrahedron, 1970, 26, 5101).

1-Methyl-3,4-dinitropyrrole reacts with sodium methoxide in boiling methanol to give *trans*-4,5-dimethoxy--1-methyl-3-nitro-2-pyrroline (1), m.p. 57-58.5°. The regiospecific acid-promoted decomposition of the 2-pyrroline (1), by treatment of the methanolic solution with one equivalent of trifluoracetic acid at room temperature yields 2-methoxy-1-methyl-4-nitropyrrole, formally the product of *cine*-substitution of 1-methyl-3,4--dinitropyrrole (2) (P. Mencarelli and F. Stegel, Chem. Comm., 1978, 564).

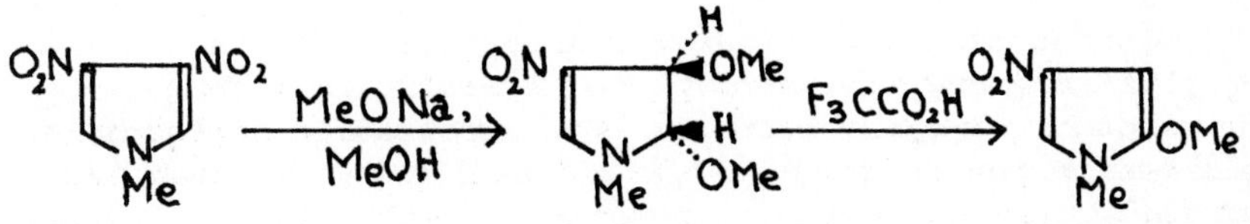

The absence of *cine*-substitution in a basic medium, as
opposed to the ready acid-promoted formation of (2),
indicates that the base-catalysed addition-elimination
mechanism envisaged for the formation of 2-arylthio-4-
nitrothiophenes from 3,4-dinitrothiophene is unlikely to
be operating in this reaction (C. Dell'Erba, D. Spinelli,
and G. Leandri, Gazz., 1969, 99, 535).

Simple 2-aminopyrroles are unstable but an electron-
withdrawing group confers stability. 1-Alkyl- and 1-aryl-
-2-amino-4-cyanopyrroles (4) are obtained by cyclization
of the appropriate aminomethylenesuccinononitrile (3),
in ethanol in the presence of potassium ethoxide
(A. Brodrick and D.G. Wibberley, J. chem. Soc., Perkin
I, 1975, 1910).

They undergo condensation with 1,3-dicarbonyl compounds
and their acetals, β-oxo-esters, and diethyl malonate in
the presence of hydrochloric acid to give 1H-pyrrolo-
[2,3-b]pyridines.

It has been shown that 2-cyanopyrrole and its
1-methyl analogue rearrange photochemically to the
corresponding 3-cyano isomers. The permutation patterns
followed in the phototransposition of 2-cyano to 3-cyano
pyrroles have been defined and show that the nitrogen atom
and C-5 interchange in the reaction. A mechanism has been
proposed involving initial 2,5-bonding to give (5),
followed by a 1,3-sigmatropic shift ('walk') of the
aziridine nitrogen atom to give (6), which can then
aromatize to the 3-cyano-isomer (J. Barltrop *et. al.*, Chem.
Comm., 1975, 786).

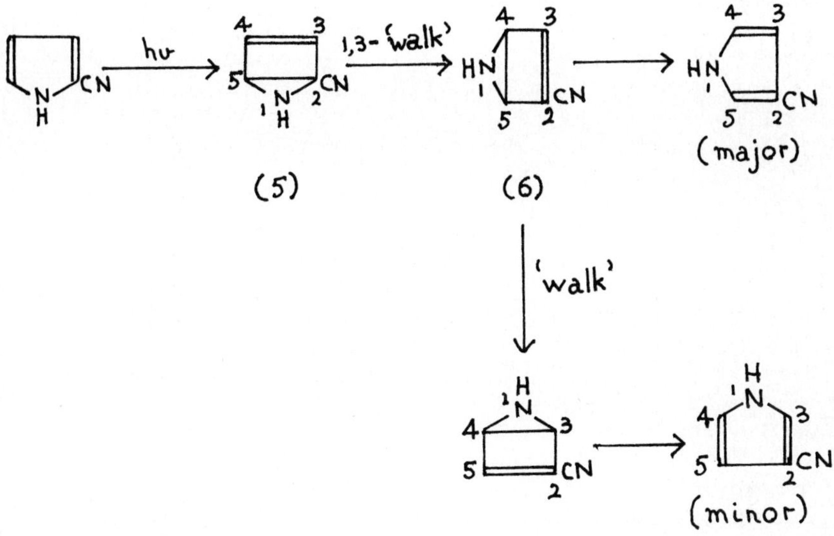

The proposed mechanism is supported by trapping experiments
with methanol and furan, which afford strong evidence for
the intermediacy of bicyclic species of the type (6)
(5-azabicyclo[2.1.0]pent-2-enes) in the rearrangement,
for instance, irradiation of 2-cyano-1-methylpyrrole in
methanol gives, besides the 3-cyano-isomer, a methanol
adduct (7), and 2-cyanopyrrole and 2-cyano-1-methylpyrrole
on irradiation in furan, besides the 3-cyano-isomers,
yield adducts (8) and (9) (Barltrop, A.C. Day and
R.W. Ward, *ibid.*, 1978, 131).

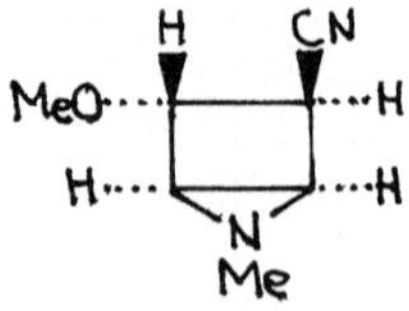

(7) 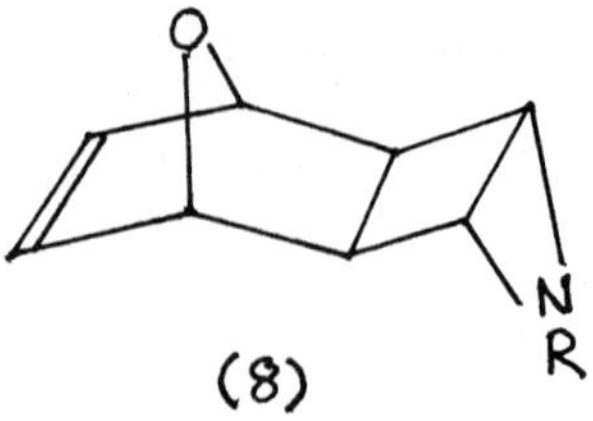   (8) 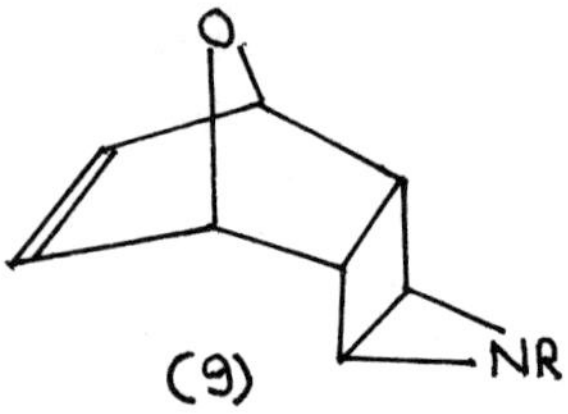   (9)

R = H, Me

For synthetic approaches to cyano-pyrroles and
1-methylpyrroles see G.H. Barnett, H.J. Anderson, and
C.E. Loader (Canad. J. Chem., 1980, 58, 409). Pyrrole
and 1-methylpyrrole are cyanated by the combined reagent,
triphenylphosphine-thiocyanogen in methylene chloride at
-40° to give preferentially the 2-cyano derivative.
If the 2-position is already substituted then the 3-cyano
derivative is formed (Y. Tamura *et. al.*, J. chem. Soc.,
Perkin I, 1980, 1132).

Modifications to the synthesis of porphobilinogen
(10), the biosynthetic precursor of porphyrins,
chlorophylls, and vitamin B12 have been described
(G.W. Kenner, K.M. Smith, and J.F. Unsworth, Chem. Comm.,
1973, 43).

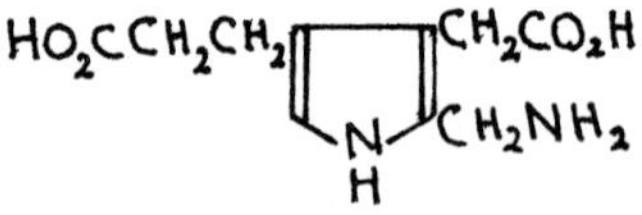

(10)

(vi) Aldehydes and ketones

Pyrrole-2-carboxaldehyde may be obtained by the formylation of pyrrole using triethyl orthoformate in the presence of trifluoroacetic acid (P.S. Clezy, C.J.R. Fookes and A.J. Liepa, Austral. J. chem., 1972, 25, 1979). 2,4-Dimethylpyrrole in ethanol reacts with carbon monoxide under pressure in the presence of sodium ethoxide to give 3,5-dimethylpyrrole-2-carboxaldehyde (A. Treibs and R. Wilhelm, Ann., 1979, 11).

Pyrrole-2-carboxaldehyde *via* its anion reacts with the bisphosphonium iodide (1), in he presence of sodium hydride in boiling xylene, to give pyrrolo[1,2-g]-5--azacyclo[3.2.2]azine (5). Initially the vinylphosphonium salt (2) is formed by a substitution-elimination process, further reaction with pyrrole-2-carboxaldehyde anion gives the pyrrolizine (3), which isomerizes to (4) and cyclizes to (5) (V.W. Flitsch and E.R. Gesing, Tetrahedron Letters, 1976, 1997).

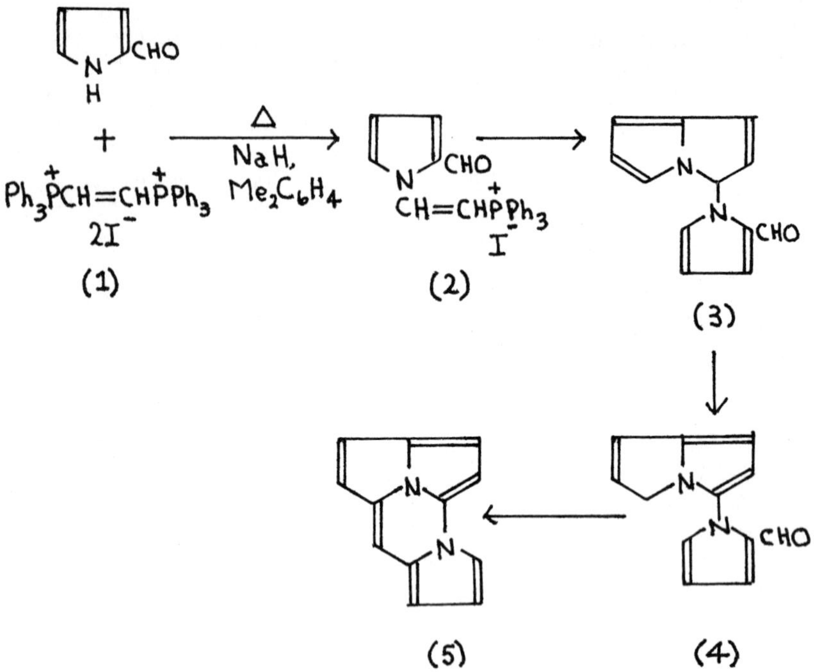

Further to the fact that direct acetylation of pyrrole gives 3-acetylpyrrole as the minor product, it has been found that trifluoroacetic anhydride catalyzed acetylation of 2-acetylpyrrole yields the hitherto unreported 2,4-diacetylpyrrole, m.p. 158–158.5° (A.G. Anderson, Jr. and M.M. Exner, J. org. Chem., 1977, 42, 3952). The photolysis of the acyloins obtained by reaction of biacetyl with pyrroles in the presence of formic acid, readily affords, in high yield 2-acetyl-pyrroles (H.-S. Ryang and H. Sakurai, Chem. Comm., 1972, 77).

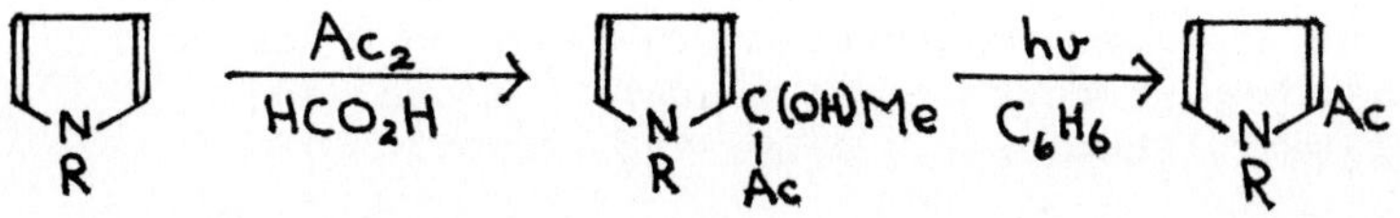

R = H, Me

N-Tosylmethylimino derivatives (6) react with Michael acceptors (7) to give 3-pyrryl phenyl ketones (8) (H.A. Houwing, J. Wildeman, and A.M. van Leusen, Tetrahedron Letters, 1976, 143).

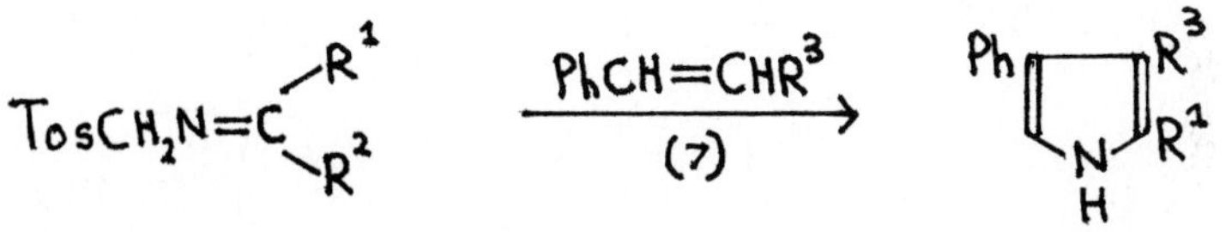

(6)   (8)

R^1 = Me, Ph, MeS, MeO R^3 = COPh
R^2 = MeS, MeO

If compounds (7) (R^3 = CO_2Me and CN) are used then methyl pyrrole-3-carboxylates (8) (R^3 = CO_2Me) and 3-cyanopyrroles (8) (R^3 = CN), respectively are obtained. The acylation of the pyrrolyl ambident anion has been investigated. The metal cation, solvent or complexing agent, halide of the pyrrole Grignard reagent, and temperature were varied. The reactions of acylating agents of the carbonyl, cyanide, and carbimine type with pyrrole Grignard reagent have also been studied to determine N/C acylation ratios under the same condition (N.-C. Wang, and H.J. Anderson, Canad. J. Chem., 1977, 55, 4103).

A number of substituted pyrroles are acetylated in methylene chloride at position-1, with acetic anhydride

and triethylamine in the presence of 4-(dimethylamino)-
pyridine at room temperature. There is no trace of
C-acetylated product (K. Nickisch, W. Klose, and
F. Bohlmann, Chem. Ber., 1980, __113__, 2036).

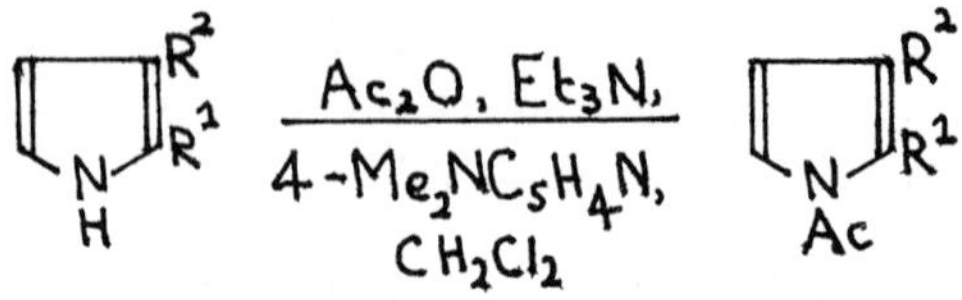

R^1 = CHO, R^2 = H , CO_2Et;
R^1 = Me, R^2 = CO_2Et

For 3,4-diacetyl-2,5-dimethylpyrrole, m.p. 178-179°,
see (5), p. 4.

The Vilsmeier-Haack acylation of pyrroles has been
re-examined and it has been found that despite the
widespread use of N,N-dimethylamides, morpholides are
better reagents except when the phenyl nucleus carried
a strongly electron withdrawing group such as a nitro
group, which renders reaction with phosphoryl chloride
incomplete. In such cases, the dimethylamide analogues
give better results (J. White and C. McGillivray, J. org.
Chem., 1977, __42__, 4248).

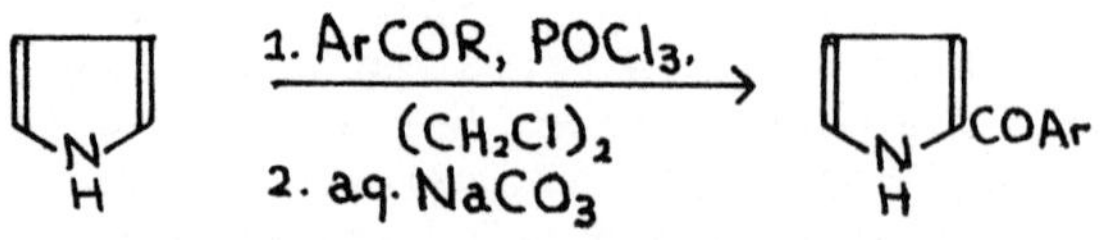

Ar = $4-NO_2-$, $-Cl-$, $-Me-$, $-MeO-C_6H_4$, Ph
R = Me_2N, $-CH_2CH_2OCH_2CH_2-$

2-Benzoyl-, m.p. 77.5-78°, 2-(4-nitrobenzoyl)-, m.p.
160-162°, 2-(4-chlorobenzoyl)-, m.p. 118.5-119.5°,
2-(4-methylbenzoyl)-, m.p. 118-119°, 2-(4-methoxy-
benzoyl)pyrrole, m.p. 112.5-113.5°.

The reaction of pyrrolylmagnesium iodide with
phthalic anhydride affords a mixture of 2- and 3-
(2-carboxybenzoyl)pyrrole, and acylation of pyrrole with
succinic anhydride gives γ-(2-pyrrolyl)oxobutyric acid
(35%) (V.N. Eraksina *et. al.*, Chem. Abs., 1979, 91,
175117z). The reaction between 2-chloroacetyl pyrrole and
triethyl phosphite affords only β-ketophosphonate esters
(Arcoria *et. al.*, J. heterocyclic Chem., 1975, 12, 215).

(vii) Pyrrolecarboxylic acids

Pyrrole esters are most generally prepared by ring
synthesis, but esterification of pyrrole acids with
reagents other than diazomethane frequently proves
troublesome. It has been shown that trichloroacetylation
of pyrroles using trichloroacetyl chloride and appropriate
bases as acid scavengers, for example, 2,6-lutidine, and
subsequent treatment with alcohol and mild base, is a
convenient method for directly introducing an ester group
into the pyrrole nucleus. This method has been applied
to alkylpyrroles, α-ethoxycarbonylpyrroles, and
dipyrrolylmethanes (J.W. Harbuck and H. Rapoport, J. org.
Chem., 1972, 37, 3618). Alkaline hydrolysis of
2-trihalogenoacetylpyrrole gives pyrrole-2-carboxylic acid
(S. Clementi and G. Marino, Tetrahedron, 1969, 25, 4599);
J. chem. Soc., Perkin II, 1972, 71; Gazz., 1970, 100,
556).

For the preparation of dimethyl 5-phenylpyrrole-2,3-
dicarboxylate, m.p. 140° see (1) p.1; for dimethyl
2,5-dimethylpyrrole-3,4-dicarboxylate, m.p. 120-121°,
dimethyl 1,2,5-trimethylpyrrole-3,4-dicarboxylate, m.p.
149.5-150°, dimethyl 1-benzyl-2,5-dimethylpyrrole-3,4-
-dicarboxylate, m.p. 105-106°, and dimethyl 2,5-dimethyl-
-1-phenylpyrrole-3,4-dicarboxylate, m.p. 86-87°, see
(5) p.4; for dimethyl 1-cyclohexyl-2-phenylpyrrole-3,4-
-dicarboxylate, oil, amide, m.p. 131-132.5°, dimethyl-1-
-cyclododecyl-2-phenylpyrrole-3,4-dicarboxylate, m.p. 106-
107.5°, amide, m.p. 142-143.5° and dimethyl 2-phenyl-1-
-*tert*-butylpyrrole-3,4-dicarboxylate, m.p. 92-93°, amide

m.p. 135-136.5°, see (6) p. 5. Ethyl 2,3-dichloro-
pyrrole-4-carboxylate, m.p. 141-144°, and ethyl
4,5-dichloropyrrole-2-carboxylic acid-3-carboxylate,
m.p. 193-195° have been prepared (p. 20) (Durham, Hughes,
and Rees, *loc. cit.*).

The reaction of pyrrole with ethoxycarbonyl
isocyanate in toluene yields N-ethoxycarbonylpyrrole-2-
-carboxamide (1), which is readily hydrolyzed to either
pyrrole-2-carboxamide or pyrrole-2-carboxylic acid and
cyclized to pyrrole-1,2-dicarboximide (2). Pyrrolyl-
potassium reacts with ethoxycarbonyl isocyanate in tetrahydrofuran to give, after acidification *N*-ethoxycarbonyl-
pyrrole-1-carboxamide (3) (E.P. Papadopoulos, J. org.
Chem., 1972, **37**, 351).

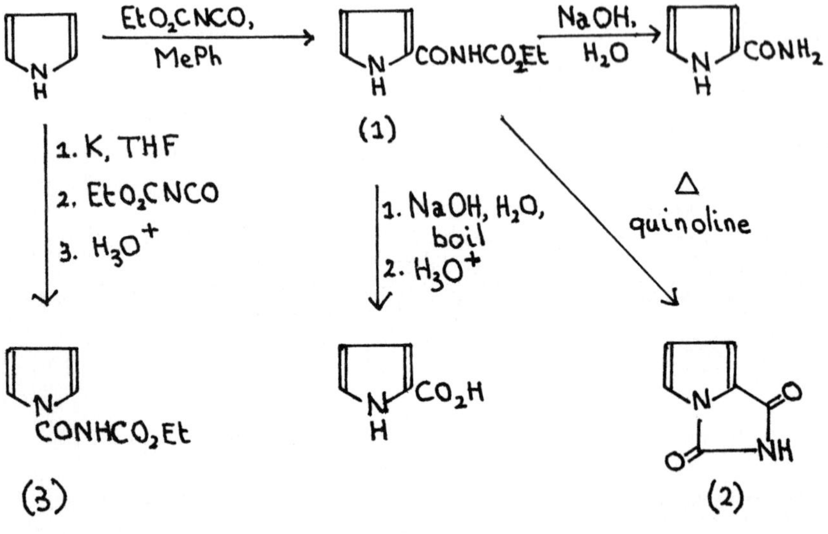

2. *Pyrrolines, Dihydropyrroles*

(a) *1-Pyrrolines*

The reaction of pyrrole with 5-(2-pyrrolyl)pyrrolidin-
-2-one in tetrahydrofuran, in the presence of phosphoryl
chloride yields 2,5-di(2-pyrrolyl)-1-pyrroline (1) m.p.
140-141° and a small amount of 2,5-di(2-pyrrolyl)-
pyrrolidine (G.P. Gardini and V. Bocchi, Ateneo Parmense,
Sez. 1, Suppl., 1968, 39, 22):

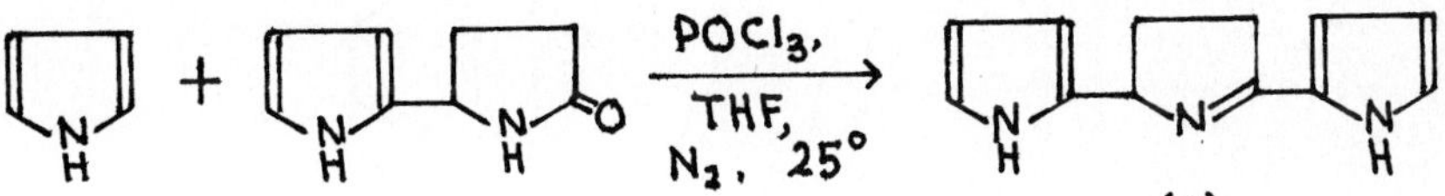

$$(1)$$

2,4-Dimethylpyrrole and 5-(2-pyrrolyl)pyrrolidin-2-one
give 2-(2,4-dimethyl-5-pyrrolyl)-5-(2-pyrrolyl)-1-pyrroline,
m.p. 168-170°.

(b) *2-Pyrrolines*

Reduction of 1-methylpyrrolidin-2-one using lithium
tetrahydridoaluminate or sodium bis-(2-methoxyethoxy)-
aluminium hydride gives 1-methyl-3-(1-methylpyrrolidin-2-
-yl)-2-pyrroline (1) and 1-methyl-3-[1-methylpyrrolidin-2-
-yl)pyrrolidin-2-yl]-2-pyrroline (2).

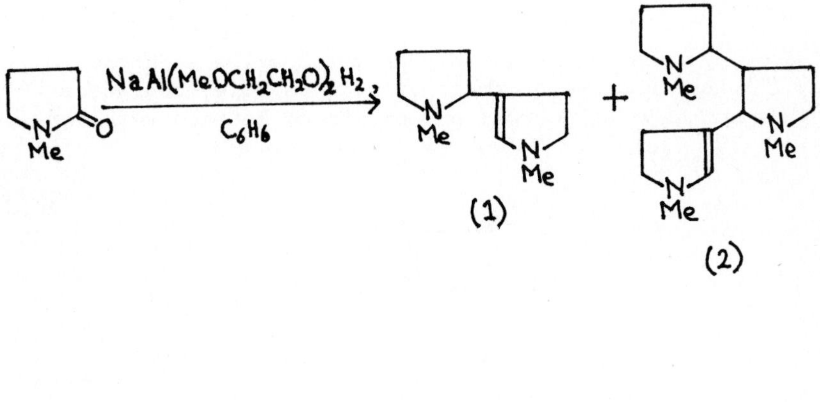

Reduction of 1-*p*-tolypyrrolidin-2-one (3) yields a mixture of two stereoisomeric forms of the dipyrrolo[1,2-a:3',2'- -c]quinoline (4) (G.A. Swan and J.D. Wilcock, J. chem. Soc., Perkin, I, 1974, 885):

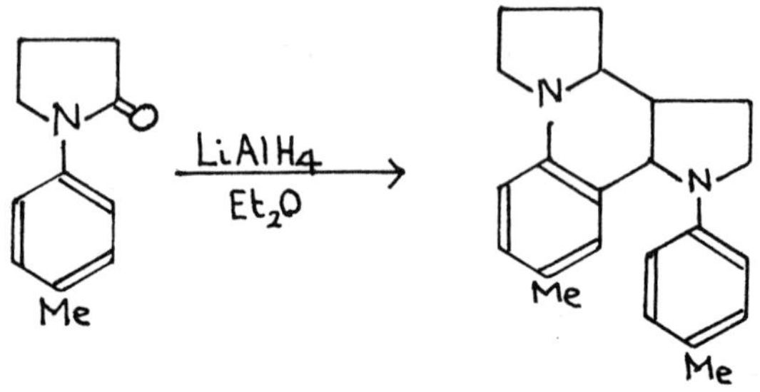

(3)   (4)

The cyclisation of acylsuccinic esters with ammonia or primary amines leads to 2-pyrrolin-5-ones (M. Pesson *et. al.*, Comp. rend., 1971, 272C, 478).

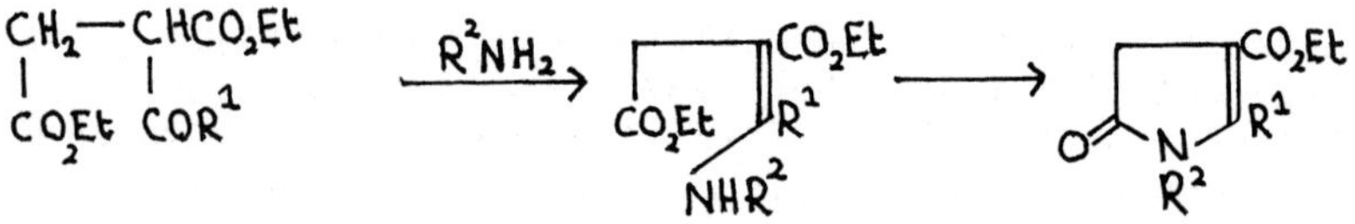

3-Hydroxypyrroles exist largely in the 2-pyrrolin-4-one form (5), but because of their vinylogous amide character they do not show the normal properties of ketones (H.Bauer, Ann., 1970, 736, 1).

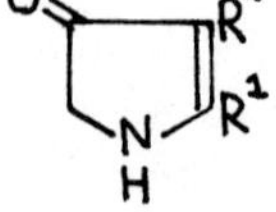

(5)

(c) 3-Pyrrolines

The reaction between imines of α-amino acid esters
and dimethyl acetylenedicarboxylate in toluene at 110°
affords 3-pyrrolines (R. Grigg *et. al.*, Chem. Comm., 1978,
109).

$$R^1CH=NCH(R^2)CO_2Me \xrightarrow[\substack{MePh \\ 110°}]{MeO_2CC\equiv CCO_2Me,}$$

Trimethyl 2,5-diphenyl-3-pyrroline-2,3,4-tricarboxylate
($R^1 = R^2 = Ph$), m.p. 135-137°.

The photo-oxidation of 3,4-diethylpyrrole in methanol
yields mainly diethylmaleimide and 3,4-diethyl-5-methoxy-3-
-pyrrolin-2-one. Under similar conditions the photolysis
of pyrrole gives mainly 5-methoxy-3-pyrrolin-2-one with
a low yield of maleimide, but in aqueous solution, only
5-hydroxy-3-pyrrolin-2-one is formed and no maleimide
(G.B. Quistad and D.A. Lightner, Chem. Comm., 1971, 1099).
The photo-oxygenation of 3,4-diethyl-2-methylpyrrole gives
3,4-diethyl-5-methoxy-5-methyl-3-pyrrolin-2-one (11%),
3,4-diethyl-5-hydroxy-5-methyl-3-pyrrolin-2-one (40%),
m.p. 121-123°, and diethylmaleimide, which involves
demethylation of the α-methyl substituent *(idem,*
Tetrahedron Letters, 1971, 4417; Lightner, G.S. Bisacchi,
and R.D. Norris, J. Amer. chem. Soc., 1976, 98, 802).
For 3-pyrrolin-2-one see p. .

3. Pyrrolidines, Tetrahydropyrroles

The reaction between butan-1,4-diol and ammonia
in the presence of 5% ferric oxide — bentonite gives
pyrrolidine (47.8%), tetrahydrofuran (1.2%), and but-3-en-
-1-ol (1.9%) (Kh. I. Areshidze and G.O. Chivadze, Chem.
Abs., 1973, 79, 126213w.

Imines of α-amino acid esters undergo cycloaddition
reactions with acrylonitrile and ethyl acrylate to give
pyrrolidine derivatives (1) and (2). Adducts (1) are

produced as single diastereoisomers and adducts (2) show clear evidence through nmr spectral data of being mixtures (Grigg *et. al., loc. cit.*).

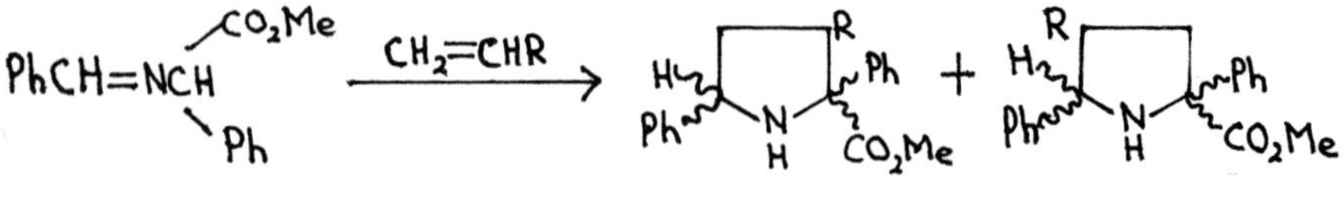

(1)   (2)

R = CN, CO$_2$Et

Gamma radiolysis of, or reaction of either *tert*--butyoxy-radicals or dibenzoyl peroxide with 1-phenyl-pyrrolidine gives two stereoisomeric forms of 2,3,3a,3b,4,5,6,11b-octahydro-1-phenyl-1H-dipyrrolo[1,2-a:-3',2'-c]quinoline (5), which presumably arises by dimerization of 1-phenyl-2-pyrroline (4), formed by disproportionation of 1-phenylpyrrolidin-2-yl radicals (3) (D.K. Ghanshyam, Swan, and R.B. Roy, J. chem. Soc., Perkin I, 1974, 891).

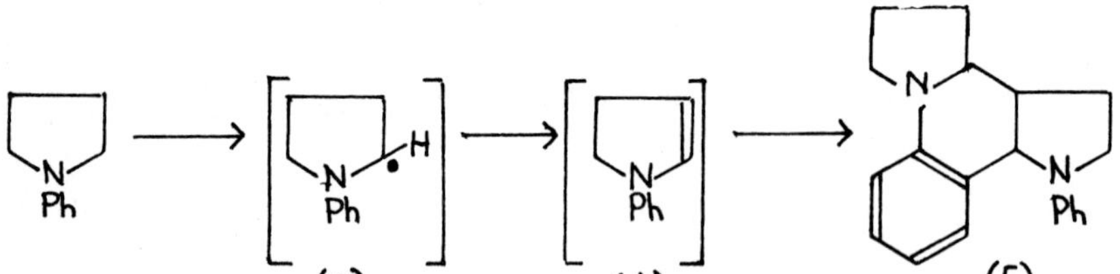

(3)   (4)   (5)

For the reduction of 1-methylpyrrolidin-2-one see p.31. Thermolysis of the adduct (6) formed from 1-phenylpyrrolidine and diethyl azodicarboxylate leads to a mixture of isomeric dimers same as (5) above (G.H. Kerr *et al., ibid.,* p.1614).

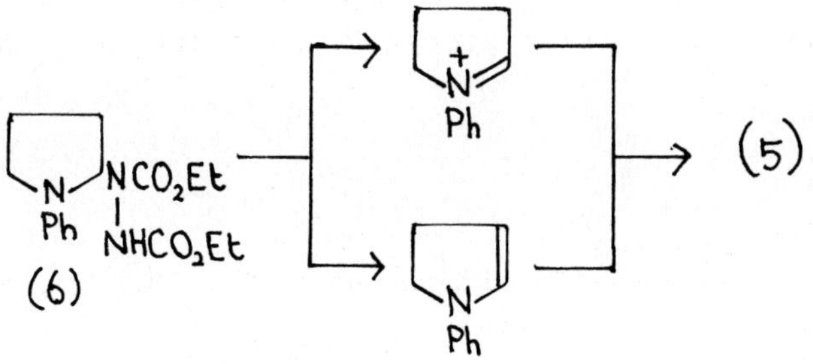

Similar dimers are obtained by the reduction of 1-p-tolyl-pyrrolidin-2-one see p. 32.

(1-Aminopyrrolidin-2-yl)methyl methyl ether (7) has been used as a reagent in a regiospecific and enantioselective aldol reaction. It first transforms the ketone into the chiral hydrazones (8), which are metallated with n-butyllithium in tetrahydrofuran, treated with carbonyl compounds and the resulting adducts silyated by chlorotrimethylsilane to form the doubly protected ketols (9). Oxidative hydrolysis of the protected ketols (9) yield the chiral ketols (10) (H. Eichenauer *et. al.*, Angew. Chem. internat. Edn., 1978, 17, 206):

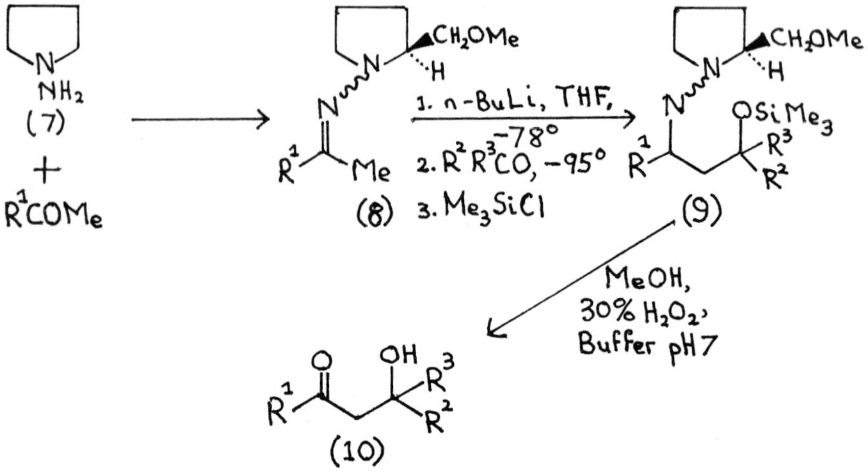

L-proline may be conveniently synthesized from L-pyroglutamic acid (5-oxopyrrolidine-2-carboxylic acid) (H.J. Monteiro, Synth., 1974, 137).

The properties and reaction of pyrrolidin-2,5-dione (succinimide) have been reviewed (M.K. Hargreaves, J.G. Pritchard, and H.R. Dave, Chem. Rev., 1970, 70, 439). 7-Azabicyclo[2.2.1]heptane (11) has been obtained starting from 4-acetylaminophenol (R.R. Fraser and R.B. Swingle, Canad. J. Chem., 1970, 48, 2065).

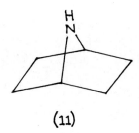

(11)

4. *Compounds Having Two Or More Independent Five-Membered Rings*

(a) Bipyrrolyls

Raney nickel reduction of the diesters of dipyrrolo-thiophenes results in desulphurization and production of bipyrrolyl diesters, which on hydrolysis and decarboxylation afford the related 2,2'-, 2,3'- and 3,3'bipyrrolyls (M. Farnier, S. Soth, and P. Fournari, Canad. J. Chem., 1976, <u>54</u>, 1083, A. Gossauer 'Die Chemie der Pyrrole', Springer, Verlag. Berlin, 1974);

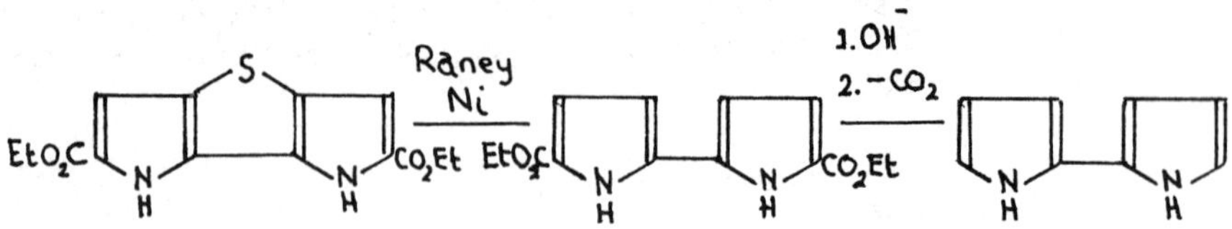

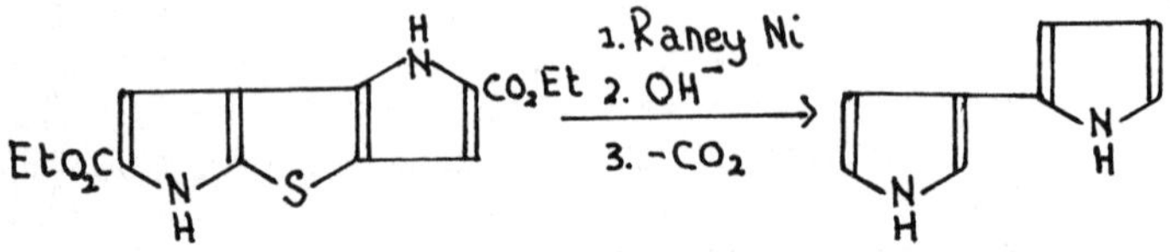

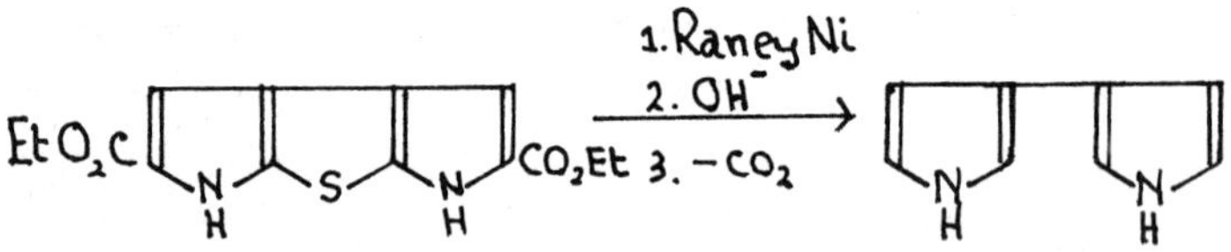

(b) *Compounds having two or more pyrrole rings linked through methylene or methine groups*

(i) *Dipyrrolymethanes*

2-Acetoxymethylpyrroles on treatment with 2-unsubstituted pyrroles in either methanol acetic acid containing a catalytic amount of p-toluenesulphonic acid hydrate, affords the corresponding dipyrrolylmethane. The reaction in methanol may be modified to give tripyrranes, for example (1) (J.A.S. Cavaleiro *et. al.*, J. chem. Soc., Perkin I, 1973, 2471).

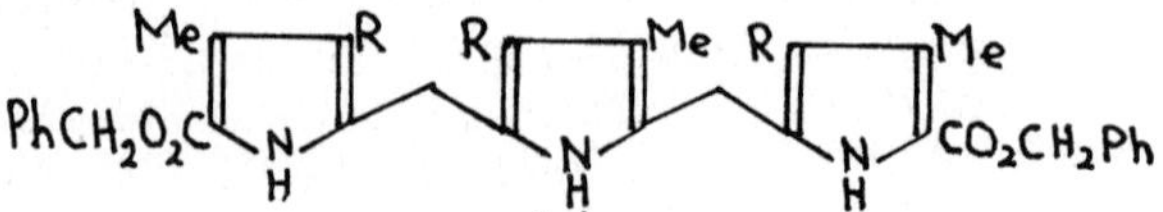

(1)

$$R = (CH_2)_2CO_2Me$$

Some 2-aminomethyldipyrrolylmethanes have been synthesized
(A.R. Battersby, J.F. Beck, and E. McDonald, *ibid.*, 1974,
160; A. Valasinas, E.S. Levy, and B. Frydman, J. org.
Chem., 1974, <u>39</u>, 2872).

5,5'-Diformyldipyrrolylmethanes on treatment with
bromine followed by sulphuryl chloride yield 5,5'-
-diformyldipyrrolyl ketones (P.S. Clezy *et. al.*, Austral.
J. Chem., 1970, <u>23</u>, 589).

(ii) Pyrromethenes

Bromination of the pyrrolecarboxylic acid (2) gives
the pyrrolylmethene hydrobromide (3) (A.H. Jackson,
G.W. Kenner, and J. Wass, J. chem. Soc., Perkin I, 1972,
1475).

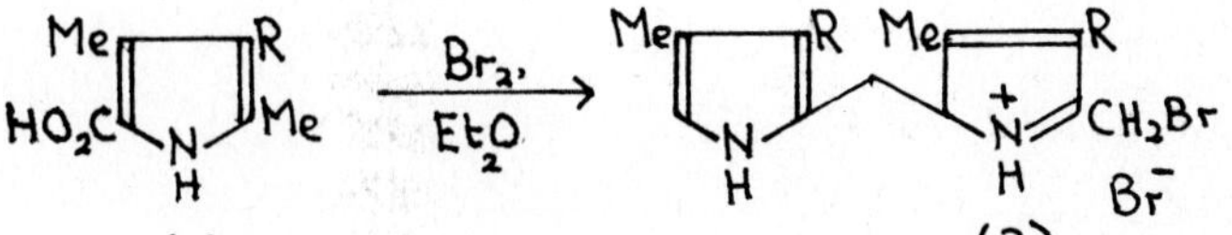

$$R = (CH_2)_2CO_2Et$$

(iii) Prodigiosins

The basic pyrrolyl-dipyrrolylmethene unit (4)
contained in the prodigiosins has been synthesized
(W.R. Hearn *et. al.*, J. org. Chem., 1970, <u>35</u>, 142).

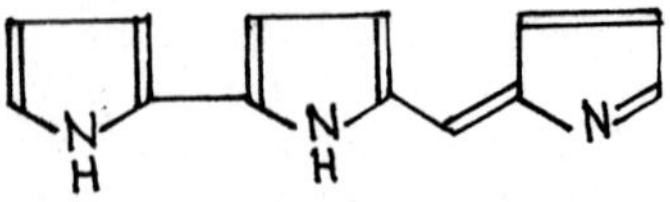

(4)

Prodigiosins additional to those described in Vol. IV A
include, nonylprodigiosin and a 2,10-nonamethylene-
prodigiosin from *Actinomadura madural* (N.N. Gerber, Appl.
Microbiol., 1969, 18, 1; Tetrahedron Letters, 1970, 809)
and a 2,10(1'-methyldecamethylene)prodigiosin from
Actinomadura pelletieri (idem, J. Antibiot., 1974, 24,
636).

Chapter 5

COMPOUNDS CONTAINING FIVE-MEMBERED RINGS WITH ONE HETERO
ATOM FROM GROUP V: NITROGEN: FUSED RING COMPOUNDS

R. LIVINGSTONE

1. *Indoles (2,3-Benzopyrroles) and 3H-Indoles*

A number of new methods of synthesizing indoles have
appeared, along with modifications and further
investigations of well established methods. A number of
new derivatives have also been described.

(a) General synthetic methods[1]

(i) The Fischer indole synthesis

It is generally assumed that in Fischer the indole
synthesis the last intermediate in the mechanistic sequence
is a 2-aminoindoline, which eliminates an amine to form the
indole. It has now been shown that 1-acetyl-2-(2-
-toluidino)indoline eliminates 2-toluidine to form
1-acetylindole, a reaction equivalent to the suggested last
step of the Fischer synthesis (T.P. Forrest and
F.M.F. Chen, Chem. Comm., 1972, 1067).

[1] R.J. Sundberg, "The Chemistry of Indoles", Academic
Press, New York, 1970; R.K. Bown, "Indole", Part I,
W.J. Houlihan, Ed., Wiley-Interscience, New York, 1972;
G.I. Zhungietu, V.A. Budylin, and A.N. Kost, "Preparative
Chemistry of Indole". Shtiintas: Kishinev, Mold. SSR,
1975.

The migration of a methyl group, involving a double 1:2 shift has been observed during the indolization of the hydrazone (1), which yields the indole derivative (2) (R. Fusco and F. Sannicolò, Tetrahedron Letters, 1975, 3351).

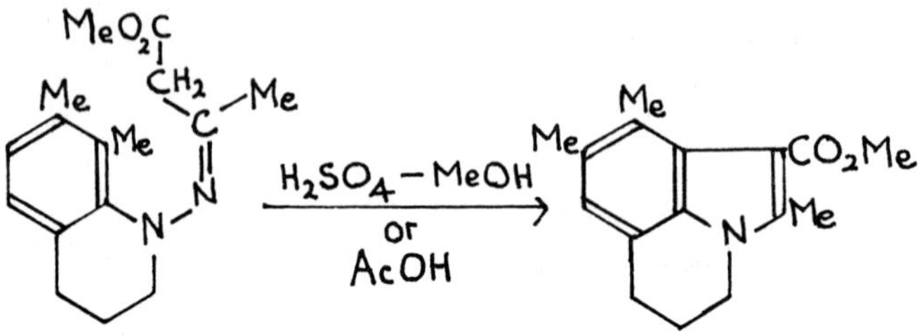

(1) $H_2SO_4 - MeOH$ or AcOH   (2)

Evidence has been reported that the overall [1,4] methyl migration in Fischer indole rearrangements previously described, actually proceeds by a series of [1,2] shifts (B. Miller and E.R. Matjeka, *ibid.*, 1977, 131). Stereochemical effects on the direction of cyclization in the Fischer indole synthesis have been discussed (R.A. Lohr, Jr., Diss. Abs. Int. B, 1978, **38**, 3697).

(ii) Other syntheses of indoles

(1) Indoles are otained by treating 2-acylanilines in ether-tetrahydrofuran with diazomethane and magnesium iodide or lithium iodide or with dibromomethane and lithium-mercury (M. Panunzio, N. Tangari, and U. Umani Ronchi, Chem. Comm., 1972, 415).

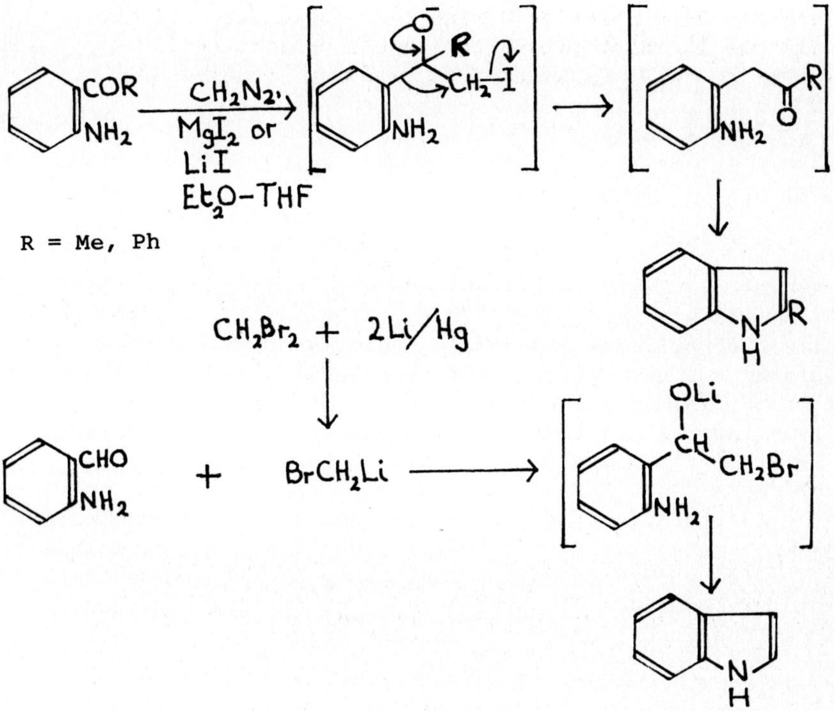

R = Me, Ph

(2) The addition of *tert*-butyl hypochlorite to an
arylamine -65° followed by methylthiopropan-2-one at
-65° with the mixture being allowed to rise to room
temperature after the addition of triethylamine gives an
indole derivative. Reduction with Raney nickel at 25°
removes the thiomethyl group. (P.G. Gassman and
T.J. van Bergen, J. Amer. chem. Soc., 1973, 95, 590).

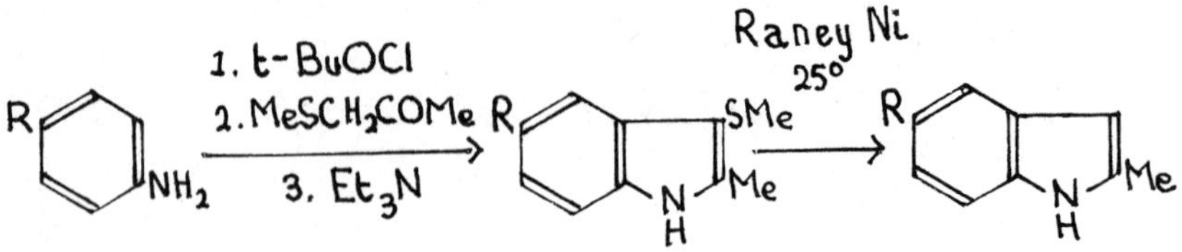

R = Me, H, Cl, CO_2Et, OCOMe

2-Alkyl- and 2-aryl- indoles have also been synthesized
by the above method (*idem*, *ibid.*, p.591) and it is stated
that it offers many advantages over the Fischer method
(Gassman *et. al.*, *ibid.*, 1974, **96**, 5495). The method has
been modified to allow the synthesis of 3-alkyl- and 3-
-aryl- indoles and thus 2,3-disubstituted indoles (Gassman,
D.P. Gilbert and van Bergen, Chem. Comm., 1974, 201).

(3) Treatment of the *N*-nitrosomethylaminobenzophenone
(1) with potassium *tert*-butoxide gives the nitrosoindoline
(2), which on hydrogenation over Raney nickel yields the
3-hydroxyindoline (3), stable under neutral or alkaline
conditions, but converted with catalytic amounts of acid
into 5-chloro-3-phenylindole (A. Walser and G. Silverman,
J. heterocyclic Chem., 1973, **10**, 883).

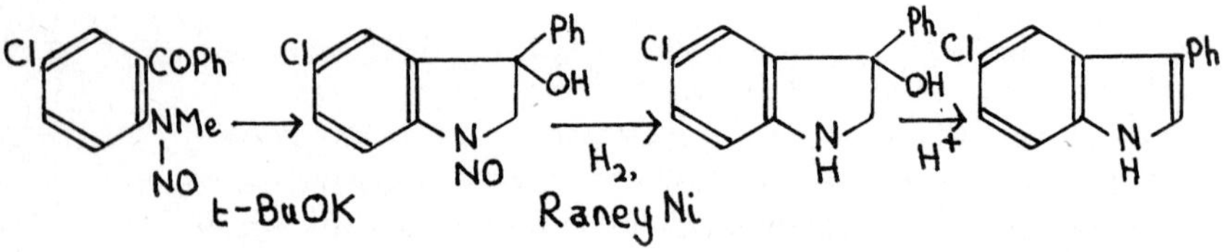

(1)   (2)   (3)

(4) The reaction of the substituted vinyl bromide
(4) containing a dimethylamino group at the *ortho* position
of the β-aryl substituent, in 80% ethanol affords a 2,3-
-diaryl-1-methylindole (T. Sonoda *et. al.*, Chem. Comm., 1976,
612).

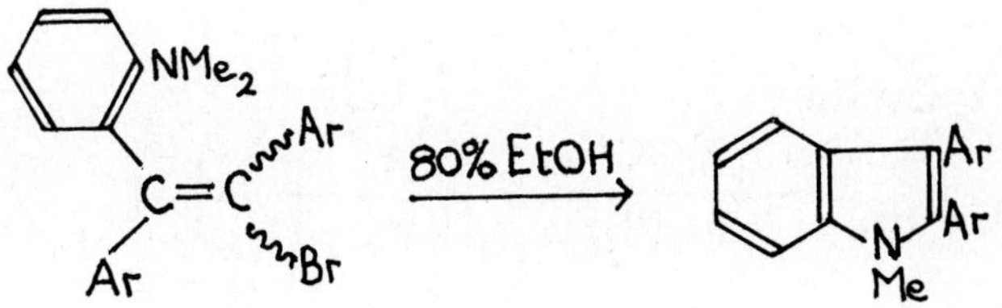

(4)

$$Ar = 4\text{-}MeOC_6H_4$$

(5) Indoles may be obtained directly in high yields from the appropriate potassium enolate in liquid ammonia and 2-iodoaniline and irradiating for a short time (R. Beugelmans and G. Roussi, *ibid.*, 1979, 950).

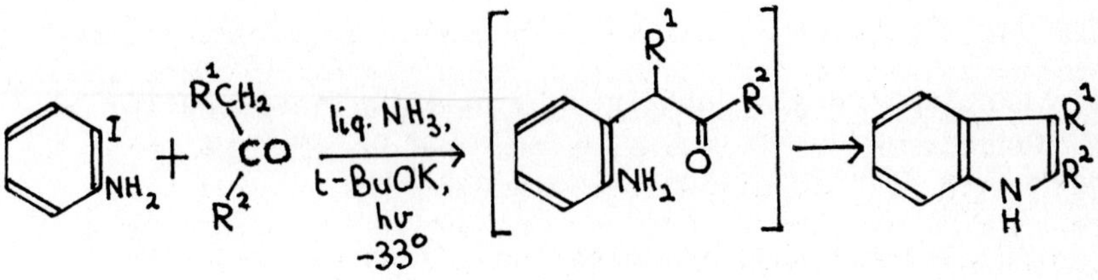

$$R^1 = H, \quad R^2 = H, \text{ Me, } Pr^i; \quad R^1R^2 = -(CH_2)_4-$$

2-Amino-1-(2- or 3- bromoaryl) ethanols (5 or 6) on treatment with potassamide in liquid ammonia give indoles (I. Fleming and M. Woolias, J. chem. Soc., Perkin I, 1979, 827).

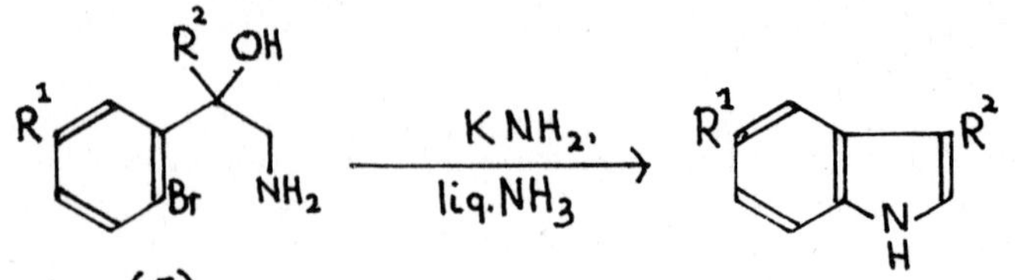

(5)

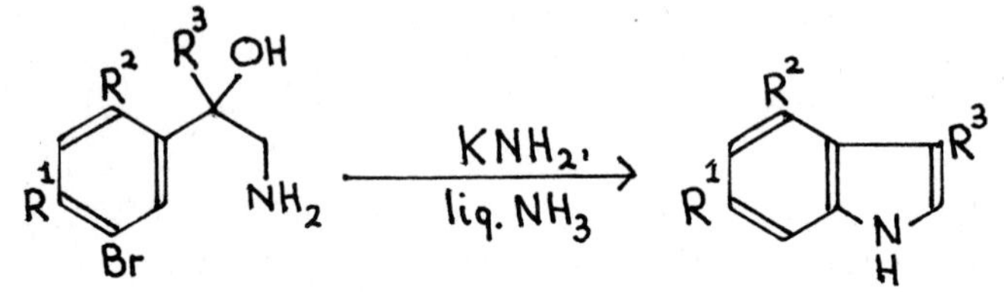

(6)

Indoles are also obtained by heating 2-amino-1-(2-
-bromophenyl)ethanols (5) in a solution of ammonia in
methanol to 140-170° (*idem, ibid.*, p.829).

(6) The *NN'*-diarylformamidine (7) on heating with
2-toluidine and sodamide gives the indole (8) (A. Tamaoki,
K. Yamamoto, and K. Maeda, Japan Kokai Tokkyo Koho, 80
49,353/1980).

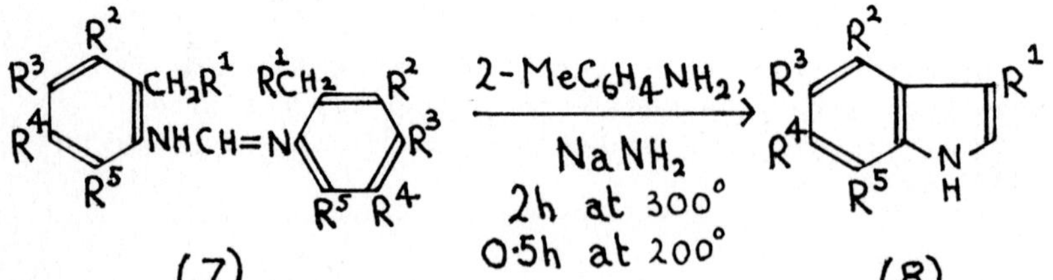

(7)   2-MeC$_6$H$_4$NH$_2$,
        NaNH$_2$
        2h at 300°
        0.5h at 200°   (8)

R^1-R^4 = H, alkyl;   R^5 = H, CH$_2$R

(b) *General properties and reactions of indoles*

(i) *Reduction*

Stereospecific hydrogenation of 2,3-disubstituted indoles occurs with sodium in liquid ammonia to give cis addition and borane in tetrahydrofuran provides trans addition (J.G. Berger *et. al.,* Tetrahedron Letters, 1975, 1807). Sodium tetrahydridoborate reduction of indole in neat carboxylic acid results in reduction of the double bond and alkylation at the nitrogen atom (G.W. Gribble *et. al.,* J. Amer. chem. Soc., 1974, **96**, 7812).

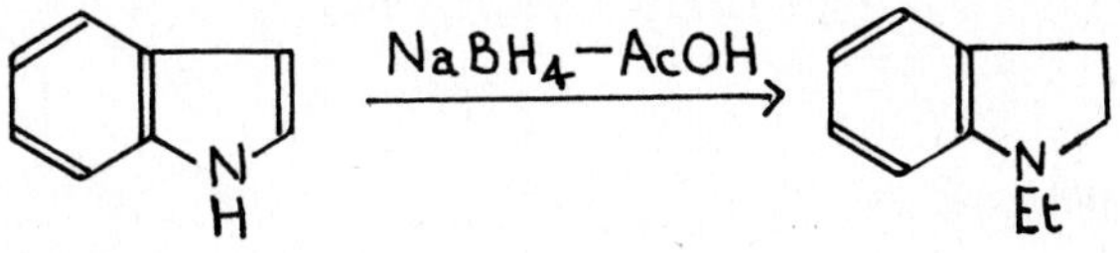

The yield of 4,7-dihydroindole by the Birch reduction of indole has been increased (J.W. Ashmore and G.K. Helmkamp, Org. prep. Proced. Int., 1976, **8**, 223).

(ii) *Addition reactions*

Indole gives 1:1, (1) and (2), and 1:2 adducts (3) and (4) with 1,4-benzoquinone and 1,4-naphthoquinone (K.K. Prasad, Tetrahedron Letters, 1974, 1361).

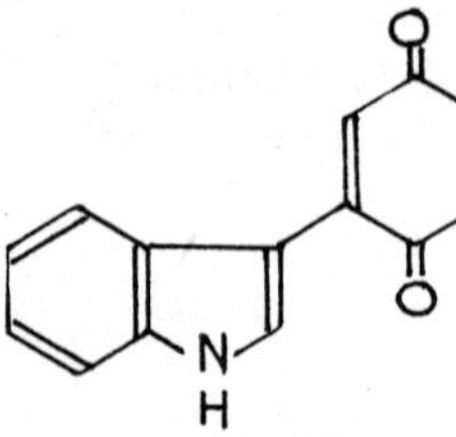

(1)

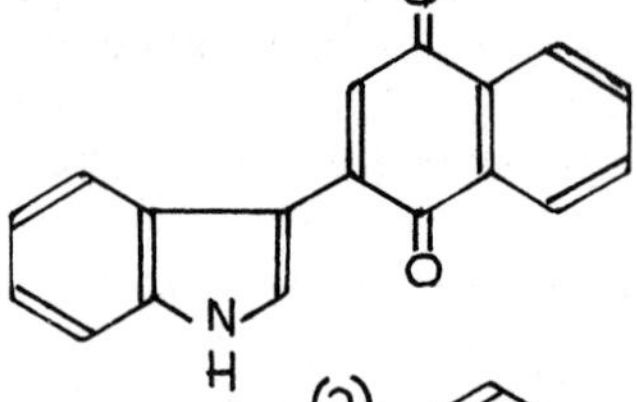

(2)

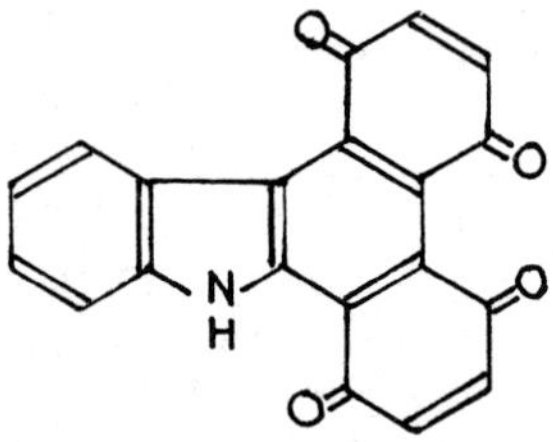

(3)

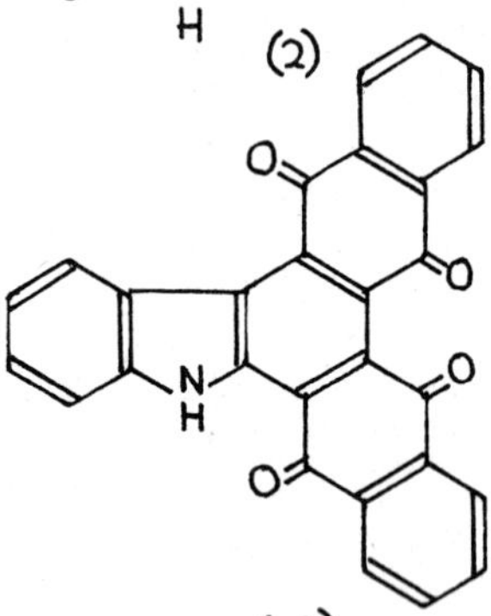

(4)

7,7,8,8-Tetracyanoquinodimethane with indole in
acetonitrile affords an adduct (5), which on irradiation
in acetonitrile gives the trinitrile (6). 2-Methylindole
behaves in a similar manner (B.P. Bespalov, E.V. Getmanova,
and V.V. Titov, *ibid.*, 1976, 1867).

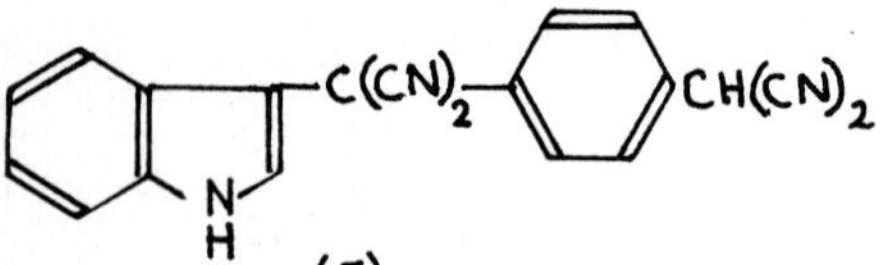

(5)

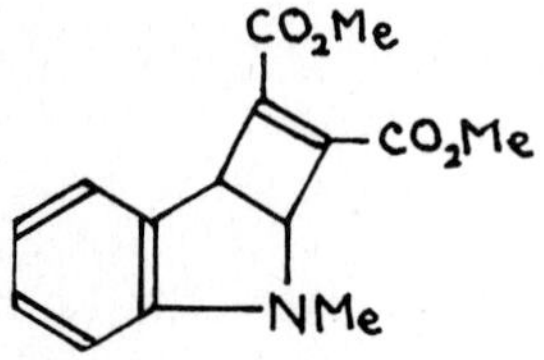

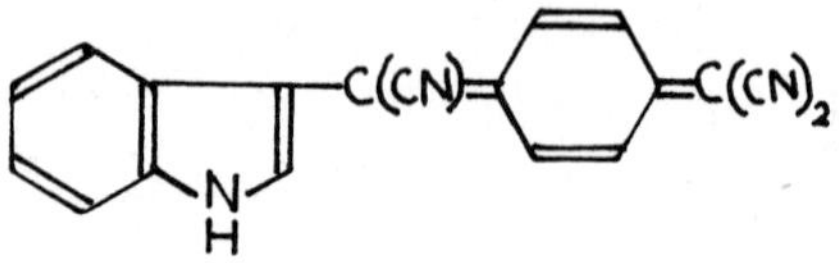

(6)

(7)

Photocycloaddition of dimethyl acetylenedicarboxylate to
1-methylindoles yields cyclobutene derivatives, for
example, 9,10-bis(methoxycarbonyl)-2-methyl-3,4-benzo-2-
-azatricyclo[3.3.1.0]deca-3,9-diene (8) (P.D. Davis and
D.C. Neckers, J. org. Chem., 1980, 45, 456).

(iii) Substitution reactions

Irradiation of 1-ethoxycarbonylindole in ethanol
results in a Fries-type photochemical rearrangement to
give 3-(46.5%), 4-(23.1%), and 6-(8.8%)-ethoxycarbonyl-
indole and diindolylmethane (1.4%) (M. Somei and
M. Natsume, Tetrahedron Letters, 1973, 2451). The chromium
tricarbonyl complex of 1-methylindole undergoes
nucleophilic substitution reactions with generation of
7-substituted indoles (A.P. Kozirowski and K. Isobe, Chem.
Comm., 1978, 1076). Indole reacts with acyl-adducts of
quinoline 1-oxide and ethyl nicotinate 1-oxide to yield
3-(2-quinolyl)indole and ethyl 6-(3-indolyl)nicotinate,
respectively (M. Hamana and I. Kumadaki, Chem. pharm. Bull.
1970, 18, 1742).

Indole and 2,3-dichloro-5,6-dicyano-1,4-benzoquinone
form a stable crystalline 1:1 donor-acceptor complex in
methylene chloride, but in dioxane the complex affords
the substitution product (1), which on heating eliminates
hydrogen cyanide to give 6-cyano-2,3-dichloro-5-(3-indolyl)-
-1,4-benzoquione (2) (J. Bergman, R. Carlsson, and S.
Misztal, Acta Chem. Scand., 1976, B30, 853). 3-Alkylindoles
and dichlorodicyanobenzoquinone give 3-alkylidene-3H-indoles.

The photo-induced reaction of indole or 3-methyl indole with N-methyl-2,3-dibromomaleimide in dioxane gives the corresponding 2-indolylmaleimide (3) (T. Matsuo, S. Mihara, and I. Ueda, Tetrahedron Letters, 1976, 4581). With dibromomaleic anhydride no reaction occurs.

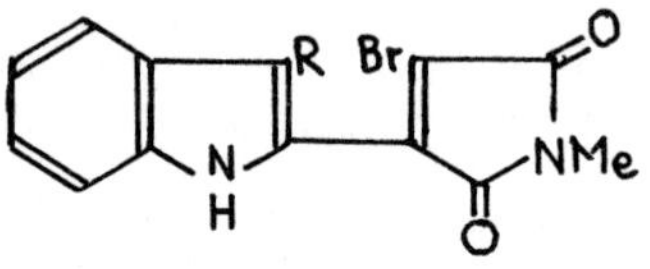

R = H, Me

(3)

(iv) Metal compounds

1-Methoxymethyl and 1-benzenesulphonyl derivatives of indole are satisfactorily lithiated in the 2-position and these derivatives may subsequently undergo addition reactions with carbonyl and cyano compounds. The 1-benzene-sulphonyl group is removed by mild alkaline hydrolysis and is therefore a good N-protecting group in synthesis via 2-lithioindoles. A number of 2-acylindoles and 2-indolylcarbinols have been prepared using the above reactions (R.J. Sundberg and H.F. Russell, J. org. Chem., 1973, 38, 3324).

(v) Oxidation

The oxidative coupling of indole with homocatechol in methanol-water, in the presence of ferric chloride yields 5-(3-indolyl)-4-methylcatechol together with a dimer of homocatechol. Similarly 2,3-dihydroxynaphthalene with indole affords 1,2-dihydro-1,1-di(3-indolyl)-3-hydroxy-2--oxonaphthalene with a trace of dimer (Y. Takizawa and T. Mitsuhashi, Chem. Abs., 1973, 78, 15948z).

In the presence of chloroperoxidase indole is oxidised by hydrogen peroxide to give mainly oxindole (M.D. Corbett and B.R. Chipko, Biochem, J., 1979, 183, 269).

(vi) Alkylation

Alkylation of indole with 5-chloropentan-2-one
ethylene ketal affords 5-(1-indolyl)pentan-2-one ethylene
ketal, which yields (1) on hydrolysis. 3-Methylindole
on similar treatment and hydrolysis gives pyrido[1,2-a]-
indole (2) (M.K. Eberle, J. org. Chem., 1976, **41**, 633).

(1)   (2)

Indole with tetramethylallene in the presence of acetic
acid gives 2-(3-indolyl)-2,4-dimethylpent-3-ene
(G. Nagendrappa and J.G. Hiriyakkanaver, Indian J. Chem.,
1975, **13**, 1124).

(vii) Miscellaneous reactions

Both indole and 1-methylindole react with
formaldehyde under strongly acid conditions to give small
yields of cyclooligomeric products, also formed by simple
acid treatment of 3,3'-diindolylmethane and 3,3'-di(1-
-methylindolyl)methane. The reaction between indole and
aromatic aldehydes yields dimeric products (J. Bergman,
S. Högberg, and J.-O. Lindström, Tetrahedron, 1970, **26**,
3347). The 2:1, 1:1, and 2:2 condensation products of
3-methylindole with metaldehyde in sulphuric acid have
been identified (H. Auterhoff and K.J. Aymanns, Arch.
Pharm., 1974, **307**, 885).

2-Methylene-oxoquinuclidine reacts with indole to

give 2-(3-indolylmethyl)-3-oxoquinuclidine (1) (V.A. Bondarenko *et. al.*, Chem. Abs., 1979, <u>91</u>, 5093w).

(1)   (2)

Phosphorylation occurs on heating indole with tri-(diethylamido)phosphite at 135-140° to give the 1-phosphorylated derivative (2) (A.I. Razumov *et. al.*, Zh. obshch. Khim., 1980, <u>50</u>, 778).

4-(1-alkyl-3-indolyl)pyrylium and thiopyrylium salts are obtained by the reaction of 1-alkylindoles with 4-chloropyrylium or 4-chlorothiopyrylium salts (S.N. Baranov *et. al.*, Chem. Abs., 1977, <u>86</u>, 16514g). The preparations of η^6-indole-(pentamethylcyclopentadienyl)-rhodium (III) and - iridium (III) hexafluorophosphates have been described (C. White, S.J. Thompson, and P.M. Maitlis, Chem., Comm., 1976, 409). Fluorescence of indole derivatives has been induced by ketone-sensitized photoreactions (E. Fujimori, Photochem. Photobiol. 1972, <u>16</u>, 61).

(viii) Characterisation of indoles

The preparative separation of indole and related products and the analysis of the indole fraction have been discussed [O.N. Karpov, Zh. Prikl. Khim. (Leningrad), 1975, <u>48</u>, 1175].

(c) Indole and its substitution products

Indole has been prepared by gas phase cyclization of 2-(2-aminophenyl)ethanol in presence of a copper catalyst (J. Bakke, H. Heikman, and E.B. Hellgren, Acta

Chem. Scand., 1974, B28, 393) and also by the initial
reduction of 2-(2-nitrophenyl)ethanol followed by catalytic
formation of the heterocyclic ring (Bakke and Heikman, Ger.
Pat. 2,052,678/1971). It has been obtained by the Fischer
method by using a flow method, with a stream of nitrogen
to remove the indole from the contact catalyst immediately
after its formation (M. Nakazaki and K. Yamamoto, J. org.
Chem., 1976, 41, 1877) and by passing acetaldehyde phenylhydrazone with benzene and carbon dioxide through a
catalytic oven at 320-330°. The latter method has
been used to obtain 1-, 2- and 5 methyl-, 3-ethyl-, 5-
chloro-, and 5-bromo- indole (N.N. Suvorov *et. al.*, Fr.
Pat. 1,562,253/1969).

2-Lithiomethylphenyl isocyanide, obtained by treating
2-methylphenyl isocyanide with lithium diisopropylamide
(LDA) in diglyme, when allowed to warm up to room
temperature produces, after water workup, indole (Y. Ito,
K. Kobayashi, and T. Saegusa, J. Amer. chem. Soc., 1977,
99, 3532),

The reduction of oxinole with lithium tetrahydridoaluminate [P.N. Stefanescu, Rev. Chim. (Bucharest), 1971,
22, 370] and the thermolysis of 1,2,3,4-tetrahydroquinoline
at 650-750° in the presence of water vapour affords
indole (K. Handrick and G. Koelling, Ger. Pat.
2,822,907/1979). 2-Nitro-β-aminostyrene obtained from
2-nitrotoluene and dimethylformamide acetal in dimethylformamide at 145-150°, gives indole on hydrogenation under
pressure over palladium-carbon (A.D. Barcho and
W. Leimgruber, Ger. Pat. 2,057,840/1971). Indole is
obtained by the catalytic dehydrogenation of indoline over
a copper or a chromium catalyst at high temperature or
in toluene using a palladium-carbon catalyst at 110°
(J.M. Bakke, Fr. Pat. 1,576,807/1969).

Indole labelled with hydrogen isotopes in the pyrrole

and benzene rings has been synthesized (L.I. Dmitrievskaya
et. al., Chem. Abs., 1980, 92, 215197e). The following
deuteriated indoles have been prepared, $[1-^2H]$, $[1,3-^2H_2]$,
$[3-^2H]$, $[2-^2H]$, $[5,7-^2H_2]$, $[4,5,5,7-^2H_4]$
(M.F. Lautie, J. labelled Compd. Radiopharm., 1979, 16,
735).

Indole is prepared by cyclodehydrogenation of
2-ethylaniline at 710-750° over a porous catalyst
prepared by heating silica and alumina (A. Gerces *et. al.*,
Hung. Pat. 16,696/1979) or in the presence of water vapour
over γ-alumina at 550° (A. Mathe, Gerecs, and T. Toth,
Magy. Kem. Poly., 1972, 78, 247); by the cracking of the
boric ester of *N*-(2-hydroxyethyl)aniline at 750°
(J. Lieto *et. al.*, Bull. Soc. chim. Fr., 1976, 1246;
M. Petinaux and J.P. Aune, *ibid.*, 1973, 2485); by
oxidation of 2-naphthol to cinnamic acid, followed by
conversion into cinnamamide and subsequent Hofmann
rearrangement (G.N. Petrova *et. al.*, Khim. Geterotiskl.
Soedin., 1973, 753); and by treating 2-carbamoyl
cinnamamide with aqueous sodium hypochlorite (V.F. Shner,
A.P. Levashova, and N.N. Suvorov, U.S.S.R. Pat. 306,126
/1971). The use of zinc chloride in the cyclization
of acetaldehyde phenylhydrazone to indole has been
investigated (N.N. Surorov, Tr. Mosk. Khim.-Tekhnol. Inst.,
1974, 80, 63). Substituted phenylhydrazones afford
substituted indoles (*idem*, Brit. Pat. 1,174,034/1969).
Metal oxides have also been used as catalysts (Suvorov,
V.G. Avramenko, and V.N. Shkil'kova, U.S.S.R. Pat. 262,904
/1970). Indole has been obtained from *N*-ethylaniline
and iodine in the vapour phase at 300-500° (W.H. Bell
and R.A.C. Rennie, Brit. Pat. 1,184,242/1970).

Indole and benzofuran have been studied by an
ab initio LCAO-SCF-MO procedure [J. Koller, A. Azman, and
N. Trinajstic, Experientia, Suppl. 1976, 23, (Quant.
Struct.-Act. Relat.), 205].

Indole and its 3-carboxaldehyde and 3-diethylamido
derivative react with tri(diethylamido)phosphine to yield
the corresponding 1-indolyldi(diethylamido)phosphine
(A.I. Razumov, P.A. Gurevich, and S.A. Muslimov, Zh.
obshch. Khim., 1976, 46, 2749).

(i) N-Derivatives

Thermolysis of certain fused β-lactams of type (1) gives 1-alkylindoles with elimination of phenyl isocynate (R.L. Bentley and H. Suschitzky, J. chem. Soc., Perkin I, 1976, 1725). The β-lactams may also be converted to indole-2-carboxylic acids (p.33).

(1)

Alkylation of indole by benzyl chloride in 50% aqueous sodium hydroxide containing benzyltrimethylammonium chloride gives 1-benzylindole (N.M. Suvorov *et. al.,* Khim. Geterotsikl. Soedin., 1976, 191), also obtained from 1-indolylsodium and benzyl chloride (G.M. Rubottom and J.C. Chabala, Org. Synth., 1974, 54, 60) and from indole and benzyl bromide in dimethyl sulphoxide containing potassium hydroxide (H. Heaney and S.V. Ley, *ibid.,* p.58). In contrast to the exclusive 1-benzylation of indole, the reaction of benzyl chloride with 2-methylindole gives a mixture of mainly 1-benzyl-2-methylindole, and some 3-benzyl- and 1,3-dibenzyl-2-methylindole (R. Garner *et. al.,* Chem. and Ind., 1974, 110). A number of 1-alkyl-indoles have been obtained by this method (Heaney and Ley, J. chem Soc., Perkin I, 1973, 499). Treatment of indole with alkyl sulphates, iodides, or bromides in a two-phase system of 50% aqueous sodium hydroxide and benzene containing tetrabutylammonium hydrogen sulphate affords 1-alkylindoles (A. Barco *et. al.,* Synth., 1976, 124). Other catalytic alkylation of indole in aqueous medium have been reported (A. Jonczyk and M. Makosza, Rocz. Chem., 1975, 49, 1203). Treatment of indole and 3-methylindole with tetrabutylammonium bromide in sulpholane containing potassium carbonate at 170° affords the respective

1-butylindoles (Suvorov, D.N. Plutitskii, and
Yu. I. Smushkevich, Khim. Geterotsikl. Soedin., 1980, 275).
Indole is also alkylated in the 1-position by methyl
iodide, benzyl bromide or isopropyl iodide in ether-
-dimethylformamide containing thallium (I) ethoxide
(A. Banerji and J. Banerji, Indian J. Chem., 1975, 13,
945), by heating with Raney nickel in primary or secondary
alcohols (H. Plieninger, H.P. Kraemer, and C. Roth, Ber.,
1975, 108, 1776), or in alcohols and aluminium alkoxides
(M. Botta, F. De Angelis, and R. Nicoletti, J. heterocyclic
Chem., 1979, 16, 501), and by alkyl halides in the presence
of 18-crown-6 ether (E. Santaniello, C. Farachi, and
F. Ponti, Synth., 1979, 617). Treating indole with sodium
hydride in hexamethylphosphoramide and then with methyl
iodide affords 1-methylindole in high yield (Rubottom and
Chabala, Org. Synth., 1973, 53, 1871). Other 1-alkylindoles
have been obtained by this method (*idem*, Synth.1972, 566).

2-Chloro-N-methyl-N-allylaniline on boiling with
tetrakis(triphenylphosphine)nickel in ether, followed by
cooling and bubbling oxygen through the mixture to convert
the triphenylphosphine to the oxide, affords 1,3-dimethyl-
indole (M. Mori and Y. Ban, Tetrahedron Letters, 1976,
1803). Similarly 2-chloro-N-methyl-N-phenylallylaniline
affords 3-benzyl-1-methylindole.

The radical copolymerisation of 1-vinylindole with
electron-accepting monomers has been discussed (Y. Oshiro,
Y. Shirota, and H. Mikawa, J. polym. Sci., polym. Chem.
Ed., 1974, 12, 2199).

(ii) C-Alkyl-, C-aryl- and related compounds

Moderate yields of 2-methylindoles are obtained by
heating N-2-chloroallylanilines with polyphosphoric acid
at 100° (B. McDonald, A. McLean, and G.R. Proctor, Chem.

1973, 208).

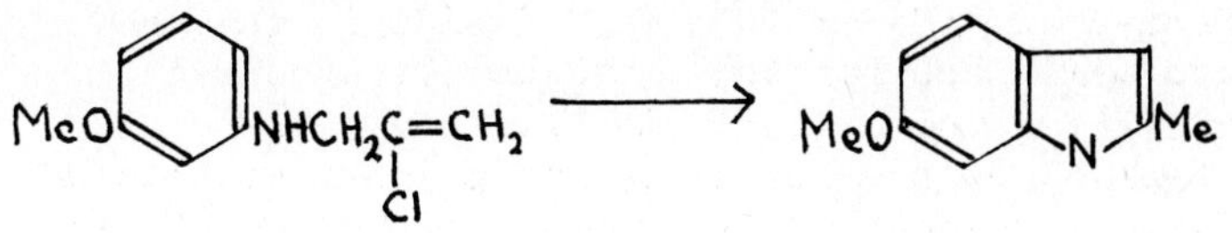

The addition of 2-allylaniline to a tetrahydrofuran
solution of $PdCl_2(MeCN)_2$ gives a complex, which on
treatment with triethylamine yields 2-methylindole
(L.S. Hegedus, G.F. Allen, and E.L. Waterman, J. Amer.
chem. Soc., 1976, 98, 2674).

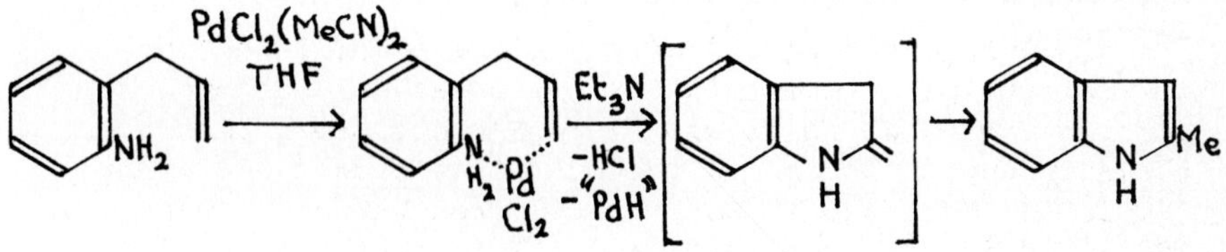

1-Substituted 2-lithioindoles react with
trialkylboranes or B-alkyl-9-borabicyclo[3.3.1]nonane
derivatives to give, after iodination, 2-alkylindoles
(A.B.L evy, J. org. Chem., 1978, 43, 4684).

Photo-induced addition of indoline-2-thiones (1)
to methyl acrylate gives the 2-substituted indoles (2)
and (3) (C. Marazano, J.-L. Fourrey, and B.C. Das, Chem.
Comm., 1977, 742).

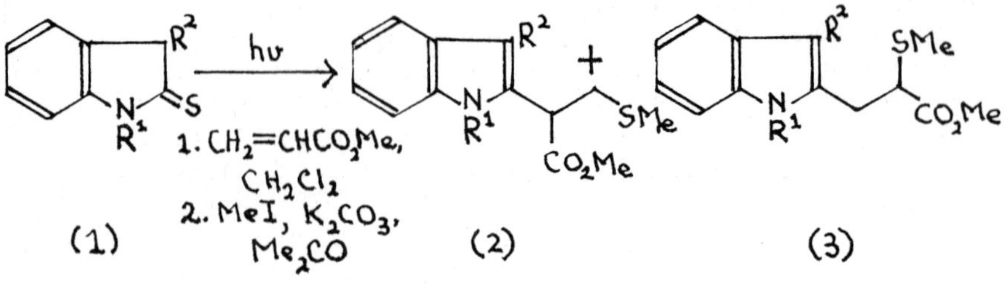

(1)   (2)   (3)

Reaction of indole with succinimidodiallyl- or
succinimidoallylalkyl-sulphonium chloride yields 2-alkyl-
-3-(allylthio)- or 3-(alkylthio)2-allyl- indole,
respectively, which on reductive desulphurization with
Raney nickel or zinc-acetic acid afford 2-alkyl- or 2-allyl-
-indoles (K. Tomita, A. Terada, and R. Tachikawa,
Heterocycl., 1976, 4, 729, 733).

3-Alkylindoles are obtained by treating 2-lithio-
methylphenyl isocyanide with alkyl halides followed by
lithium 2,2,6,6-tetramethylpiperidide (LTMP) (Y. Ito,
K. Kobayashi, and T. Saegusa, J. Amer. chem. Soc., 1977,
99, 3532).

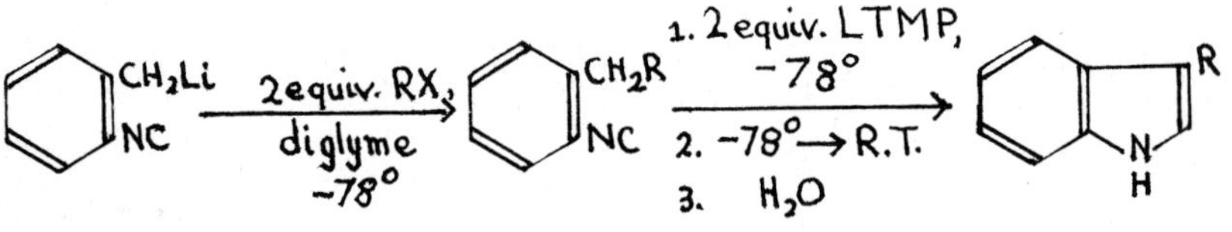

Indole on boiling with triphenylmethyl chloride in pyridine
gives 3-triphenylmethylindole (P. Buckus and N. Raguotiene,
Khim. Geterotsikl. Soedin., 1970, 1056). 3-Alkylindoles
with a branched alkyl group have been prepared by reacting
indole with a mixture of the appropriate alcohol and
derived sodium alkoxide. The 3-alkylindoles rearrange
in polyphosphoric acid to yield the corresponding

2-alkylindoles (A.N. Kost, V.A. Budylin, and
O.A. Mochalova, Vestn. Mosk. Univ., Khim., 1975, 16, 467).
3-Alkyl and 3-aryl-indoles may be obtained from the
reaction of aldehydes, in alcoholic or aqueous-alcoholic
solution, with indole in the presence of alkali metal
tetracarbonylhydridoferrate (G.P. Boldrini, M. Panunzio
and A. Umani-Ronchi, Chem.Comm., 1974, 359). Some 3-alkyl-
and 3-alkyl-2-methyl-indoles are obtained from the
appropriate indole with the necessary alkyl chloride or
bromide in the presence of zinc chloride — dipyridine zinc
chloride (Budylin, M.S. Ermolenko, and Kost, Khim.
Geterotsikl. Soedin., 1978, 921). 3-Benzylindole may be
obtained by the reaction of indole with 2-methylthio-2-
-phenyl-1,3-dithiane and boiling the product (4) with
lithium tetrahydridoaluminate, zinc chloride, and copper
(II) chloride in tetrahydrofuran (P. Stuetz and
P.A. Stadler, Org. Synth., 1977, 56, 8).

(4)

The methyl pyrrole-1-carboxylate (5) on treatment
with an alkylmagnesium halide, followed by oxidation of
the resulting secondary alcohol with pyridinium
chlorochromate, gives a 4-alkylindole on subsequent
reaction with stannic chloride (M. Natsume and H. Muratake,
Chem. Abs., 1980, 93, 71455h).

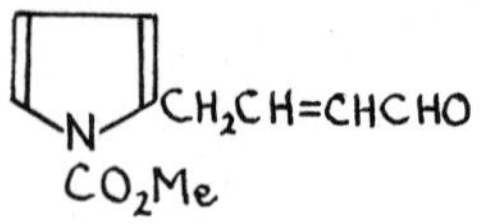

$$\text{structure with } CH_2CH=CHCHO, \ CO_2Me$$

(5)

A convenient synthesis of 4-alkylindoles is achieved by utilizing the stannous chloride - effected reaction of the endoperoxide of 1-methoxycarbonylpyrrole with carbon nucleophiles. The endoperoxide is obtained by photo-oxidation of the pyrrole (Natsume and Muratake, Tetrahedron Letters, 1979, 3477).

Indole with allyl acetate in acetic acid containing palladium acetylacetonate-triphenylphosphine yields 1-allylindole (7%), 3-allylindole (54%) and 1,3-diallylindole (11%) (W.E. Billups, R.S. Erkes, and L.E. Reed, Synth. Comm., 1980, 10, 147).

1-Allyl- and 1-(*trans*-2-butenyl)- indole with aluminium trichloride rearrange to give 3-allyl- and 3-(1-methylallyl)- indole, respectively (S. Inada *et. al.*, Bull. chem. Soc. Japan, 1976, 49, 833).

(iii) Halogeno-indoles

1-Chloroindole is prepared by treating indole in pentane, hexane, chloroform, or carbon tetrachloride with an aqueous solution of sodium hypochlorite. On boiling in butanol containing potassium carbonate it rearranges to 3-chloroindole (M. De Rosa, Chem. Comm., 1975, 482).

Both 4- and 6-chloroindole have been obtained by cyclizing
the appropriate chloro-1-(2-chloroethyl)-2-nitrobenzene
with iron and dehydrogenating the resulting indoline with
palladium-carbon (J. Bakke, Acta Chem. Scand., 1974, B28,
134).

When indole is treated with 1-chloropyrrolidine in
petroleum ether-benzene or acetonitrile, or with
N-chlorodibutylamine in petroleum ether-benzene in the
dark at room temperature, 3-chloroindole is obtained
(V. Snieckus and M.-S. Lin. J. org. Chem., 1970, 35, 3995).
It is also prepared by heating indole in diethyl
chloroacetamide at 60-70° (V.N. Rusinova, L.M. Orlova,
and N.N. Suvorov, Zh. Vses. Khim. Obsch., 1974, 19, 116).

Bromination of indole with excess bromine in acetic
anhydride at -15° affords 1-acetyl-2,3,5,6-tetrabromo-
indole. Similarly 3-methylindole with 3 moles bromine
gives 1-acetyl-3-methyl-2,5,6-tribromo-indole and 2-methyl-
indole with 1.5 moles bromine gives 3,6-dibromo-2-
methylindole and with 6 moles, 3,5,6-tribromoindole-2-
-carboxaldehyde [A. Da Settimo et. al., Chim., Ind. (Milan),
1976, 58, 220]. 5-Bromoindole has been prepared from
6-bromonaphth-2-ol (G.N. Petrova et. al., Khim.
Geterotsikl. Soedin., 1973, 753).

Reaction of indole with iodobenzene diacetate in
methanol containing potassium hydroxide yields the betaine
(1), which on treatment with toluene-4-sulphonic acid gives
the 3-indolyliodonium derivative (2) (B. Karele et. al.,
ibid., 1974, 214).

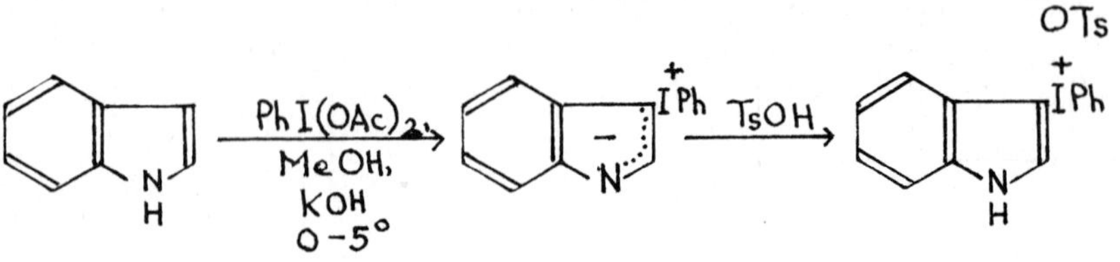

(1)   (2)

The isolation and properties of 3-bromoindolenines (T. Hino and M. Nakagawa, Symp. Heterocycl., [Pap.], 1977, 182; Heterocycl., 1977, 6, 1680) and the usefulness of 3-halogenoindolenines in indole chemistry have been reviewed (M. Ikeda and Y. Tamura, Yuki Gosei Kagaku Kyokaishi, 1979, 37, 568).

(iv) Sulphur compounds

Phase transfer-catalyzed 1-sulphonation of indole occurs when it is treated with sulphonyl chlorides in the presence of tetrabutylammonium hydrogen sulphate and sodium hydroxide in benzene (V.O. Illi, Synth., 1979, 136).

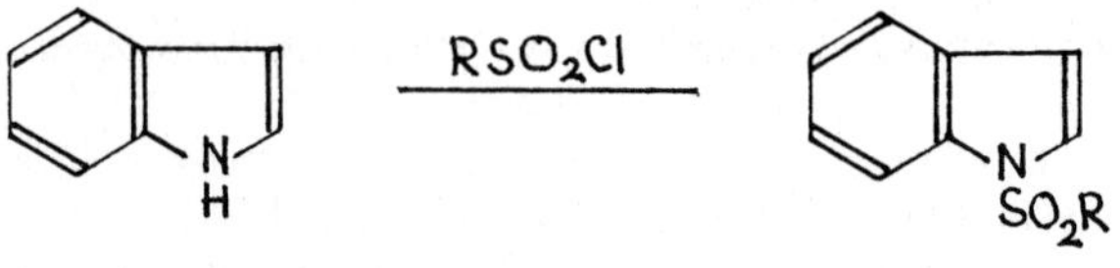

R = Ph, 4-tolyl, 2,4,6-$(Me_2CH)_3C_6H_2$, Me

Treatment of 1-benzenesulphonyl-2-lithioindole with triethylborane and iodine leads to 1-benzenesulphonyl-2-ethylindole, which on hydrolysis gives the 2-alkylated indole (A.B. Levy, J. org. Chem., 1978, 43, 4684).

The sulphonation of 1- and 2- methyl-, and 1,3-

-dimethylindole by pyridinium-1-sulphonate in boiling
pyridine gives the corresponding pyridinium indole-3-
-sulphonate, which on conversion into the silver salt and
treatment with methyl iodide affords the methyl ester.

Methyl 1-methylindole-3-sulphonate, m.p. 127-128°;
pyridinium 2-methylindole-3-sulphonate, m.p. 136-138°,
pyridinium 1,2-dimethylindole-3-sulphonate; methyl 1,2-
dimethylindole-3-sulphonate, m.p. 148°; pyridinium 3-
methylindole-2-sulphonate; pyridinium 1,3-dimethylindole-
-2-sulphonate; methyl 1,3-dimethylindole-2-sulphonate,
m.p. 102-104°; methyl 1-methylindole-2-sulphonate
m.p. 66-68° (G.F. Smith and D.A. Taylor, Tetrahedron,
1973 29, 669).

Thiocyanation of 2-phenyl-, 2-methyl-indole and ethyl
indole-2-carboxylate with thiocyanogen in acetic acid at
0° gives the corresponding 3-thiocyanoindoles, whereas
similar treatment of 3-methyl-2-phenylindoles and 2,3-
-diphenylindoles yield the corresponding 6-thiocyanoindoles.
2-Methyl-3-thiocyanoindole, m.p. 102-103°; 2-phenyl-3-
-thiocyanoindole, m.p. 166-168°; ethyl 3-thiocyanoindole-
-2-carboxylate, m.p. 133-134° (Y. Tamura et. al.,
J. Heterocycl. Chem., 1978, 15, 425).

(v) Nitro-, amino-, cyano- and related indoles

4-Aminoindole has been prepared by cyclizing 1-(2-
-chloroethyl)-2,6-dinitrobenzene with iron and

dehydrogenating the resulting dihydroindole with palladium-carbon (J. Bakke, Acta chem. Scand., 1974, B28, 134).

Benzenediazonium chlorides afford different products with indole at different pH, for example, at pH 8, a mixture of derivatives (1) and (2) are obtained, whereas at pH 1-2 the main product is (3) with minor amounts of (1), (2) and (4) (V.G. Avramenko *et. al.*, Tr. Mosk. Khim.-Tekhnol. Inst. 1970, 66, 132).

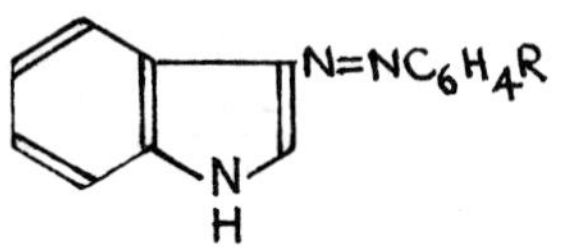

(1)

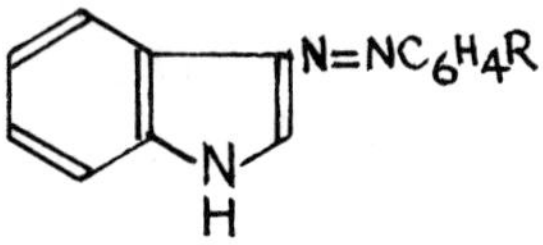

(2)

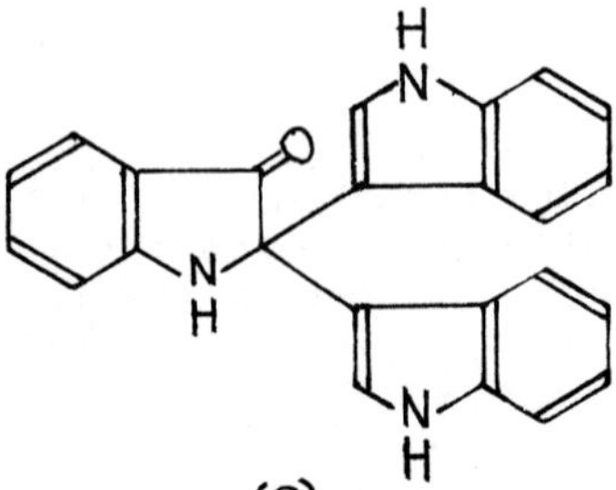

(3)

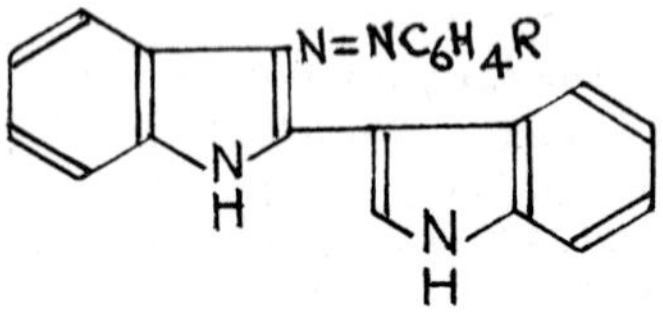

(4)

R = H, 2-, 3-, and 4-Me, 2- and 4-MeO

A number of 3-arylazoindoles have been prepared and their tuberculostatic activity determined (*idem*, Khim.-Farm. Zh., 1974, 8, 10) and the diazo coupling of indole with 2-aminophenyl methyl sulphide has been investigated (P.K. Sarma, F.S. Ahmed, and S.K. Barooah, Indian J. Chem., 1978, 16B, 530; Sarma and Barooah, *ibid.*, 1979, 17B, 274).

Indoles on reacting with iodine azide give different

kinds of products, *i.e.*, 3-azidoindolenines, 2-azidomethyl-indoles, and 3a-azido-furo- and -pyrrolo-[2,3-b]indoles, depending on the position and nature of the substituents. 2,3-Dimethyl- and 2-phenyl-3-substituted indoles yield the 2-azidomethyl derivative (5) and the 3-azidoindolenine (6), respectively, and 2-methyl-3-(2-hydroxyethyl)indole gives the azidofuroindole (7) (M. Ikeda Tetrahedron Letters, 1976, 2347).

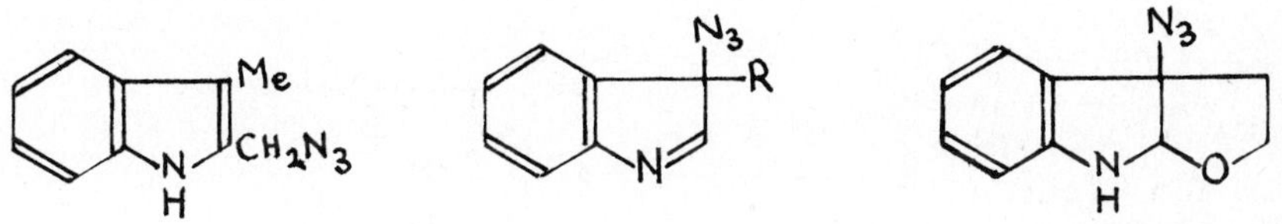

R = Me, Ph

A combined reagent of triphenylphosphine and thiocyanogen reacts with indole to give the 3-cyanoindole in good yield (Tamura *et. al.*, Tetrahedron Letters, 1977, 4417). Similarly pyrrole affords 2-cyanopyrrole.

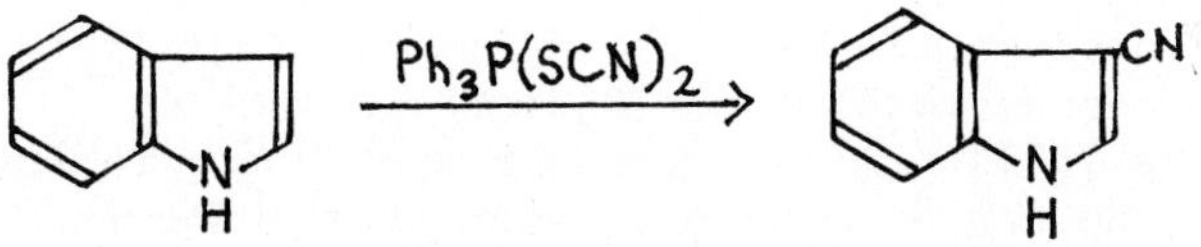

Irradiation of ethyl 2-cyano-1,2-dihydroquinoline-1--carboxylate gives ethyl 2-cyanomethylindole-1-carboxylate, m.p.121° (Ikeda *et. al.*, Chem, Comm., 1973, 922).

(vi) Hydroxyindoles

Hydroxyindoles, indole alcohols, and indolethiols have been reviewed [T.F. Spande, Chem. Heterocycl. Compd., 1979, 25 (Indole, Pt. 3), 1].

(vii) Indolylcarbinols, aminoalkylindoles and related compounds

3-Dimethylaminomethylindoles on treatment with thionyl chloride and dimethylformamide yield the immonium salts (1), which react with the appropriate nitroalkane to yield the 3-(2-alkyl-2-nitrovinyl)indole (2) (K.K. Babievskii, K.A. Kochetkov, and V.M. Belikov, Invest. Akad. Nauk SSSR, Ser. Khim., 1977, 2310).

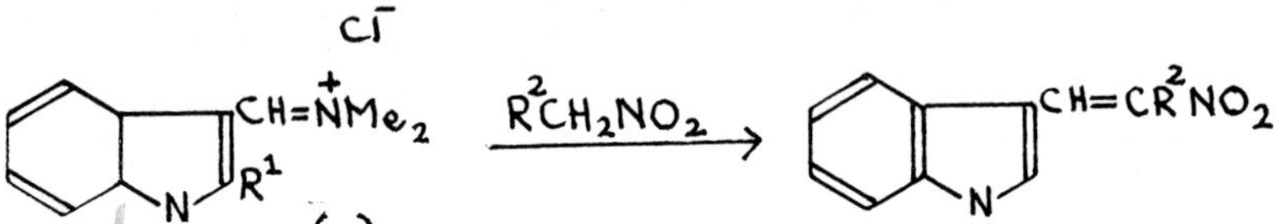

$$R^2 = H, Me, Et$$

3-(2-Nitrovinyl)indoles are also obtained by condensing indoles with 1-dimethylamino-2-nitroethene in trifluoroacetic acid.   1-Methyl-3-(2-nitrovinyl)indole, m.p. 167-168°;   6-benzoloxy-1-methyl-3-(2-nitrovinyl)indole, m.p. 194-195°;   5-methoxy-3-(2-nitrovinyl)indole m.p. 162-165° (G. Büchi and C.-P. Mak, J. org. Chem., 1977, $\underline{42}$, 1784).

1-(Dialkylaminoalkyl)indoles (3) are prepared by the reaction of indole with the appropriate dialkylamino-alkyl chloride hydrochloride [$R_2N(CH_2)_nCl.HCl$ (R = alkyl, n = 2,3)]   (A.N. Prilepskaya *et. al.*, Tezisy Doklady - Simp. Khim. Tekhnol. Geterotsikl. Soedin. Goryuch. Iskop, 2nd, 1973, 92).

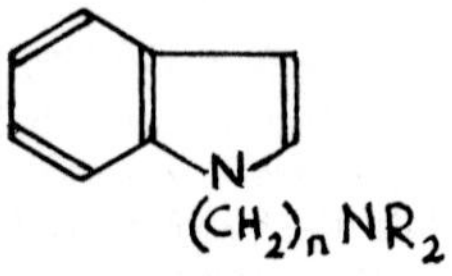

$$(3)$$

3-(2-Aminoethyl)indoles are obtained by treatment of indole with an aziridine (E.P. Styngach *et. al.*, Khim. Geterotsikl. Soedin., 1974, 1066). 5-Chloroacetylaminomethylindoles are prepared by the reaction of the appropriate indole with chloroacetylaminomethanol in sulphuric acid - phosphoric anhydride (A. Muminov, A.N. Kost, and L.G. Yudin, U.S.S.R. Pat. 697,507/1979).

2-Substituted indoles react with some methyl anils on heating at 40-70° for several hours to give 3-(methylaminomethyl)indoles (4). At higher temperatures di(3--indolyl)methanes (5) are formed (V.N. Borisova, E.N. Gordeev, and N.N. Suvorov, Khim. Geterotsikl. Soedin., 1977, 357).

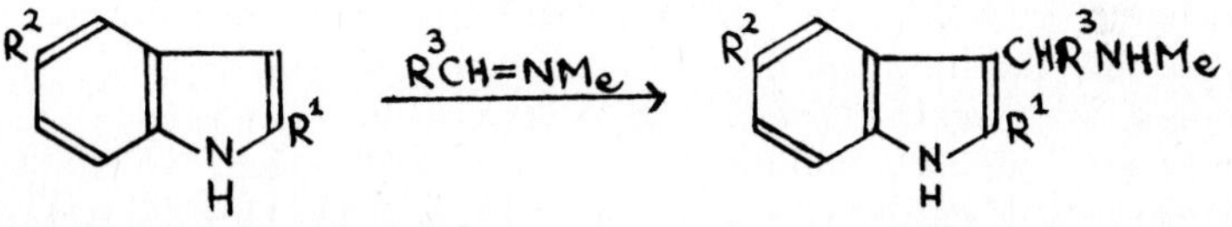

(4)

$R^1 = R^2 = H$; $R^3 = Ph, 4-MeOC_6H_4$;
$R^1 = H, R^2 = MeO$; $4-NO_2C_6H_4$, 2-thienyl
$R^1 = Me, R^2 = H$

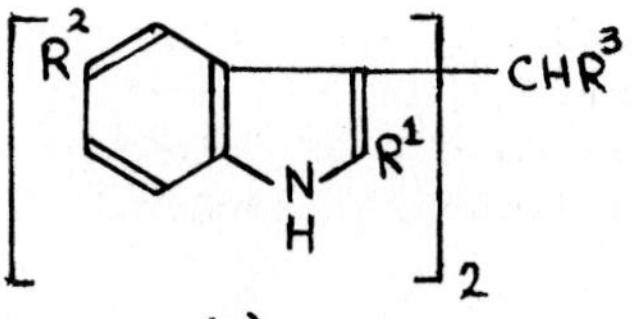

(5)

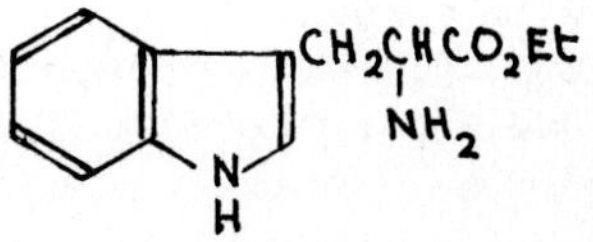

(6)

Ethyl 3-bromo-2-hydroxyiminopropanoate reacts with indole in the presence of sodium carbonate to give an α-hydroxyimino ester, which can be reduced to the ethyl 2-amino-3-(3-indolyl)propanoate (6) (T.L. Gilchrist, D.A. Lingham, and T.G. Roberts, Chem. Comm., 1979, 1089). Some 3-amidoalkylindoles have been prepared (V.P. Marshtupa and A.P. Kucherenko, Chem. Abs., 1977, <u>86</u>, 5258a).

(viii) Aldehydes and ketones

Indoles react with hexamethylenetetramine to give
indole-3-carboxaldehyde. 3-Methylindole on formylation
with dimethylformamide and phosphoryl chloride affords
3-methylindole-1-carboxaldehyde, 3-methylindole-2-
-carboxaldehyde, and 2-formamidoacetopheone, and 2,3-
-dimethylindole gives 2,3-dimethylindole-1-carboxaldehyde.
Indole-1-carboxaldehyde exists in two conformations in
carbon tetrachloride at 30°, while 2,3-dimethylindole-1-
-carboxaldehyde exists solely in one conformation
(A. Chatterjee and K.M. Biswas, J. org. Chem., 1973, <u>38</u>,
4002). Small amounts of indole-3-carboxaldehyde are formed
along with other products on irradiation of quinoline
N-oxide in ethanol. Indole-3-carboxaldehyde-2-d_1 is
obtained from quinoline N-oxide-2-d_1 (O. Buchardt,
K.B. Tomer, and V. Madsen, Tetrahedron Letter, 1971, 1311).
Indole-3-carboxaldehyde may be obtained by the reaction
of indole with chloromethylenemalononitrile and treating
the resulting dinitrile (1) with aqueous potassium
hydroxide (W. Ertel and K. Friedrich, Ber., 1977, <u>110</u>,
86); by treating indole with dimethylformamide

(1)

and sulphuryl chloride and hydrolysing the product
(K.K. Babievskii, K.A. Kochetkov, and V.M. Belikov,
U.S.S.R. Pat. 704,941/1979), and by the reaction of indole
with 2-oxazolines followed by hydrolysis of the product
(Y. Motoyama, Y. Furuya, and N. Hirose, Japan Kokai Tokkyo
Koho 79 90,173/1979).

Treatment of indole-3-carboxaldehyde with
phenylacetic acid chloride and potassium carbonate in

dimethylformamide affords the 1-phenylacetylindole-3-
-carboxaldehyde. Reacting indole-3-carboxaldehyde under
the same conditions with 1,2-dibromoethane gives 1,2-bis-
(3-formyl-1-indolyl)ethane (G.L. Papayan and L.S. Galstyan,
Arm. Khim. Zh., 1972, 25, 814).

The addition of the chromium tricarbonyl complex
of 1-methylindole in tetrahydrofuran, containing
hexamethylphosphoric triamide to 2-lithio-1,3-dithian in
tetrahydrofuran, containing hexamethylphosphoric triamide
furnishes exclusively the 7-substituted indole (2). Copper
(II) catalysed hydrolysis of dithian (2) affords 1-methyl-
indole-7-carbaldehyde.

(2)

A significant downfield shift is observed for the 1-methyl
group in the nmr spectrum of 1-methylindole-7-
carboxaldehyde relative to the 1-methyl group of
1- methylindole, owing to the deshielding effect of the
formyl substituent (A.P. Kozikowski and K. Isobe, Chem.
Comm., 1978, 1076).

1-Acylindoles are obtained by the reaction of indole
with acid chlorides in a mixture of methylene chloride
and powdered sodium hydroxide in the presence of tetrabutylammonium hydrogen sulphate (V.O. Illi, Synth., 1979,
387). Appropriate 1-aroylindoles may be cyclized in the
presence of palladium acetate to give isoindolo[2,1-a]-

indol-6-ones (T. Itahara, *ibid.*, p.151).

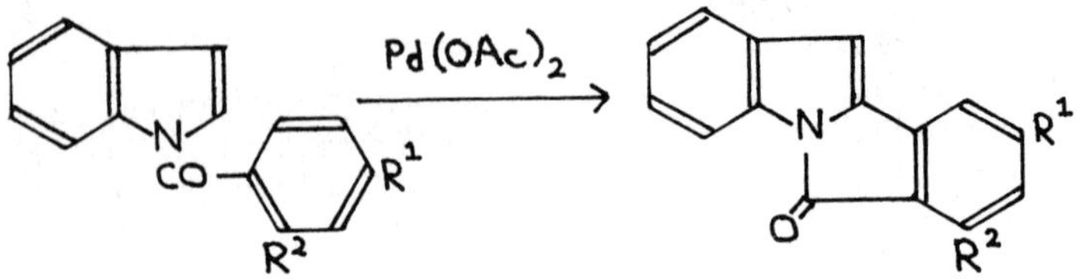

$$R^1, R^2 = H, Me, Cl$$

1-Acetylindoles may be prepared by treating the indole with acetic anhydride — potassium hydroxide in dimethyl sulphoxide (G.W. Gribble, L.W. Reilly, Jr., and J.L. Johnson, Org. Prep. Proced. Int., 1977, 9, 271). They may also be obtained by transacylation on boiling the indole with *N*-acetylimidazole, but slow addition of indole in acetic anhydride to a solution of *N*-acetyl-imidazole in acetic anhydride at 125° gives 3-(1,3--diacetyl-1,2,-dihydro-2-imidazolyl)indole (3), as the major product (J. Bergman, Tetrahedron Letters, 1972, 4273).

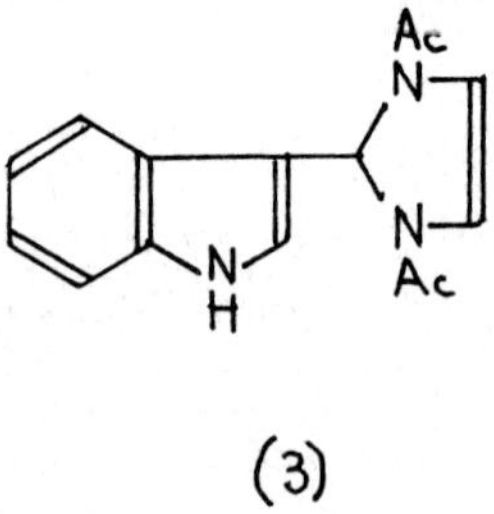

(3)

1-Acylindoles react with iodine azide to give remarkable stable 1-acyl-*cis*- and -*trans*-2,3-diazido-indolines in high yield, which may be useful in the synthesis of indole derivatives of potential pharmalogical activity (Tamura *et. al.*, *ibid.*, 1975, 3291).

R^1 = COPh, R^2 = H, R^3 = Me; R^1 = Bz, R^2 = R^3 = H;

R^1 = Bz, R^2 = R^3 = Me; R^1 = $SO_2C_6H_4Me$-4, R^2 = H,

R^3 = Me; R^1 = CO_2Et, R^2 = H, R^3 = Me

2-Methylindole reacts with chloroacetic acid chloride in the presence of pyridine to yield 3-chloroacetyl-2-
-methylindole (V.P. Arya *et. al.*, Indian J. Chem., 1977, 15B, 473).

3-Benzoylindole may be obtained by boiling the product from indole and 2-methylthio-2-phenyl-1,3-dithiane (p. 19) with copper (II) chloride and oxide in acetone (Stuetz and Stadler, *loc. cit.*). 3-Benzoylindole with phosphorus pentachloride in dimethylformamide affords 3-chloro-3-(3-indolyl)-1-oxo-2-phenylprop-2-ene which can be converted to 3-(phenylethynyl)indole (A.B. Kamenskii *et. al.*, Khim. Geterotsikl. Soedin., 1980, 956). 3-(α-
-Halogenoacyl)indoles rearrange under the influence of various bases, for example, hydroxide hydride, and Grignard reagents (see p. 34) (Bergman, J.-E. Bäckvall, and J.-O. Lindström, Tetrahedron, 1973, 29, 971; Bergman and Bäckvall, *ibid.*, 1975, 31, 2063). 3-Acetoacetylindoles are prepared by the reaction of 3-unsubstituted indoles with diketene (P. De Cointet *et. al.*, Eur. J. med. Chem.-
Chim. Ther,. 1976, 11, 471). The reaction between 1-methylindole and the selenol ester, prepared from methyl heptanoate, in benzene in the presence of cuprous triflate—
benzene complex gives 1-methyl-3-indolyl hexyl ketone (Kozikowski and A. Ames, J. Amer. chem. Soc., 1980, 102, 860).

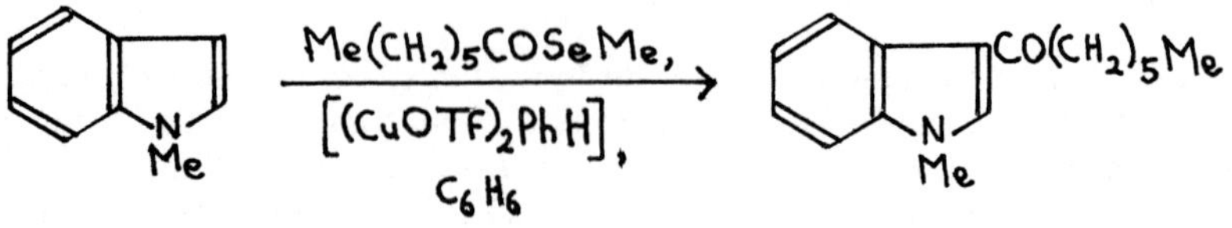

Trifluoroacetylation of indole affords a mixture of 1-trifluoroacetylindole (50%), 3-trifluoroacetylindole (20%) and 2-(3-indolyl)-1-trifluoroacetylindoline (4) (15%) (A. Cipiciani *et. al.*, Tetrahedron, 1976, 32, 2595).

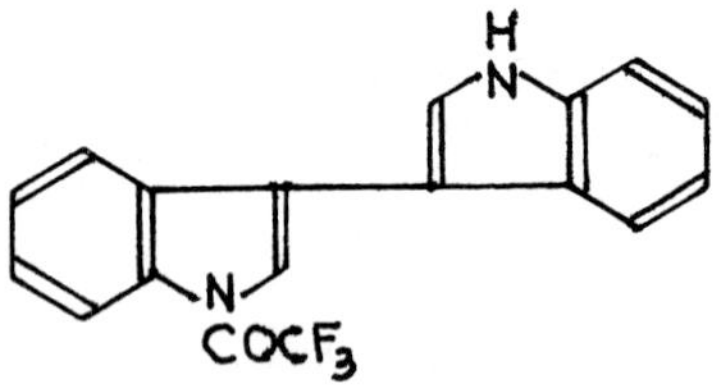

(4)

Studies relating to the synthesis of 3-acylindoles have been reported (O.O. De Sousa Campos, Diss. Abs. Int. B, 1980, 40, 3166). For a review of indole aldehydes and ketones see W.A. Remers [Chem. heterocycl. Compd., 1979, 25, (Indoles Pt. 3), 357].

Indole on boiling with diacetyl in acetic acid yields 2,2-di(3-indolyl)butan-3-one, m.p.197°. Similarly benzil gives di(3-indolyl)phenylbenzoylmethane, m.p.316° (G.I. Zhungietu and F.N. Chukhrii, Zh. Vses. Khim. obschch., 1970, 15, 228). The reaction between indole and phenylglyoxal in acetic acid affords di(3-indolyl)-benzoylmethane, m.p.210° *(idem,* Khim. Geterotsikl. Soedin., 1969, 952).

(ix) Carboxylic acids

1-(ω-Chloroalkyl)indole-2-carboxylic acids (3) are formed by treating the indolinium-β-lactam (1) with aqueous sodium hydroxide and then hydrolysing the resulting amides (2) in either boiling sodium hydroxide solution or aqueous 15% hydrochloric acid (R.L. Bentley and H. Suschitzky, J. chem. Soc., Perkin I, 1976, 1725).

$R^1 R^1 = [CH_2]_{5-6}$

$R^2 = Ph$

$R^1 = [CH_2]_{5-6}OH$

or $[CH_2]_{5-6}Cl$

Amide (2) (R^1 = Me, R^2 = H) on boiling with hydrochloric acid affords 1-methylindole-2-carboxylic acid.

3-Indolylacetonitrile obtained by treating the Mannich base from indole, formaldehyde, and methylaniline with sodium cyanide, on hydrolysis gives indole-3-acetic acid (C. Leonte, A. Sauciuc, and E. Carp, Rom. 64,006/1978). The reaction of 3-unsubstituted indoles with chlorosulphonyl isocyanate in ether or acetonitrile at 0-5° yields indole-3-carboxamide sulphonyl chlorides, which on hydrolysis with alkali in aqueous acetone give indole--3-carboxamides. Similar treatment of 2,3-dimethylindole affords 2,3-dimethylindole-1-carboxamide (D.N. Dhar and S.C. Suri, Synth., 1978, 374). Indole-3-carboxylic acid derivatives are readily prepared from indoles and oxalyl chloride (Bergman, R. Carlsson, and B. Sjöberg,

434

J. heterocyclic Chem., 1977, 14, 1123).

    Copper-induced thermolysis of ethyl diazoacetate
in 1,3-dimethylindole yields ethyl 1,3-dimethylindole-2-
-acetate in low yield (E. Wenkert *et. al.*, J. org. Chem.,
1977, 42, 3945). Indole-3-acetic acids are obtained *via*
the Favorskii rearrangement on boiling 3-(α-halogenoacyl)-
indoles with sodium hydroxide in 80% ethanol (Bergman and
Backvall, Tetrahedron Letters, 1973, 2899; Tetrahedron,
1975, 31, 2063).

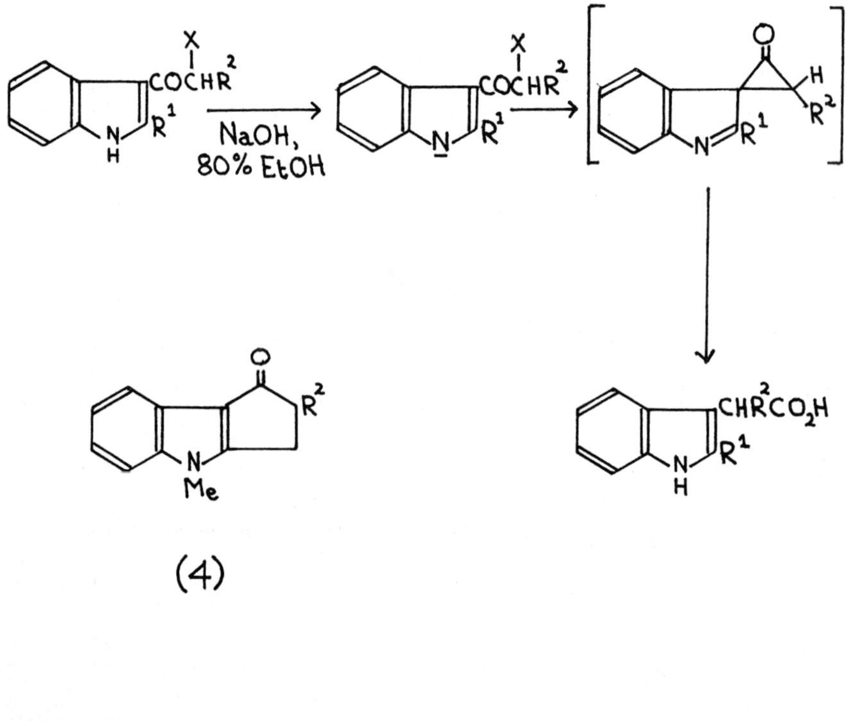

(4)

1,2-Dimethyl-3-(α-halogenoacyl)indoles under the above
conditions afford the N-methylindolocyclopentanone (4).
Indoles condense with chloracetic acid in aqueous potassium
hydroxide at 240-250° and 5-atmospheres pressure to yield
indole-3-acetic acids (5) (V.G. Avramenko *et. al.*, Khim.,
Geterotsikl. Soedin., 1974, 1375). At 245-250° and 45-50
atmospheres indoles and chloroacetic acid give indole-3-
-acetic acid (90-97%) (N.N. Suvorov *et. al.*, Sin.

Geterotsikl. Soedin., 1972, 47).

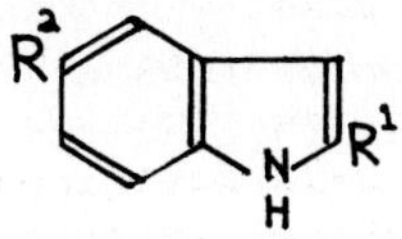

$$R^1 = H, Me, Ph;$$
$$R^2 = H, Cl, MeO, HO$$

(5)

Irradiation of a mixture of indole and methyl chloroacetate
gives methyl indole-1-acetate and six other methyl indoleacetates, which can be separated and identified (S. Naruto
and O. Yonemitsu, Tetrahedron Letters, 1971, 2297; Chem.
pharm. Bull., 1972, 20, 2161). Methyl indole-1-acetate
on irradiation affords an isomeric mixture of methyl
indoleacetates (*idem, ibid.*, p.2272).

Ethyl 3-indolylpropionates (7) are obtained in good
yields by the decarboxylative ethanolysis of the product
(6) formed by a novel Mannich-type condensation between
indole and Meldrum's acid (Y. Oikawa, H. Hirasawam and
Yonemitsu, Tetrahedron Letters, 1978, 1759).

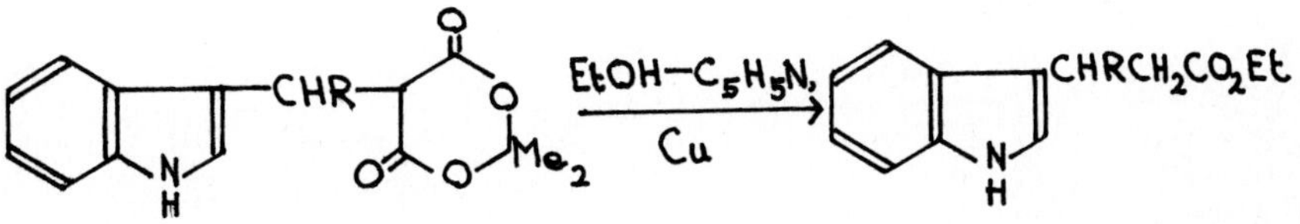

(6)   (7)

R = alkyl, Ph, 3-$NO_2C_6H_4$,
2,5-$(MeO)_2C_6H_3$, 3,4-$(OCH_2O)C_6H_3$

3-Tritylindole on boiling with acrylonitrile in benzene
in the presence of potassium hydroxide gives
3-indolylpropionitrile, which on boiling with potassium

hydroxide — ethanol followed by acidification affords
3-indolylpropionic acid (P. Buckus and N. Raguotiene, Khim.
Geterotsikl. Soedin., 1970, 1056).

The reaction between equimolar proportions of 2,5-
-dichlorothiophenium bismethoxycarbonylmethylide and indole
in boiling toluene in the presence of bis(acetylacetonato)-
copper affords dimethyl indole-3-malonate (R.J. Gillespie
and A.E.A. Porter, Chem. Comm., 1978, 50). Indole reacts
with diethyl chloromethylenemalonate in the presence of
aluminium chloride to give diethyl 3-indolylmethylene-
malonate (W. Ertel and K. Friedrich, Ber., 1977, 110, 86).
A number of indole-3-ketoacids hve been synthesized
(A.N. Kost *et. al.*, Vestn. Mosk. Univ., Khim., 1974, 15,
323). For 3-indolyl(α-amino)propionic esters see p. .

(d) 3H-Indoles (indolenines)

3,3-Dimethyl-3H-indole reacts with indoles of type
(1) in acetic acid to give 3,3-dimethyl-2-(3-indolyl)-
indoline (2), whereas with 3-methylindole the 3,3-dimethyl-
-2-(1-indolyl)indoline (3) is formed (V. Bocchi,
R. Marchelli, and V. Zanni, Synth., 1977, 343).

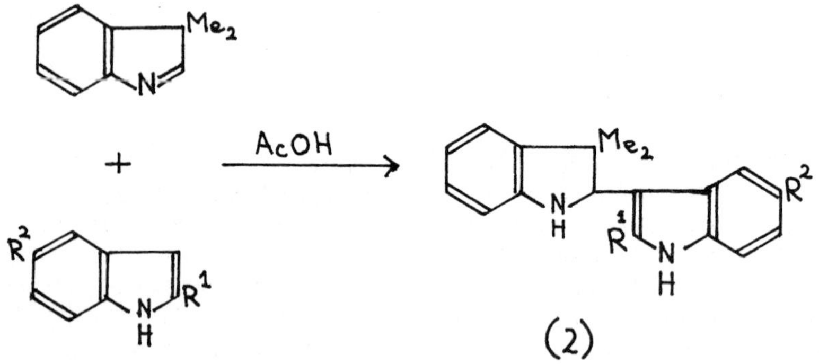

R^1 = H, Me; R^2 = H, OH, Cl

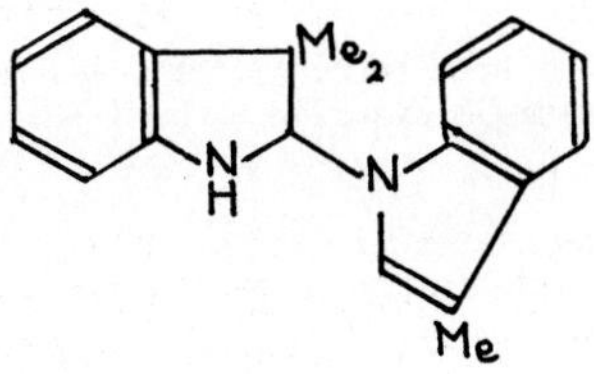

(3)

2. Hydroindoles

(a) Indolines (dihydroindoles)

(i) Synthesis

(1) Indoline and 2,3-dimethylindoline (*cis* and *trans*) are obtained by reduction of the corresponding indoles with borane and trimethylamine (J.G. Berger, Synth., 1974, 508). Certain 2,3-disubstituted indoles, where the substituents constitute a ring, undergo a stereospecific hydrogenation with sodium in liquid ammonia to give the *cis*-disubstituted indoline and with borane in tetrahydrofuran to give the *trans*-isomer (Berger *et. al.*, Tetrahedron Letters, 1975, 1807). Indolines are also obtained by treating indoles and fused indoles with borane in trifluoroacetic acid (B.E. Maryanoff and D.F. McComsey,

U.S. Pat., 4,210,590/1980). Selective reduction of the
double bond to give indolines, occurs when indoles
containing reducible groups are treated with borane-pyridine
[Y. Kikugawa, J. chem. Res. S (Synopses), 1977, 212].

   (2) N-Allylanilines rearrange under acid conditions
to give 2-allylanilines, which photocyclize to yield
indolines (K. Krowicki et. al., J. heterocyclic Chem.,
1976, 13, 555).

   (3) Treatment of indole with sodium tetrahydridoborate in neat carboxylic acid results in reduction of
the double bond and alkylation of the nitrogen atom to
yield 1-alkylindolines. Sodium cyanoborohydride—acetic
acid affords indoline (G.W. Gribble et. al., J. Amer. chem.
Soc., 1974, 96, 7812).

A number of indolines have been obtained by the latter
method, but 5-nitroindole and 2,3-diphenylindole are not
reduced under these conditions (Gribble and J.H. Hoffman,
Synth., 1977, 859). Indoles are reduced by sodium
tetrahydridoborate—aluminium chloride in pyridine to yield
the corresponding indolines (Kikugawa, Chem. pharm. Bull.,
1978, 26, 108). Catalytic hydrogenation of indole and
some of its derivatives in the presence of Raney nickel
at room temperature and 1 atmosphere pressure affords
indolines [T. Toth and A. Gerecs, Acta Chim. (Budapest),
1971, 67, 229].

(ii) General properties and reactions

   Dehydrogenation of 5-iodo- and 5-acetamido-indoline
with 2,3-dichloro-5,6-dicyano-1,4-benzoquinone affords
the corresponding indoles (Bergman, Carlsson, and

S. Misztal, Acta Chem. Scand., 1976, B30, 853). The
oxidation of indoline in water by palladium dichloride
and gold trichloride gives indole in yields of 0-83%
depending on reaction condition, with the optimum reaction
in methanol and trimethylamine at room temperature
M.E. Kuehne and T.C. Hall, J. org. Chem., 1976, 41, 2742).

(iii) Derivatives and indolines

Heating indole with arenesulphonyl azides yields
2-arylsulphonyliminoindolines and on similar treatment
1-methylindole affords mainly the 2-arylsulphonylimino-1-
-methylindolines with, as minor products, the 3-aryl-
sulphonylaminoindoles. In the presence of pyridine,
2-arylsulphonylimino-3-diazoindolines are formed
(A.S. Bailey *et. al.*, J, chem. Soc., Perkin I, 1972,
2411).

The reaction of ω-chloroalkylamine hydrochlorides
with indolines gives the 1-(ω-aminoalkyl)indolines
(V.P. Marshtupa and A.P. Kurcherenko, Tezisy Doklady-Simp.
Khim. Tekhnol. Geterotsikl. Soedin. Goryuch. Iskop., 2nd,
1973, 95).

R = alkyl, n = 2,3

The corresponding derivatives of indole can be prepared
in a similar manner. 4,5,6,7-Tetrafluoroindole has been
synthesized (R. Filler, S.M. Woods, and A.F. Freudenthal,
J. org. Chem., 1973, 38, 811). For 1-acyl-*cis*- and -*trans*-
-2,3-diazidoindolines see p. 30.

(b) Oxoindolines (indolinones) and hydroxyindoles

(i) 2-Oxoindolines (oxindoles)

(1) *Synthesis.* Oxindole is obtained by the reaction of aniline in methylene chloride at -65° with *tert*-butyl hypochlorite, then with ethyl methylthioacetate and finally with triethylamine to yield the very unstable amino ester (1), which on treatment with dilute acid gives the oxindole (2). Raney nickel reduction of derivative (2) results in the formation of oxindole (P.G. Gassman and T.J. van Bergen, J. Amer. chem. Soc., 1973, <u>95</u>, 2718; 1974, <u>96</u>, 5508).

5-Methoxyindole has been prepared by a related method (Gassman, van Bergen, and G. Gruetmacher, *ibid.*, 1973, <u>95</u>, 6508). Oxindole (75-80%) is obtained by the Willgerodt-Kindler reaction from 2-nitroacetophenone [P.N. Stefabescu, Rev. Chim. (Bucharest), 1971, <u>22</u>, 370].

Treating 2-chloro-*N*-methyl-*N*-acrylanilide (3) with tetrakis(triphenylphosphine)nickel in toluene at 50-60° gives 1,3-dimethyloxindole; also obtained by this method from appropriate starting materials are methyl 1-methyloxindolyl-3-acetate and 1-methyl-3-benzyloxindole (M. Mori and Y. Ban, Tetrahedron Letters, 1976, 1807).

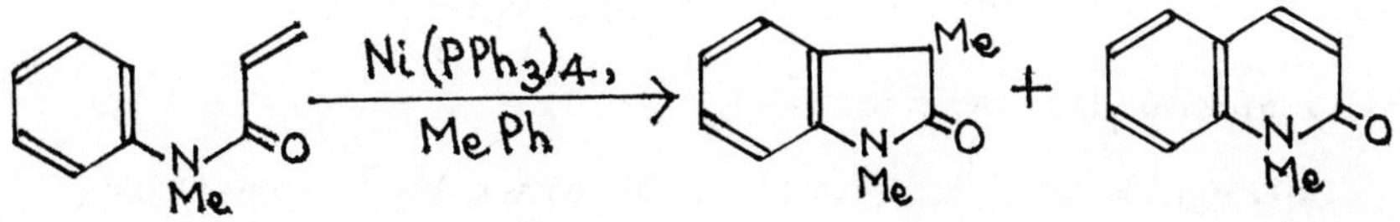

$$(3) \qquad (43\cdot5\%) \qquad (7\cdot3\%)$$

β-Nitrostyrenes react with acetyl chloride in the presence of ferric chloride in methylene chloride at 0° to give 3-chlorooxindoles (P. Demerseman *et. al., ibid.,* 1978, 2011). The reaction of indole with 1-chloropyrrolidine under strongly acidic Hofmann-Loeffler-Freytag conditions gives 3,3-dichlorooxindole (26-44%) (V. Snieckus and M.-S. Lin, J. org. Chem., 1970, <u>35</u>, 3994).

(2) *Properties and reactions.* The pyrolysis of 3,3-dimethyloxindole at 850°/0.2-0.4mm gives 2-quinolone (carbostyril) (52%) and indole (25%). Similarly, 1,3,3--trimethyloxindole affords 1-methyl-2-quinolone (30%), but 1- and 3-methyloxindoles give complex mixtures of decarbonylated and rearranged products (R.F.C. Brown and M. Butcher, Austral. J. Chem., 1973, <u>26</u>, 369).

Michael addition of indoles to oxindole-3-ylidene ketones affords adducts (4) and (5) (A. Kubo, T. Naki, and T. Nozoye, Heterocycl., 1974, <u>4</u>, 1675).

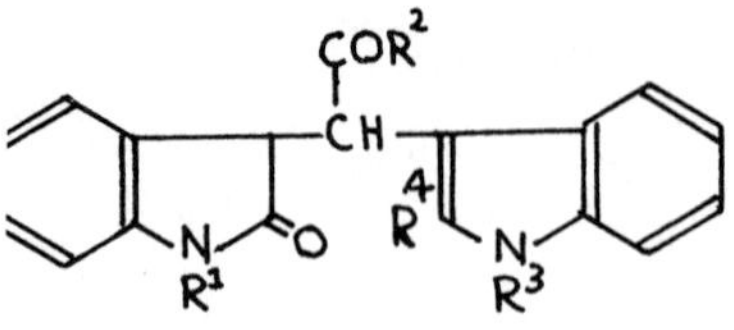

(4)

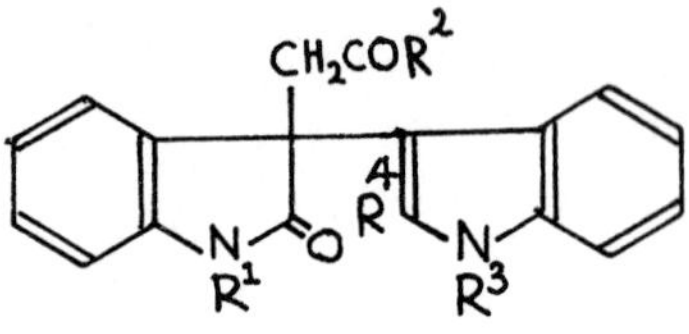

(5)

$$R^1 = H, Me, COMe;$$
$$R^2 = Ph, Me;$$
$$R^3, R^4 = H, Me$$

Bubbling air through a solution of 3-methylthiooxindole in ether in the presence of an equimolar amount of potassium *tert*-butoxide results in the formation of isatin (Gassman, Ger. Pat., 3,000,338/1980).

Oxindolylidenenitromethane (7) is obtained by the dehydrochlorination of 3-chloro-3-nitromethyloxindole (6) with triethylamine. Compound (7) on treatment with lithium chloride-dimethylformamide affords oxindolylidenechloromethane (8) (D.R. Long and C.G. Richards, Tetrahedron Letters, 1975, 1603).

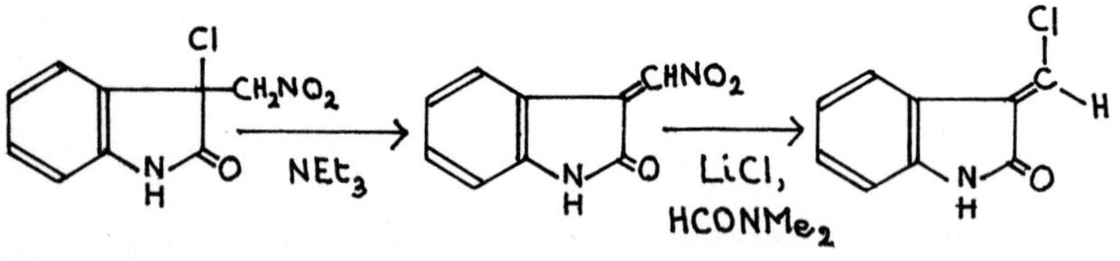

(6)

(7)

(8)

Both compound (7) and (8) exist as the *E*-isomer, and the nitro compound is an excellent dienophile, reacting very rapidly with cyclopentadiene to give the nitro adduct (9), which on treatment with alkali yields 3-phenyloxindole.

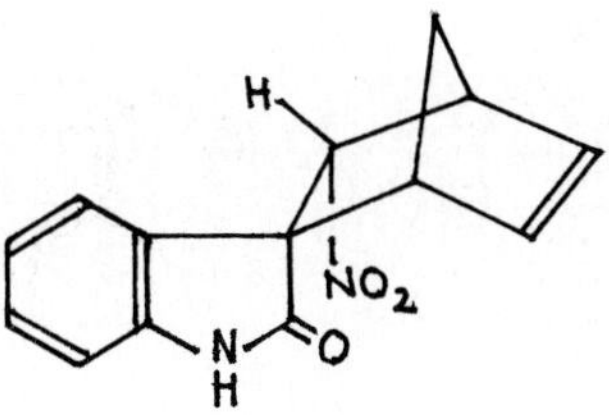

(9)

(ii) 3-Oxoindoline; indoxyl

Ring contraction occurs when 4H-3,1-benzoazines (1) are reacted with potassium amide in liquid ammonia to give 2,3-disubstituted 3-hydroxy-3H-indoles (2). The latter on treatment with hot base reaarange to indoxyls (D. Lednicer and D.E. Emmert, J. heterocyclic Chem., 1970, 7, 575).

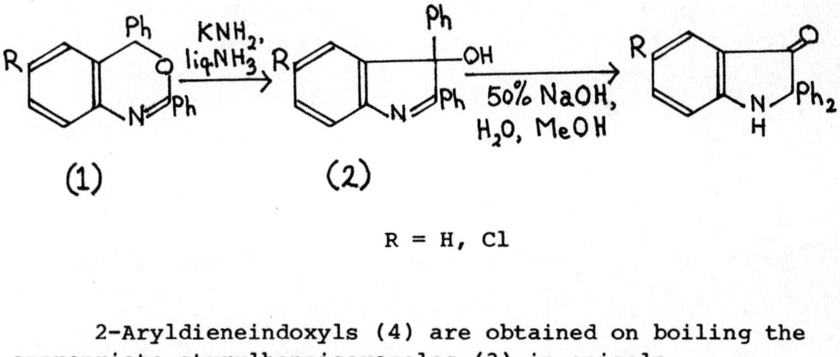

R = H, Cl

2-Aryldieneindoxyls (4) are obtained on boiling the appropriate styrylbenzisoxazoles (3) in anisole (R.K. Smalley, R. Smith, and H. Suschitzky, Tetrahedron Letters, 1978, 2309).

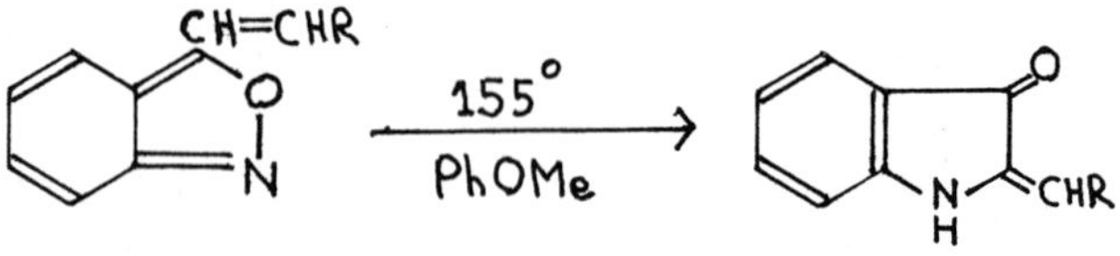

(3)   (4.)

$R = Ph, 4-MeOC_6H_4, 3,4-(OCH_2O)C_6H_3$

It was previously believed that indigo reacted with hydrazine and sodium hydroxide in aqueous ethanol to give desoxyindigo, but it has now been shown that the product possesses structure (5). Treatment of indigo with anhydrous hydrazine at 35° yields a quinazoline derivative (6) (Bergman, B. Egestad, and N. Eklund, *ibid.*, p.3147).

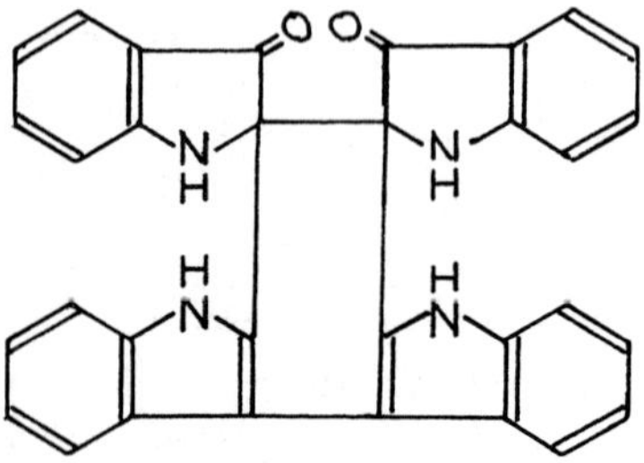

(5)

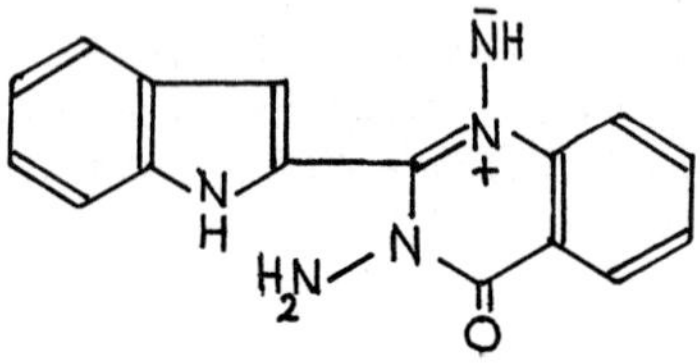

(6)

(iii) 2,3-Dioxoindolines, isatins

The reaction of 1-substituted isatins with potassium cyanide and ammonium carbonate gives 1'-substituted spiro-(imidazolidine-4,3'-indoline)2,2',5-trione (1) (H. Otomasu, K. Natori, and H. Takahasi, Chem. Pharm.

Bull., 1975, <u>23</u>, 1431).

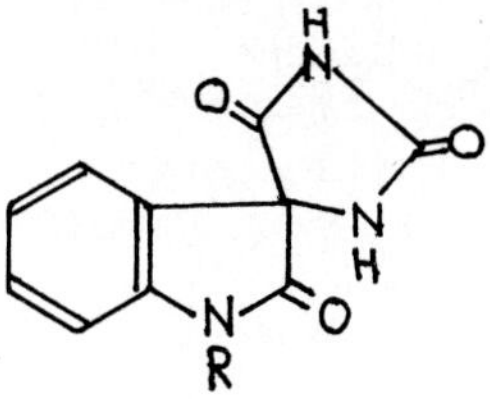

(1)

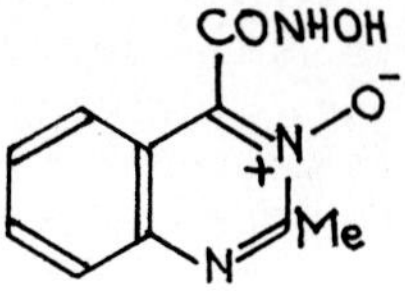

(2)

1-Acetylisatin on treatment with two moles of
hydroxylamine gives 2-methyl-3-N-oxidoquinazoline-4-
-hydroxamic acid (2) (Bergman, Carlsson, and
J.-O Lindström, Tetrahedron Letters, 1976, 3611, 3615)
and not the isatin dioxime as originally described. The
chemistry based on the supposed dioxime has, therefore,
been modified (M. Takahashi, Bull. chem. Soc., Japan, 1970,
<u>43</u>, 2986).   Isamic acid obtained from the reaction between
isatin and ammonia is 1,2-dihydro-2-oxospiro-[3H-indole-3,-
2'(1'H)-quinazoline]-4'-carboxylic acid (3)
(J.W. Cornforth, J. chem. Soc., Perkin 1, 1976, 2004).

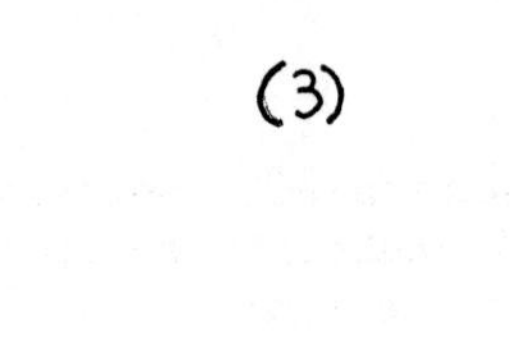

(3)

(4)

R^1, R^2 = H or Me

   The reaction 1-chloromethylisatin or 1-chloromethyl-5-
-methylisatin with indole or 2-methylindole gives the
indole derivative (4) (G.I. Zhungietu, L.P. Sinyavskaya,
and T. Ya. Filipenko, Khim. Geterotsikl. Soedin., 1977,
217). 1-Aminomethylisatin (5) condenses with acetone
in the presence of diethylamine to give the 3-acetyl-
methyl-1-aminomethyl-3-hydroxyoxindole (6) (O.M. Radul
and Zhungietu, *ibid.*, 1976, 1211).

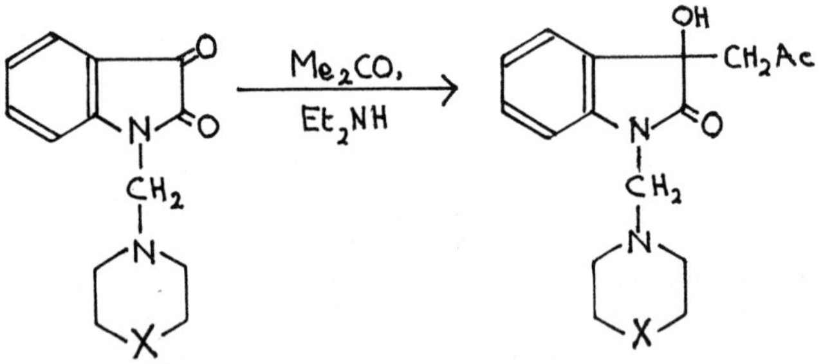

(5)   (6)

X = 0, CH$_2$, NPh

Condensation of isatin with indole in ethanol, containing
diethylamine at 60° gives 3-hydroxy-3-(3-indolyl)oxindole
(7), which on treatment with hydrazine hydrate affords
2-hydroxy-2-(3-indolyl)indoxyl (8), (Zhungietu and
Sinyavskaya, *ibid.*, p.204).

(7)   (8)

(iv) Isatogens

Isatogens (1) are obtained when 2-nitrotolans are heated in dry pyridine (C.C. Bond and M. Hooper, J. chem. Soc., C, 1969, 2453).

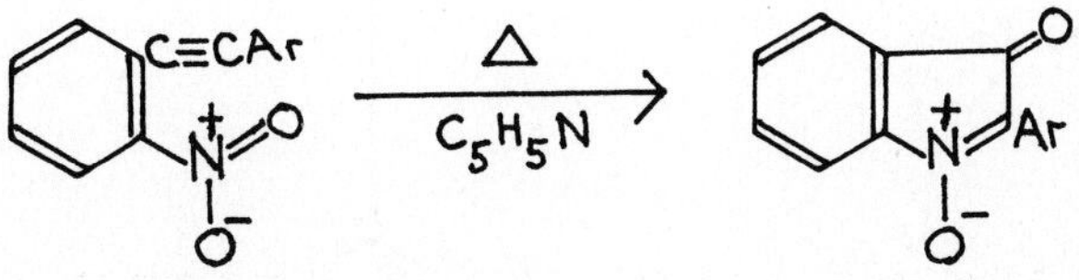

(1)

The oxidation of indolines with 3-chloroperbenzoic acid affords isatogens. This method appears to be a general route to isatogens including the previously unknown 2-alkylisatogens (T.H.C. Bristow, H.E. Foster, and Hooper, Chem. Comm., 1974, 677).

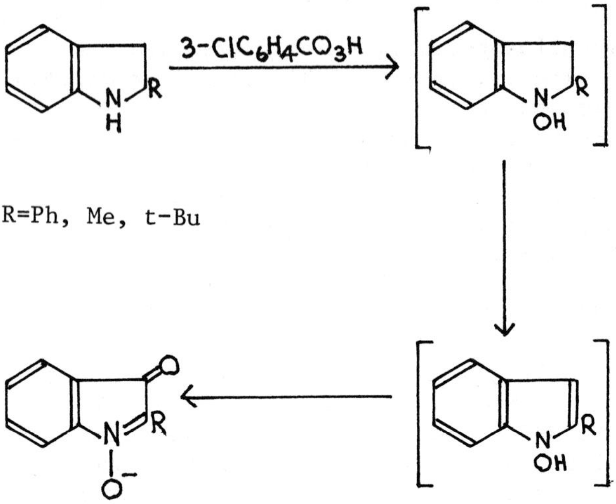

R=Ph, Me, t-Bu

2-Arylisatogens photoisomerizes to 2-aryl-4H-3,1-
-benzoxazin-4-ones (D.R. Eckroth and R.H. Squire, J. org.
Chem., 1971, $\underline{36}$, 224).

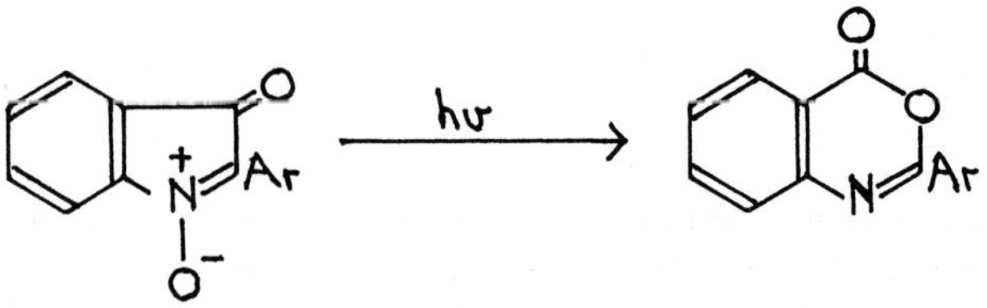

Ar = C_6H_5, 4-BrC_6H_4

(v) *4,7-Dihydroindole*

For preparation see p. 7.

(c) Tetra-, hexa-, and octa-hydroindoles

The condensation product of aminoacetaldehyde dimethyl acetal with cyclohexane-1,3-diones is cyclized by toluene-4-sulphonic acid to 4-oxo-4,5,6,7-tetrahydro-indoles. 4-Oxo-4,5,6,7-tetrahydroindole, m.p. 184-186°, 6,6-dimethyl-, m.p. 182-183, 1-benzyl-, m.p. 81-82°, 1-methyl- m.p. 84-86°, and 1,6,6-trimethyl-4-oxo-4,5,6,7--tetrahydroindole, m.p. 109-110° (J.M. Bobbitt *et. al.*, *ibid.*, 1978, *43*, 3541).

(d) Compounds having more than one indole nucleus

Indole on boiling with propiolic acid in methanol gives a 2:1 adduct, which hydrolyzes to give the acid and decarboxylates to afford di(3-indolyl)methylmethane, also obtained by boiling indole with indole-3-acrylic acid [S.H. Zee and C.S. Chen, J. Chin. chem. Soc. (Taipei), 1974, *21*, 229].

The reaction between indole and protocatechualdehyde in methanol-water-sodium hydroxide under oxygen gives 1,1--di(3-indolyl)-1-(3,4-dihydroxyphenyl)methane, m.p. 151° (Y. Takizawa and T. Mitsuhashi, Tokyo, Gakugei Daigaku Kiyo, Dai-4-Bu, 1972, *24*, 130).

3. Isoindoles and Isoindolines

(a) Isoindoles

Isoindole (2) has been obtained as a white solid on a surface at *ca.* 77K following pyrolysis of 2-(methoxy-carbonyloxy)isoindoline (1) at 500° and 0.01mm. It decomposes rather rapidly at room temperature.

Decomposition of the carbonate (1) in boiling xylene in the presence of *N*-phenylmaleimide gives the *exo*-adduct (3), m.p.208-209° (R. Bonnett and R.F.C. Brown, Chem. Comm., 1972, 393).

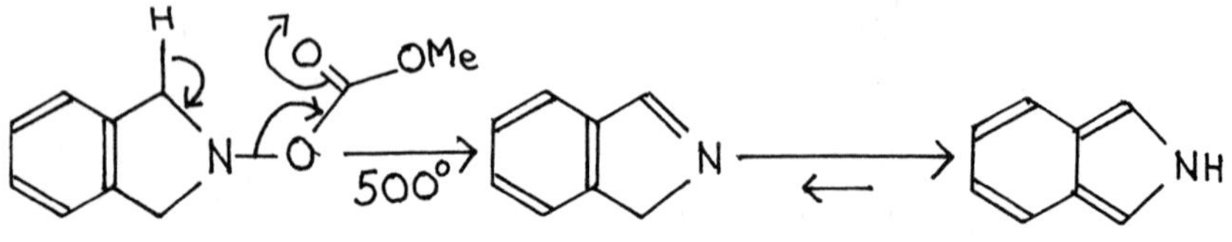

Isoindole is also obtained by the vacuum thermolysis of 1,2,3,4-tetrahydro-1,4-epiminonaphthalene (4) in essentially quantitative yield (J. Bornstein and D.E. Remy, *ibid.*, p.1149) and from the dihydropyridazine (5). 2-Ethoxycarbonylisoindole, m.p. *ca.* 35°, rapidly polymerizes on melting, 2-methyl-3,4,5,6-tetrafluoroiso-indole, m.p.178° (G.M. Priestley and R.N. Warrener, Tetrahedron Letters, 1972, 4295). Electronic and nmr spectroscopy suggests that the NH-tautomer (2) predominates, in accord with its oxygen and sulphur analogues (Bonnett and Brown, *loc. cit.*).

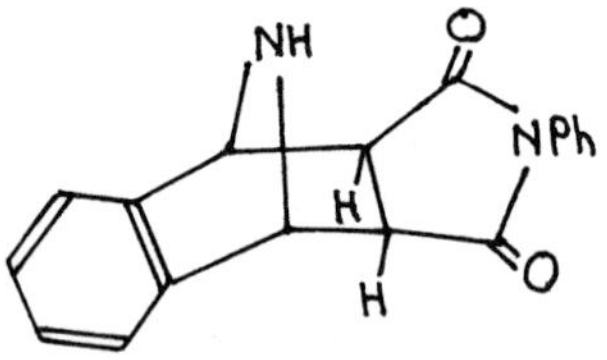

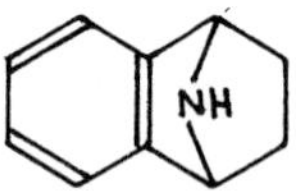

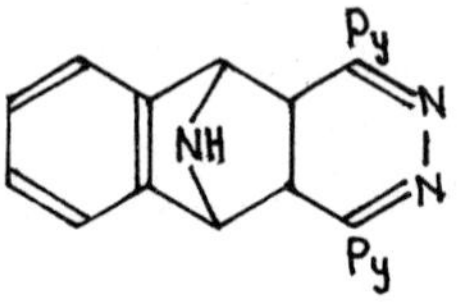

A simple one-step synthesis of isoindoles and related heterocycles involves treatment of an ethanolic solution

of a 1,2-dibenzoylbenzene with a primary amine followed
by addition of sodium tetrahydridoborate, for example,
the preparation of 1,3-diphenyl-2-methylisoindole and 1,3-
-diphenyl-2,5,6-trimethylisoindole (M.J. Haddadin and
N.C. Chelhot, Tetrahedron Letters, 1973, 5185).

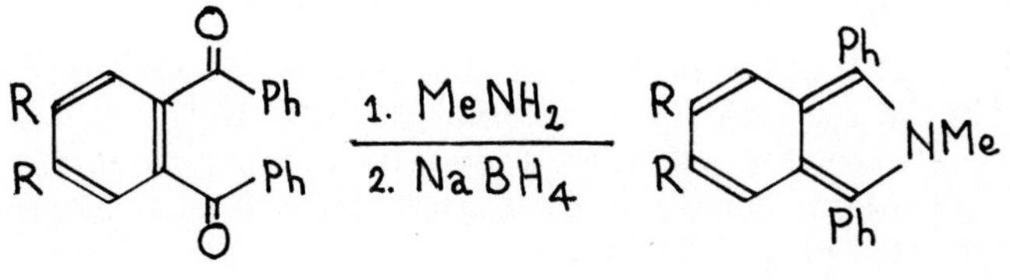

R = H, Me

2-Ethyl- and 2-n-butyl-isoindole react exothermally
with dimethyl acetylenedicarboxylate to give directly the
corresponding 1:2 adducts (6). Even at $0°$ no
intermediate 1:1 adduct is isolable. Several 1,2,3,4,7-
pentasubstituted isoindoles give 1:1 adducts with dimethyl
acetylenedicarboxylate, and their failure to form 1:2
adducts is ascribed to steric effects. Isoindole does
not react with ethyl propiolate in the expected manner
and the only product obtained is diethyl, 3,3'-
oxydiacrylate (L.J. Kricka and J.M. Vernon, J. chem. Soc.,
Perkin I, 1972, 904).

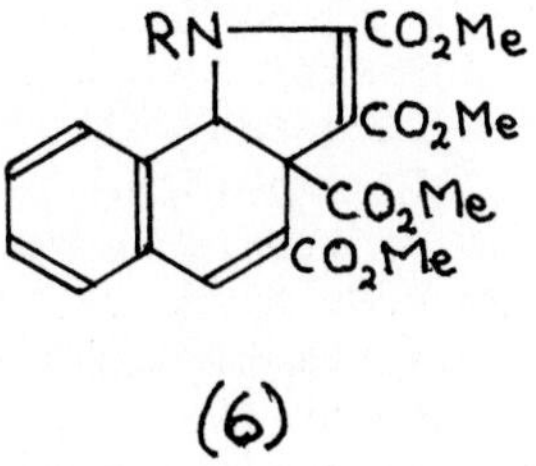

(6)

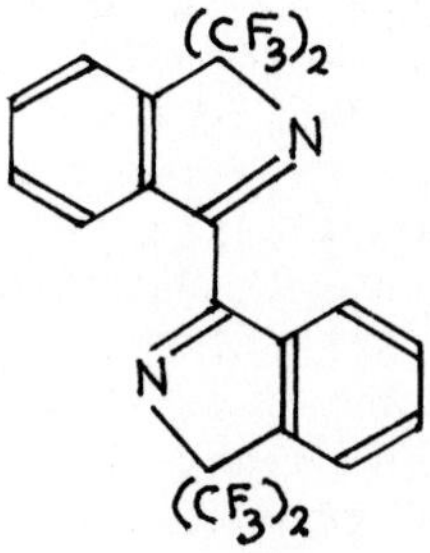

R - Et, n-Bu

(7)

The structure of 1,1,1',1'-tetrakis(trifluoromethyl)-3,3'-
-bi(1H-isoindole) (7) has been determined by X-ray
analysis (A. Gieren, K. Burger, and K. Einhellig, Angew.
Chem. internat. Edn., 1973, 12, 157). For the synthesis
and properties of isoindole and its derivatives see E.
Chacko (Diss. Abs. Int. B, 1977, 37, 4460) and D.E. Remy
(*ibid.*, 38, 697).

Structural effects on the isoindole-isoindolenine
equilibrium by substitution in the pyrrole ring, *i.e.*, at
positions -1 and -3 have been investigated using nmr
spectroscopic data (Chacko, J. Bornstein and D.J. Sardella,
Tetrahedron, 1979, 35, 1055).

3-(Dimethylamino)-1-ethoxy-2-azapentalene (10), m.p.
78-79° is obtained, by converting 3-(dimethylamino)-2H-2-
-azapentalen-1-one (8) by O-alkylation into the tetrafluoroborate (9), using triethyloxonium tetrafluoroborate. The
deprotonation of (9) gives (10) (K. Hafner and F. Schmidt,
Angew. Chem. internat. Edn., 1973, 12, 418).

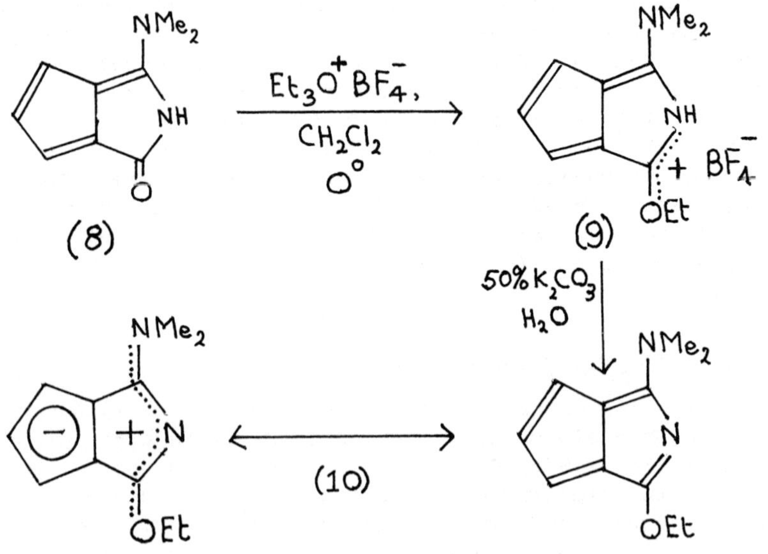

(b) Isoindolines

2-(Methoxycarbonyloxy)isoindole is prepared by treating 2-hydroxyisoindoline with methyl 4-nitrophenyl carbonate (Bonnett and Brown, *loc. cit.*).

4. Compounds Containing the Carbazole Nucleus

(a) Carbazole and its derivatives

(i) Synthetic methods

Carbazole is obtained by the deoxygenation of 2-nitrosobiphenyl *via* transition metal atom condensation: nitrene or nitrenoid intermediates are implicated (S. Togashi *et. al.*, J. org. Chem., 1980, **45**, 3044), by deoxygenation of 2-nitrobiphenyl using tris(trimethylsilyl)-phosphite (yield 84%) (M. Sekine, H. Yamagata, and T. Hata, Tetrahedron Letters, 1979, 375), by intramolecular dehydrogenative cyclization of diphenylamine in the vapour phase on platinum at 450-540° (M. Naito, K. Murayama, and M. Matsumoto, Aromarikkusu, 1980, **23**, 39), and by the photo-sensitized oxidation of diphenylamine using methylene blue, rose bengal, or eosin (N.R.K. Raju *et. al.*, Indian J. Chem., 1974, **12**, 422). The iodine-promoted thermal cyclization of anthranilic acid gives a mixture of carbazole, 2-benzoylaniline, and diphenylamine (P. Bhattacharyya *et. al.*, J. Indian chem. Soc., 1979, **56**, 328). Carbazole has been conveniently obtained from 2-nitro-4-*tert*-butylbiphenyl, following cyclization with triethyl phosphite and de-*tert*-butylation by aluminium chloride in benzene (M. Tashiro and T. Yamato, Synth., 1979, 48).

(ii) General properties and reactions

Oxidation of carbazole in the liquid phase with ozone gives a high yield of tetraozonide (N.F. Tyupalo *et. al.*, Ukr. Khim. Zh., 1976, **42**, 394). Photooxidation of carbazole in sulphuric acid—ethanol at 40° by passage of air affords 1-hydroxy-3-oxo-3H-carbazole (1), also obtained on similar treatment of 9-acetyl- and 9-methyl- -carbazole. 3-Acetyl-, 3-acetyl-9-methyl-, 3,6-diacetyl-, and 3,6-diacetyl-9-methyl-carbazoles all yield the 6-acetyl derivatives of (1) (G.N. Ivanov *et. al.*, Izv. Tomsk.

454

Politekh. Inst., 1973, <u>257</u>, 101).

(1)

The ir spectra of carbazole and some of its
derivatives, including 9-ethylcarbazole (S.S. Rogacheva
and V.I. Danilova, Izv. Vyssh. Uchebn. Zaved., Fiz., 1974,
<u>17</u>, 156) and of complexes with pyromellitic dianhydride
or antimony trihalides have been studied (V.K. Kondratov,
Zh. obshch. Khim., 1979, <u>49</u>, 1832). ^{13}C-nmr spectra
of carbazole and some of its derivatives (I. Mester,
D. Bergenthal, and J. Reisch, Z. Naturforsch., B: Anorg.
Chem. Org. Chem., 1979, <u>34B</u>, 650; A. Ahond, C. Poupat,
and P. Potier, Tetrahedron, 1978, <u>34</u>, 2385) and the [1]H-
and ^{13}C-nmr spectra of some carbazole derivatives have
been reported [G. -Z. Xu *et. al.*, Sci. Sin. (Engl. Edn.),
1980, <u>23</u>, 74]. Comparative ^{13}C-nmr studies have been
made of carbazoles, deuteriated carbazoles and related
heterocycles (J. Giraud and C. Marzin, Org. Mag. Res.,
1979, <u>12</u>, 647). [1]H-nmr spectra of a number of carbazole
derivatives have been determined (J. Polaczek, Zesz. Nauk.,
Inst. Ciekiej Syn. Org. Blachowni Slask., 1971, <u>3</u>, 65;
F. Balkau and M.L. Heffernan, Austral. J. Chem., 1973,
<u>26</u>, 1501). The pKa values of carbazole and its 1,2,3,4-
-tetrahydro, 3-methyl, 3-iodo, 3-chloro, 3,6-dichloro, and
3,6-dibromo derivatives have been determined in dimethylformamide and dimethyl sulphoxide [I.P. Zherebtsov,
V.P. Lopatinskii, and A.A. Zhiganova, Mater. Obl. Nauchn.
Konf. Vses. Khim. O-Va., Posuyashch. 75-Letiyu Khim.-
Tekhnol. Fak. Tomsk. Politeckh. Inst., 3rd 1972 (Pub.
1973), 22; M.I. Terekhova *et. al.*, Khim. Geterotsikl.
Soedin., 1979, 1104].

(iii) Carbazole

For review of carbazole see G. Collin (Ullmanns
Encykl. Tech. Chem., 4.Aufl., 1975, 9, 96). The aluminium
chloride catalyzed carbonization of carbazole has been
studied kinetically and by structural analysis of
intermediates (I. Mochida *et. al.*, Carbon, 1978, 16, 453).

A study has been made by the CNDO/S method, of the
(SO-S1) electron spectra and dipole moment variations of
the carbazole molecule under excitation (M. Deumie,
P. Viallet, and O. Chalvet, J. Photochem., 1979, 10, 365).
For the application of self-consistent HMO theory to the
carbazole molecule see T. Kakitani and H. Kakitani (Theor.
Chim. Acta, 1977, 46, 259), and for configuration
interaction calculations on the ground and lower excited
states of carbazole see L.E. Nitzche, C. Chabalowski, and
R.E. Christoffersen (J. Amer. chem. Soc., 1976, 98, 4794).
A variant of the Hartree-Fock perturbation theory has been
used to calculate π-electron currents in carbazole and
related heterocycles (Yu. B. Vysotskii, Zh. Strukt. Khim.,
1978, 19, 605; E. Corradi, P. Lazzeretti, and F. Taddei,
Mol. Phys., 1973, 26, 41). An esr signal, indicating
radical formation has been observed during the reaction
of carbazole with cobalt acetate and bromide (Kondratov
et. al., Zh. fiz. Khim., 1978, 52, 1485). The uv spectrum
of carbazole (K.R. Popov and N.V. Platonova, Zh. Prikl.
Spektrosk., 1978, 28, 717), and the ^{14}N-nmr spectra of
carbazole and 9-methylcarbazole have been reported
(M. Witanowski *et. al.*, Tetrahedron, 1972, 28, 637). A
new method of predicting aromaticity (B.A. Hess, Jr., and
L.J. Schaad, J. Amer. chem. Soc., 1971, 93, 305) has been
extended to conjugated heterocyclic systems, including
carbazole (Hess, Jr., Schaad, and C.W. Holyoke, Jr.,
Tetrahedron, 1972, 28, 3657). Aromatic resonance energy
for carbazole has been obtained from equilibrium data
(M.J. Cook *et. al.*, Tetrahedron Letters, 1972, 5019).

(iv) N-substituted carbazoles

The preparation of 9-carbazolyl alkali salts by
heating carbazole and caustic alkali, with or without
adding a little water, in xylene with azeotropic removal
of the water formed has been discussed (H. Otsuki,
K. Sakuma, and I. Matsuzawa, Japan Kokai 74 66,674/1974).

9-Methylcarbazole or its 3-substituted or 3,6-
-disubstituted derivatives are prepared by heating
carbazole or its derivatives with (2-chloroethyl)tri-
methylammonium chloride in a dipolar aprotic solvent in
the presence of strong alkali (Zherebtsov, V.P. Lopatinskii,
and N.I. Likhtarovich, U.S.S.R. Pat. 591/465/1978).
Carbazole on treatment with alkyl halides in 50% sodium
hydroxide in the presence of benzyltriethylammonium
chloride gives 9-alkylcarbazoles. Similarly pyrrole and
indole may be N-alkylated (A. Jonczyk and M. Makosza, Rocz.
Chem., 1975, 49, 1203). Treatment of carbazole with sodium
tetrahydridoborate in acetic acid gives 9-ethylcarbazole
(92%) (G.W. Gribble *et. al.*, J. Amer. chem. Soc., 1974,
96, 7812). Methyl 2-(carbazol-9-yl)benzoate, m.p.150-151°,
is obtained from carbazole and methyl 2-iodobenzoate in
the presence of copper—bronze powder and anhydrous
potassium carbonate in nitrobenzene at 170-180°
R. Glaser, J.F. Blount, and K. Mislow, *ibid.*, 1980, 102,
2777). 9-Acetylcarbazole on heating with ethyl oxalate
acid chloride affords ethyl 9-acetylcarbazole-2-oxalate,
which has been converted to 6-chlorocarbazole-2-oxalic
acid (L. Berger *et. al.*, U.S. Pat. 4.150.031/1979).
Carbazole on cyanoethylation gives 9-(2-cyanoethyl)carba-
zole (M.K. Murshtein *et. al.*, Chem. Abs., 1973, 78, 84170b).
Irradiation of carbazole with acrylonitrile affords 9-(1-
-cyanoethyl)carbazole (K. Yanasaki, I. Saito, and
T. Matsuura, Tetrahedron Letters, 1975, 313). Trifluoroacetylation of carbazole in 1,2-dichloroethane gives
9-trifluoroacetylcarbazole, m.p.61° (A. Cipiciani *et.
al.*, Tetrahedron, 1976, 32, 2595).

Hydrogen bonding and N-alkylation effects on the
electronic structure of carbazole have been discussed
(R.W. Bigelow and G.P. Caesar, J. phys. Chem., 1979, 83,
1790).

(v) C-Alkylcarbazoles

4-Methylcarbazole may be obtained from the 2- or
2'-nitro derivative of 4,4'-di-*tert*-butyl-6-methylbiphenyl
(p. 53) (Tashiro and Yamato, *loc. cit.*). The rearrangement
of cyclohexanone 2-ethyl-6-methylphenylhydrazone in acetic
acid, followed by dehydrogenation with chloranil, gives
2-ethyl-1-methylcarbazole (B. Miller and E.R. Matjeka,
Tetrahedron Letters, 1977, 131).

(vi) Halogenocarbazoles

Methods of synthesis and chemical properties of chloro-, bromo-, and iodo-carbazoles have been reviewed and discussed (J. Kyziol and J. Pielichowski, Zesz. Nauk. Politech. Krakow., Chem., 1978, 3).

3-Chlorocarbazole has been obtained from carbazole using *N*-chlorobenzotriazole units on homo- or co-polymers of *p*-aminostyrene [G. Manecke *et. al.*, Isr. J. Chem., 1978 (Pub. 1979), 17, 257]. Chlorination of carbazole by disulphur dichloride and aluminium chloride in sulphuryl chloride gives octachlorocarbazole [A.F. Andrews, C. Glidewell, and J.C. Walton, J. chem. Res. (S), 1978, 294]. Chlorination by two equivalents of sulphuryl chloride affords 3,6- and 1,3-dichlorocarbazole. The latter derivative is also obtained on treating 3-chloro-carbazole with sulphuryl chloride (Lopatinskii and Zherebtsov, Izv. Tomsk. Politekh. Inst., 1974, 198, 73). Bromination of carbazole with bromine-pyridine yields the 3-bromo, 3,6-dibromo, 1,3,6-tribromo, or 1,3,6,8-tetrabromo derivative depending on the amount of bromine used. Similarly 9-vinylcarbazole affords 9-(1,2-dibromoethyl)-carbazole and 3-bromo-9-(1,2-dibromoethyl)carbazole, which can be hydrolyzed to 3-bromocarbazole (J. Pielichowski and J. Kyziol, Monatsh., 1974, 105, 1306). 1,3,6,8-Tetra-bromocarbazole has been prepared by heating carbazole and dibromomethane at 50° (S. Narusawa and Y. Ohtsuka, Japan Kokai Pat. 95,899/1977), and by treating carbazole with the appropriate amount of bromine in methylene bromide. The tetrabromocarbazole on treatment with sodium hydride in dimethylformamide, followed by ethyl iodide gives 9-ethyl-1,3,6,8-tetrabromocarbazole. Related 9-alkyl derivatives and other derivatives have been prepared (Narisawa and H. Kawahara, Japan Kokai Pat. 151,162/1977).

Photobromination of 9-benzenesulphonyl-3-methylcarba-zole gives 9-benzenesulphonyl-3-bromoethylcarbazole (80%) m.p.144°, and 9-benzenesulphonyl-3-dibromomethylcarbazole (10%) (M.T. Goetz, J. hetereocyclic Chem., 1974, 11, 445).

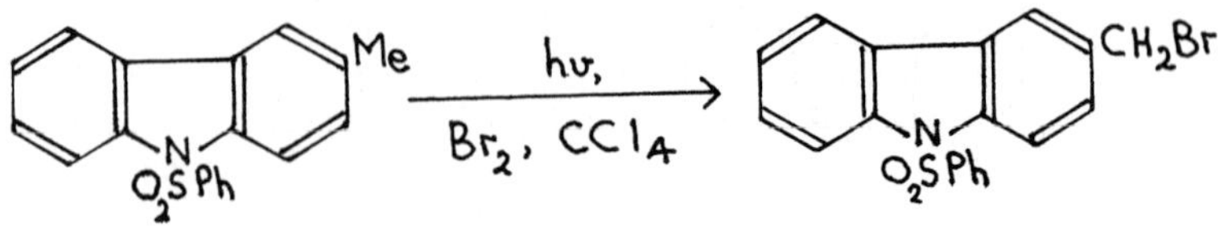

(vii) Nitro-, nitroso-, and amino-carbazoles

Nitration of carbazole using nitric acid-acetic acid at 40° gives a mixture of 1- and 3-nitrocarbazole. Uv--spectral data indicates that both isomers have a quinoid structure (V.I. Shishkina, Chem. Abs., 1973, <u>78</u>, 43188v).

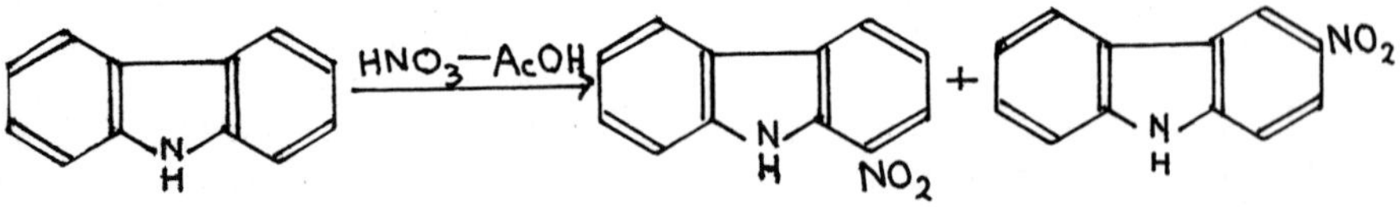

The kinetics of the nitration of carbazole, 9-methyl-carbazole, and 3-nitrocarbazole have been discussed (G.M. Novikova, V.F. Degtyarev, and Shishkina, Zh. fiz. Khim., 1973 <u>47</u>, 2178). It has been found that the copper (II) nitrate-acetic acid-acetic anhydride nitration of carbazole yields only 3-nitrocarbazole, whereas 9-ethyl-, 9-propyl-, 9-butyl- and 9-phenyl-carbazole afford the 3,6--dinitro derivatives as the sole products (Pielichowski and A. Puszynski, Monatsh., 1973, <u>105</u>, 772).

3-Nitrosocarbazole is obtained by treating carbazole in acetic acid with sodium nitrite, or by the addition of hydrochloric acid to 9-nitrosocarbazole in acetic acid. The addition of water over 4 hours to the solution of 3-nitrosocarbazole in acetic acid affords 3-(3-carbazolyl)--9-quinonediiminocarbazole (1) (Shishkina, *loc. cit.*).

(1)

(viii) Sulphonic acids and related compounds

The reaction of carbazole with thiocyanogen in acetic acid at 0-20° gives 1- and 3-carbazolyl thiocyanate, with two moles thiocyanogen to one mole carbazole a mixture of 1-carbazolyl thiocyanate and 1,3-carbazolyl dithiocynate is obtained (N.I. Baranova and Shishkina, Izv. Vyssh. Ucheb. Zaved. Khim. Tekhnol., 1972, **15**, 1678).

Thiocyanation of 3-nitrocarbazole and its 9-methyl derivative gives 3-nitro-6-carbazolyl thiocyanate and 9-methyl-3-nitro-6-carbazolyl thiocyanate , respectively. 1-Nitrocarbazole and its 9-methyl derivative afford no products (Baranova, L.N. Pushkina, and Shishkina, Zh. org. Khim., 1978, **14**, 192). Photo-bromination of 9-benzene-sulphonyl-3-methylcarbazole see (p. 57).

(ix) Alcohols, aldehydes and ketones derived from carbazole

Acetylation of carbazole by the Vilsmeier-Haack reagent (*N*,*N*-dimethylacetamide-carbonyl chloride) gives 93.1% of the 9-acetyl and 6.9% of the 3-acetyl derivatives. The observed order of reactivity in *N*-acetylation by the above reagent is pyrrole<<indole<carbazole (A. Cipiciani *et. al.*, J. chem. Soc., Perkin II, 1977, 1284). 9-Acetylcarbazole is obtained by treating carbazole with ketene in ethyl acetate containing methanesulphonic acid at -5 to 0° (M.N. Modi *et. al.*, Indian J. Chem. 1973, **11**, 1049), by boiling carbazole in benzene or xylene with acetic anhydride and a catalytic amount of phosphorus pentoxide (A.V. Spasov and A. Aleksiev, Z. Chem., 1974, **14**, 58) and on treating carbazole with acetic anhydride

and a small amount of zinc chloride or ferric chloride
at 130°. The latter method using benzoyl chloride at
130° gives 9-benzoylcarbazole (93%), whereas at 170°,
3-benzoylcarbazole (60%) is obtained. Heating 9-benzoyl-
carbazole at 170° affords 3-benzoylcarbazole (55%)
(Kh. Yu. Yuldashev, N.G. Sidorova, and A. Khaitbaeva,
Chem., Abs., 1976, 85, 46291d). The acetylation of
carbazole with two equivalents of acetyl chloride in
nitromethane containing one equivalent of aluminium chloride
at 90° for 5 minutes yields 3-acetyl (76.9%), 1-acetyl
(3.2%), 9-acetyl (4.3%) and 3,6-diacetylcarbazole (10%)
(Yu. G. Yur'ev, V.L. Ivasenko, and Lopatinskii, Izv. Tomsk.
Politekh. Inst., 1973, 257, 50).

9-Alkoxyalkylcarbazoles are obtained by the
condensation of 9-unsubstituted carbazoles with appropriate
aldehydes and alcohols in the presence of concentrated
hydrochloric or sulphuric acid (V.A. Anfinogenov,
V.D. Filimonov, and E.E. Sirotkina, Zh. org. Khim., 1978,
14, 1723).

(x)   Carbazolecarboxylic acids

3-(3-Carbazolyl)propionic acid, m.p.224°, is
obtained by the reaction of 9-benzenesulphonyl-3-bromo-
methylcarbazole with the sodium salt of dimethyl malonate
and subjecting the product to hydrolysis, acidification
and decarboxylation (Goetz, *loc. cit.*).

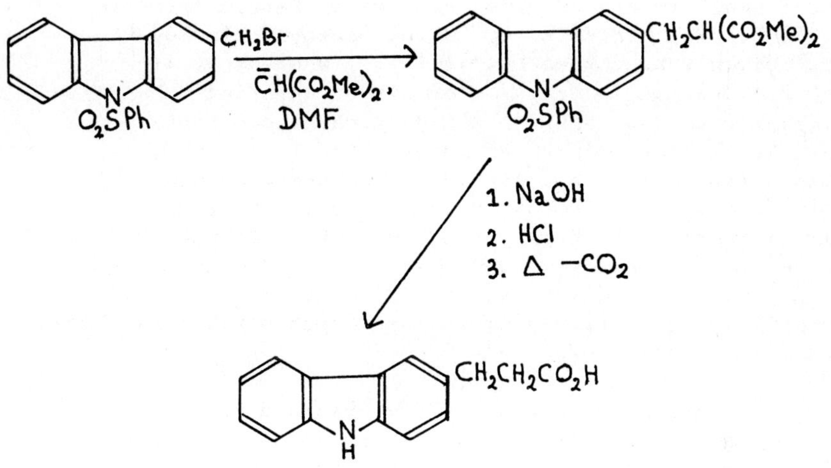

(b) Hydrocarbazoles

(i) Dihydrocarbazole

The Birch reduction of carbazole affords 1,4-
-dihydrocarbazole (J.W. Ashmore and G.K. Helmkamp, Org.
Prep. Proced. Int., 1976, $\underline{8}$, 223).

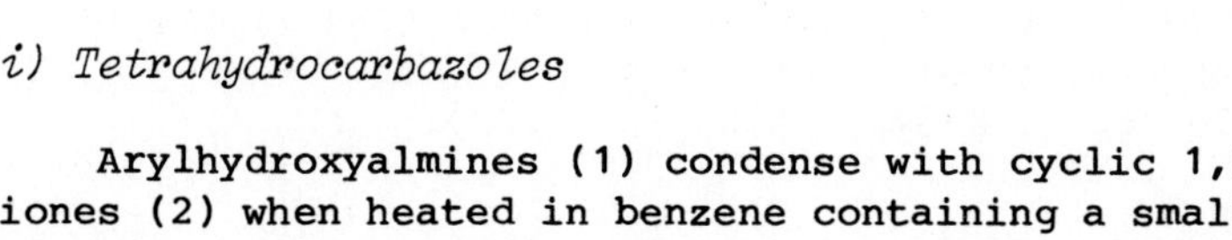

(ii) Tetrahydrocarbazoles

Arylhydroxyalmines (1) condense with cyclic 1,3-
-diones (2) when heated in benzene containing a small amount
of ascorbic acid to give products (3), whose acetates (4)

in trifluoroacetic acid easily cyclize to the 4-oxo-1,2,3,-
4-tetrahydrocarbazoles (5) (T. Okamoto and K. Shudo,
Tetrahedron Letters, 1973, 4533).

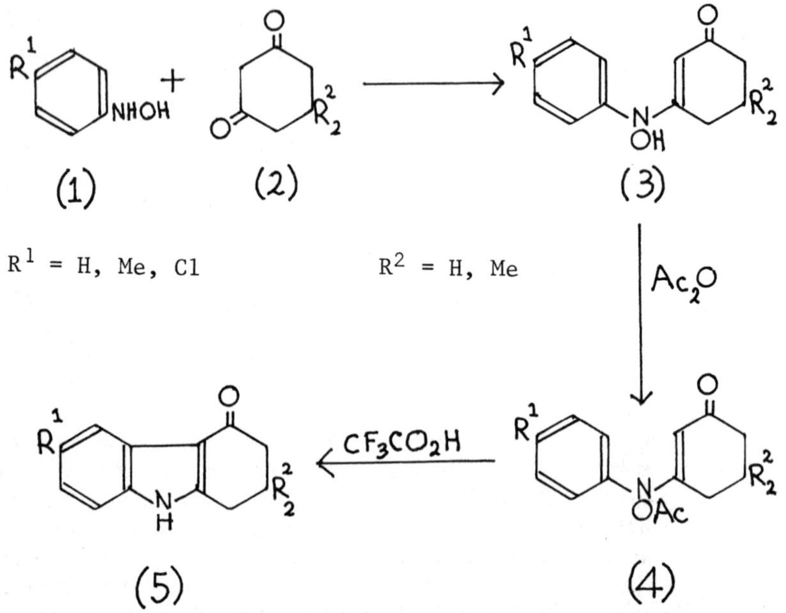

R^1 = H, Me, Cl R^2 = H, Me

(c) Carbazoles with additional fused rings

Indole and substituted indoles with thallium
triacetate yield both monomeric and dimeric oxidation
products, for example, indole (1) gives 6,6a,12,12a
-tetrahydroindolo[3,2-b]carbazole (2), which on sodium
tetrahydridoborate reduction yields 5,5a,6,6a,11,11a,12,-
12a,-octahydroindolo[3,2-b]carbazole (3) (A. Banerji and
R. Ray, Indian J. Chem., 1978, 16B, 422).

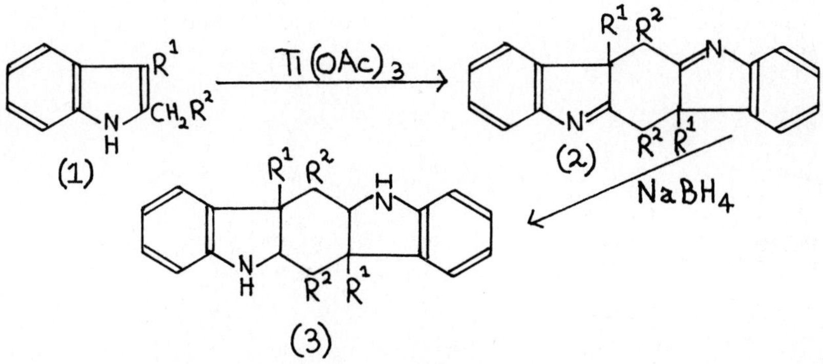

R^1R^2 = Me, H; CH$_2$Ph, H; Me, Me

5. Other Tricyclic Pyrrole Systems

7H-Pyrrolo[3,4-e]isoindole (I), m.p.>120°
(decomp.), sublimes at 112-117°/0.01mm, is prepared by
the reaction of 1,2,3,4-tetrakis(bromomethyl)benzene with
methanesulphonamide followed by a twofold elimination of
methanesulphonic acid leading to tautomeric forms, which
rearrange spontaneously to the desired product. Its
chemical reactivity resembles that of isoindole, as shown
by its sensitivity towards oxygen and the formation of
1:2 adduct with N-phenylmaleimide (R. Kreher and K.J. Herd,
Angew. Chem. internat. Edn., 1978, 17, 68).

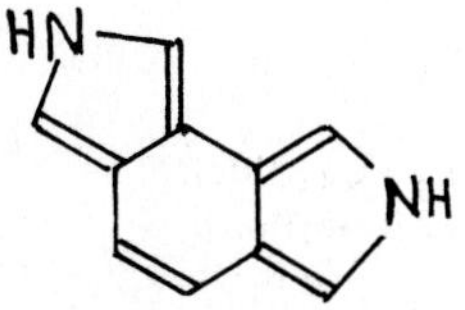

(1)

Chapter 6

OTHER FIVE-MEMBERED RING COMPOUNDS WITH ONE HETERO ATOM IN
THE RING FROM GROUPS III, IV AND V

R LIVINGSTONE

1. Phosphorus Compounds

(a) Mononuclear compounds

(i) Phospholes[1]

Newer techniques of phosphole synthesis, particularly
those which lead to simple phospholes substituted with
active functional groups, have been briefly surveyed and a
detailed account of chemical, physico-chemical, and
spectroscopic studies given. Theoretical studies relating
to the phosphole aromaticity problem and the present
conflicting position have been discussed (A.N. Hughes and
D. Kleemola, J. heterocyclic Chem., 1976. 13, 1). The
deductions from recent photoelectron spectroscopic and
theoretical studies are in considerable disagreement
although the LCGO studies of M.H. Palmer and R.H. Findlay
(J. chem. Soc., Perkin II, 1975, 974) provide a powerful
argument for a non-armotic phosphole system in either the
pyramidal or planar states. A one-electron MO analysis,
aided by explicit SCF-MO-CNDO/2 calculations, leads to the
conclusion that the nonplanar geometry of phosphole is due
to σ interactions and not to lack of aromaticity of the π
system, as suggested by Palmer and Findlay   (N.D. Epiotis
and W. Cherry, J. Amer. chem. Soc., 1976, 98, 4365).
Reference is also made to arsoles. The electronic
structure of phospholes and arsoles has been investigated
using an extended CNDO/2 approach (H.L. Hase *et. al.*,
Tetrahedron 1973, 29, 469). Nmr spectral data of phosphole
(C.L. Khetrapel, A.C. Kunwar, and K.P. Sinha, Chem. phys.
Letters, 1979, 67, 444). Dipole and second moments have
been reported for a number of five- and six-membered ring

---

[1]   Recent developments in phosphole and phosphoindole
chemistry, A.N. Hughes, Stud, Org. Chem. (Amsterdam),
1979, 3 (New Trends Heterocycle. Chem.), 216.

heterocycles containing one hetero atom, phosphorus, nitrogen, oxygen, or sulphur (Palmer, Findlay, and A.J. Gaskell, J. chem. Soc., Perkin II, 1974, 420).

1-Phenyl-, 3-methyl-1-phenyl-, and 3,4-dimethyl-phosphole readily form bis-complexes, L_2MCl_2 [L = ligand and M = Ni(π), Pd(π), or Pt(π)], or tris-complexes, L_3M Cl_2 with the chlorides of Ni(π), Pd(π), and Pt(π). Comparisons have been made with other simple phospholes which do not form Ni(π) complexes (G. Holah *et. al.*, J. heterocylic Chem., 1978, 15, 89).

1-Phenylphospholes react with *tert*-butyllithium to give 1-*tert*-butylphosphole, for example, 1-*tert*-butyl-3--methylphosphole and 1-*tert*-butyl-3,4-dimethylphosphole. The reaction between phospholes and trifluoroacetic acid, followed by neutralisation affords 3-phospholene oxides (F. Mathey, Tetrahedron, 1972, 28, 4171).

Reaction of lithium derivatives of 3-phospholene oxides with nitriles gives 1-aza-2-phospha-4,6-cycloheptadiene 2-oxides (Mathey and J.P. Lampin, Ger. Pat. 2,305,922/1973). The polymerization of phospholene glycol acrylates and some of their α-substituted analogues has been discussed (B.A. Arbuzov *et. al.*, Vysokomol. Soedin., Ser., B, 1980, 22, 95).

1-Methyl-3-oxophospholane 1-oxide (1) is in tautomeric equilibrium with 1-methyl-3-hydroxy-2--phospholene 1-oxide, permitting uncatalyzed exchange with deuterium oxide at the 2-position.

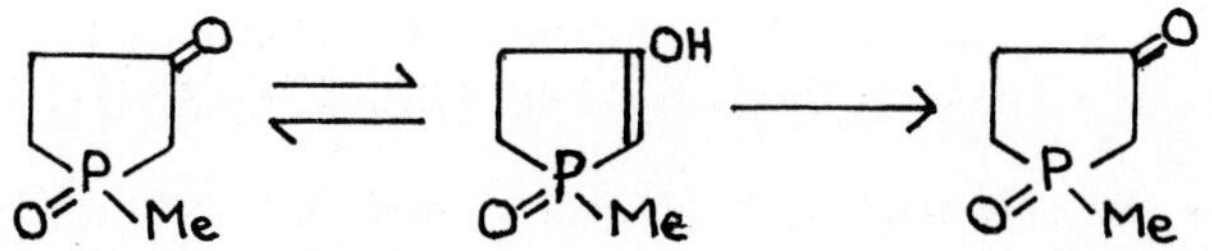

(1)

Some properties and reactions of (1) have been discussed
(L.D. Quin and R.C. Stocks, J. org. Chem., 1974, 39, 686).

Cycloaddition occurs when 1-methyl-3-phospholene 1-
oxide is irradiated in the presence of dichlorovinylene
carbonate to give the cyclobutane derivative (2)
(Von G. Märkl, G. Dannhardt, and J. Siller, Tetrahedron
Letters, 1979, 2979).

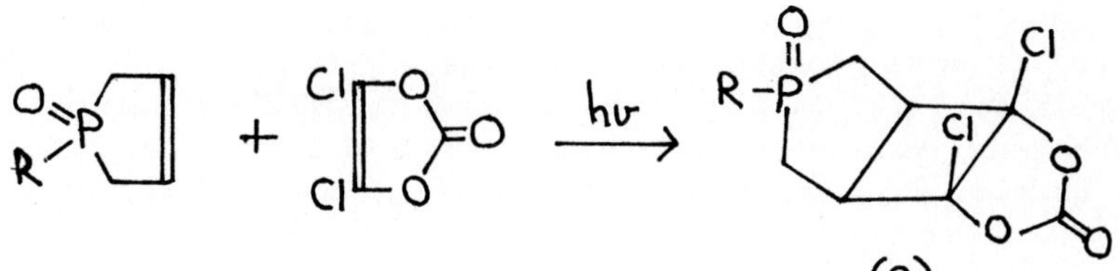

R = Me, OEt, Cl

The hydrolysis of 1-(substituted amino)-3-phospholene
1-oxides (phospholinic acid amides) is a 1st-order reaction
at constant pH (M. Kenn, W. Jansen, and O. Schmitz-DuMont,
Z. anorg. allg. Chem., 1979, 452, 176). A strategy module
has been developed for the PASCOP program, which performs
computer design of synthesis in organophosophorus chemistry
(F. Chloplin et. al., Nouv. J. Chim., 1979, 3, 223).

Phospholanes (3) (R = Me, Et, cyclohexyl, Ph, 1-
naphthyl) may be prepared by chlorinating diethylaminophos-
pholane (3) (R=NEt$_2$) and reacting the product (3) (R=Cl)
with the appropriate Grignard reagent (B. Fell and
H. Bahrmann, Synth., 1974, 119).

(3)

Phospholanes may be synthesized by the reaction of
secondary phosphine oxides containing benzyl groups with
α,β-unsaturated esters in the presence of stoichiometric
amounts of sodium hydride in tetrahydrofuran (R. Bodalski
and M. Pietrusiewicz, Zesz. Nauk. Politech. Lodz., Chem.,
1973, 27, 158).

The synthesis of 3-oxophospholane oxides (β-
-ketophospholane oxides) has been reviewed [K. Forner and
H.G. Henning, Khim. Primen. Fosfororg. Soedin., Tr. Kinf.,
5th 1972 (Pub. 1974), 290]. ^{13}C nmr (F.J. Weigert and
J.D. Roberts, Inorg. Chem., 1973, 12, 313), and ^{31}P nmr
spectra of some phospholanes (J.J. Breen *et. al.*,
Phosphorus, 1972, 2, 55), and the mass spectra of some
phospholane 1-oxides have been reported (G.L. Kenyon,
D.H. Eargle, Jr., and C.W. Koch, J. org. Chem., 1976, 41,
2417).

1,1-Diethylphospholanium bromide on treatment with
alkali undergoes a high degree of ring opening to give
diethyl-n-butylphosphine oxide. Some 1-ethylphospholane
1-oxide is also formed (K.L. Marsi and J.E. Oberlander,
J. Amer. chem. Soc., 1973, 95, 200). Nmr shift reagents
have been used to identify the above phosphine oxide
(B.D. Cuddy, K. Treon, and B.J. Walker, Tetrahedron
Letters, 1971, 4433). 1,1-Dimethylphospholanium chloride
inhibits the growth of wheat and other related
phospholanium salts are also plant-growth regulators
(A.-G. Bayer Fr. Pat. 2,274,218/1976).

1-Phosphabicyclo[2.2.1]heptane 1-oxide (4) is
prepared by Grignard cyclization of diethyl 5-bromo-3-
-bromoethylpentylphosphonate (R.B. Wetzl and Kenyon,
J. Amer. chem. Soc., 1974, 96, 5189):

$$(4)$$

(b) Polynuclear, fused ring compounds

(i) Phosphindole (benzo[b]phosphole)

Developments in the chemistry of phosphole and phos-
phindole have been reviewed [A.H. Hughes, Stud. org. Chem.
(Amsterdam), 1979, 3 (New Trends Heterocycl. Chem.) 216].
Treatment of 2-phenylethylphosphonous dichloride with zinc
chloride at 170°, followed by hydrolysis with
hydrochloric acid and oxidation affords 1-hydroxyphosphin-
doline 1-oxide (2,3-dihydro-1-hydroxyphosphindole 1-oxide).
1-Chlorophosphindoline 1-oxide may be converted into
1-ethyl- and 1-phenylphosphindoline 1-oxide and the
bromination of 1-methoxyphosphindoline 1-oxide with
N-bromosuccinimide followed by dehydrobromination gives
1-methoxyphosphindole 1-oxide (D.J. Collins, L.E. Rowley,
and J.M. Swan, Austral. J. Chem., 1974, 27, 831).
Phosphindoline 1-oxides (Hughes, S. Phisithkul, and
T. Rukachaisirikul, J. heterocyclic Chem., 1979, 16, 1417);
1-oxides and sulphides (Swan, U.S. Pat. 3,931,196/1976).
The reaction between phenylethyne and phosphorus
pentachloride in benzene gives an adduct, which on heating
affords 1,1,1,2,3-pentachlorophosphindole, converted on
heating with an alcohol into 1-alkoxy-2,3-dichlorophos-
phindole 1-oxide. Some 1-dialkylamino derivatives have
also been prepared (S.V. Fridland, Yu. K. Malkov, and
A.I. Efremov, Zh. obshch. Khim., 1978, 48, 361; U.S.S.R.
535,310/1976). The pyramidal stability of substituted
phosphindoles and dibenzophospholes has been discussed
[G. Zon, Diss. Abs. Int. B, 1972, 32, (Pt.1), 6939].

1-Phenyl-4,5,6,7-tetrahydrophosphindole (2) and

4,5-dihydro-3-phenylnaptho[2,1-b]phosphole (4) are obtained
by the reaction of 1,5-diazobicyclo[5.4.0]undec-5-ene with
the cycloaddition products (1) and (3) derived from the
reaction of phenylphosphorus dibromide with
1-vinylcyclohexene and 1-vinyl-3,4-dihydronaphthalene,
respectively. The ^{31}P nmr spectral data show that the
phosphorus nucleus in the respective potassium phospholides
is strongly deshielded, consistent with considerable
double-bond character for the di-co-ordinate phosphorus
(L.D. Quin and W.L. Orton, Chem. Comm., 1979, 401).

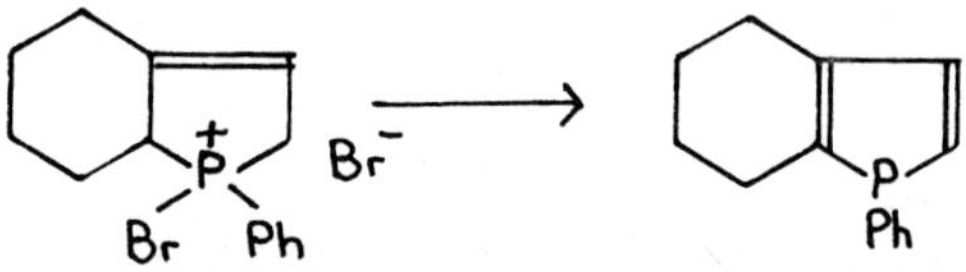

(1)   (2)

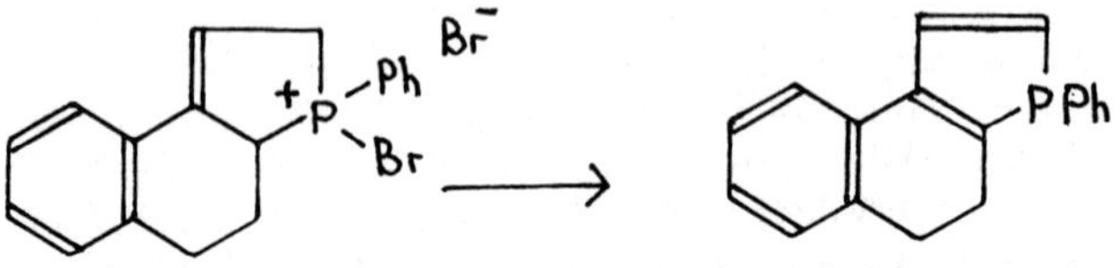

(3)   (4)

(ii) Dibenzophospholes

Dibenzophosphole, m.p.47-48°, is obtained by the
hydrolysis of the lithium salt, formed on treating
5-phenyldibenzophosphole with lithium in an inert solvent
(E.H. Braye, I. Caplier, and R. Saussez, Tetrahedron, 1971,
27, 5523).

The observed phosphorus coupling constants of a number of radical anions obtained from phosphine derivatives, including 5-phenyldibenzophosphole have been correlated with the calculated spin densities (A.G. Evans, J.C. Evans, and D. Sheppard, J. chem. Soc., Perkin II, 1976, 492). The use of the sodium salt of dibenzophosphole in the transition-metal-catalysed asymmetric organic synthesis *via* polymer-attached optically active phosphine ligands has been described (S.J. Fritschel *et. al.*, J. org. Chem., 1979, **44**, 3152).

5,5-Dimethyldibenzosilole reacts with phosphorus trichloride in the presence of aluminium chloride to yield 5-chlorodibenzophosphole (E.A. Chernyshev *et. al.*, Zh. obshch., Khim., 1977, **47**, 2572). For the mass spectra of 5-chloro-, 5-methyl-, and 5-phenyl-dibenzophosphole see V.N. Bochkarev *et. al.* (*ibid.*, 1974, **44**, 1273). 5-Substituted dibenzophospholes, on treatment with benzoyl chloride in the presence of triethylamine, followed by hydrolysis, undergo ring expansion to form 5,6-dihydro-dibenzo[b,d]phosphorin 5-oxides (D.W. Allen and A.C. Oades, J. chem. Soc., Perkin I, 1976, 2050):

R = Ph, Me, Et, Pr[i]

5-Phenyl-1,2,3,4,6,7,8,9-octahydrodibenzophosphole, b.p.165-170°/0.1mm, is obtained by treating the cyclo-adduct from phenylphosphorus dibromide and 1,1'-bicyclo-hexenyl with 1,5-diazabicylco[5.4.0]undec-5-ene (DBU) (F. Mathey and D. Thavard, Canad. J. Chem., 1978, 56, 1952).

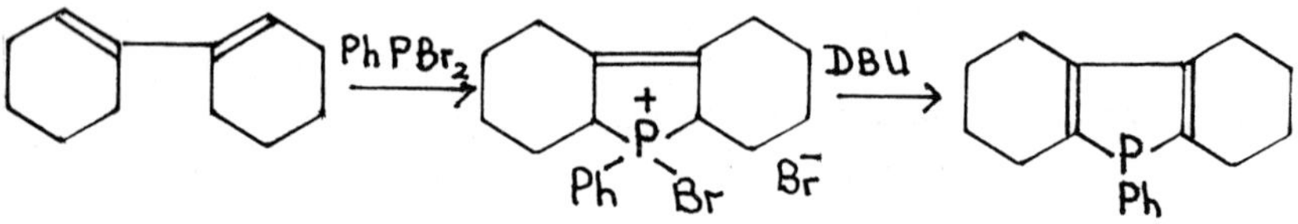

2. *Arsenic and Antimony Compounds*

(a) Mononuclear compounds

(i) Arsoles

1,2,5-Triphenylarsole, m.p. 186.5-187.5°, 2,5-di(4-
-methylphenyl)-1-phenylarsole, m.p. 179-180.5°, 2,5-di(4 -
-chlorophenyl)-1-phenylarsole, m.p. 160-161°, 2,5-di(1-
-naphthyl)-1-phenylarsole, m.p. 203.5-205°, and 2,5-
-dimethyl-1-phenylarsole are obtained by base-catalysed cycloaddition of phenylarsine to the appropriate buta-1,3-
-diyne (G. Märkl and H. Hauptmann, Tetrahedron Letters, 1968, 3257).

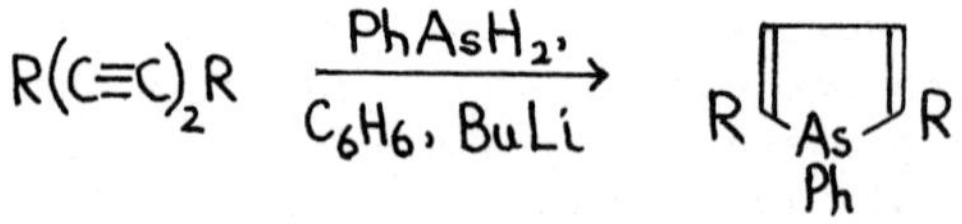

The properties of the arsoles resemble those of pyrroles
rather than phospholes. 1,2,5-Triphenylarsole on oxidation
with hydrogen peroxide gives 1,2,5-triphenylarsole 1-oxide
and on treatment with potassium in boiling dimethoxyethane
it affords 2,5-diphenyl-1-potassioarsole, which on
treatment with alkyl halides yields 1-alkyl-2,5-diphenyl-
arsoles. The latter unlike 1,2,5-triphenylarsole produces
a stable dianion on reaction with two equivalents of alkali
metal, which with ethyl iodide gives a mixture of 1-methyl-
and 1-ethyl-2,5-diphenylarsole, probably *via* the penta-
-coordinated species (1)(Märkl and Hauptmann, Angew. Chem.
internat. Edn., 1972, 11, 439).

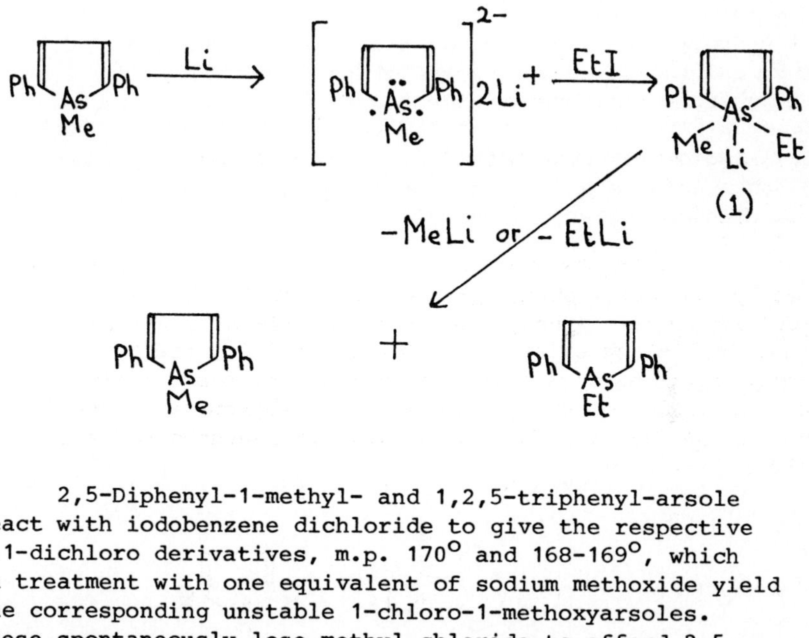

2,5-Diphenyl-1-methyl- and 1,2,5-triphenyl-arsole
react with iodobenzene dichloride to give the respective
1,1-dichloro derivatives, m.p. 170° and 168-169°, which
on treatment with one equivalent of sodium methoxide yield
the corresponding unstable 1-chloro-1-methoxyarsoles.
These spontaneously lose methyl chloride to afford 2,5-
-diphenyl-1-methyl-, and 1,2,5-triphenyl-arsole 1-oxide,
respectively.

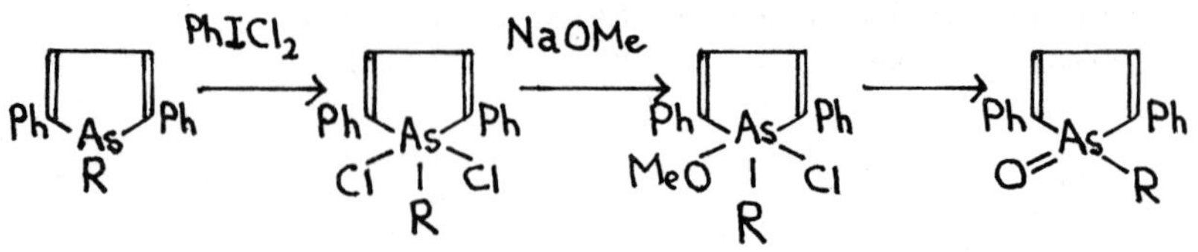

R = Me, Ph

1-Benzyl-1,1-dichloro-2,5-diphenylarsole decomposes at room temperature to give 1-chloro-2,5-diphenylarsole, m.p. 136-137°C (*idem. ibid.*, p.441). The reaction between 2,5-diphenyl-1-*tert*-butylarsole and phenylchlorocarbene or dichlorocarbene yields an arsenin (arsabenzene) (P. Jutzi, *ibid.*, 1975, 14, 232). 2,5-Diphenyl-1-methyl- and 1,2,5--triphenyl-arsole react with dimethyl acetylenedicarboxylate to give 3,6-disubstituted phthalates *via* the adduct (2), and a stable adduct (3) is obtained from 2,5-dimethyl--1-phenylarsole and tetracyanoethene (Märkl and Hauptmann, Tetrahedron Letters, 1968, 3257).

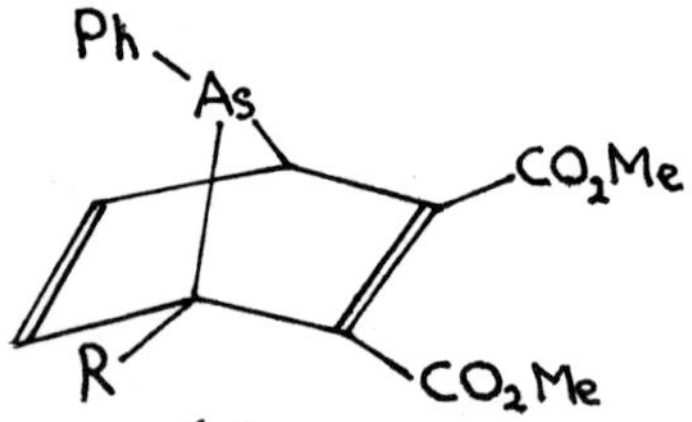

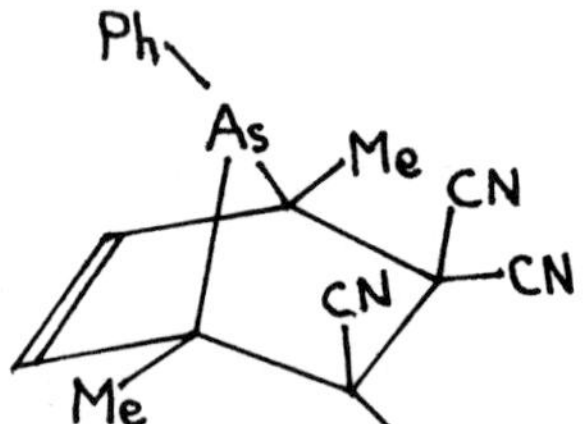

R = Me, Ph

1-Phenylarsolane (1-phenylarsacyclopentane, 1-phenyl-arsolidine) is obtained by the reaction between the di-Grignard reagent from 1,4-dibromobutane and phenyl-dichloroarsine. On oxidation with 30% hydrogen peroxide

it affords 1-phenylarsolane 1-oxide, m.p. 99-108°
(J.J. Monagle, J. org. Chem., 1962, <u>27</u>, 3851).

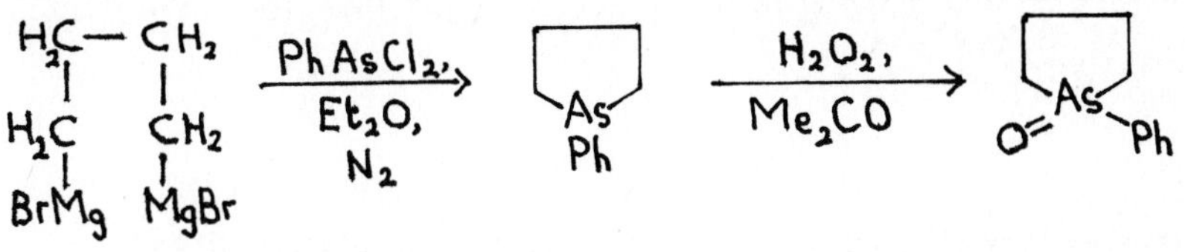

Pentaphenylstibole, m.p. 162-170°, is obtained
on reacting 1,4-dilithiotetraphenylbutadiene with
phenyldichlorostibine. Air oxidation converts it to
pentaphenylstibole 1-oxide, m.p. 250-255° (decomp.)
(E.H. Braye, W. Hübel, and I. Caplier, J. Amer. chem. Soc.,
1961, <u>83</u>, 4406).

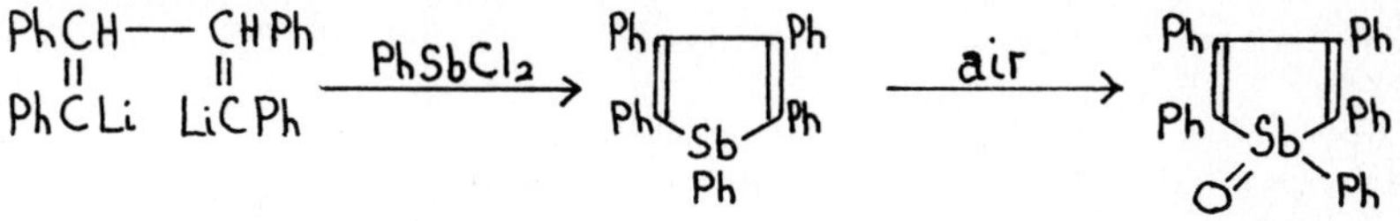

(b) Fused ring compounds

(i) Arsindolines

1-Methylarsindole is formed on cyclizing the
chloroarsine (1) by using aluminium chloride in carbon
disulphide (E.E. Turner and F.W. Bury, J. chem. Soc., 1923,
<u>123</u>, 2489).

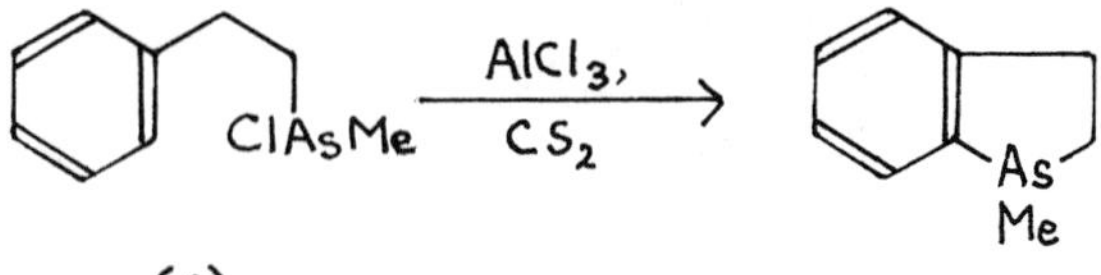

(1)

1-Phenylarsindole, b.p. 126-128°/0.6mm, is obtained by
heating a mixture of phenylethylphenylarsinic acid and
concentrated sulphuric acid and reducing the resulting
arsindoline oxide.

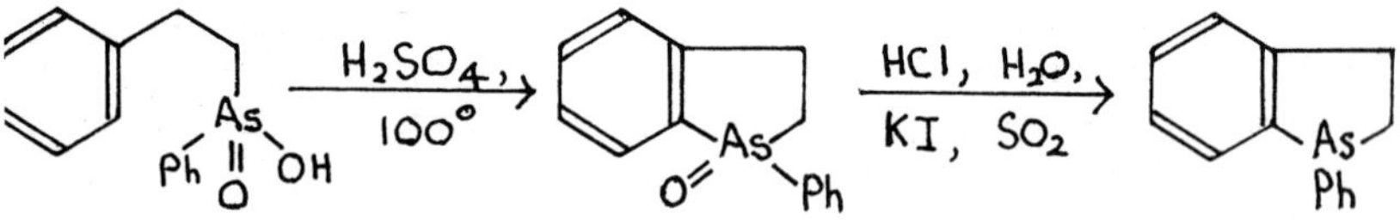

Attempts to dehydrogenate the arsinoline to 1-phenylarsin-
dole fail, and attempts to oxidise to a 2- or 3-oxo-
-derivative give mainly the 1-oxide (E.R.H. Jones and
F.G. Mann, *ibid.*, 1958, 1719).   A number of 2-aryliso-
arsindolines (2) have been prepared by the action of
2-xylylene dibromide and sodium on the corresponding
aryldichloroarsines.   When heated with hydriodic acid these
2-arylisoarsindolines are readily converted into 2-iodoiso-
arsindoline, m.p. 107-108°, which on treatment with the
appropriate Grignard reagent affords the corresponding
2-alkyl- or 2-aryl-isoarsindoline;   2-chloroisoarsindoline,
m.p. 73-74° (D.R. Lyon, Mann, and G.H. Cookson, *ibid.*,
1947, 662).

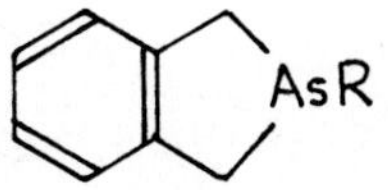

(2)

(ii) Dibenzoarsoles (9-arsafluorenes)

5-Methyldibenzarsole (1), b.p. 125°/0.05mm, m.p.
41-41.5°, obtained on heating the cyclic diquaternary
dibromides formed from 2,2'-biphenylenebis(dimethylarsine)
and either 1,2-dibromoethane or 1,3-dibromopropane
(H. Heaney *et. al., ibid.,* 1958, 3838), also prepared with
m.p. 46° by converting 5-hydroxydibenzarole 5-oxide to
the 5-iodo-derivative, and treating with methylmagnesium
iodide (J.A. Aeschlimann *et. al., ibid.,* 1925, 127, 66);
5,5-dimethyldibenzarsolium iodide monohydrate, m.p. 206-
207° (decomp.).

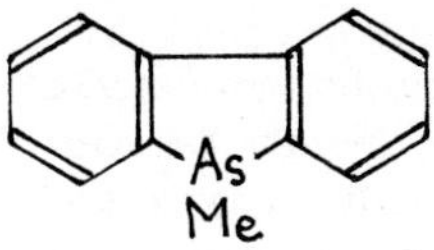

(1) 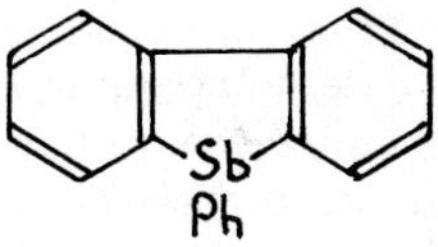   (2)

5-Methyldibenzarsole and 5-phenyldibenzostibole (9-phenyl-9-
-stibiafluorene (2), m.p. 101°, are prepared by the
interaction of 2,2'-dilithiobiphenyl and the appropriate
dihalogeno-arsine or -stibine. This method offers a route
for the preparation of other 5-alkyl- and 5-aryl-
-dibenzoarsoles and -dibenzostiboles (D.M. Heinekey and
I.T. Miller, *ibid.,* 1959, 3101).

3. Silicon Compounds

(a) Mononuclear compounds

(i) Silacyclopentadienes; siloles

Fast-neutron irradiation of phosphine-butadiene mixtures produces [^{31}Si]-1-silacyclopent-3-ene and [^{31}Si]-1-silapentadiene (R.-J. Hwang and P.P. Gaspar, J. Amer. chem. Soc., 1978, 100, 6626), identified by catalytic hydrogenation to [^{31}Si]-1-silacyclopent-3-ene. Silacyclopentadiene is sensitive to γ-ray irradiation and is also thermally unstable above 100° (E.E. Siefert et. al., ibid., 1980, 102, 2285). Also reported ^{31}Si nmr spectral data. Ionization potentials and one-electron properties of cyclopentadiene and silacyclopentadiene (W. Von Niessen, W.P. Kraemer, and L.S. Cederbaum, Chem. Phys., 1975, 11, 385). 1,1-Diethynyl-2,3,4,5-tetraphenyl-silacyclopentadiene (N.A. Vasneva, O.N. Gavrilova, and A.M. Sladkov, Izv. Akad. Nauk SSSR, Ser. Khim., 1978, 2149). 1-Chloro-1,2,3,4,5-pentaphenylsilacyclopentadiene, m.p. 181-183°, 1-chloro-1-methyl-2,3,4,5-tetraphenyl-silacyclopentadiene, m.p. 194-195°, 1-hydroxy-1-methyl- -2,3,4,5-tetraphenylsilacyclopentadiene, m.p. 199-202°, 1-methyl-2,3,4,5-tetramethylsilacyclopentadiene, m.p. 223-224°, 1,2,3,4,5-pentaphenylsilacyclopentadiene, m.p. 198-199° (M.D. Curtis, J, Amer. chem. Soc., 1969, 91, 6011) 1,1-diethynyl-2,3,4,5 -tetraphenylsilacyclopentadiene (Vasneva, Gavrilova, and Sladkov, loc. cit.). 1,1-Dimethyl-2,3,4,5-tetraphenylsilacyclopentadiene undergoes ready addition to hexafluorobut-2-yne to give trifluoromethyl-substituted silanorbornadiene (N.K. Hota and C.J. Willis, J. organometal. Chem., 1968, 15, 89).

(ii) Silacyclopentenes

1,1-Diphenylsilacyclopent-2-ene is prepared by brominating the corresponding silacyclopent-3-ene using N-bromosuccinimide and reducing the resulting 4-bromo-1,1- -diphenylsilacyclopent-2-ene with lithium tetrahydrido-aluminate (G. Manuel, P. Mazerolles, and J.M. Darbon, ibid., 1973, 59, C7).

$$\underset{\underset{Ph_2}{Si}}{\boxed{}} \xrightarrow[\substack{CCl_4, \\ Bz_2O_2}]{NBS,} \underset{\underset{Ph_2}{Si}}{\boxed{}}^{Br} + \underset{\underset{Ph_2}{Si}}{\boxed{}}_{Br} + \overset{Br}{Ph_2SiCH=CHCH=CH_2}$$

$$\Big\downarrow \substack{LiAlH_4, \\ Et_2O}$$

$$\underset{\underset{Ph_2}{Si}}{\boxed{}} + \underset{\underset{Ph_2}{Si}}{\boxed{}} + \overset{H}{Ph_2SiCH=CHCH=CH_2}$$

(70%)   (10%)   (20%)

The related 1,1-dimethyl compounds have also been prepared. 1,1-Dimethyl-, 1,1-diphenyl-, and 1,1,3-trimethyl-silacyclo--pent-2-ene-4-thiol are obtained from the respective silacyclopent-3-enes (A. Laporterie *et. al.*, Tetrahedron, 1978, **34**, 2669). 1,1-Dimethyl-2-methoxysilacyclopent-2-ene on hydrolysis affords 1,1-dimethylsilacyclopentan-2-one (J.A. Soderguist and A. Hassner, J. org. Chem., 1980, **45**, 541).

Flow pyrolysis of 1-alkyl-1-methyl-silacyclopent-3--ene affords 1-methylsilacyclopentadiene (T.J. Barton and G.T. Burns, J.organometal. Chem., 1979, **179**, C17). The electron diffraction determination of the gas-phase molecular structures of silacyclopent-3-ene, 1,1-difluoro- and 1,1--dichloro- silacyclopent-3-ene (S. Cradock *et. al.*, J. mol. Struct., 1979, **57**, 123) and the electronic structure of silacyclopent-3-ene have been reported (M Eckert-Maksic, K. Kovacevic, and Z.B. Maksic, J. organometal. Chem., 1979, **169**, 295). 1,1-Dimethylsilacyclopent-3-ene is obtained by the cyclization of diallyldimethyl silane in the presence of the catalyst system Re_2O_7/Al_2O_3-$SnBu_4$ (E.Sh. Finkel'shtein *et. al.*, Izv. Akad. Nauk SSSR, Ser. Khim., 1979, 474). Air oxidation of 1,1-diphenylsila-

cyclopent-3-ene at 100° yields 1,1-diphenyl-4-
-hydroperoxysilacyclopent-2-ene, which on reduction with
triphenylphosphine or sodium tetrahydridoborate affords
1,1-diphenyl-4-hydroxysilacyclopen-2-ene
(L.N. Khabibullina *et. al., ibid.,* p.416). Nitration of
silacyclopent-3-enes by dinitrogen tetroxide gives
predominantly dinitrosilacyclopentanes (E.V. Trukhin,
V.M. Berestovitskaya, and V.V. Perekalin, Chem. Abs.,
1979, <u>91</u>, 20584a). The ^{13}C- and ^{1}H- nmr spectral data
of some silacyclopent-3-ene derivatives have been reported
(M.L. Filleux-Blanchard, Nguyen Dinh An, and G. Manuel,
J. organometal. Chem., 1977, <u>137</u>, 11).

(iii) Silacyclopentanes

The ^{13}C nmr spectral data of 1,2-dimethylsilacyclo-
pentanes (J. Dubac *et. al., ibid.,* <u>127</u>, C69) and the
^{1}H nmr spectra of 1,1-dimethoxysilacyclopentane and
related derivatives of silacyclopentane and silacyclopent-
-3-ene have been reported (A.D. Naumov *et. al.,* Zh. obshch.
Khim., 1976, <u>46</u>, 1808). The electronic structure (Eckert-
Masksic, Kovacevic, and Maksic, *loc. cit.),* the Raman
spectrum (J.R. Durig, W.J. Natter, and V.F. Kalasinsky,
J. chem. Phys., 1977, <u>67</u>, 4756), the microwave spectrum
(Durig, W.J. Lafferty, and Kalasinsky, J. phys. Chem.,
1976, <u>80</u>, 1199), the electron diffraction (V.S. Mastryukov
et. al., Zh. struckt. Khim., 1979, <u>20</u>, 726; J. mol.
Struct., 1979, <u>54</u>, 121), and the photosensitized
decomposition of silacyclopentance (R.Y.S. Huo, Diss.
Abs. Int. B, 1973, <u>34</u>, 612) have been discussed.

1,1-Dichloro-2-methylsilacyclopentane (R.A. Benkeser
et. al., J. org. Chem., 1979, <u>44</u>, 1370). 1-Chloro-1,2-
dimethylsilacyclopentane undergoes *cis-trans*-isomerisation
catalyed by a variety of nucleophilic species, including
polar aprotic solvents. Kinetics of the process have been
investigated for hexamethylphosphorus triamide in carbon
tetrachloride (F.K. Cartledge, B.G. McKinne, and
J.M. Wolcott, J. organometal. Chem., 1976, <u>118</u>, 7).
Alcoholysis in the presence of amines and catalytic
alcoholysis of 1-chloro-1,2-dimethylsilacylopentane
predominantly affords the thermodynamically more stable
E-isomers of 1-alkoxy-1,2-dimethylsilacyclopentane, but
more *Z*-isomer is obtained on methanolysis in aniline,
pyridine, or ispropylamine (Dubac Mazerolles, and M. Joly,

ibid., 1977, <u>128</u>, C21). Brominolysis of 1,2-dimethyl-1-
-phenylsilacyclopentanes occurs with predominant inversion
of configuration at the silicon atom to give 1-bromo-1,2-
-dimethylsilacyclopentanes (*idem. ibid.*, p.C18). The
stereochemistry of reactions of 1,2-dimethylsilacyclo-
pentane derivatives has been discussed (Cartledge *et. al.*,
ibid., 1978, <u>154</u>, 187, 203) and 1-diethylphosphino-1,2-
-dimethylsilacyclopentane has been prepared (Dubac *et. al.*,
ibid., 1979, <u>174</u>, 263). Chlorination of 1,2,5-
-trimethylsilacyclopentane using benzoyl peroxide-carbon
tetrachloride proceeds with 100% retention of configuration
at silicon, but treatment with triphenylmethyl chloride
in benzene gives two-thirds retention and one-third
inversion (F. Franke and P.R. Wells, J. org. Chem., 1979,
<u>44</u>, 4055).

1,1-Dimethylsilacyclopentane on heating with ethyl
diazoacetate in the presence of $Rh_2(OAc)_4$ gives ethyl
(1,1-dimethylsilacylopent-3-yl) acetate (L.P.Danilkina
and G.V. Sidorenko, Zh. obshch. Khim., 1979, <u>49</u>, 2776).

1,1-Dimethyl-2-hydroxysilacyclopentane, b.p. 114-
117°/100mm, obtained on treating 1,1-dimethylsilacyclo-
pent-2-ene with diborane in the presence of sodium
hydroxide and hydrogen peroxide.

2-Hydroxy-1-methyl-1-phenylsilacyclopentane, 1,1-diphenyl-
2-hydroxysilacyclopentance, b.p. 162-163°/0.8mm, n_D^{20}
1.5984, 3-hydroxy-1-methyl-1-phenylsilacyclopentance, b.p.
117-120°/1mm, n_D^{20} 1.5497 (Manuel, Mazerolles, and J.
Gril, J. organometal. Chem., 1976, <u>122</u>, 335).

(b) Fused ring compounds

(i) Silaindenes and silaindanes

The reaction between hexachlorodisilane and cyclooctatetraene at 550° gives a mixture of 1,1-dichloro-1-silaindene (17%), 1,1-dichloro-1-silaindane (13%), 2,2-dichloro-2-silaindane(3,4-benzo-1,1-dichloro-silacyclopentane) (7%) and 1,1-dichloro-1-sila-8,9-dihydro-indene (~8%) (E.A. Chernyshev *et. al., ibid.,* 1975, <u>45</u>, 2221).

1,1-Dimethyl-1-silaindane, b.p. 80-88°/14mm (R.A. Benkeser *et. al.,* J. org. Chem., 1979, <u>44</u>, 1370). (4-Trimethylsilyl)phenyl carbene rearranges to give 1,1-dimethyl-1-silaindane as the major product (G.R. Chambers and M. Jones, Jr., Tetrahedron Letters, 1978, 5193; A. Sekiguchi and W. Ando, Bull. chem. Soc., Japan, 1977, <u>50</u>, 3067). The pyrolysis of bis(trimethyl-silyl)phenylmethanol (*idem,* Tetrahedron Letters, 1979, 4077) and phenyltrimethylsilyldiazomethane (Ando *et. al.,* J. Amer. chem. Soc., 1977, <u>99</u>, 6995) also afford 1,1-
-dimethyl-1-silaindane (27% and 15-20%, respectively) besides other products. The ^{13}C-nmr spectral data of 2,2-dimethyl-2-silaindane and 2,2-dimethyl-5-fluoro-2-
-silaindane (S.Q.A. Rizvi *et. al.,* J. organometal. Chem., 1973, <u>63</u>, 67) and the mass spectra of 2,2-dimethyl-2-
-silaindane have been reported (V.G. Zaikin *et. al.,* Neftekhimiya, 1975, <u>15</u>, 212).

1,1-Dichlorooctahydro-1-silaindene (mixture of 20% *trans* and 80% *cis),* b.p. 51-54°/0.2mm on treatment with methylmagnesium bromide affords 1,1-dimethyloctahydro-1-
-silanindene (20% *trans* and 80% *cis)* (Benkeser *et. al., loc. cit.),* also obtained by catalytic hydrogenation of 1,1-dimethyl-1-silanindane over nickel.

(ii) Dibenzosiloles (silafluorenes)

Dibenzosilole m.p. 36.6°, b.p. 86°/0.31mm, is prepared by the lithium tetrahydridoaluminate reduction of 5,5-difluorodibenzosilole. On treatment with ozone it gives 5-hydroxydibenzosilole (L. Spialter *et. al.,* J. Amer. chem. Soc., 1971, <u>93</u>, 5682). The electronic spectra and structure of 5,5-dimethyl-, 5,5-diethyl-, and

5,5-diphenyl-dibenzosilole and related heterocycles (Si replaced by Ge, Sn, N, P, Sb, Bi) (S.N. Davydov *et. al.*, Zh. fiz. Khim., 1980, **54**, 506), and the ir spectra of dibenzosilole and 5-methyl-, 5-phenyl-, and 5-chloro-dibenzosilole (A.N. Egorochkin *et. al.*, Izv. Akad. Nauk SSSR, Ser. Khim., 1974, 1604), and the mass spectra of dibenzosilole and some of its derivatives have been discussed (V.N. Bochkarev *et. al.*, Zh. obshch. Khim., 1977, **47**, 1799) and a X-ray structural study of 5,5-dichlorodi-benzosilole carried out [V.A. Sharapov *et. al.*, Tezisy Doklady- Vses. Soveshch. Org. Kristallokhim., 1st, 1974 (Pub. 1975), 33]. Thermodynamic parameters have been reported for the ethanolysis of 5-chloro-5-phenyldibenzo-silole (Chernshev *et. al.*, Zh. obshch. Khim., 1979, **49**, 139) and a study made of the ^{35}Cl nuclear quadruple resonances of 5,5-dichlorodibenzosilole and related compounds (V.L. Rogachevskii *et. al.*, *ibid.*, p.352). 2,2-Dimethyldibenzosilole on reaction with germanium tetrachloride in the presence of aluminium chloride yields 5,5-dichlorodibenzogermole and with phosphorus trichloride yields 5-chlorodibenzophosphole (Chernyshev *et. al.*, *ibid.*, 1977, **47**, 2572).

4. *Germanium Compounds*

(a) *Five-membered mononuclear compounds*

(i) *Germacyclopentadiene; germole*

1,1-Dichloro-2,3,4,5-tetraphenylgermacyclopentadiene, m.p. 197-199°, obtained from 1,4-dilithiotetraphenyl-butadiene and tetrachlorogermane; 1-chloro-1,2,3,4,5--pentaphenylgermacyclopentadiene, m.p. 210-211°, obtained by the addition of diphenylacetylene and excess lithium to phenyltrichlorogermane in tetrahydrofuran; 1-hydroxy--1,2,3,4,5-pentaphenylgermacyclopentadiene, m.p. 256-257°, 2,3,4,5-tetraphenylgermacyclopentadiene, m.p. 192-193°, 1,2,3,4,5-pentaphenylgermacyclopentadine, m.p. 187-188°, 1,1,2,3,4,5-hexaphenylgermacyclopentadiene, m.p. 198-199°, octaphenyl-1,1'-spirobigermacyclo-pentadiene (1), m.p. 258-260° (M.D. Curtis, J. Amer. chem. Soc., 1969, **91**, 6011).

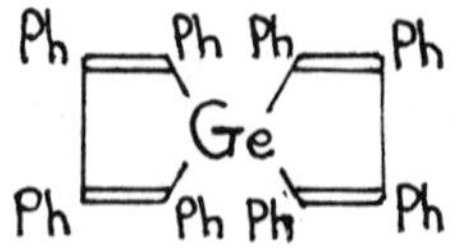

(1)

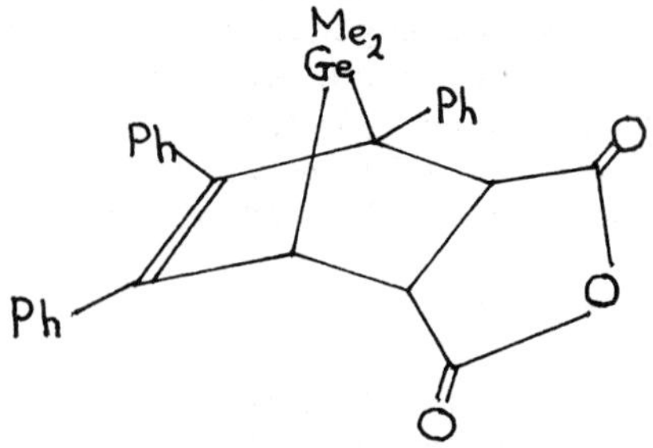

(2)

1,1-Dimethyl-2,3,4,5-tetraphenylgermacyclopentadiene, m.p. 179-181°, from 1,4-dilithiotetraphenylbutadiene and dimethyldichlorogermane, does not undergo simple Diels—Alder addition to various acetylenes, but on heating with maleic anhydride in benzene in a sealed tube affords a 1:1 adduct, 7,7-dimethyl-1,4,5,6-tetraphenyl-7-germa-bicyclo-[2.2.1]hept-5-ene-2,3-dicarboxylic anhydride (2) (N.K. Hota and C.J. Willis, J. organometal. Chem., 1968, 15, 89). Pentaphenylgermacyclopentadiene reacts with diiron enneacarbonyl to give μ-[1-(1,2,3,4-tetraphenyl-butadienyl)phenylgermylene]octacarbonyldiiron (Curtis, W.M. Butler, and J. Scibelli, *ibid.*, 1980, 192, 209). The photochemical reaction of iron pentacarbonyl with 1,1-dimethyl-, 1,1-dibenzyl- and 1-chloro-1-methyl-2,3,4,5-tetraphenylgermacyclopentadiene affords the corresponding diene-iron tricarbonyl complex.

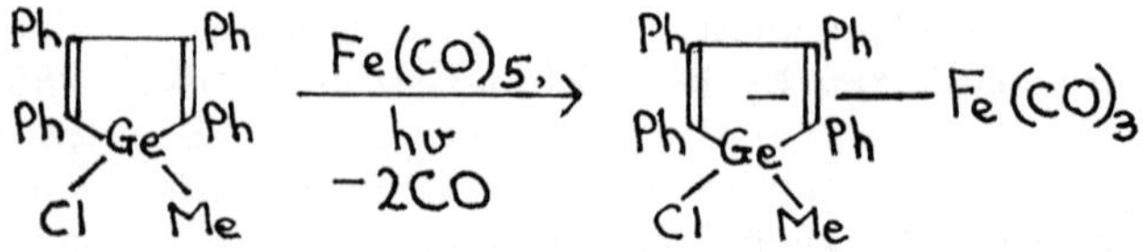

Related silacyclopentadienes behave in a similar manner

(P. Jutzi and A. Karl, *ibid.*, 1977, **128**, 57). 1,1-Dichloro-
-2,3,4,5-tetraphenylgermacyclopentadiene on treatment with
magnesium ethyl bromide, and acetylene in tetrahydrofuran
yields 1,1-diethynyl-2,3,4,5-tetaphenylgermacyclo-
pentadiene (N.A. Vasneva, O.N. Gavrilova, and A.M. Sladkov,
Izv. Akad. Nauk SSR, Ser. Khim., 1978, 2149).

(ii) Germacyclopentenes

1,1-Diphenylgermacyclopent-2-ene is obtained from
its cyclopent-3-ene isomer and on hydroboration-oxidation
it affords 1,1-diphenyl-2-hydroxygermacyclopentane, which
on treatment with sodium gives 1-oxa-2,2-diphenyl-2-germa-
cyclohexane (G. Manuel, G. Bertrand, and P. Mazerolles,
J. organometal. Chem., 1978, **146**, 7). 1,1-Dimethyl- and
1,1-diphenyl- germacyclopent-3-ene react with $PhSO_2NSO$
to give the corresponding germacyclopent-2-ene derivatives,
which on reduction are converted into 1,1-dimethyl- and
1,1-diphenyl-germacyclopent-2-ene-4-thiols respectively
(A. Laporterie *et. al.*, Tetrahedron, 1978, **34**, 2669).

R = Me, Ph

1,1-Dimethyl-2-methoxygermacyclopent-2-ene on hydrolysis
affords 1,1-dimethylgermacyclopentan-2-one
(J.A. Soderquist and A. Hassner, J. org. Chem., 1980, **45**,
541.

Complexes of phenylchlorogermylene and tertiary
phosphines on heating with 2,3-dimethylbutadiene give
1-chloro-3,4-dimethyl-1-phenylgermacyclopent-3-ene
(J. Escudie *et. al.*, J. organometal. Chem., 1977, **124**,
C45); 1,1-diiodogermalcylopent-3-ene (Manuel, Bertrand,
and Mazerolles, *loc. cit.*). The ir and Raman spectra of
of 1,1-dichloro-, 1,1-dimethoxy-, and 1,1-dimethyl-germa-

cyclopent-3-ene (P.W. Jagodzinski, J. Laane, and Manuel,
J. mol. Struct., 1978, 49, 239), and of some sila- and
germacyclopent-3-enes (A. Marchand *et. al.*, J. organometal.
Chem., 1977, 135, 23), uv spectral data of 1,1-dimethyl-
germacyclopent-3-ene (J. Reffy, T. Veszpremi, and J. Nagy,
Period. Polytech. Chem. Eng., 1976, 20, 223), and [13]C-
and [1]H-nmr spectral data of some germacyclopent-3-ene
derivatives (M.L. Filleux-Blanchard, Nguyen Dinh An, and
G. Manuel, J. organometal. Chem., 1977, 137, 11.) have
been reported.

(iii) Germacyclopentane

2-Methylgermacyclopentane, b.p. 70°/400mm, obtained
by lithium tetrahydridoaluminate reduction of 1,1-dichloro-
-2-methylgermacyclopentane, prepared from pentyl-1,4-
dimagnesium dibromide and tetrachlorogermane. Similarly
the reaction between the above Grignard reagent and phenyl-
or methyl-trichlorogermane affords 1-chloro-2-methyl-1-
-phenyl-, *(Z,E)*, b.p. 95-100°/0.4mm, or 1-chloro-1,2-
-dimethyl-germacyclopentane, *(Z/E,* 50/50), b.p. 100-105°/
65mm, respectively, and with diphenyldibromogermane it
gives 1,1-diphenyl-2-methylgermacyclopentane, b.p. 130°/
0.05mm; 1,2-dimethyl-, *(Z/E,* 55/45), b.p. 78-82°/200mm,
1-bromo- 1,2-dimethyl-, *(Z/E,* 55/45), b.p. 82-88°/20mm,
and 1,2-dimethyl-1-phenyl-germacyclopentane, *(Z/E,* 45/55),
b.p. 85-90°/2.4mm (Dubac *et. al.*, J. organometal. Chem.,
1977, 127, C69). The stereoselective alcoholysis,
aminolysis, and thiolysis of 1-bromo-1,2-dimethylgerma-
cyclopentane and the formation of other geometric isomers
have been discussed *(idem, ibid.,* 1979, 165, 175; Dubac,
Mazerolles, and J. Cavezzan, Tetrahedron Letters, 1978,
3255), and 1-diethylphosphino-1,2-dimethylgermacyclopentane
has been synthesized (Dubac *et. al.*, J. organometal. Chem.,
1979, 174, 263). 1,1-Diphenyl-2-hydroxygermacyclopentane
(G. Manuel, G. Bertrand, and P. Mazeroles, J. organometal.
Chem., 1978, 146, 7), 1,1-dimethylgermacyclopentan-2-one
(J.A. Soderquist and A. Hassner, J. org. Chem., 1980, 45,
541), the pyrolysis and photolysis of the latter compound
and the related sila compound have been studied *(idem,*
Tetrahedron Letters, 1980, 429). The mechanics, and
stereochemistry of the reactions of 1,2-dimethyl-1-methoxy-
germa (and sila) cyclopentane have been discussed (Dubac
et. al., ibid., 1978, 4499).

(b) Polycyclic compounds with germanium in a five-membered ring

(i) Benzogermole (germaindene)

1,1-Dimethyl-2,3-dihydrobenzogermole is formed by the flash vacuum pyrolysis of the lithium salt (1) of the tosylhydrazone obtained from phenyltrimethylgermyl ketone, or by gas-phase decomposition of the diazo compound (2) prepared by gently heating (1) under vacuum (E.B. Norsoph, B. Coleman, and Jones, Jr., J. Amer. chem. Soc., 1978, 100, 994).

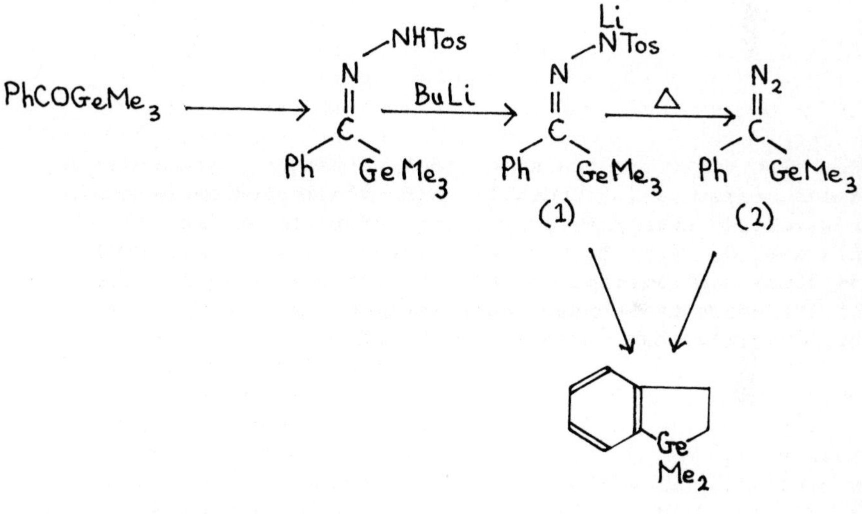

The gas-phase rearrangement of 4-(trimethylgermyl)phenyl-carbene to give 1,1-dimethyl-2,3-dihydrobenzogermole has been discussed (G.R. Chambers and M. Jones, Tetrahedron Letters, 1978, 5193).

(ii) Dibenzogermole

The reaction of 5,5-dimethyldibenzosilole with tetrachlorogermane in the presence of aluminium chloride give 5,5-dichlorodibenzogermole (Chernyshev *et. al.*, Zh. obshch. Khim., 1977, 47, 2572), also obtained along with

compound (1) on passing 2-chlorobiphenyl over
germanium—copper—zinc chloride at 520° (*idem, ibid.,*
1978, 48, 636).

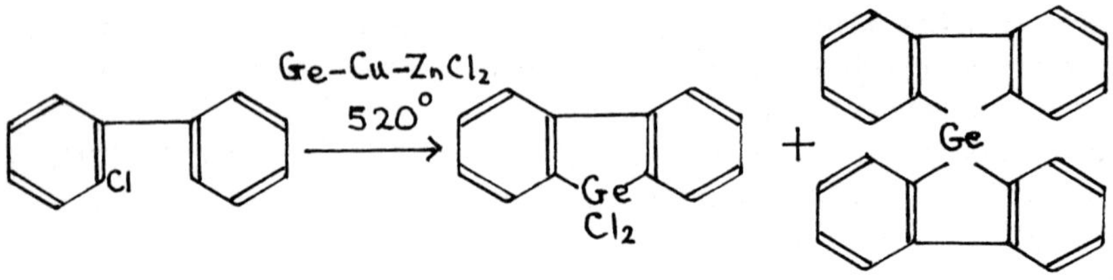

(1)

The electronic spectra of 4-ethyl-, -methyl-, and -phenyl-,
and 5,5-diethyl-, -dimethyl-, and -diphenyl-dibenzogermole
and related compounds containing C, Si, Sn, N, Sb, Bi in
place of Ge (S.N. Davydov *et. al.,* Zh. fiz. Khim., 1980,
54 506), and the mass spectra of dibenzogermole and some
of its derivatives have been discussed (V.N. Bochkarev
et. al., Zh. obshch. Khim., 1977, 47, 1799.

Guide to the Index

This index is constructed in a similar manner to the volume indexes of the first edition of the Chemistry of Carbon Compounds. However, to make the index easier to use, more descriptive entries have been made for the commonly occurring individual, and groups of chemicals.

The indexes cover primarily the chemical compounds mentioned in the text, and also include reactions and techniques, where named, and some sources of chemical compounds such as plant and animal species, oils, etc.

Chemical compounds have been indexed alphabetically under the names used by authors, editing being restricted to ensuring uniformity of entries under the same heading. In view of the alternative nomenclature that can often be used, a limited amount of cross-referencing has been done where it is considered to be helpful, but attention is particularly drawn to Convention 2 below.

For this and the succeeding volumes, the indexing conventions listed below have been adopted.

1. *Alphabetisation*

(a) The following prefixes have not been counted for alphabetising:

n-	*o-*	*as-*	*meso-*	D	*C*
sec-	*m-*	*sym-*	*cis-*	DL	*O-*
tert-	*p-*	*gem-*	*trans-*	L	*N-*
	vic-				*S-*
		lin-			*Bz-*
					Py-

Some prefixes and numbering have been omitted in the index, where they do not usefully contribute to the reference.

(b) The following prefixes have been alphabetised:

Allo	Epi	Neo
Anti	Hetero	Nor
Cyclo	Homo	Pseudo
	Iso	

(c) A letter by letter alphabetical sequence is followed for entries, firstly for the main entry, followed by the descriptive entry. The only exception to this sequence is the placing of plural entries in front of the corresponding individual entries to prevent these being overlooked by a strict alphabetical sequence which could lead to a considerable separation of plural from individual entries. Thus "butanes" will come before *n*-butane, "butenes" before 1-butene, and 2-butene, etc.

2. *Cross references*

In view of the many alternative trivial and systematic names for chemical compounds, the indexes should be searched under any alternative names which may be indicated in the main body of the text. Only a limited amount of cross-referencing has been carried out, where it is considered that it would be helpful to the user.

3. *Esters*

In the case of lower alcohols esters are indexed only under the acid, e.g. propionic methyl ester, not methyl propionate. Ethyl is normally omitted e.g. acetic ester.

4. *Derivatives*

Simple derivatives are not normally indexed if they follow in the same short section of the text.

5. *Collective and plural entries*

In place of "– derivatives" or "– compounds" the plural entry has normally been used. Plural entries have occasionally been used where compiunds of the same name but differing numbering appear in the same section of the text.

6. *Main entries*

The main entry of the more common individual compounds is indicated by heavy type. Multiple entries, such as headings and sub-headings over several pages are shown by "–", e.g., 67–74, 137–139, etc.

INDEX

494

504

508